# Manual of
# Electrophysiology

# Manual of Electrophysiology

**Editor**
**Kanu Chatterjee** MBBS
Clinical Professor of Medicine
The Carver College of Medicine
University of Iowa
United States of America

Emeritus Professor of Medicine
University of California, San Francisco
United States of America

JAYPEE *The Health Sciences Publisher*
New Delhi | London | Philadelphia | Panama

 **Jaypee Brothers Medical Publishers (P) Ltd**

**Headquarters**

Jaypee Brothers Medical Publishers (P) Ltd
4838/24, Ansari Road, Daryaganj
New Delhi 110 002, India
Phone: +91-11-43574357
Fax: +91-11-43574314
Email: jaypee@jaypeebrothers.com

**Overseas Offices**

J.P. Medical Ltd
83 Victoria Street, London
SW1H 0HW (UK)
Phone: +44 20 3170 8910
Fax: +44 (0)20 3008 6180
Email: info@jpmedpub.com

Jaypee-Highlights Medical Publishers Inc
City of Knowledge, Bld. 237, Clayton
Panama City, Panama
Phone: +1 507-301-0496
Fax: +1 507-301-0499
Email: cservice@jphmedical.com

Jaypee Medical Inc
The Bourse
111 South Independence Mall East
Suite 835, Philadelphia, PA 19106, USA
Phone: +1 267-519-9789
Email: jpmed.us@gmail.com

Jaypee Brothers Medical Publishers (P) Ltd
17/1-B Babar Road, Block-B, Shaymali
Mohammadpur, Dhaka-1207
Bangladesh
Mobile: +08801912003485
Email: jaypeedhaka@gmail.com

Jaypee Brothers Medical Publishers (P) Ltd
Bhotahity, Kathmandu, Nepal
Phone: +977-9741283608
Email: kathmandu@jaypeebrothers.com

Website: www.jaypeebrothers.com
Website: www.jaypeedigital.com

© 2015, Jaypee Brothers Medical Publishers

The views and opinions expressed in this book are solely those of the original contributor(s)/author(s) and do not necessarily represent those of editor(s) of the book.

All rights reserved. No part of this publication may be reproduced, stored or transmitted in any form or by any means, electronic, mechanical, photocopying, recording or otherwise, without the prior permission in writing of the publishers.

All brand names and product names used in this book are trade names, service marks, trademarks or registered trademarks of their respective owners. The publisher is not associated with any product or vendor mentioned in this book.

Medical knowledge and practice change constantly. This book is designed to provide accurate, authoritative information about the subject matter in question. However, readers are advised to check the most current information available on procedures included and check information from the manufacturer of each product to be administered, to verify the recommended dose, formula, method and duration of administration, adverse effects and contraindications. It is the responsibility of the practitioner to take all appropriate safety precautions. Neither the publisher nor the author(s)/editor(s) assume any liability for any injury and/or damage to persons or property arising from or related to use of material in this book.

This book is sold on the understanding that the publisher is not engaged in providing professional medical services. If such advice or services are required, the services of a competent medical professional should be sought.

Every effort has been made where necessary to contact holders of copyright to obtain permission to reproduce copyright material. If any have been inadvertently overlooked, the publisher will be pleased to make the necessary arrangements at the first opportunity.

**Inquiries for bulk sales may be solicited at:** jaypee@jaypeebrothers.com

*Manual of Electrophysiology*

First Edition: **2015**

ISBN 978-93-5152-664-3

Printed at : Samrat Offset Pvt. Ltd.

# Contributors

**Alexander Mazur** MD
Associate Professor of Medicine
The Carver College of Medicine
University of Iowa, USA

**Arthur C Kendig** MD
Associate Professor of Medicine
The Carver College of Medicine
University of Iowa, USA

**Brian Olshansky** MD
Professor of Medicine
The Carver College of Medicine
University of Iowa, USA

**Christine Miyake** MD
The Carver College of Medicine
University of Iowa, USA

**David Singh** MD
Department of Cardiology
University of California
San Francisco, USA

**Dwayne N Campbell** MD
The Carver College of Medicine
University of Iowa, USA

**Frank I Marcus** MD
Professor of Medicine
University of Arizona School of Medicine
Tucson, Arizona, USA

**Fred Kusumoto** MD
Professor of Medicine
Mayo Clinic
Jacksonville, Florida, USA

**Gordon A Ewy** MD
Professor of Medicine
University of Arizona College of Medicine
Director, University of Arizona Sarver Heart Center
Tucson, Arizona, USA

**Indrajit Choudhuri** MD
University of Wisconsin Medical School and Public Health
Department of Medicine
Cardiovascular Disease Section
Sinai/St Lukes Medical Centers
Milwaukee, Wisconsin, USA

**James B Martins** MD
Professor of Medicine
The Carver College of Medicine
University of Iowa, USA

**Jeffrey E Olgin** MD
Ernest Gallo-Kanu Chatterjee Distinguished
Professor of Medicine
Director, Chatterjee Center for Cardiac Research
Professor of Medicine
University of California
San Francisco, USA

**Jooby John** MD
Interventional Cardiology
Lenox Hill Hospital
New York, USA

**Mark Anderson** MD PhD
Professor, Departments of Internal Medicine and Molecular Physiology and Biophysics
Head, Department of Internal Medicine
Francois M Abboud Chair in Internal Medicine
The Carver College of Medicine
University of Iowa, USA

**Masood Akhtar** MD
Clinical Professor of Medicine
University of Wisconsin Medical School and Public Health
Department of Medicine
Cardiovascular Disease Section
Electrophysiology
Sinai/St Luke's Medical Centers
Milwaukee, Wisconsin, USA

**Melvin Scheinman** MD
Professor of Medicine
University of California
San Francisco, USA

**Moniek GJP Cox**
University of Arizona College of Medicine
Tucson Arizona, USA

**Nitish Badhwar** MD
Associate Professor of Medicine
University of California
San Francisco, USA

**Nora Goldschlager** MD
Professor of Medicine
University of California
San Francisco, USA

**Peter J Mohler** PhD
Professor of Medicine
The Ohio State University of
Medicine and Public Health
Columbus, Ohio, USA

**Rakesh Gopinathannair** MD MA
University of Kentucky
Kentucky, USA

**Renee M Sullivan** MD
Department of Cardiology
The Carver College of Medicine
University of Iowa, USA

**Richard E Kerber** MD
Professor of Medicine
The Carver College of Medicine
University of Iowa, USA

**Richard NW Hauer** MD
University of Arizona School of
Medicine
Tucson, Arizona, USA

**Seyed Hashemi** MD
Division of Cardiology
The Carver College of Medicine
University of Iowa, USA

**Vasanth Vedantham** MD PhD
Division of Cardiology
Electrophysiology Section
University of California
San Francisco, USA

**Vijay Ramu** MD
Mayo Clinic Medical Center
Jacksonville, Florida, USA

**Wei Wei Li** MD PhD
Fellow in Cardiology
Electrophysiology Section
The Carver College of Medicine
University of Iowa, USA

**Yanfei Yang** MD
Department of Cardiology
Electrophysiology Section
University of California
San Francisco, USA

# Preface

There have been revolutionary changes in the field of pathophysiologic mechanisms, diagnostic modalities, and management of heart diseases. Electrophysiology has a very important role in ensuring accurate clinical diagnoses of heart diseases. Many neurological diseases cause symptoms that manifest far from the injured or deceased tissues. Locating and treating all the affected areas of the body is essential for proper patient care. Cardiac electrophysiology allows for the investigation of abnormal electrical signals in the heart tissues. It provides quantitative data to clinicians, supporting diagnostic processes, and evaluating treatment success. *Manual of Electrophysiology* has been designed for the readers seeking a comprehensive overview of all the aspects of electrophysiological studies of the heart. Written by star-studded authors of international repute, the book focuses on the current understanding and the recent advances that are taking place at a fast pace in the field.

The book covers detail discussion on electrophysiological studies, arrhythmia mechanisms, syncope, atrial fibrillation, antiarrhythmic drugs, ventricular and supraventricular tachycardia, bradycardia and heart block, arrhythmogenic right ventricular dysplasia/cardiomyopathy, long QT (LQT) syndrome, short QT (SQT) syndrome, and Brugada syndrome, cardiac resynchronization therapy, cardiac arrest and resuscitation, ambulatory electrocardiographic monitoring, risk stratification of sudden cardiac death, and cardiocerebral resuscitation. The book provides an easy-to-follow format containing practical advice to correctly diagnose the disease with a focus on hands-on therapeutic guidance to the clinicians.

I sincerely thank Shri Jitendar P Vij (Group Chairman), Mr Ankit Vij (Group President), Mr Tarun Duneja (Director-Publishing), Ms Samina Khan (PA to Director-Publishing), Dr Richa Saxena, and the expert team of M/s Jaypee Brothers Medical Publishers (P) Ltd., New Delhi, India for their hard work and professional expertise, without which the book could not have been published.

**Kanu Chatterjee**

# Contents

## 1. Arrhythmia Mechanisms — 1
Mark Anderson
- Arrhythmia Initiation   *2*

## 2. Antiarrhythmic Drugs — 28
Rakesh Gopinathannair, Brian Olshansky
- Arrhythmia Mechanisms and Antiarrhythmic Drugs   *29*
- Indications for Antiarrhythmic Drug Therapy   *30*
- Proarrhythmia   *30*
- Classification Scheme   *31*
- Vaughan-Williams Classification   *31*
- Miscellaneous Drugs   *66*
- Newer Drugs   *67*
- Emerging Antiarrhythmic Drugs   *70*
- Antiarrhythmic Drug Selection in Atrial Fibrillation   *70*
- Outpatient versus In-Hospital Initiation for Antiarrhythmic Drug Therapy   *71*
- Antiarrhythmic Drugs in Pregnancy and Lactation   *72*
- Comparing Antiarrhythmic Drugs to Implantable Cardioverter Defibrillators in Patients at Risk of Arrhythmic Death   *73*
- Antiarrhythmic Drug-device Interactions   *74*

## 3. Electrophysiology Studies — 84
Indrajit Choudhuri, Masood Akhtar
- Cardiac Electrophysiology Study: Philosophy, Requirements, and Basic Techniques   *85*
- Fundamentals of the Cardiac Electrophysiology Study   *93*
- Programmed Electrical Stimulation and Associated Electrophysiology   *101*
- Cardiac Electrophysiology Study for Evaluation of Drug Therapy   *123*
- Electrophysiology Study to Guide Ablative Therapy   *123*
- Complications   *126*

## 4. Syncope — 130
Vijay Ramu, Fred Kusumoto, Nora Goldschlager
- Epidemiology   *131*
- Diagnostic Tests   *135*
- Approach to the Evaluation of Syncope   *149*
- Specific Patient Groups   *151*
- Syncope and Driving   *157*

## 5. Atrial Fibrillation — 169
Vasanth Vedantham, Jeffrey E Olgin
- Definition and Classification   *169*
- Epidemiology   *170*
- Etiology and Pathogenesis   *172*
- Diagnosis   *177*
- Management   *181*

## 6. Supraventricular Tachycardia — 206
*Renee M Sullivan, Wei Wei Li, Brian Olshansky*

- Classification  *207*
- Diagnosis  *222*
- Treatments  *226*

## 7. Clinical Spectrum of Ventricular Tachycardia — 243
*Masood Akhtar*

- Monomorphic Ventricular Tachycardia  *245*
- Polymorphic Ventricular Tachycardia  *254*

## 8. Bradycardia and Heart Block — 266
*Arthur C Kendig, James B Martins*

- Conduction System Anatomy and Development  *266*
- Bradycardia Syndromes/Diseases  *267*
- Clinical Presentation  *270*
- Measurement/Diagnosis  *270*
- Sinus Node Disease  *271*
- AV Node Disease  *272*
- Hemiblock  *274*
- Bundle Branch Block  *276*
- Treatment  *277*

## 9. Arrhythmogenic Right Ventricular Dysplasia/Cardiomyopathy — 281
*Richard NW Hauer, Frank I Marcus, Moniek GJP Cox*

- Molecular and Genetic Background  *283*
- Epidemiology  *289*
- Clinical Presentation  *289*
- Clinical Diagnosis  *290*
- Non-classical ARVD/C Subtypes  *299*
- Differential Diagnosis  *300*
- Molecular Genetic Analysis  *301*
- Prognosis and Therapy  *302*

## 10. Long QT, Short QT and Brugada Syndromes — 310
*Seyed Hashemi, Peter J Mohler*

- LQT Syndrome  *310*
- SQT Syndrome  *320*
- Brugada Syndrome  *322*

## 11. Surgical and Catheter Ablation of Cardiac Arrhythmias — 331
*Yanfei Yang, David Singh, Nitish Badhwar, Melvin Scheinman*

- Supraventricular Tachycardia  *331*
- Atrioventricular Nodal Re-entrant Tachycardia  *333*
- Wolff-Parkinson-White Syndrome and Atrioventricular Re-entrant Tachycardia  *336*
- Focal Atrial Tachycardia  *339*
- Atrial Flutter  *342*
- Ablation of Ventricular Tachycardia in Patients with Structural Cardiac Disease  *348*
- Idiopathic Ventricular Tachycardia  *364*

## 12. Cardiac Resynchronization Therapy 390
*David Singh, Nitish Badhwar*

- *CRT: Rationale for Use  391*
- *CRT in Practice  391*
- *Summary of CRT Benefit  398*
- *Prediction of Response to CRT Therapy  398*
- *Role of Dyssynchrony Imaging  402*
- *Dyssynchrony Summary  409*
- *LV Lead Placement  409*
- *CRT Complications  409*
- *Emerging CRT Indications  411*

## 13. Ambulatory Electrocardiographic Monitoring 431
*Renee M Sullivan, Brian Olshansky, James B Martins, Alexander Mazur*

- *Holter Monitoring  432*
- *Event Recorders  437*
- *Mobile Cardiac Outpatient Telemetry  440*
- *Implantable Loop Recorders  441*
- *Key Considerations in Selecting a Monitoring Modality  444*

## 14. Cardiac Arrest and Resuscitation 450
*Christine Miyake, Richard E Kerber*

- *Overview or Background  450*
- *Basic Life Support  459*
- *Advanced Cardiac Life Support  467*
- *Cessation of Resuscitation  478*
- *Post-resuscitation Care  479*

## 15. Risk Stratification for Sudden Cardiac Death 488
*Dwayne N Campbell, James B Martins*

- *Healthy Athletes  488*
- *Brugada Syndrome  490*
- *Long QT Interval Syndrome  490*
- *Early Repolarization  491*
- *Short QT Syndrome  491*
- *Catecholamine Polymorphic Ventricular Tachycardia  491*
- *Wolff-Parkinson-White Syndrome  491*
- *Arrhythmogenic Right Ventricular Cardiomyopathy  492*
- *Hypertrophic Cardiomyopathy  493*
- *Marfan Syndrome  493*
- *Noncompaction  494*
- *Congenital Heart Disease  494*
- *Non-ischemic Cardiomyopathy  495*
- *Coronary Artery Disease  496*

## 16. Cardiocerebral Resuscitation for Primary Cardiac Arrest 503
*Jooby John, Gordon A Ewy*

- *Etiology and Pathophysiology of Cardiac Arrest  505*
- *Drug Therapy in Cardiac Resuscitation  525*
- *Cardiac Resuscitation Centers  526*
- *Ending Resuscitative Efforts  529*

*Index* 537

CHAPTER 1

# Arrhythmia Mechanisms

*Mark Anderson*

## Chapter Outline

- Arrhythmia Initiation
  - Molecular and Cellular Mechanisms
  - Action Potentials Require Orchestrated Ion Channel Opening and Inactivation
  - Action Potential Physiology is a Consequence of Ion Channel and Cellular Properties
  - Action Potentials are Designed for Automaticity and to Initiate Contraction
  - Action Potential Physiology is Reflected by the Surface Electrocardiogram
  - Afterdepolarizations and Triggered Arrhythmias
  - Proarrhythmic Substrates
  - Proarrhythmic Triggers and Substrates are Promoted in Failing Hearts

## INTRODUCTION

Arrhythmias require initiating conditions and a hospitable substrate for perpetuation. Triggers and substrates are often considered as unrelated or independent events. However, new findings suggest that triggers and substrates may be connected, particularly in structural heart disease, by hyperactivity of signaling molecules, intracellular $Ca^{2+}$ and reactive oxygen species (ROS).[1] There is now a body of evidence to support a view that the increased ROS and disturbed intracellular $Ca^{2+}$ homeostasis that mark structural heart disease contribute to arrhythmia initiation, while actively promoting a proarrhythmic substrate. Ion channels are the fundamental effectors that determine membrane currents and arrhythmias, but ion channels are regulated by multiple factors in myocardium, including intracellular $Ca^{2+}$, phosphorylation and ROS. These same factors participate in responses to common forms of myocardial injury, including ischemia and infarction, which lead to proarrhythmic adaptations in myocardium. This chapter will briefly review ion channel biology, genetic diseases of ion channels, and cellular and tissue arrhythmia mechanisms in an effort to present a broad, but comprehensible, approach to understanding arrhythmia mechanisms.

At a basic level, much of our understanding is due to studies in reduced systems (e.g. isolated heart muscle cells or non-cardiac cells heterologously expressing ion channel proteins) and animal models. However, many key arrhythmia mechanisms, including afterdepolarizations[2,3] and reentry[4] have been identified in patients. In fact, clinical studies and therapies, particularly ablation of focal and reentrant arrhythmias have provided strong evidence for fundamental concepts first formulated from analysis of animal studies. However, not all basic knowledge supporting discussion in this chapter has been translated to and validated in patients.

## ARRHYTHMIA INITIATION

### Molecular and Cellular Mechanisms

Ion channels and exchangers are the fundamental units directing physiological and pathological membrane excitability and conduction.

Equation 1:

$$E = \frac{RT}{zF} \ln \frac{[\text{ion outside cell}]}{[\text{ion inside cell}]} = 2.303 \frac{RT}{zF} \log_{10} \frac{[\text{ion outside cell}]}{[\text{ion inside cell}]}$$

Nernst equation E-equilibrium potential or Nernst potential is the cell membrane potential that is necessary to oppose the diffusion of an ion across the cell membrane as motivated by the concentration gradient of each ion (R—universal gas constant; T—temperature in degrees kelvin; z—valence: F—Faraday's constant). At 25°C, RT/F = 25.693 mV.

Selective membrane permeability coupled with active pumps (ATPases) allow for an electrochemical gradient across cell membranes. The Nernst equation[5] is a powerful, but simplified (i.e. relies exclusively on two ions), description of a half cell that predicts how ionic gradients determine cell membrane potential. The maintenance of $Na^+$ and $K^+$ gradients under conditions of selective membrane permeability requires a $Na^+$ and $K^+$ 'pump'—the $Na^+/K^+$ ATPase. The $Na^+/K^+$ pump transports extracellular $Na^+$ $[Na^+]_o$ and intracellular $K^+$ $[K^+]_i$ against their concentration gradients, a process that requires energy input from ATP hydrolysis. The $Na^+/K^+$ ATPase is required to maintain physiological $[Na^+]_o$ (~ 145 mM), $[K^+]_o$ (~ 4 mM) and $[Na^+]_i$ (~ 10 mM), $[K^+]_i$ (~ 140 mM) in the face of the tendency of these gradients to dissipate with repetitive opening of $Na^+$ and $K^+$ channel proteins. Under resting conditions myocardial cell membrane potentials approximate the equilibrium potential for $K^+$, ~ –90 mV, where the cytosolic side of the membrane is negative and the extracellular side of the membrane is positive, because the cell membrane permeability is greatest for $K^+$ under resting conditions. The resting membrane permeability to $K^+$ occurs because a particular ion channel, the inward rectifier, opens at the negative potentials present in resting membranes.

Equation 2:

$$E_{eq, K^+} = \frac{RT}{zF} \ln \frac{[K^+]_o}{[K^+]_i},$$

*Nernst Equation for $K^+$*

The resting membrane potential is highly dependent upon $[K^+]_o$ and the resting membrane potential determines membrane excitability in part because voltage-gated Na$^+$ channels (mostly Na$_V$1.5) begin to inactivate at membrane potentials more positive than –100 mV. At 37 °C (~ 310 °K) the equilibrium potential for K$^+$ (Eeq, K$^+$) is –91 mV for [K$^+$]$_o$ = 4.5 mM and [K$^+$]$_i$ = 140 mM. If the [K$^+$]$_o$ is reduced to 2.5 mM the Eeq, K$^+$ is –107.5 mV (and more Na$_V$1.5 channels are available to activate), and if the [K$^+$]$_o$ = 6.5 mM, the Eeq, K$^+$ is –82 mV (with reduced Na$_V$1.5 channel availability). Thus, the Nernst equation provides quantitative insight into the importance of K$^+$ homeostasis for normal cardiac electrophysiology.

Ion channels are protein complexes embedded in cell membranes **(Figs 1A to D)**. All ion channels consist of a pore forming α subunit **(Figs 1A to C)**. Some α subunits (e.g. K$^+$ channels) aggregate with identical or similar α subunits to form a cell membrane spanning pore. This pore is the conductance pathway that allows individual ions to cross lipid bilayer membranes with high throughput. Ion channels are configured for relative ion selectivity. The specific amino acids lining the pore create a 'filter' that selects ionic species for conductance based on ionic size and charge. In solution ions are effectively larger due to a sphere of hydration that is a result of charge-associated water molecules. The selectivity filter in ion channels may remove water (dehydrate) from permeant ions as a requirement for passage through the conductance pore. Other α subunits are formed from a single large protein (e.g. Na$^+$ and Ca$^{2+}$ channels). Ion channels open and close in response to a blend of various stimuli. In contracting atrial and ventricular myocardium and in specialized pacemaking [sinoatrial node (SAN)] and conduction tissue (atrioventricular node and His-Purkinje system) the most important and best understood ion channels are primarily opened by changes in membrane potential. These so-called 'voltage-gated ion channels' all contain a cell membrane spanning domain enriched in charged amino acids that act as a membrane voltage sensor **(Figs 1C and D)**. The voltage sensor moves in response to changes in the membrane potential, and these movements are allosterically coupled to the pore domain. Voltage-gated ion channels open and close in response to a change in membrane potential, but also inactivate. Inactivation appears to be the result of various protein conformations that hinder the availability of the pore domain to open in response to a voltage stimulus, before the ion channel is 'reset' by recovering from the state of inactivation. Importantly, voltage-gated ion channels respond to additional factors, including amino acid phosphorylation and oxidation, which influence the probability of ion channels to open **(Fig. 2A)**.

The voltage dependence of ionic current carried by voltage-dependent ion channels and exchangers is often presented as a current-voltage (I-V) relationship **(Figs 2B and C)**. The I-V

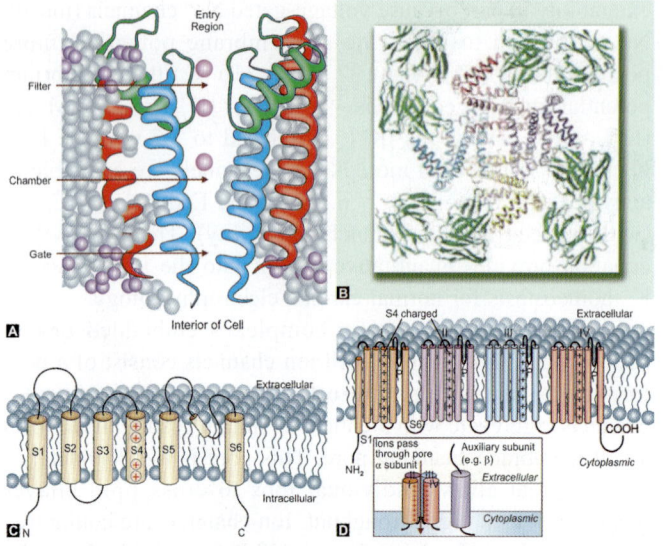

**FIGURES 1A TO D:** Ion channels are proteins that form a conductance pore through bilayer lipid cell membranes. (A) A ribbon diagram representation of the pore forming α subunit for a bacterial voltage-gated K⁺ channel viewed from the side. (B) Ribbon diagram of a voltage-gated K⁺ channel viewed from above. This view shows the fourfold symmetry of α subunit proteins that assemble to form a conductance pore for K⁺ (center). (C) Schematic representation of a voltage-gated K⁺ channel α subunit showing the voltage sensor (S4) and the pore (P) loop between S5 and S6. (D) A schematic representation of a voltage-gated Na⁺ or Ca²⁺ channel that is similar to four concatenated K⁺ channel α subunits

relationship is obtained in voltage-clamped cells or tissue, typically under conditions designed to isolate individual currents (e.g. by controlling the ionic constituents in the intracellular and extracellular solutions, addition of antagonist drugs or pore blocking ions, or by heterologous expression of individual ionic channels in non-excitable cells by gene transfection). The I-V relationships can reveal important ion channel behaviors such as the voltage dependence of activation and inactivation, ion selectivity, rectification and conductance. Voltage-gated ion channels activate and inactivate over a range of membrane potentials. In some cases, the voltage-range of activation and inactivation permits a 'window current' where ion channels can reactivate **(Fig. 2D)**. An important window for voltage-gated Ca²⁺ channel ($Ca_V1$) currents ($I_{Ca}$) occurs during the membrane potentials present during the AP plateau. Excessive $Ca_V1$ window currents are a cause of triggered arrhythmias. Many ion channels (e.g. $Na_V$, $K_V$ and $Ca_V$) have a very high selectivity for their namesake ions under physiological conditions. For example, K⁺ channels are greater than 1,000 times more likely to conduct K⁺ compared to Na⁺. A simple, Ohmic, I-V relationship is linear with the line crossing through

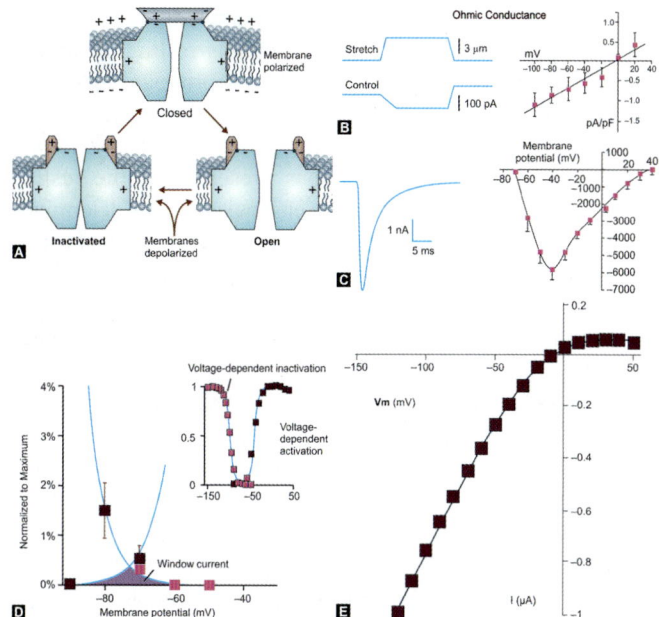

**FIGURES 2A TO E:** Ion channel gating is the process that determines the probability of an α subunit being available to conduct ionic current. (A) A schematic representation of basic gating states: open; closed and inactivated for a voltage-gated ion channel. (B) Examples of a non-rectifying, stretch-activated ionic current (left). The current, normalized to membrane surface area, (pA/pF)-voltage (mV) relationship for this current shows an Ohmic conductance that is linear and passes through zero. (C) The left panel shows an example of a voltage-gated $Na^+$ current that activates rapidly (inward deflection) and then rapidly inactivates (resolution of the inward current back to baseline within a few milliseconds). The right panel shows the parabolic current-voltage relationship that is characteristic of voltage-gated $Na^+$ current in myocardium. (D) An example of a 'window current' for voltage-gated $Na^+$ channels. The shaded overlap between the voltage-dependent loss of $Na^+$ channel availability to open (inactivation, pink boxes) and voltage-dependent $Na^+$ channel activation (purple boxes) is the window current. (E) An example of a current-voltage relationship for an inwardly rectifying $K^+$ channel current ($I_{K1}$).

the zero point **(Fig. 2B)**. However, the I-V relationship of most ion channels in heart is complex, and curvilinear **(Fig. 2C)**. The point of current reversal, or equilibrium potential (mV), can be calculated by the Nernst equation: ∼ +60 for $Na^+$, ∼ −98 for $K^+$ and ∼ +130 for $Ca^{2+}$ under physiological conditions. The I-V relationship is influenced by the electrochemical gradient, which determines where a current transitions from inward to outward (as referenced to the cell membrane and cytoplasm). Convention holds that inward currents are negative and outward currents are positive. The I-V relationship is also affected by a property of some ion channels called rectification. Rectification is the tendency of a current to conduct preferentially inwardly or outwardly. A prominent example is the inwardly rectifying $K^+$ current ($I_{K1}$) that is crucial for determining resting membrane potential in myocardium. $I_{K1}$ exhibits a pronounced inward rectification that is most evident at very negative membrane potentials. However, the physiologically relevant outward current is relatively small and is present near the resting membrane potential **(Fig. 2E)**. Ion channel current is determined by gating properties, including opening probability, conductance, rectification, the electrochemical gradient of a particular ion and ion selectivity. Some ion channels may assume more than a single conductance (i.e. a subconductance state). The $Ca^{2+}$-gated ryanodine receptor $Ca^{2+}$ channel has multiple subconductance states. Ion channel activity is also regulated by ions (e.g. $Ca^{2+}$ and $H^+$), oxidation and phosphorylation.

Ion channel $\alpha$ subunits do not exist or operate in isolation. Accessory subunit proteins, often labeled as $\beta$, $\delta$ and $\gamma$, comprise the ion channel macromolecular complex. These accessory subunits may serve as chaperones to increase expression of $\alpha$ subunit proteins on the cell membrane. Accessory subunits are also targets for regulatory proteins, such as kinases and phosphatases, and may influence the probability of $\alpha$ subunits to open in response to a voltage stimulus. Ion channel macromolecular complexes require precise localization in the cellular ultrastructure to function properly. For example, voltage-gated $Ca^{2+}$ channels, $Ca_V1$, are enriched in T-tubular membranes across from intracellular $Ca^{2+}$ channels called ryanodine receptors (RyR2) that control $Ca^{2+}$ release from the sarcoplasmic reticulum (SR) **(Fig. 3)**.[6,7] Distortion of the relationship of $Ca_V1$ and RyR channels occurs in heart failure and contributes to loss of normal intracellular $Ca^{2+}$ homeostasis, mechanical dysfunction and promotes arrhythmia-initiating afterdepolarizations.[8] Cytoskeletal proteins also contribute to ion channel disposition and localization, and cytoskeletal diseases, such as the ankyrin syndromes,[9,10] cause arrhythmias and other pathological phenotypes in excitable cells in brain and pancreas.

The current view of ion channel structure and function arose using three fundamental investigational approaches. The first was a combination of voltage clamp and mathematical modeling. Voltage clamp uses an operational amplifier with feedback control to 'clamp' a cell membrane at a command potential. By controlling cell membrane potential and the concentration of ions in the cell interior and exterior, it was possible to study individual macroscopic currents that arose from all the ion channels of a particular type operating together on the cell membrane. Originally, voltage clamp studies were focused on very large excitable cells, such as the squid giant axon, which were amenable to early techniques such as Vaseline gap and intracellular electrodes. Hodgkin and Huxley used data obtained

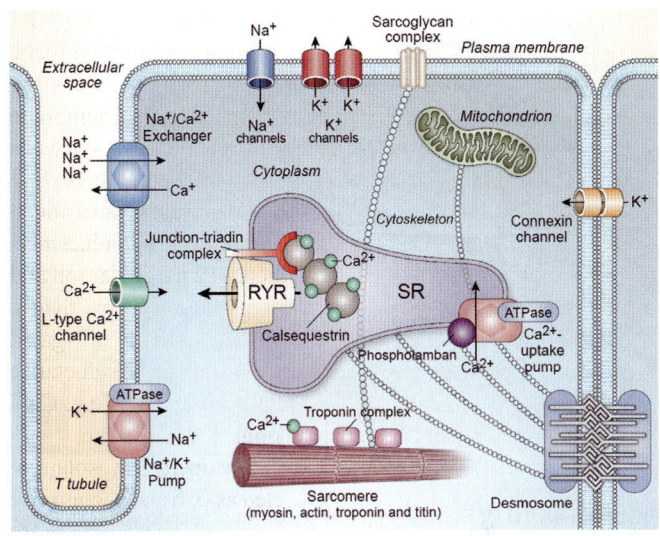

**FIGURE 3:** Myocardial cells are designed for excitation-contraction coupling, the process whereby action potentials generate inward $Ca^{2+}$ current that triggers myofilament-activating $Ca^{2+}$ release from ryanodine receptors (RyR) on the sarcoplasmic reticulum (SR) to cause contraction. The cell membrane ultrastructure formed by T tubules allows $Ca^{2+}$ channels and RyR to face one another across a narrow (~ 10 nm) cytoplasmic space

in squid axon to develop a model of ion channel physiology that postulated 'gates' for activation and inactivation.[11] Their studies provided a conceptual and quantitative framework for understanding ion channels that has endured, albeit with modifications, into the modern era. In 1981, Hammell et al. published the first description of voltage clamp studies using the patch clamp technique **(Figs 4A to D)**.[12] Cardiac myocytes were the subject of one of the first studies using patch clamp that described currents flowing through individual ion channels.[13] Patch clamp allowed for high resistance, giga-Ohm, seals between a glass microelectrode and the cell membrane. This high resistance seal allowed resolution of the extremely small currents associated with individual ion channels (in the pico-Ampere range for $Ca_V$). Patch clamp used in the whole cell mode allowed investigators to measure macroscopic currents in single cells grown in culture or isolated from tissue, and to control intracellular contents by dialysis of an investigator-selected solution. Modern molecular biology techniques of gene cloning and expression were developed after voltage clamp.[14] Expression of wild type and mutant ion channels studied in non-native and native cells allowed investigators to determine the biophysical purpose of various ion channel domains such as the voltage sensor.[15] These 'structure-function' studies provided highly detailed information that led to more complete

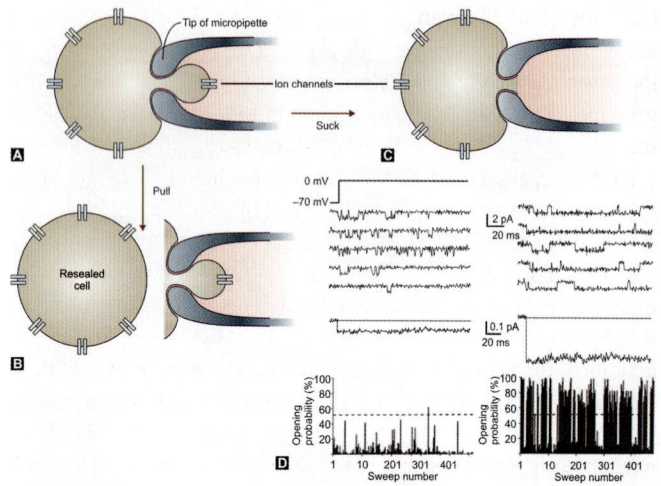

**FIGURES 4A TO D:** Patch clamp is a flexible approach to voltage clamp single cells or cell membrane patches. The high resistance seals (giga Ohm) between the glass micro-pipette and the cell membrane allow for resolution of very small (pA) currents. (A) On cell configuration for recording a subset of ion channels on a cell membrane. (B) Excised membrane patch for recording a subset of ion channels on a cell membrane under conditions where the cytoplasmic constituents can be easily manipulated. (C) Whole cell mode configuration for recording all the ion channels on a cell membrane and where the pipette solution can be dialyzed into the cell. (D) Examples of single $Ca^{2+}$ channel recordings ($Ca_V1.2$) using excised cell membrane patches (as in panel B) at baseline (left panels) and after application of calmodulin kinase II to the cytoplasmic face of the membrane. The top panels show ionic currents from single $Ca_V1.2$ channels in response to a voltage clamp command from –70 to 0 mV. The downward deflections indicate channel openings. The middle tracing is an ensemble current averaged from multiple 'sweeps', as shown in the top five tracings. The bottom panels show a diary plot that indicates the opening probability of the single channel in the recording for each sweep. Panel D is adapted from Dzhura et al. 2000

understanding of ion channel molecular physiology in health and disease. Because ion channel proteins are expressed in cells at relatively low copy number, have prominent lipophilic regions (that allow for membrane insertion) and are large, they are difficult to crystallize. However, the MacKinnon laboratory overcame many of these obstacles by over-expressing bacterial $K^+$ channels,[16,17] which have served as a structural model for many of the voltage-gated cation channels present in heart. The combination of voltage clamp, molecular biology and high resolution structural information form the modern tool kit for understanding cardiac ion channels.

Ion channels are not the only source of ionic membrane currents. In myocardium, the $Na^+/Ca^{2+}$ exchanger is the predominant mechanism for removing $Ca^{2+}$ from the cytoplasm to the extracellular space. The $Na^+/Ca^{2+}$ exchanger transfers a

Ca$^{2+}$ for 3Na$^+$ (forward exchange mode). Because there is a single net positive charge moved to exchange a Ca$^{2+}$ ion from the cytoplasm to the extracellular space, the Na$^+$/Ca$^{2+}$ exchanger produces a small inward Na$^+$ current in forward mode. Although the Na$^+$/Ca$^{2+}$ exchanger does not directly require ATP, the Na$^+$ gradient necessary for forward mode exchange depends upon the ATP-requiring Na$^+$/K$^+$ ATPase. The Na$^+$/K$^+$ ATPase and a sarcolemmal Ca$^{2+}$ ATPase produce small, but measurable currents. The Na$^+$/Ca$^{2+}$ exchanger current, although small in magnitude compared to Na$_V$ or Ca$_V$ channel currents, contributes to AP duration. It is essential for the direct myocardial inotropic actions of digitalis glycosides, which inhibit the Na$^+$/K$^+$ ATPase leading to accumulation of [Na$^+$]$_i$ and consequent increase in [Ca$^{2+}$]$_i$, because the gradient for Ca$^{2+}$ extrusion by Na$^+$/Ca$^{2+}$ exchanger is less favorable than when [Na$^+$]$_i$ is lower. The Na$^+$/Ca$^{2+}$ exchanger is a source of inward currents for arrhythmia triggering afterdepolarizations, as will be discussed below.

## Action Potentials Require Orchestrated Ion Channel Opening and Inactivation

Action potentials are the fundamental unit of membrane excitability **(Fig. 5)**. In most myocardial cells action potentials are initiated by opening of voltage-gated Na$^+$ channels, Na$_V$1.5. The inward Na$_V$1.5 current ($I_{Na}$) depolarizes atrial and ventricular myocytes in a few milliseconds. The brevity of $I_{Na}$ is due to the rapidity of the inactivation process, which competes with activation to modulate the peak current. The membrane potential depolarizes (becomes more positive) from the negative resting potential ($\sim$ –80 mV) to approach the reversal potential for Na$^+$, estimated by the Nernst equation ($\sim$ +50 mV). Specialized myocytes that are dedicated more to automaticity (i.e. SAN) and conduction (i.e. the atrioventricular node) than contraction rely on $I_{Ca}$ for their (phase 0) action potential upstroke. Membrane depolarization activates a combination of voltage-gated ion channels, but the most prominent are depolarizing inward Ca$_V$1.2/1.3 currents ($I_{Ca}$) and several distinct, but structurally related repolarizing inward K$^+$ channel (K$_V$x) currents ($I_K$). The interplay between $I_{Ca}$ and $I_K$ largely determines the duration of the myocardial action potentials, which last hundreds of milliseconds. Atrial and ventricular myocardial action potentials have different shapes and electrophysiological properties. In fact, there are important heterogeneities in action potential configuration within the atrium and ventricle. The ventricular endocardium, mid-myocardium and epicardium show prominent differences in action potential configuration, due to variability in expression of repolarizing K$^+$ currents **(Fig. 6)**. While the physiological

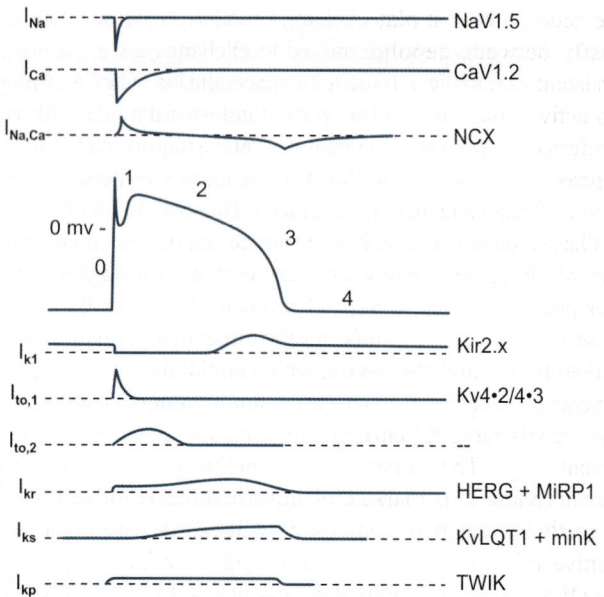

**FIGURE 5:** The action potential duration and configuration is shaped by the interplay between inward and outward-going ionic currents. The top two tracings represent $Na_V1.5$ and $Ca_V1.2$ inward currents that initiate and sustain action potential depolarization. The third tracing from the top is the $Na^+/Ca^{2+}$ exchanger (NCX) that can produce inward (forward mode) and outward (reverse mode) currents at various action potential phases. The ventricular action potential is labeled by phase (0–4). The lower six tracings represent some of the $K^+$ currents that contribute to action potential repolarization

benefit of action potential heterogeneity is unknown, the heterogeneities are affected by $K^+$ channel antagonist drugs and by electrical remodeling during heart failure, where expression of various repolarizing $K^+$ channels is reduced.[18] In addition to voltage-gated ion channels and exchangers, there is an increasing recognition that other non-voltage-gated ion channels contribute to action potential configuration. A more complete discussion of these channels is reviewed elsewhere.[19,20]

## Action Potential Physiology Is a Consequence of Ion Channel and Cellular Properties

Myocardial action potentials are distinguished from action potentials in other excitable tissues by their extreme length, lasting up to hundreds of milliseconds. In contrast, action potentials in most neurons last only a few milliseconds. Cardiac action potentials are often described in phases **(Fig. 5)**. Phase 0 marks the abrupt depolarization from the resting potential and is attributable to $Na_V1.5$ current in most myocardial cells. Cardiac action potentials are long because of their plateau.

The action potential plateau occurs because of a fine balance, mostly between depolarizing inward $Ca_V$ current, a small persistent (slowly inactivating) component of $Na_V1.5$ current, and activation of repolarizing $K^+$ currents. The initial plateau is referred to as phase 2, while the later plateau is referred to as phase 3. In electrically healthy myocardium phase 3 is the period of repolarization to resting membrane potential (phase 4). Phase 3 occurs as inward currents inactivate and repolarizing currents become preeminent. Phase 1 occurs immediately after peak membrane potential depolarization (i.e. the end of phase 0) and where prominent (e.g. ventricular epicardium) is marked by a 'notch' that is due to a combination of $K_V$ channel currents that support a transient inward current ($I_{to}$) and a more rapid repolarizing $K^+$ current (the ultrarapid transient outward current, $I_{Kur}$). The initial component of the action potential plateau (phase 2) is marked by high membrane resistance (R), so small increases in net inward current lead to prominent positive increases in membrane voltage, according to Ohm's law ($V = I \times R$). In automatic cells phase 4 is not stable, but instead consists of an increasing positive membrane potential in late diastole that leads to activation of $Ca_V$ channel currents to initiate phase 1 AP depolarization. Thus, a rich diversity of ion channels contributes to various AP configurations. These AP configurations are matched to the purpose of particular myocardial cells (e.g. pacing or contraction), but in disease AP parameters are directly relevant to arrhythmia initiation and perpetuation.

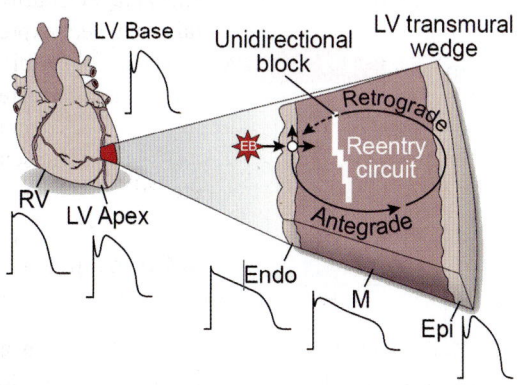

**FIGURE 6:** Ventricular action potentials are heterogenous and vary between base and apex and across the myocardium from endocardium to epicardium. M cells in the mid-myocardium have characteristically long action potentials with a reduced phase 1. Structural defects, such as scar tissue, can serve as a structural barrier that supports a reentry circuit for arrhythmias. Exaggeration of action potential heterogeneities, by genetic disease or acquired disease, can also support a reentry circuit, even in the absence of scar

Action potentials can be repetitively initiated in atrial and ventricular myocardium within the time constraints of the tissue refractory period **(Figs 7A and B)**. The refractory period is determined in large part by the duration of the cardiac action potential. Action potentials are initiated by positive (inward) current sufficient to depolarize the membrane potential to the threshold for activation of $Na_V1.5$ in contracting myocardium or $Ca_V1$ in specialized conduction tissue. During phase 2 of the action potential plateau myocardial cells are absolutely refractory, meaning that no amount of inward current is adequate

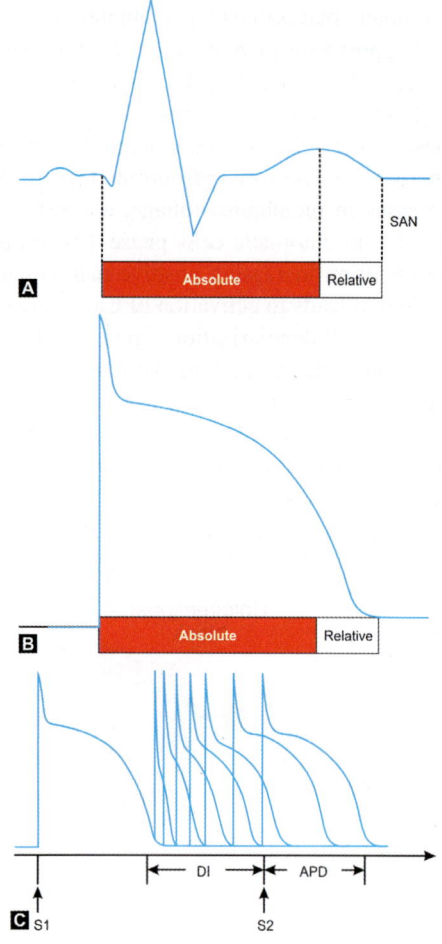

**FIGURES 7A TO C:** Tissue refractoriness to excitation is determined by action potential repolarization and reflected in the surface ECG. (A) A schematic ECG tracing. (B) The surface ECG is a reflection of many action potentials. Myocardial tissue is absolutely refractory to repeat stimulation (dark bars) until late in repolarization. Tissue is potentially excitable prior to completion of repolarization, but initiation of excitation requires a supranormal depolarizing current, a state of relative refractoriness (light bars). (C) Action potential restitution is revealed by a premature stimulus (S2) deployed over a range of coupling intervals to elicit an action potential. Later in the course of action potential repolarization (phase 3) an action potential can be stimulated, but only by a larger inward current than would be necessary after completion of action potential repolarization. Tissue where an action potential can only be stimulated by a supranormal current is said to be relatively refractory. Under physiological conditions action potentials shorten in response to shorter stimulation intervals (i.e. faster rates), due to a process called restitution **(Fig. 7C)**. Action potential restitution occurs, in part, because rapid simulation enhances net outward repolarizing current. Action potential restitution is impaired in genetic long QT syndromes (LQTS), where repolarizing currents are defective, or in common forms of heart failure where reduction in repolarizing currents is a signature event in the proarrhythmic electrical remodeling process. Tissue refractoriness can persist after action potential repolarization under conditions of reduced availability of inward currents responsible for phase 0 depolarization (i.e. $Na_V1.5$ in contracting myocardium and $Ca_V1.2$ and $Ca_V1.3$ in specialized conduction tissue). Various factors contribute to availability of these channels to open, including cell membrane potential (e.g. fewer $Na_V$ and $Ca_V$ channels are available to open at depolarized potentials because membrane depolarization favors inactivation), oxidation, pH, $[Ca^{2+}]_i$, ischemia and autonomic tone. Thus, cell membrane excitability depends on multiple input variables that ultimately converge on ion channels and APs.

The rate that APs are conducted across myocardium (i.e. the conduction velocity) is determined by two principle factors. The first are determined by inputs that affect phase 0: availability of $Na_V$ currents in contracting myocardium and $Ca_V$ currents in specialized conducting and automatic tissue. The second is the efficacy of electrical coupling between myocardial cells. Myocardial cells are electrically coupled by connexin hemichannels that cooperate to form a conductance pore between adjacent cells. The predominant connexin (Cx) type is specific to atrium, ventricle and specialized conduction tissue. Cx 40 and 43 are the major forms in atrium, Cx 43 is the major form in ventricle and Cx 45 is the major form in sinus node, AV node and His-Purkinje cells. Longitudinal intercellular coupling is favored in ventricular myocardium, based on the greater density of Cx 43, compared to side-to-side connections. Conduction velocity is more rapid in the longitudinal direction, due to the greater density of Cx 43 and because $Na_V1.5$ is enriched at the longitudinal junctions, analogous to Nodes of Ranvier in neurons.[21] Like voltage-gated ion channels, Cxs are part of a substantial macromolecular complex that influences intercellular conduction. Altered Cx behavior, localization and expression[22] contributes to conduction velocity dispersion and slowing that are critical components of the proarrhythmic substrate in rare genetic diseases and common forms of structural heart disease.

## Action Potentials Are Designed for Automaticity and to Initiate Contraction

Myocardial action potentials are committed to the major tasks of myocardium: rhythmic, repetitive beating and mechanical work that propels blood through the circulatory system. Sinoatrial node (SAN) action potentials have a specialized, late diastolic component or phase 4 where membrane depolarization leads to activation of $Ca_V$ channel currents to drive phase 0 depolarization. The slope of phase 4 is the membrane potential mechanism for increasing (steeper slope) or decreasing (shallower slope) heart rate **(Fig. 8)**. In healthy hearts, the activity of phase 4 is largely confined to the SAN, where the steady increase in net inward current during late diastolic depolarization is augmented by β adrenergic receptor stimulation and reduced by muscarinic receptor stimulation. Multiple currents likely contribute to physiological phase 4 depolarization in SAN, but recent evidence suggests that two currents play a critical role in physiological pacing. The classical 'pacemaker' current is a $Na^+/K^+$ selective cation current carried by an *HCN4* gene encoded channel. The *HCN4* current, also called the funny current ($I_f$) is enhanced

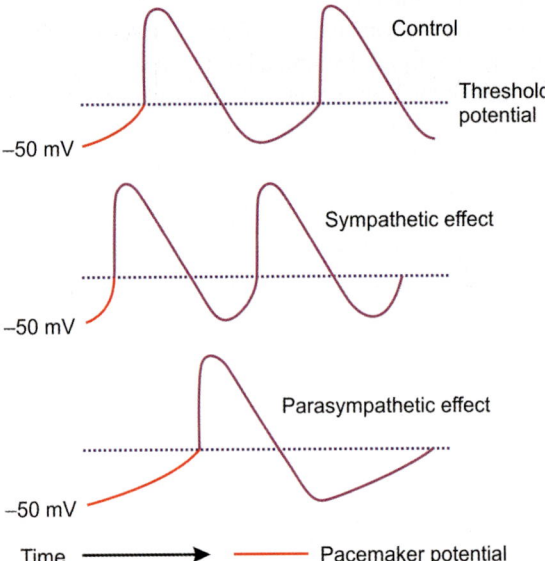

**FIGURE 8:** The cell membrane potential for determining heart rate in sinoatrial nodal cells is set by the steepness of phase 4 (pacemaker) potential. Steeper phase 4 allows the membrane potential to reach the threshold for action potential initiation more rapidly than shallow phase 4 depolarization

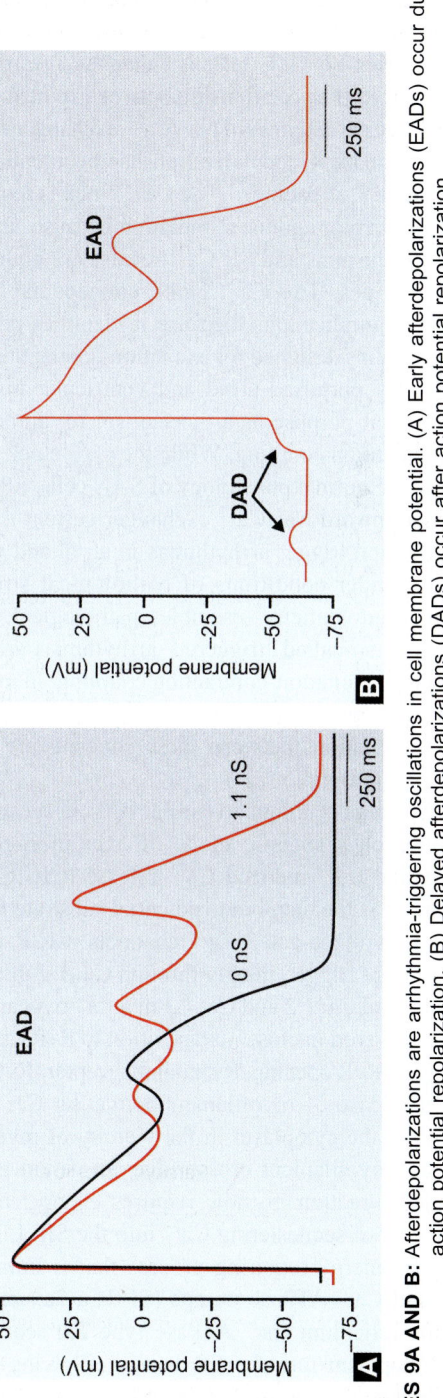

**FIGURES 9A AND B:** Afterdepolarizations are arrhythmia-triggering oscillations in cell membrane potential. (A) Early afterdepolarizations (EADs) occur during action potential repolarization. (B) Delayed afterdepolarizations (DADs) occur after action potential repolarization

by cyclic AMP, which confers increased activity (and steeper phase 4) with α adrenergic receptor agonist stimulation.[23,24] More recent understanding of physiological automaticity in SAN cells suggests that SR $Ca^{2+}$ release enhances inward $Na^+/Ca^{2+}$ exchanger current. The relationship between spontaneous SAN cell SR $Ca^{2+}$ release and inward $Na^+/Ca^{2+}$ exchanger current that contributes to phase 4 depolarization has been called a '$Ca^{2+}$ clock mechanism' of pacing.[25] The $Ca^{2+}$ clock is responsive to α adrenergic receptor agonist stimulation because cellular $Ca^{2+}$ entry by $Ca_V1$ currents and SR $Ca^{2+}$ release are both increased by catecholamines. The $Ca^{2+}$ clock concept has important and interesting implications, because it identifies proteins and subcellular systems designed for excitation-contraction coupling in mechanically purposed atrial and ventricular myocardium as serving a dual purpose as a mechanism for automaticity—excitation-excitation coupling. While the $Ca^{2+}$ clock appears to contribute to the normal physiology of SAN cells, SR $Ca^{2+}$ leak and increased inward $Na^+/Ca^{2+}$ exchanger current is known to induce DADs and trigger arrhythmias in atrial and ventricular myocardium under conditions of pathological stress. Thus, physiological automaticity resembles pathological triggering, suggesting that so-called 'triggered' arrhythmias are a natural consequence of excitation-contraction coupling. In my opinion, the similarities between automaticity and triggering suggest that bright line distinctions between these concepts are no longer warranted or appropriate.

The AP plateau is unique to cardiac muscle because cardiac muscle relies on a specific mode of excitation-contraction coupling called $Ca^{2+}$-induced $Ca^{2+}$ release (CICR, Fig. 3).[26] The AP plateau is the membrane potential substrate for grading $Ca^{2+}$ entry by voltage-gated $Ca^{2+}$ channels. CICR is initiated by a $Ca^{2+}$ current trigger, mostly through $Ca_V1.2$ in ventricular myocardium, and $Ca_V1.2$ and $Ca_V1.3$ in atrial myocardium. $Ca_V$ channels are arrayed in close juxtaposition to RyRs and the $Ca_V$ current triggers RyR opening. Ryanodine receptor (RyR) opening results in a release of myofilament-activating $Ca^{2+}$ from the SR lumen into the cytoplasm in the vicinity of myofilaments. $Ca^{2+}$ triggers myofilament crossbridge formation that causes myocardial contraction. Systole requires energy, in part, due to the ATP cost of sequestering $Ca^{2+}$ into the SR. Like systole, diastole is an energy requiring process that is initiated when the SR bound $Ca^{2+}$ ATPase pumps (SERCa 2a: sarcoplasmic endoplasmic reticulum $Ca^{2+}$ ATPase type 2a) sequester $Ca^{2+}$ from the cytoplasm into the SR lumen, allowing release of myofilament crossbridge formation and myocardial relaxation. SR $Ca^{2+}$ release occurs in a highly structured subcellular domain, resulting in very high local $[Ca^{2+}]_i$. SR $Ca^{2+}$ affects myocardial ion channels, particularly $Ca_V1$ and the $Na^+/Ca^{2+}$ exchanger. The actions at $Ca_V1$ currents are complex, and include conflicting processes called facilitation (peak current is increased and inactivation is reduced) and $Ca^{2+}$ dependent inactivation (peak current is reduced and inactivation is increased). These processes are labile and may have marked influence on the shape, duration and stability of the AP plateau. In our opinion, the best available evidence suggests that $Ca_V1$ channel current facilitation is due to phosphorylation of a specific residue on the $Ca_V1$ β subunit by the multifunctional $Ca^{2+}$ and calmodulin-dependent protein kinase II (CaMKII).[27] $Ca_V1$ channels current facilitation occurs because $Ca_V1$ channels enter a highly active gating mode after CaMKII phosphorylation where the probability of channel opening rises significantly above baseline.[28] CaMKII actions on $Ca_V1$ channels cause proarrhythmic afterdepolarizations and arrhythmias.[29-31]

## Action Potential Physiology Is Reflected by the Surface Electrocardiogram

The electrocardiogram (ECG) is one of the most commonly ordered medical tests in most hospitals. The ECG is a surface report on myocardial electrical activity. Although multiple factors influence ECG parameters, the basic intervals (PR, QRS, QT) reflect ion channel-directed AP parameters **(Figs 7A to C)**. The PR interval is the duration required for an electrical impulse to conduct from the point of 'break out' near the SAN, through atrial myocardium and AVN to the ventricle. In healthy myocardium, this interval will be dominated by the slowest conducting segment, which is in the AVN. In diseased myocardium, impaired atrial and His-Purkinje conduction may contribute to PR prolongation. The QRS interval reflects the speed of conduction and depolarization through the right and left ventricles. The QRS interval can be prolonged by $Na_V$ or Cx gene defects or antagonist drugs, injury or disease in the His-Purkinje system or myocardial injury, including myocardial ischemia, infarction and scar. The QT interval corresponds to ventricular repolarization. Ventricular repolarization is complex, due to the physiological variation in repolarizing ionic currents in endocardium, mid-myocardium and epicardium, as well as between the ventricular apex and base. QT interval prolongation can occur in long QT syndromes that are due to intrinsic defects in repolarizing ionic currents or their cellular localization (LQTS). Ion channel antagonist drugs are the most common reason for QT interval prolongation. Importantly, a wide variety of drugs are antagonists of the hERG (human ether-a-go-go related gene)[32,33] or *KCNH2* encoded $K_V11.1$ $K^+$ channel α subunit protein that conducts the rapid delayed rectifier current ($I_{Kr}$).[34] Rectifier current antagonist properties are a major obstacle for drug development because of the link between QT prolongation, Torsade de Pointes ventricular arrhythmia and sudden death.[35] Diseases of ion channel encoding genes that alter membrane repolarization **(Table 1)** can result in AP and QT interval lengthening (Long QT syndromes) or AP and QT interval shortening (Short QT syndromes).[36] Failing myocardium from a variety of causes (e.g. myocardial infarction, valvular disease, genetic disease) undergoes a proarrhythmic electrical remodeling process where repolarizing $K^+$ currents are reduced resulting in AP and QT interval prolongation.[18] Understanding basic electrophysiological principles constitutes the foundation for understanding arrhythmia mechanisms and for interpreting ECGs.

## Afterdepolarizations and Triggered Arrhythmias

Afterdepolarizations are arrhythmia-initiating oscillations in cell membrane potential. Early afterdepolarizations (EADs) occur during the plateau phases (2 and 3) of AP repolarization. Delayed afterdepolarizations (DADs) occur after AP repolarization, during phase 4 **(Figs 9A and B)**. EADs and DADs can trigger an arrhythmia by propagating to adjacent tissue under favorable source-sink conditions. In theory, EAD and DADs can emerge from an essentially limitless set of conditions, sharing a common requirement that net inward current is enhanced to initiate a depolarizing oscillation in membrane potential. EADs and DADs of sufficient magnitude depolarize the cell membrane to reach the threshold for activation of $Na_V$ and/or $Ca_V$ channel currents to initiate AP phase 0. EADs and DADs that occur at the same time in a sufficient number of cells can lead to a premature AP. One or more premature APs can trigger an arrhythmia by engaging a proarrhythmic substrate supporting reentry. Although there are many potential scenarios for increasing net inward current to initiate EADs or DADs, there is an emerging body of experimental evidence that a common pathway for promoting EADs is reactivation of $Ca_V$ channel currents, while a common pathway favoring DADs is loss of synchronous SR $Ca^{2+}$ release leading to inward $Na^+/Ca^{2+}$ exchanger current. Thus, both EADs and DADs can be thought to arise as a consequence of corruption of key components of CICR.

EADs and DADs are hypothesized to initiate life-threatening arrhythmias in long QT syndromes, catecholaminergic polymorphic VT, atrial fibrillation, and ventricular arrhythmias in heart failure. Long QT syndromes are mostly the result of dominant or dominant negative mutations that cause a defect

## TABLE 1
### A compilation of genetic arrhythmia syndromes due to mutation in ion channel proteins

| | Rhythm | Inheritance | Locus | Ion channel | Gene |
|---|---|---|---|---|---|
| **Long QT syndrome (RW)** | TdP | AD | | | |
| LQT1 | | | 11p15 | $I_{Ks}$ | *KCNQ1, KvLQT1* |
| LQT2 | | | 7q35 | $I_{Kr}$ | *KCNH2, HERG* |
| LQT3 | | | 3p21 | $I_{Na}$ | *SCN5A, Na$_V$1.5* |
| LQT4 | | | 4q25 | | *ANKB, ANK2* |
| LQT5 | | | 21q22 | $I_{Ks}$ | *KCNE1, minK* |
| LQT6 | | | 21q22 | $I_{Kr}$ | *KCNE2, MiRP1* |
| LQT7 (Andersen-Tawil syndrome) | | | 17q23 | $I_{K1}$ | *KCNJ2, Kir 2.1* |
| LQT8 (Timothy syndrome) | | | 6q8A | $I_{Ca}$ | *CACNA1C, Ca$_V$1.2* |
| LQT9 | | | 3p25 | $I_{Na}$ | *CAV3, caveolin-3* |
| LQT10 | | | 11q23.3 | $I_{Na}$ | *SCN4B, Na$_v$b4* |
| **Long QT syndrome (JLN)** | TdP | AR | | | |
| | | | 11p15 | $I_{Ks}$ | *KCNQ1, KvLQT1* |
| | | | 21q22 | $I_{Ks}$ | *KCNE1, minK* |
| **Brugada syndrome** | | | | | |
| BrS1 | PVT | AD | 3p21 | $I_{Na}$ | *SCN5A, Na$_V$1.5* |

*Contd...*

*Contd...*

|  | Rhythm | Inheritance | Locus | Ion channel | Gene |
|---|---|---|---|---|---|
| BrS2 | PVT | AD | 3p24 | $I_{Na}$ | GPD1L |
| BrS3 | PVT | AD | 12p13.3 | $I_{Ca}$ | CACNA1C, $Ca_v1.2$ |
| BrS4 | PVT | AD | 10p12.33 | $I_{Ca}$ | CACNB2b, $Ca_v\beta_{2b}$ |
| **Short QT syndrome** | | | | | |
| SQT1 | VT/VF | AD | 7q35 | $I_{Kr}$ | KCNH2, HERG |
| SQT2 |  | AD | 11p15 | $I_{Ks}$ | KCNQ1, KvLQT1 |
| SQT3 |  | AD | 17q23.1–24.2 | $I_{K1}$ | KCNJ2, Kir2.1 |
| SQT4 |  | AD | 12p13.3 | $I_{Ca}$ | CACNA1C, $Ca_v1.2$ |
| SQT5 |  | AD | 10p12.33 | $I_{Ca}$ | CACNB2b, $Ca_v\beta_{2b}$ |
| **Catecholaminergic VT** | | | | | |
| CPVT1 | VT | AD | 1q42–43 |  | RyR2 |
| CPVT2 | VT | AR | 1p13–21 |  | CASQ2 |

(Abbreviations: AD: Autosomal dominant; AR: Autosomal recessive; JLN: Jervell and Lange-Nielsen; RW: Romano-Ward; TdP: Torsade de pointes; VF: Ventricular fibrillation; VT: Ventricular tachycardia; PVT: Polymorphic VT; $I_{Ks}$: Slowly activating delayed rectifier current; $I_{Kr}$: Rapidly activating delayed rectifier current; $I_{Na}$: Na$^+$ channel current; $I_{K1}$: Inward rectifier current; $I_{Ca}$: Ca$^{2+}$ channel current; GPDIL: Glycerol-3-phosphate dehydrogenase 1-like gene; RyR2: Ryanodine receptor 2 gene; CASQ2: Calsequestrin 2 gene. (*Source:* Antzelevitch 2007)

in depolarization that results in AP prolongation **(Table 1)**, secondary increases in $Ca_V1$ current and afterdepolarizations. CaMKII is activated in atrial fibrillation[37,38] and during AP prolongation,[39] due to enhanced $Ca^{2+}$ entry, and is thought to promote arrhythmias by enhancing $Ca_V1$ current facilitation,[29] the non-inactivating component of $Na_V1.5$[40] and SR $Ca^{2+}$ leak[41] in animal and cellular models. CaMKII inhibition can suppress afterdepolarizations[29,30,39] and arrhythmias[31] without AP or QT interval shortening, suggesting that CaMKII contributes to a critical proarrhythmic connection between AP prolongation and afterdepolarizations. EADs and DADs are also implicated in arrhythmogenesis in heart failure, due to a proarrhythmic electrical remodeling process where $K^+$ current expression is reduced—leading to AP prolongation and increased activity and expression of CaMKII in failing myocardium.[42] CaMKII activity and/or expression are increased in failing myocardium from animal models and from patients.[43] Thus, emerging concepts suggest that afterdepolarizations and excessive CaMKII activity constitute a unified mechanism for arrhythmia triggering in genetic and structural forms of heart disease.[1,44,45] CaMKII may contribute to other competing concepts favoring afterdepolarizations, including RyR2 $Ca^{2+}$ leak due to ROS[46] and hyperphosphorylation by protein kinase A.[47]

## Proarrhythmic Substrates

Cardiac arrhythmias are often initiated by afterdepolarizations, but sustained by a mechanism called reentry **(Fig. 10)**. Reentry can occur over a large tissue domain (e.g. typical atrial flutter, bundle branch reentry ventricular tachycardia, the atrioventricular reciprocating tachycardia), or in a small volume of tissue (e.g. atrioventricular nodal tachycardia, fasicular ventricular tachycardia). Processes that lead to myocardial scar formation, such as myocardial infarction, can favor reentry by producing regions of slowed conduction.[4] Reentry can be supported by an anatomically defined pathway involving scar, specialized conduction tissue, or both. However, functional reentry can occur in structurally normal tissue due to exaggerated electrical inhomogeneities of activation[48,49] or repolarization.

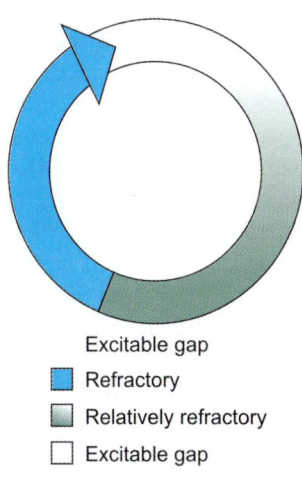

**FIGURE 10:** A simplified reentrant circuit with core components indicated by color coding

- Excitable gap
- ■ Refractory
- ■ Relatively refractory
- □ Excitable gap

Physiological electrical heterogeneity is exaggerated by proarrhythmic drugs, and in animal models of mycoardial hypertrophy.[50] Enhanced dispersion of repolarization is thought to support a voltage gradient that constitutes a functional reentrant circuit. Reduced $I_{Na}$, as occurs in the Brugada Syndrome, can also induce a functional reentrant circuit by unmasking enhanced transient outward $K^+$ current in AP phase 1.[51] In cases of structural heart disease where scar and fibrosis contribute to anatomical reentrant pathways, the exaggeration of heterogeneity of repolarization may also contribute to creation of a sustainable arrhythmia circuit. It is likely that failing human hearts exhibit focal and reentrant arrhythmias,[52,53] with the caveat that an apparent arrhythmia focus could be a 'microreentrant' circuit. Programmed electrical stimulation (discussed in another chapter) can be used to distinguish between reentry and focal arrhythmia mechanisms.

## Proarrhythmic Triggers and Substrates Are Promoted in Failing Hearts

Although afterdepolarizations and reentry are distinct entities, there is a growing appreciation that common biological factors can promote development of proarrhythmic triggers and substrates in heart failure. CaMKII has emerged as a signal that drives structural and electrical components of myocardial injury, providing a molecular rationale to explain why failing hearts are prone to arrhythmias. While it is likely that many signaling molecules participate in promoting afterdepolarizations and proarrhythmic tissue substrates, this concept is best developed for CaMKII. Failing myocardium is consistently marked by AP prolongation, loss of normal intracellular $Ca^{2+}$ homeostasis, increased ROS and increased expression of CaMKII. These factors favor EADs because the prolonged AP plateau occurs over a membrane potential window permissive for $Ca_V1.2$ opening.[54,55,56] CaMKII is activated by $Ca^{2+}$ bound calmodulin and by ROS,[57] and CaMKII mediated phosphorylation leads to high $Ca_V1.2$ activity (so-called mode 2 gating)[28] and afterdepolarizations.[29,31,58] CaMKII actions at a specific site on a $Ca_V1.2$ β subunit (Thr 498)[27] lead to increased cellular $Ca^{2+}$ entry and increased SR $Ca^{2+}$ filling.[29] CaMKII also phosphorylates RyR2 (at Ser 2814)[59] leading to increased RyR2 opening, SR $Ca^{2+}$ leak and afterdepolarizations that promote ventricular arrhythmia in failing hearts.[41] A similar mechanism may also favor atrial fibrillation.[38] RyR2 $Ca^{2+}$ leak can trigger inward $Na^+/Ca^{2+}$ exchanger current[60] that promotes DADs and phase 3 EADs. CaMKII activity at key $Ca^{2+}$ homeostatic proteins ($Ca_V1.2$ and RyR2) promotes loss of normal intracellular $Ca^{2+}$ homeostasis, which may reduce the efficacy of CICR resulting in reduced mechanical performance.[61]

After myocardial infarction the borderzone tissue between non-living scar and normal myocytes serves as a substrate for reentry. Surviving borderzone tissue undergoes electrical remodeling marked by reduced $Na_V1.5$ expression that is due, at least in part, to reduction in ion channel-targeting ankyrin G expression.[21] Loss of $Na_V1.5$ current contributes to conduction slowing. In addition, borderzone tissue is enriched in ROS and ROS activated CaMKII is increased in the MI borderzone,[62] where it may contribute to conduction slowing by effects, at least in part, on $Na_V$ channels.[63] CaMKII activation contributes to scar formation by increasing myocardial death in response to ischemic injury.[64] The pro-survival effects of CaMKII inhibition are likely multifactorial, and have been mapped to $Ca_V1.2$,[29,65] SR $Ca^{2+}$,[64] and mitochondria.[65,66] CaMKII activation after MI results in activation of inflammatory signaling by increased nuclear factor for κB (NF-κB) transcription.[67] Thus, understanding CaMKII signaling provides insight into how a properly positioned nodal signal can produce the twin phenotypes of heart failure and arrhythmias. CaMKII resides at an intersection of the β adrenergic receptor and angiotensin II signaling pathways,[57] both of which are extensively therapeutically validated to improve heart failure symptoms and reduce sudden death after MI. Improved understanding of cellular signaling important for arrhythmias has the potential to lead to more effective and novel non-invasive antiarrhythmic treatments.

## REFERENCES

1. Tomaselli GF, Barth AS. Sudden cardio arrest: oxidative stress irritates the heart. Nat Med. 2010;16:648-9.
2. Kurita T, Ohe T, Shimizu W, et al. Early afterdepolarizationlike activity in patients with class IA induced long QT syndrome and torsades de pointes. Pacing and Clinical Electrophysiology. 1997;20:695-705.
3. Shimizu W, Ohe T, Kurita T, et al. Effects of verapamil and propranalol on early afterdepolarizations and ventricular arrhythmias induced by epinephrine in congenital long QT syndrome. J Am Coll Cardiol. 1995;26:1299-309.
4. Brunckhorst CB, Delacretaz E, Soejima K, et al. Identification of the ventricular tachycardia isthmus after infarction by pace mapping. Circ. 2004;110:652-9.
5. Nernst R. Die elektromotorische wirksamkeit der ionen. Z Phys Chem. 1889;4:129-81.
6. Tanabe T, Mikami A, Numa S, et al. Cardiac-type excitation-contraction coupling in dysgenic skeletal muscle injected with cardiac dihydropyridine receptor cDNA. Nature. 1990;344:451-3.
7. Sun XH, Protasi F, Takahashi M, et al. Molecular architecture of membranes involved in excitation-contraction coupling of cardiac muscle. Journal of Cell Biology. 1995;129:659-71.
8. Song LS, Sobie EA, McCulle S, et al. Orphaned ryanodine receptors in the failing heart 3. Proc Natl Acad Sci USA. 2006;103:4305-10.

9. Mohler PJ, Splawski I, Napolitano C, et al. A cardiac arrhythmia syndrome caused by loss of ankyrin-B function. Proc Natl Acad Sci USA. 2004;101:9137-42.
10. Mohler PJ, Schott JJ, Gramolini AO, et al. Ankyrin-B mutation causes type 4 long-QT cardiac arrhythmia and sudden cardiac death [comment]. Nature. 2003;421:634-9.
11. Hodgkin AL, Huxley AF. A quantitative description of membrane current and its application to conduction and excitation in nerve. Journal of Physiology. 1952;117:500-44.
12. Hamill OP, Marty A, Neher E, et al. Improved patch-clamp techniques for high-resolution current recording from cells and cell-free membrane patches. Pflugers Arch. 1981;391:85-100.
13. Colquhoun D, Neher E, Reuter H, et al. Inward current channels activated by intracellular Ca in cultured cardiac cells. Nature. 1981;294:752-4.
14. Papazian DM, Schwarz TL, Tempel BL, et al. Cloning of genomic and complementary DNA from Shaker: a putative potassium channel gene from Drosophila. Science. 1987;237:749-53.
15. Timpe LC, Schwarz TL, Tempel BL, et al. Expression of functional potassium channels from Shaker cDNA in Xenopus oocytes. Nature. 1988;331:143-5.
16. Doyle DA, Morais CJ, Pfuetzner RA, et al. The structure of the potassium channel: molecular basis of $K^+$ conduction and selectivity. Science. 1998;280:69-77.
17. MacKinnon R, Cohen SL, Kuo A, et al. Structural conservation in prokaryotic and eukaryotic potassium channels. Science. 1998; 280:106-9.
18. Tomaselli GF, Zipes DP. What causes sudden death in heart failure? Circ Res. 2004;95:754-63.
19. Grant AO. Cardiac ion channels. Circ Arrhythm Electrophysiol. 2009;2:185-94.
20. Roden DM, Balser JR, George AL Jr., et al. Cardiac ion channels. Annu Rev Physiol. 2002;64:431-75.
21. Lowe JS, Palygin O, Bhasin N, et al. Voltage-gated $Na_v$ channel targeting in the heart requires an ankyrin-G dependent cellular pathway 1. J Cell Biol. 2008;180:173-86.
22. Hesketh GG, Shah MH, Halperin VL, et al. Ultrastructure and regulation of lateralized connexin43 in the failing heart. Circ Res. 2010;106:1153-63.
23. Ludwig A, Zong X, Stieber J, et al. Two pacemaker channels from human heart with profoundly different activation kinetics. EMBO J. 1999;18:2323-9.
24. Seifert R, Scholten A, Gauss R, et al. Molecular characterization of a slowly gating human hyperpolarization-activated channel predominantly expressed in thalamus, heart, and testis. Proc Natl Acad Sci USA. 1999;96:9391-6.
25. Maltsev VA, Vinogradova TM, Lakatta EG. The emergence of a general theory of the initiation and strength of the heartbeat 1. J Pharmacol Sci. 2006;100:338-69.
26. Fabiato A, Fabiato F. Contractions induced by a calcium-triggered release of calcium from the sarcoplasmic reticulum of single skinned cardiac cells. J Physiol. 1975;249:469-95.
27. Grueter CE, Abiria SA, Dzhura I, et al. L-Type Ca(2+) channel facilitation mediated by phosphorylation of the beta subunit by CaMKII 1. Mol Cell. 2006;23:641-50.

28. Dzhura I, Wu Y, Colbran RJ, et al. Calmodulin kinase determines calcium-dependent facilitation of L-type calcium channels. Nat Cell Biol. 2000;2:173-7.
29. Koval OM, Guan X, Wu Y, et al. $Ca_V1.2$ beta-subunit coordinates CaMKII-triggered cardiomyocyte death and afterdepolarizations. Proc Natl Acad Sci USA. 2010;107:4996-5000.
30. Wu Y, MacMillan LB, McNeill RB, et al. CaM kinase augments cardiac L-type Ca2+ current: a cellular mechanism for long Q-T arrhythmias. American Journal of Physiology. 1999;276:H2168-H2178.
31. Wu Y, Temple J, Zhang R, et al. Calmodulin kinase II and arrhythmias in a mouse model of cardiac hypertrophy. Circ. 2002;106:1288-93.
32. Warmke JW, Ganetzky B. A family of potassium channel genes related to eag in Drosophila and mammals. Proc Natl Acad Sci USA. 1994;91:3438-42.
33. Trudeau MC, Warmke JW, Ganetzky B, et al. HERG, a human inward rectifier in the voltage-gated potassium channel family. Science. 1995;269:92-5.
34. Sanguinetti MC, Jiang C, Curran ME, et al. A mechanistic link between an inherited and an acquired cardiac arrhythmia: HERG encodes the IKr potassium channel. Cell. 1995;81:299-307.
35. Anderson ME, Al Khatib SM, Roden DM, et al. Cardiac repolarization: current knowledge, critical gaps, and new approaches to drug development and patient management. Am Heart J. 2002;144:769-81.
36. Morita H, Wu J, Zipes DP. The QT syndromes: long and short. Lancet. 2008;372:750-63.
37. Tessier S, Karczewski P, Krause EG, et al. Regulation of the transient outward K(+) current by Ca(2+)/calmodulin-dependent protein kinases II in human atrial myocytes. Circulation Research. 1999;85:810-9.
38. Chelu MG, Sarma S, Sood S, et al. Calmodulin kinase II-mediated sarcoplasmic reticulum Ca2+ leak promotes atrial fibrillation in mice. J Clin Invest. 2009;119:1940-51.
39. Anderson ME, Braun AP, Wu Y, et al. KN-93, an inhibitor of multifunctional Ca++/calmodulin-dependent protein kinase, decreases early afterdepolarizations in rabbit heart. J Pharm Exp Ther. 1998;287:996-1006.
40. Wagner S, Dybkova N, Rasenack EC, et al. Ca2+/calmodulin-dependent protein kinase II regulates cardiac Na+ channels. J Clin Invest. 2006;116:3127-38.
41. Ai X, Curran JW, Shannon TR, et al. $Ca^{2+}$/calmodulin-dependent protein kinase modulates cardiac ryanodine receptor phosphorylation and sarcoplasmic reticulum $Ca^{2+}$ leak in heart failure. Circ Res. 2005;97:1314-22.
42. Sag CM, Wadsack DP, Khabbazzadeh S, et al. Calcium/calmodulin-dependent protein kinase II contributes to cardiac arrhythmogenesis in heart failure. Circ Heart Fail. 2009;2:664-75.
43. Zhang T, Brown JH. Role of $Ca^{2+}$/calmodulin-dependent protein kinase II in cardiac hypertrophy and heart failure. Cardiovasc Res 2004;63:476-86.
44. Anderson ME. CaMKII and a failing strategy for growth in heart. J Clin Invest. 2009;119:1082-5.
45. Qi X, Yeh YH, Chartier D, et al. The calcium/calmodulin/kinase system and arrhythmogenic afterdepolarizations in bradycardia-

related acquired long-QT syndrome. Circ Arrhythm Electrophysiol. 2009;2:295-304.
46. Belevych AE, Terentyev D, Viatchenko-Karpinski S, et al. Redox modification of ryanodine receptors underlies calcium alternans in a canine model of sudden cardiac death. Cardiovasc Res. 2009;84: 387-95.
47. Marx SO, Reiken S, Hisamatsu Y, et al. PKA phosphorylation dissociates FKBP12.6 from the calcium release channel (ryanodine receptor): defective regulation in failing hearts [In Process Citation]. Cell. 2000;101:365-76.
48. Ziv O, Morales E, Song YK, et al. Origin of complex behaviour of spatially discordant alternans in a transgenic rabbit model of type 2 long QT syndrome. J Physiol. 2009;587:4661-80.
49. Antzelevitch C. Role of spatial dispersion of repolarization in inherited and acquired sudden cardiac death syndromes. Am J Physiol Heart Circ Physiol. 2007;293:H2024-H38.
50. Volders PG, Sipido KR, Vos MA, et al. Cellular basis of biventricular hypertrophy and arrhythmogenesis in dogs with chronic complete atrioventricular block and acquired torsade de pointes. Circ. 1998;98:1136-47.
51. Yan GX, Antzelevitch C. Cellular basis for the Brugada syndrome and other mechanisms of arrhythmogenesis associated with ST-segment elevation. Circ. 1999;100:1660-6.
52. Pogwizd SM, McKenzie JP, Cain ME. Mechanisms underlying spontaneous and induced ventricular arrhythmias in patients with idiopathic dilated cardiomyopathy. Circ. 1998;98:2404-14.
53. Pogwizd SM, Chung MK, Cain ME. Termination of ventricular tachycardia in the human heart. Insights from three-dimensional mapping of nonsustained and sustained ventricular tachycardias. Circ. 1997;95:2528-40.
54. Antoons G, Volders PG, Stankovicova T, et al. Window $Ca^{2+}$ current and its modulation by $Ca^{2+}$ release in hypertrophied cardiac myocytes from dogs with chronic atrioventricular block 8. J Physiol. 2007;579: 147-60.
55. January CT, Riddle JM, Salata JJ. A model for early after-depolarizations: induction with the $Ca^{2+}$ channel agonist Bay K 8644. Circulation Research. 1988;62:563-71.
56. Wu Y, Kimbrough JT, Colbran RJ, et al. Calmodulin kinase is functionally targeted to the action potential plateau for regulation of L-type $Ca^{2+}$ current in rabbit cardiomyocytes. J Physiol. 2004;554: 145-55.
57. Erickson JR, Joiner ML, Guan X, et al. A dynamic pathway for calcium-independent activation of CaMKII by methionine oxidation. Cell. 2008;133:462-74.
58. Xie LH, Chen F, Karagueuzian HS, et al. Oxidative-stress-induced afterdepolarizations and calmodulin kinase II signaling. Circ Res. 2009;104:79-86.
59. Wehrens XH, Lehnart SE, Reiken SR, et al. $Ca^{2+}$/calmodulin-dependent protein kinase II phosphorylation regulates the cardiac ryanodine receptor. Circ Res. 2004;94:e61-e70.
60. Wu Y, Roden DM, Anderson ME. Calmodulin kinase inhibition prevents development of the arrhythmogenic transient inward current. Circulation Research. 1999;84:906-12.

61. Couchonnal LF, Anderson ME. The role of calmodulin kinase II in myocardial physiology and disease. Physiology (Bethesda). 2008;23: 151-9.
62. Christensen MD, Dun W, Boyden PA, et al. Oxidized calmodulin kinase II regulates conduction following myocardial infarction: a computational analysis. PLoS Comput Biol. 2009;5:e1000583.
63. Hund TJ, Decker KF, Kanter E, et al. Role of activated CaMKII in abnormal calcium homeostasis and I(Na) remodeling after myocardial infarction: insights from mathematical modeling. J Mol Cell Cardiol. 2008;45:420-8.
64. Yang Y, Zhu WZ, Joiner ML, et al. Calmodulin kinase II inhibition protects against myocardial cell apoptosis in vivo 3. Am J Physiol Heart Circ Physiol. 2006;291:H3065–H75.
65. Chen X, Zhang X, Kubo H, et al. $Ca^{2+}$ influx-induced sarcoplasmic reticulum $Ca^{2+}$ overload causes mitochondrial-dependent apoptosis in ventricular myocytes 1. Circ Res. 2005;97:1009-17.
66. Timmins JM, Ozcan L, Seimon TA, et al. Calcium/calmodulin-dependent protein kinase II links ER stress with Fas and mitochondrial apoptosis pathways. J Clin Invest. 2009;119:2925-41.
67. Singh MV, Kapoun A, Higgins L, et al. $Ca^{2+}$/calmodulin-dependent kinase II triggers cell membrane injury by inducing complement factor B gene expression in the mouse heart. J Clin Invest. 2009;119:986-96.

**CHAPTER 2**

# Antiarrhythmic Drugs

*Rakesh Gopinathannair, Brian Olshansky*

## Chapter Outline

- Arrhythmia Mechanisms and Antiarrhythmic Drugs
- Indications for Antiarrhythmic Drug Therapy
- Proarrhythmia
- Classification Scheme
- Vaughan-Williams Classification
  - Class I Antiarrhythmic Drugs: Sodium Channel Blockers
  - Class II Antiarrhythmic Drugs: Beta-Adrenoceptor Blockers
  - Class III Antiarrhythmic Drugs: Drugs that Prolong Repolarization
  - Class IV Antiarrhythmic Drugs: Calcium Channel Antagonists
- Miscellaneous Drugs
  - Adenosine
- Newer Drugs
  - Tedisamil
  - Vernakalant
  - Ivabradine
  - Ranolazine
- Emerging Antiarrhythmic Drugs
- Antiarrhythmic Drug Selection in Atrial Fibrillation
- Outpatient versus in-Hospital Initiation for Antiarrhythmic Drug Therapy
- Antiarrhythmic Drugs in Pregnancy and Lactation
- Comparing Antiarrhythmic Drugs to Implantable Cardioverter Defibrillators in Patients at Risk of Arrhythmic Death
- Antiarrhythmic Drug-device Interactions

## INTRODUCTION

Antiarrhythmic drugs (AADs) were developed to suppress cardiac arrhythmias, and therefore improve survival, symptoms and morbidity. Much of the original data were based on studies performed on cellular preparations and in vivo animal models. Despite a surfeit of supporting data demonstrating that AADs can have potent impact on various cardiac ion channels and receptors to affect arrhythmias, the lofty goal of improving survival and outcomes in patients with cardiovascular disease and arrhythmias have been less than anticipated based on results from large long-term randomized controlled clinical trials.

The AAD therapy has undergone constant evolution as new therapies have emerged and the risk benefit profile of these drugs on major clinical endpoints is better understood. The AAD therapy continues to have a critical role in the management of patients with cardiac arrhythmias, but its place is now better appreciated and understood in light of other advancements, including radiofrequency catheter ablation and implantable devices. The role has transformed, as it is now realized that AADs are often not perfectly effective under all circumstances and there is risk for proarrhythmia. Many older AADs, considered the staple of arrhythmia management for years, have begun to disappear with the emergence of several purportedly safer and potentially more effective therapies.

The history of AAD therapy can be best described as a somewhat sobering transition from "panacea" to "Pandora's box". Currently, AADs, for the most part, are used as an adjunct to therapies that target and cure the rhythm like catheter ablation or those directed against the underlying structural heart disease. This role reversal has resulted from superior efficacy of newer therapies, as well as concerns over the safety and effectiveness of AADs.

Perhaps the biggest concern, notwithstanding mediocre efficacy, is the proarrhythmic, as well as systemic, side effects of AADs. Proarrhythmia may have contributed to the lack of benefit from AADs on hard clinical endpoints. The AADs are among the most complex to prescribe and monitor. Now, with a better understanding of the risks and benefits, AADs are used in a much more regulated and rigorous fashion. Several drugs disappearing from the scenery include quinidine, procainamide, phenytoin, tocainide and bretylium. Others (mexiletine and disopyramide) are used infrequently. Now, there is a better understanding of the proarrhythmic and toxic effects of these AADs. Even though guidelines are developed for their use, AADs is still often used indiscriminately without careful observation for adverse effects. AADs are available, being used and being developed. Proper and effective AAD therapy continues to play an important role to treat symptomatic and potentially life-threatening arrhythmias. The drugs are used to treat a wide variety of sustained and nonsustained atrial and ventricular tachyarrhythmias, as well as atrial and ventricular ectopy. As these drugs play an important role to treat a wide variety of arrhythmias, clinicians who use these drugs must be familiar with their indications, pharmacology, mechanisms of action, dosing, adverse effects, proarrhythmic effects, and interactions with other drugs.

This chapter describes the classification schema as well as clinical pharmacology, adverse effects, and interactions of individual drugs. We will also focus on the clinical applicability of the individual agents based on available clinical data. A small section at the end of the chapter focuses on emerging and investigational AADs.

## ARRHYTHMIA MECHANISMS AND ANTIARRHYTHMIC DRUGS

Cardiac tachyarrhythmias are due to several well understood mechanisms including various forms of reentry, triggered activity and automaticity. The AADs can affect cardiac ionic channels and receptors to affect properties that alter the chance of initiation, perpetuation and termination of tachyarrhythmias. The AADs can affect cardiac excitability, conduction, and refractoriness. The AADs, depending on the type, can block the sodium channel and, therefore, slow down conduction in the myocardium by reducing the electrical gradient of cellular activation (Vmax, rate of rise of phase 0 of the action potential) to reduce the presence of reentrant ventricular and supraventricular arrhythmias. The AADs can also suppress spontaneous depolarization of cells leading to decreased automaticity. Many of the AADs are specific for certain cardiac tissue, such as atrial, AV nodal or ventricular myocardium. Some AADs affect myocardial repolarization by affecting several potassium channels. Other AADs block calcium channels to affect other forms of reentry triggered activity, as well as automaticity dependent on the tissue and the mechanism of the arrhythmia. Some newer AADs also can affect cell-to-cell communications or work by other novel mechanisms.

## INDICATIONS FOR ANTIARRHYTHMIC DRUG THERAPY

The AADs are now mainly used to treat atrial tachyarrhythmias, particularly atrial fibrillation (AF).[1] While mortality outcome with regard to rhythm control with an AAD is not superior to rate control,[2] symptom reduction and improvement in quality of life can be superior in select patients who have AF and atrial flutter. The AADs are used to treat other supraventricular tachyarrhythmias including AV node reentry, sinoatrial reentry, AV reentry tachycardia and atrial tachycardias. Occasionally, AADs are used to suppress ventricular and atrial ectopy including nonsustained and even sustained ventricular tachycardia (VT) but their use is balanced by potential adverse effects. The AADs can be used as primary therapy for patients with idiopathic VT but for patients with underlying structural heart disease and VT, AADs are not generally recommended as primary therapy unless there are specific reasons to do so in lieu of ablation therapy and/or implantable devices. The reason for this is that the proarrhythmic effects of the drugs can exceed the benefits.

## PROARRHYTHMIA

The AADs suppress, and otherwise treat, arrhythmias but they can also create new ones. In some instances, this is simply an increase in the amount of atrial or ventricular ectopy but in the worst case scenario, it can lead to ventricular fibrillation and sudden cardiac death. The proarrhythmic effects of the AADs are drug and patient specific but include the following potentially important problems:
- Sinus bradycardia
- Atrioventricular block
- Increased ventricular or atrial ectopy

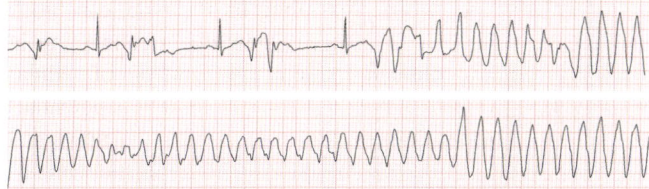

**FIGURE 1:** An example of sotalol-induced QTc prolongation resulting in torsades de pointes. This patient had a recent increase in his diuretic dosage and was hypokalemic at the time of presentation

- VT (monomorphic and polymorphic), including torsades de pointes related to QT interval prolongation **(Fig. 1)**
- Ventricular fibrillation
- Slowing of atrial tachyarrhythmias allowing one-to-one AV conduction when this was not present before the drug.

In some instances, based on the drug and the patient, a proarrhythmic response can be identified or predicted. For some drugs, starting the drug in the hospital to observe for developing proarrhythmia or the presence of QT interval prolongation that could predict proarrhythmia is effective. In other instances, this is not helpful, and only long-term monitoring can determine proarrhythmia. In some instances, it is difficult to determine if a cardiac arrest on a drug is due to drug proarrhythmia or due to lack of efficacy.

## CLASSIFICATION SCHEME

The Vaughan-Williams classification, the most commonly used and by far the most clinically relevant, classifies the drugs based on their most prominent electrophysiological action[3] **(Table 1)**. The more complex "Sicilian gambit" scheme classifies AADs based on their cellular mechanism of action and is mostly utilized for research purposes and drug development[4] **(Fig. 2)**. While the Sicilian gambit held up hope for defining the potential mechanisms of AADs better, its role has all but disappeared. There are problems with both classifications. In fact, our understanding of the mechanisms of action of AADs is at best questionable, as much of the data are from animal models and isolated muscle preparations rather than from clinical assessment. Drugs can have a multiplicity of effects by themselves and by their active metabolites that do not fit neatly into one specific classification scheme.

## VAUGHAN-WILLIAMS CLASSIFICATION

**Class I:** Sodium channel blockers:

**Class IA,** e.g. quinidine, procainamide, disopyramide

**Class IB,** e.g. lidocaine, mexiletine, phenytoin

**TABLE 1**

**The Vaughan-Williams classification of antiarrhythmic drugs**

| Class | Drug | Ion channel effect | Electrophysiological effect |
|---|---|---|---|
| I | | Block inward $Na^+$ channel and outward $K^+$ channels | |
| IA | Quinidine<br>Procainamide<br>Disopyramide | | Slow conduction velocity (predominant effect) and increase refractoriness |
| IB | Lidocaine<br>Mexiletine | | Shorten APD, especially in depolarized cells |
| IC | Flecainide<br>Propafenone | | Marked conduction slowing (minimal effect on refractoriness) |
| II | Beta-blockers | Beta-adrenoceptor blockade | Sympatholytic effect |
| III | Sotalol | Block Ikr and beta-receptors | Prolong refractoriness and APD |
| | Amiodarone | Blocks multiple potassium channels, $Na^+$ channels, $Ca^{++}$ channels, beta-receptors | Prolong refractoriness and APD |
| | Dronedarone | Blocks multiple potassium channels, $Na^+$ channels, $Ca^{++}$ channels, beta-receptors | Prolong refractoriness and APD |
| | Ibutilide | Blocks Ikr and late $Na^+$ current | Prolong refractoriness and APD |
| | Dofetilide | Blocks Ikr | Prolong refractoriness and APD |
| | Azimilide | Blocks Ikr and Iks | Prolong refractoriness and APD |
| IV | Calcium channel blockers | Blocks $Ca^{++}$ channels | Negative chronotropic and inotropic effects |

(Abbreviations: APD: Action potential duration; IKr: Rapid rectifier current; IKs: Delayed rectifier current)

# Antiarrhythmic Drugs

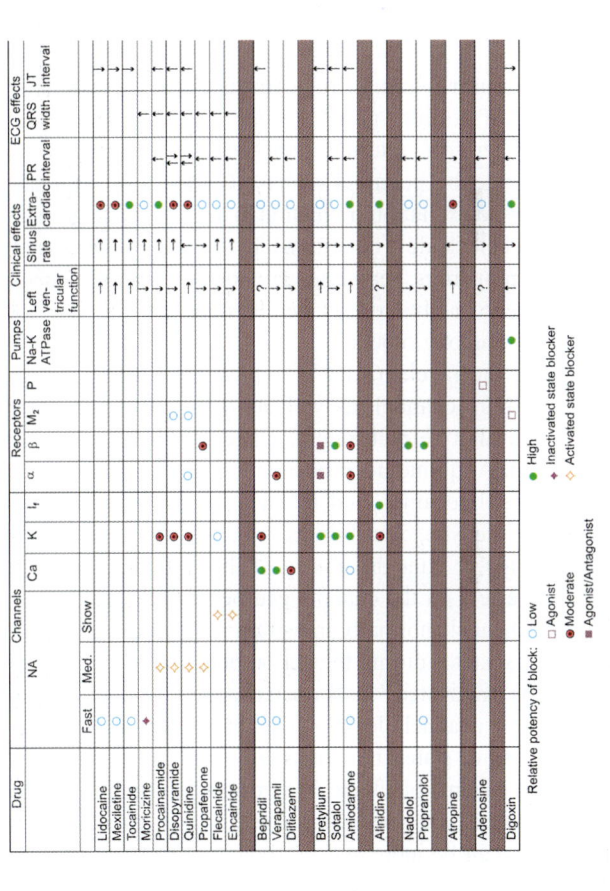

**FIGURE 2:** The Sicilian gambit scheme for classifying antiarrhythmic drugs. [*Source:* Task Force of the Working Group on Arrhythmias of the European Society of Cardiology. Circulation. 1991;84:1831-51 (Reference 4)]

**Class IC,** e.g. flecainide, propafenone

**Class II:** Sympathetic antagonists—beta-blockers

**Class III:** Prolong repolarization, e.g. sotalol, amiodarone, dofetilide, ibutilide, dronedarone, azimilide

**Class IV:** Calcium channel antagonists

The dosing, common uses and adverse effects of the orally available AADs are shown in **Table 2**. **Table 3** describes the major drug interactions of AADs.

## Class I Antiarrhythmic Drugs: Sodium Channel Blockers

The class I antiarrhythmic drugs primarily act by slowing conductance of sodium ($Na^+$) across the cell membrane. These drugs, therefore, interfere with the depolarization phase of the cardiac action potential ("phase 0") and also decrease responsiveness to excitation (reduction in $V_{max}$). The magnitude of $Na^+$ channel blockade is determined by specific cardiac tissue, specific drug properties, heart rate, autonomic (parasympathetic and sympathetic) activation, ischemic state and the state of depolarization, among others.

Based on the mechanisms by which these drugs act to block the sodium channel, as well as their effects on other channels, can alter refractoriness. Class I drugs are further classified into IA (quinidine, procainamide and disopyramide), IB (mexiletine, lidocaine), and IC (flecainide, propafenone).[5] Depending on the type, class I drugs can block sodium channels (class IC drugs) or alter the ability of the sodium channel to conduct; the effect on the channel can be short or prolonged.

Sodium channels normally transition through three distinct conformational states during the action potential: (1) open, (2) closed and (3) inactivated.[6] Only open channels conduct sodium current. Sodium channel blockers interact with open, as well as inactivated channel states, but not usually with closed channels. Thus, sodium channel blockade depends on the conformational state of the channel and blockade is phasic. The extent of sodium channel block can be increased by reducing the recovery rate of the sodium channel. This can happen in disease states, such as ischemia, or can be the property of a particular drug. For example, the class IC drugs, which unbind "very slowly" from sodium channels, are the most potent sodium channel blockers. Class I AADs exhibit use dependence. Tachycardia increases the number of sodium channels in the open and inactivated states. Since sodium channel blockers have greater affinity for the open and inactivated channels, when compared to closed channels, the extent of sodium channel blockade and consequently, conduction slowing, is greater during faster heart rates. This phenomenon is called use dependence.

**TABLE 2**

Dosing, uses, and side effects of orally available antiarrhythmic drugs

| Class | Drug | Maintenance oral dosing | Side effects | Uses |
|---|---|---|---|---|
| IA | Quinidine | 300–600 mg every 6 hours | • Nausea, vomiting, diarrhea, anorexia, abdominal pain<br>• Tinnitus, hearing loss, visual disturbance, confusion (cinchonism)<br>• Thrombocytopenia, hemolytic anemia, anaphylaxis<br>• Hypotension, QRS prolongation, syncope, torsades de pointes, QT prolongation | • PVCs<br>• Sustained VT and VF<br>• Short QT syndrome<br>• Brugada syndrome<br>• AF<br>• Atrial flutter |
|  | Procainamide | 250–1,000 mg every 4–6 hours (no longer available) | • Rash, myalgia, vasculitis, Raynaud<br>• Fever, agranulocytosis<br>• Hypotension, bradycardia, QT prolongation, torsades de pointes<br>• Drug-induced lupus | • Sustained VT<br>• Unmasking Brugada syndrome<br>• AF in WPW |
|  | Disopyramide | 100–200 mg every 6 hours | • Urinary retention, constipation, glaucoma, xerostomia<br>• QT prolongation, torsades de pointes<br>• Reduced ventricular contractility | • PVCs<br>• VT<br>• Hypertrophic CMP<br>• AF |
| IB | Mexiletine | 200–300 mg every 8 hours | • Tremor, dysarthria, dizziness, diplopia, nystagmus, anxiety<br>• Nausea, vomiting, dyspepsia<br>• Hypotension, bradycardia | • VT and VF<br>• Reduction of ICD shocks |

*Contd...*

*Contd...*

| Class | Drug | Maintenance oral dosing | Side effects | Uses |
|---|---|---|---|---|
| IC | Flecainide | 100–200 mg two times daily | • Negative inotropy, AV block, bradycardia<br>• Decreases pacing threshold<br>• Confusion, irritability | • Paroxysmal AF<br>• SVTs<br>• VT<br>• PVCs unmasking Brugada syndrome |
|  | Propafenone | 150–300 mg every 8 hours | • Dizziness, blurred vision<br>• Bronchospasm<br>• AV block, bradycardia, heart failure exacerbation<br>• Decreases pacing threshold | • Paroxysmal AF<br>• SVTs<br>• VT<br>• PVCs |
| II | Beta-blocker | Beta-blocker specific | • Hypotension, bradycardia, heart block, heart failure exacerbation<br>• Bronchospasm<br>• Depression<br>• Impairment of sexual function | • Atrial arrhythmias<br>• Rate control in AF<br>• SVTs<br>• PVCs<br>• VT<br>• VF |
| III | Amiodarone | 1,200–1,800 mg daily for the first 7–10 days, then taper gradually to 200–400 mg daily | • Pulmonary fibrosis<br>• Abnormal liver function tests<br>• Hyperthyroidism or hypothyroidism<br>• Bradycardia, heart failure exacerbation<br>• Tremor, paresthesia<br>• Photosensitivity<br>• Corneal deposits | • VT<br>• VF<br>• Reduction of ICD shocks<br>• AF<br>• Atrial flutter<br>• AF in WPW<br>• Other SVTs |

*Contd...*

## Antiarrhythmic Drugs

*Contd...*

| Class | Drug | Maintenance oral dosing | Side effects | Uses |
|---|---|---|---|---|
| | Sotalol | 80–160 mg every 12 hours | • Bradycardia, torsades de pointes | • Sustained VT/VF<br>• VT in ARVD<br>• Reduction of ICD shocks<br>• AF<br>• Atrial flutter |
| | Dofetilide | 250–500 mcg twice daily | • Torsade de pointes | • AF |
| | Dronedarone | 400 mg twice daily | • Gastrointestinal side effects | • To reduce the risk of cardiovascular hospitalization in patients with non-permanent AF and associated cardiac risk factors |
| IV | Calcium channel blocker (Verapamil) | 80–160 mg every 8 hours | • Hypotension, bradycardia, AV block | • Rhythm control in AF<br>• Idiopathic VT<br>• PVCs<br>• Rate control in AF<br>• SVTs |

(Abbreviations: AF: Atrial fibrillation; CMP: Cardiomyopathy; ICD: Implantable cardioverter defibrillator; PVCs: Premature ventricular contractions; SVTs: Supraventricular tachycardias; VF: Ventricular fibrillation; VT: Ventricular tachycardia; WPW: Wolff-Parkinson-White syndrome)

### TABLE 3
**Major drug interactions of antiarrhythmic drugs**

| Drug | Interacting drug | Interaction |
|---|---|---|
| Quinidine | Phenytoin | ↓ Quinidine levels |
| | Phenobarbital | ↓ Quinidine levels |
| | Rifampin | ↓ Quinidine levels |
| | Ketoconazole | ↑ Quinidine levels |
| | Verapamil | ↑ Quinidine levels |
| | Propafenone | ↑ Propafenone level |
| | Beta-blockers | ↑ Beta-blockade |
| | Digoxin | ↑ Digoxin concentration |
| Mexiletine | Phenytoin | ↓ Mexiletine levels |
| | Phenobarbital | ↓ Mexiletine levels |
| | Rifampin | ↓ Mexiletine levels |
| | Ketoconazole | ↑ Mexiletine levels |
| | Isoniazid | ↑ Mexiletine levels |
| | Theophylline | ↑ Theophylline levels |
| Flecainide | Digoxin | ↑ Digoxin levels |
| | Amiodarone | ↑ Flecainide levels |
| | Quinidine | ↑ Flecainide levels |
| Propafenone | Digoxin | ↑ Digoxin levels |
| | Warfarin | ↓ Warfarin clearance |
| | Cyclosporine | ↑ Cyclosporine levels |
| | Quinidine | ↑ Propafenone levels |
| Amiodarone | Digoxin | ↑ Digoxin effect |
| | Warfarin | ↑ Warfarin effect |
| | QT prolonging drugs | ↑ Risk of torsades de pointes |
| | Beta-blockers | Bradycardia and AV block |
| | Diltiazem and Verapamil | Hypotension and Bradycardia |
| | Anesthetic drugs | ↑ cyclosporine concentration |
| | Cyclosporine | |
| Sotalol | QT prolonging drugs | ↑ Risk of torsades de pointes |
| Dofetilide | QT prolonging drugs | |
| Dronedarone | | |
| Beta-blockers | Quinidine | ↑ Beta-blockade |
| | Amiodarone, Digoxin, Diltiazem, verapamil | Bradycardia |
| Calcium channel blockers (Verapamil) | Digoxin | ↑ Digoxin levels |

Class IA drugs (quinidine, procainamide, disopyramide) slow conduction in atrial and ventricular myocardium and have a moderate effect on slowing myocardial conduction by moderate effect on phase 0 ($V_{max}$) but they also have other effects. These drugs prolong repolarization by their effects on potassium channels. Disopyramide, in particular, can have an anticholinergic effect. Additionally, these drugs have vasodilatory (intravenous procainamide and quinidine), negative inotropic (disopyramide) and vagolytic (quinidine and disopyramide effects).

Quinidine, disopyramide and procainamide have active metabolites (3-hydroxyquinidine, mono-N-dealkylated disopyramide and N-acetylprocainamide respectively). These drug and metabolites can be toxic by several mechanisms. In particular, they are known to prolong action potential duration in the ventricle causing QT prolongation and torsades de pointes. These drugs can have multiple adverse side effects including negative inotropic effects (disopyramide), anticholinergic effects (disopyramide), hypotensive effects (intravenous procainamide and quinidine), autoimmune effects (procainamide in particular), agranulocytosis (disopyramide and procainamide), thrombocytopenia (quinidine) and neurological side effects with nightmares (procainamide).

Due to a wide range of side effects that cause serious problems and require termination of the drug, these drugs are used rarely as the toxicity limits their utility. Furthermore, they are not necessarily the most effective AADs to suppress atrial or ventricular arrhythmias. Due to their toxicity, they have not been well tested in controlled clinical trials, but meta-analyses[7] and other observational data would suggest that their use is limited for both atrial and ventricular arrhythmias and, therefore, these drugs have become phased out for routine use in patients. Nevertheless, there may still be a role for the use of disopyramide in particular to treat some forms of AF, particularly those suspected to be due to vagal activation.

## Class IA Antiarrhythmic Drugs

*Quinidine*: Quinidine not only blocks the rapid sodium current but also affects rapid (IKr) and slow (IKs) components of the delayed-rectifier potassium current, the inward-rectifier potassium current (IKI), the ATP-sensitive potassium channel ($IK_{ATP}$) and transient outward current (Ito). With regard to Ito, the effect is different from the other class IA AADs.

The Ito blockade is purported to reduce the disparity of repolarization in the right ventricular outflow tract and thereby attenuate anterior precordial ST-segment elevation in the Brugada syndrome. Quinidine has been shown to reduce inducibility of ventricular arrhythmias, as well as suppress electrical storm in small studies of Brugada syndrome patients.[8,9] Thus, quinidine has been proposed as an adjunct, but not an alternative, to implantable cardioverter defibrillator (ICD) therapy in high-risk patients with Brugada syndrome.[10] Otherwise, quinidine is rarely used due to its proarrhythmic effects and its adverse effects.

Diarrhea is common; thrombocytopenia can occur or tinnitus is possible. Quinidine can cause idiosyncratic QT prolongation and torsades de pointes. There is a potential interaction between digoxin and quinidine such that quinidine will increase and even double the digoxin levels. The dose of quinidine is generally 200–400 mg every six hours but can be used in a long-acting preparation that is difficult to obtain.

*Procainamide*: Similar to quinidine, the adverse effects, as well as the proarrhythmic effects, outweigh the potential benefits in many cases and therefore, this drug is rarely used. It is hardly available other than in the intravenous form. Nausea, lupus-like syndrome (positive antinuclear antibodies with antihistone antibodies) and agranulocytosis, along with proarrhythmia, are some of the reasons for not using this drug. Intravenous procainamide is very useful in the acute management of supraventricular tachycardias, in particular, rapidly conducted AF and atrial flutter in patients with Wolff-Parkinson-White syndrome. Procainamide can help to facilitate pace termination of atrial arrhythmias.[11] Procainamide can cause torsades de pointes mainly due to its active class III metabolite, N-acetylprocainamide.[12]

*Disopyramide*: Disopyramide is still available but used rarely. The primary use of disopyramide is in patients with hypertrophic cardiomyopathy and left ventricular outflow tract obstruction, and to treat AF. A multi-center study showed that disopyramide can be an effective therapy in symptomatic hypertrophic obstructive cardiomyopathy, with 66% of patients remaining asymptomatic at 3 years with a 50% reduction in outflow gradient. Although disopyramide did not show any mortality benefit in hypertrophic cardiomyopathy, it should nevertheless be considered before invasive options such as surgical myectomy.[13]

Long-term use of disopyramide, however, is limited due to its severe anticholinergic effects, including constipation and dry mouth, as well as urinary retention (as it is about 10% as potent as atropine) and cannot be used in patients with a history of ventricular dysfunction, congestive heart failure or for men with enlarged prostate. Disopyramide is usually used in a long-acting preparation. The dosing is between 400 mg and 600 mg a day in divided doses. Disopyramide can have a marked negative inotropic effect in patients with heart failure and is contraindicated under such circumstances. It can lengthen QT interval and cause torsades de pointes.

## Class IB Antiarrhythmic Drugs

Lidocaine and mexiletine are the only currently available and utilized class IB drugs. As a group, class IB drugs block sodium channels in both activated and inactivated states, but do not delay channel recovery. They affect conduction in ventricular myocardium and have little, if any, effect on atrial myocardium or on AV conduction. This results in shortening of action potential duration and refractoriness. Lidocaine may affect ischemic myocardium preferentially. Their efficacy is increased at high heart rates and also in depolarized tissues, which makes them effective in treatment of ventricular arrhythmias in the ischemic myocardium.

*Lidocaine*: Lidocaine is useful to treat patients who have had recurrent VT or ventricular fibrillation.[14] It does not appear to be effective or beneficial as a prophylactic drug for patients who have had myocardial infarction.[15-17] It has negligible effects on atrial electrophysiology. Lidocaine can suppress conduction, preferentially in ischemic myocardium and does not prolong the QT interval. The purported effect is to prevent reentrant arrhythmias, but it can also suppress automatic and escape rhythms.

Lidocaine dosing can be complex. It undergoes extensive first-pass metabolism in the liver and so can only be administered parenterally. Given rapid initial distribution (half-life of 8 minutes), lidocaine should be administered with multiple loading doses, followed by a maintenance infusion to maintain levels in therapeutic range. The drug has two active metabolites, monoethylglycinexylidide and glycinexylidide. Up to 70% of the drug is protein bound and this number increases in the acute phase of a myocardial infarction when alpha-1-acid glycoprotein increases (as such, over long time periods, lidocaine levels can increase despite a relatively short half-life). In congestive heart failure, where the volume of distribution is reduced, lidocaine achieves higher than normal initial concentration and so the initial dose should be reduced to avoid toxicity. Lidocaine is actively bound to alpha-1-acid glycoprotein, whose levels are increased in heart failure, and perhaps after myocardial infarction thereby decreasing drug availability.

Lidocaine has an elimination half-life of about 2 hours and steady state is reached in 4–5 half lives. Steady state concentration for lidocaine is determined by liver blood flow[18] and is reduced in both heart failure and liver disease. Thus, maintenance dosage of lidocaine should be reduced in both conditions. Renal dysfunction has no impact on lidocaine metabolism.

Lidocaine is most commonly used for acute suppression of potentially life-threatening ventricular arrhythmias (although little data support its role as a drug that improves survival). Lidocaine administration in this setting is based more on anecdotal experience than real data. Lidocaine is frequently ineffective, has a narrow therapeutic range and is frequently associated with neurological toxicity. There are no randomized controlled trials demonstrating benefits of lidocaine. Lidocaine has little effect on atrial tissue and has no value in treating supraventricular tachycardias. The effect of lidocaine in treating arrhythmias in Wolff-Parkinson-White syndrome is controversial[19] and other drugs, including procainamide, ibutilide or amiodarone, are preferred.

Lidocaine is usually administered as a loading dose followed by a maintenance infusion. A commonly used loading regimen is one suggested by Wyman et al., where an initial bolus of 75 mg is given, followed by 50 mg given every 5 minutes repeated three times, for a total loading dose of 225 mg.[20] This regimen usually achieves and maintains plasma concentrations in the therapeutic range of 1.5–5 mcg/mL. This is followed by a maintenance infusion at 1–4 mg/min. It should be noted that wide inter-individual variability in peak plasma concentration exists and, therefore, patients should be closely monitored for evidence of toxicity during loading. Lidocaine has little therapeutic effect at plasma concentrations below 1.5 mcg/mL, and the risk of toxicity increases above 5 mcg/mL.

Symptoms of central nervous system are the most frequent side effects associated with lidocaine administration. Symptoms include paresthesias, perioral numbness, drowsiness, diplopia, dysarthria, confusion and hallucinations. Nystagmus can be an early sign of neurological toxicity. Toxic levels can result in seizures and coma. In patients with known infranodal conduction abnormalities, lidocaine may worsen conduction and should be administered cautiously. Metoprolol, propranolol and cimetidine can reduce hepatic blood flow, decrease lidocaine clearance and can potentially result in lidocaine toxicity when administered concomitantly.[21,22]

*Mexiletine*: Mexiletine is an orally active congener of lidocaine. Mexiletine, like lidocaine, does not suppress AV conduction and has little effect on hemodynamics and ventricular function.[23] Like lidocaine, mexiletine has little effect on atrial electrophysiology. Mexiletine is almost completely absorbed orally, is primarily metabolized (90%) in the liver by the CYP2D6 system to inactive metabolites and is excreted in urine. Mexiletine has a plasma half-life of 9–12 hours. Intravenous mexiletine is not available in the United States.

Mexiletine is primarily used to suppress ventricular arrhythmias and ICD shocks in patients with structural heart disease, either as monotherapy or in combination with another AAD, such as amiodarone or, in years past, quinidine. Effectiveness of mexiletine in this setting varies widely and ranges from 6% to 60%, with majority of studies suggesting a success rate around 20%[24] depending on the condition and the type of ventricular arrhythmia being suppressed. It alone does not improve survival in a controlled trial of high-risk patients.[25] Mexiletine does not prolong and may even shorten the QT interval and can therefore be useful to suppress arrhythmias for patients with the congenital long QT syndrome type III and those with history of drug-induced torsades de pointes.[26]

Mexiletine is usually initiated at a dose of 150 mg every 8 hours. The dose can be increased at 2–3 day intervals until arrhythmia suppression or intolerable side effects develop. Suggested maximum maintenance dose is 300 mg every 6–8 hours. Patients with renal failure should be initiated at a lower dose. Dosage adjustment is also advised in patients with hepatic failure and congestive heart failure, as they impair liver blood flow and prolong elimination half-life of mexiletine.

The most common adverse events with mexiletine are gastrointestinal and neurologic. Tremor, nausea and vomiting are common; dizziness, confusion, blurred vision and ataxia are also seen. Mexiletine-induced tremor may respond to beta-blockers. Thrombocytopenia occurs infrequently.[27] Neurologic side effects are dose-dependent. Severe bradycardia and abnormal sinus node recovery times, with mexiletine have been reported in patients with otherwise or symptomatic sinus node dysfunction.

The major drug interactions of mexiletine are listed in **Table 3**. Inducers and inhibitors of the CYP2D6 system can influence mexiletine metabolism and can affect effectiveness and/or toxicity. Mexiletine decreases theophylline clearance and increases plasma theophylline concentrations.[28] Digoxin and warfarin levels are unaffected by mexiletine.

## Class IC Antiarrhythmic Drugs

The currently available class IC drugs, flecainide and propafenone, are potent sodium channel blockers and cause marked conduction slowing in cardiac tissues without exerting any effects on refractoriness. Their sodium channel blocking effects are exaggerated at high heart rates (use dependency) and in depolarized tissues. At therapeutic doses, class IC drugs prolong the PR and QRS intervals without having significant effects on the QTc interval. Class IC drugs also exert negative inotropic effects and can worsen heart failure in patients with left ventricular dysfunction. The use of these drugs is not recommended in patients who have ventricular dysfunction, who have marked left ventricular hypertrophy or who have ischemic heart disease.[29]

*Flecainide*: Oral flecainide is 90–95% bioavailable and is predominantly metabolized in the liver by CYP2D6 to inactive metabolites. Flecainide is also eliminated to some extent by the kidneys and because of this, genetic variations in CYP2D6 does not seem to significantly affect pharmacological actions of flecainide. Flecainide is eliminated slowly with a half-life of 16–20 hours.

Flecainide is highly effective in suppressing a variety of ventricular and supraventricular tachycardias.[30] It is one of the most potent drugs to suppress ventricular ectopy.[29] At the present time, flecainide is commonly used for restoration and maintenance of sinus rhythm in patients with paroxysmal AF and no structural heart disease. It can be used for maintenance therapy or as a "pill-in-the-pocket" drug for AF termination.[31,32] Flecainide is also effective for suppression of idiopathic ventricular arrhythmias of right and left ventricular outflow tract origin,[33] as well as to treat supraventricular tachycardias in patients with Wolff-Parkinson-White syndrome.

Recently, flecainide has been found to be effective in suppression of catecholaminergic polymorphic VT, which is an inherited, potentially lethal arrhythmic syndrome resulting from mutations in the ryanodine and calsequestrin receptors, causing abnormal calcium handling. Flecainide was found to completely suppress adrenergically mediated polymorphic VT in a mouse model of catecholaminergic polymorphic VT, as well as in two patients with drug-refractory catecholaminergic polymorphic VT. Flecainide was shown in the mouse model to have direct inhibitory effect on the defective ryanodine receptor-mediated calcium release.[34] Flecainide may also be beneficial for patients with long QT interval syndrome type III with a specific SCN5A (D1790G) mutation.[35]

Based on results of the Cardiac Arrhythmia Suppression Trial I (CAST I), although flecainide suppressed premature ventricular contractions (PVCs) in postmyocardial infarction patients, it increased mortality compared with placebo. The same was true of another now obsolete class IC AAD—encainide.[29] Additionally, moricizine, another class I AAD, was shown to have an early proarrhythmic effect during the loading phase in the CAST II trial, even though it did not have any long-term adverse effects compared with placebo.[36] Based on these and other similar data, class IC drugs are contraindicated in patients with advanced structural heart disease and those at risk to develop myocardial ischemia.

Oral flecainide is usually initiated at a dose of 50–100 mg twice daily and can be titrated to a maximum recommended dose of 300 mg daily. At efficacious doses, QRS widening of up to 25% is seen and this is usually evaluated by exercise treadmill testing at high heart rates.[37] A one-time dose of 300 mg or 600 mg flecainide is used when employed as a "pill-in-the-pocket" dosing.[38] To reduce the incidence of adverse effects, flecainide therapy should start with a low dosage that is maintained until steady state has been reached (at least 4 days) and altered relative to clinical response. Flecainide levels can be measured. There is little issue with regard to active metabolites. Caution should be exercised with initial dosing and up titration in patients with hepatic and renal dysfunction. Major drug interactions with flecainide are shown in **Table 3**.

Most common adverse effects of flecainide are dose-dependent and include headache, ataxia, and blurred vision. Flecainide can cause AF to convert to atrial flutter and, in the absence of AV blocking drugs, can result in rapid 1:1 AV conduction (often with aberrant conduction). In patients with depressed ventricular function, negative inotropic effects can precipitate heart failure.[39] In patients with pacemakers and ICDs, flecainide should be used with caution as it can significantly increase pacing and defibrillation thresholds.[40,41] There is a risk of incessant monomorphic VT in patients who have VT, but this is now uncommon as the drug is rarely, if ever, used under these circumstances. Flecainide is also contraindicated in patients with suspected sodium channelopathies like Brugada syndrome, as it can worsen this condition (indeed, it has been used to bring out the classic ECG abnormality). Additionally, caution is needed for patients with advanced His-Purkinje conduction system disease as infra-Hisian block can ensue. Caution about using this drug in patients with substantial left ventricular hypertrophy is recommended in the current ACC/AHA/HRS AF guidelines.[31]

*Propafenone*: Propafenone, in addition to being a potent sodium channel blocker, has beta-adrenergic blocking (about one-thirtieth of the potency of propranolol) and calcium-channel blocking properties. The drug is structurally similar to propranolol and can have significant beta-blocking properties in patients who are slow metabolizers of propafenone.[42]

Propafenone is metabolized through the hepatic CYP2D6 pathway into 5-hydroxy propafenone and this process is largely genetically determined. Approximately 7% of the US population is deficient in CYP2D6, resulting in very slow conversion of propafenone to the active metabolites, 5-hydroxypropafenone and N-depropylpropafenone. The consequent accumulation of high concentrations of propafenone leads to significant beta-adrenoceptor antagonism in poor metabolizers.[43,44] The genetic phenotype, while determining the degree of beta-blockade, does not seem to significantly affect the antiarrhythmic effects of propafenone in most patients.

Propafenone is used to help in maintaining sinus rhythm in patients with paroxysmal or persistent AF who have no associated structural heart disease. It is usually administered in doses ranging from 150 mg to 300 mg every 8–12 hours (a long-acting preparation is available). Peak plasma concentrations are achieved in 1–3 hours following an oral dose. Propafenone increases the PR and QRS intervals on the surface electrocardiogram, but it does not prolong the QT interval. Propafenone can result in acceleration of the ventricular rate in AF if it is converted to a slow atrial flutter. Therefore, administration of an AV nodal blocking drug along with propafenone is recommended. Like flecainide, propafenone is contraindicated in patients with prior myocardial infarction, known ischemic heart disease, severe ventricular hypertrophy and history of sustained VT or severe structural heart disease.[31]

The most common side effects of propafenone are nausea, dizziness and metallic taste. Neurological side effects, like paresthesias and blurred vision, are dose-dependent and are more common in poor metabolizers. Enhanced beta-blockade resulting from poor metabolism can result in bronchospasm and asthma exacerbations. Sustained VT, as a proarrhythmic effect of sodium channel blockade, has been reported and tends to occur in patients with history of VT and underlying structural heart disease.

Propafenone decreases warfarin clearance by inhibition of CYP2C9, resulting in an increased anticoagulant effect. Propafenone markedly increases digoxin levels by decreasing non-renal clearance of digoxin. Quinidine, cimetidine and antidepressants, like fluoxetine and paroxetine, can all inhibit CYP2D6; thereby increasing propafenone levels. Levels of metoprolol[45] and propranolol, which are also metabolized by CYP2D6, are increased in the presence of propafenone.

## Class II Antiarrhythmic Drugs: Beta-adrenoceptor Blockers

Beta-adrenergic blocking drugs are one of the most efficacious drugs used in clinical cardiology for a variety of purposes, including treatment of congestive heart failure and myocardial ischemia. Beta-blockers also have AAD properties and can reduce the risk of sudden cardiac death by a number of mechanisms, can reduce ventricular tachyarrhythmias in select patients, can inhibit AF, can prevent paroxysmal supraventricular tachyarrhythmias of various types and can have additive effects to other AADs. Additionally, beta-blockers can slow AV nodal conduction in patients with rapid atrial tachyarrhythmias including AF and atrial flutter.[46]

Specifically, beta-blockers can prevent catecholamine induced or modulated arrhythmias[47] that occur in catecholaminergic polymorphic VT, idiopathic exercise-induced VT, right ventricular outflow tract tachycardia and ventricular tachyarrhythmias due to the long QT interval syndrome; in particular, those patients with long QT interval syndrome type 1.[26]

Beta-blockers work by a variety of mechanisms. They can suppress automaticity and the triggers for atrial tachycardias, AF and ventricular fibrillation. They can also interfere with the reentry circuit in patients with AV node reentry and with AV reentry tachycardia (by facilitating blockade in the AV node).[48] Beta-blockers may facilitate the effects of class I AADs since their efficacy may be blunted under conditions of catecholamine excess. Additionally, recent data suggest that the combination of amiodarone and beta-blockers is most effective at preventing potentially life-threatening arrhythmias in an ICD population.[49] The mechanism by which this occurs is not completely known.

While a variety of beta-blockers are available for use, when it comes to arrhythmia management, it is important to have drug levels that persist throughout the day. Several beta-blockers do not do this when given on once-a-day basis; for example, atenolol. Specific beta-blockers may be effective for other reasons, including treatment of congestive heart failure and hypertension. When it comes to treating arrhythmias, although it is important to use a beta-blocker such that levels persist throughout the day. Sotalol is a beta-blocker, but it is actually a stereoisomer including d-sotalol and l-sotalol. While d-sotalol is a class III AAD; l-sotalol is a beta-blocker. When using sotalol at lower doses, the greatest effect is from the l-stereoisomer.

Some beta-blockers may have additional central nervous system effects and this effect may depend upon lipid solubility. Water-soluble and renally excreted beta-blockers, such as atenolol, rarely cross the blood-brain barrier, whereas lipid soluble beta-blockers, such as propranolol, cross the blood-brain barrier easily. It is likely that some of the benefits of beta-blockers are through central effects that are not well understood. Additionally, there are data to suggest that carvedilol may be more than just a beta-blocker as it can inhibit the rapid activating delayed-rectifier current, the L-type calcium current and the Ito, as well as the delayed-rectifier current (IKs).[50]

Beta-blockers can terminate specific acute arrhythmias, such as AF with rapid ventricular response rates that may occur in the period early after cardiac and noncardiac surgery. In this particular case, as in other cases whereby AF is catecholamine mediated, beta-blockers can be modestly effective.[51] Additionally, AF can be treated by a beta-blocker in the setting of thyrotoxicosis.

## Class III Antiarrhythmic Drugs: Drugs That Prolong Repolarization

All clinically available class III drugs block the rapid component of the delayed-rectifier potassium channel (Ikr), resulting in an increase in action potential duration and refractoriness in various cardiac tissues, the hallmark of a class III AAD. With class III AADs, reverse use dependence can also occur. In this situation, the AAD effect is most pronounced during slow heart rates. The class III AADs (d-sotalol and N-acetylprocainamide—a metabolite of procainamide, but not amiodarone) demonstrate reverse use dependence. Quinidine can show reverse use dependence for the potassium channel, but use dependence for the sodium channel.

### Sotalol

Sotalol is a class III AAD with beta-blocking properties. This combination results in sinus slowing, decrease in AV nodal conduction and increased refractoriness in atria, AV node, ventricle and accessory pathways. The dextrostereoisomer of sotalol (d-sotalol) is a pure class III AAD without beta-blocking properties.

Oral bioavailability of sotalol is close to 100%. Peak concentrations are seen in 2.5–4 hours following a dose. The drug has an elimination half-life of 12–16 hours and is excreted unchanged by the kidneys. Thus, drug accumulation results in the setting of renal insufficiency, increasing the risk of torsades de pointes and necessitating dose adjustment.

Sotalol is currently available in the United States only in the oral form. Usual starting dose of sotalol is 80 mg twice daily with gradual increase to 240–320 mg daily, provided the QTc is within accepted limits (<500 msec). The following dosing algorithm is proposed in patients with renal insufficiency **(Table 4A)**. No dose adjustment is needed in patients with hepatic disease.

Patients with heart failure and severe left ventricular dysfunction will fare poorly on this drug due to the substantial beta-blocking effects. Concern has been raised for those patients with marked left ventricular hypertrophy, as such patients have

**TABLE 4A**

Sotalol renal dosing algorithm

| Creatinine clearance (measured by Cockcroft-Gault method) mL/min | Dosing frequency |
|---|---|
| >60 | Every 12 hours |
| 30–60 | Every 24 hours |
| 10–30 | Every 36–48 hours |
| <10 | Individualize |

a preponderance of mid-myocardial cells and therefore can be at a greater risk for developing QT interval prolongation.

The combined class III and beta-blocking properties make sotalol effective for supraventricular and ventricular arrhythmias.[52] Sotalol is most commonly used as a rhythm control drug in AF and to suppress ventricular arrhythmias in ICD patients.

The Survival With Oral d-Sotalol (SWORD) trial evaluated the effect of d-sotalol, a pure class III drug and one that is no longer available, versus placebo on mortality in patients who had a myocardial infarction and a left ventricular ejection fraction less than or equal to 40%. The SWORD trial was stopped prematurely due to increased mortality in the d-sotalol arm which was primarily due to arrhythmic death.[53] In a multicenter, double-blind study of 1,456 patients with recent myocardial infarction randomized to d-sotalol and l-sotalol 320 mg once daily versus placebo, the mortality rate at 12-month follow-up was not significantly different between the two groups (8.9% in the sotalol group vs 7.3% in the placebo group), but the reinfarction rate was 41% lower in the sotalol group ($p < 0.05$). This beneficial effect was attributed to the beta-blocking properties of l-sotalol.[54]

In the Electrophysiologic Study Versus Electrocardiographic Monitoring (ESVEM) trial, a randomized, NIH-sponsored multicenter trial designed to determine the best method to guide drug therapy for patients who had malignant ventricular arrhythmias, sotalol was effective in 31% of the patients, which was the best among the different AADs tested.[55] It should be noted, however, that ESVEM did not test amiodarone or ICDs. Sotalol has been shown to be effective in an ICD population where, when compared to placebo, it significantly reduced the number of both appropriate and inappropriate ICD shocks.[56]

The Sotalol Amiodarone AF Efficacy Trial (SAFE-T) was a randomized, double-blind, placebo-controlled trial that compared sotalol versus amiodarone in restoration and maintenance of sinus rhythm in patients with persistent AF. A total of 665 patients were randomized to sotalol ($n = 261$), amiodarone ($n = 261$) and placebo ($n = 137$), and were monitored weekly for 1–4.5 years. The primary endpoint was time to recurrence of AF. Sotalol and amiodarone were equally efficacious in converting AF to sinus rhythm (24% in sotalol group vs 27% in amiodarone group) and both were superior to placebo. The median time to AF recurrence was 487 days in the amiodarone group when compared to 74 days in the sotalol group and 6 days in the placebo group. Amiodarone was clearly superior to sotalol and placebo for maintenance of sinus rhythm. Sotalol, however, was equally efficacious as amiodarone in maintaining sinus rhythm in the subgroup of patients with ischemic heart disease. Major adverse events were comparable among the three groups.[57]

The effects of sotalol, a class III AAD, can result in dose-dependent QTc prolongation and risk of torsades de pointes. At doses ranging from 160 mg/day to 240 mg/day, QTc prolongation of 10–40 ms was noted. Of particular concern is the situation where patients receive concomitant diuretics with frequent dose changes and inadequate potassium replacement. The overall incidence of torsades de pointes appears to be 2% and is more common in females, structural heart disease, and is exacerbated by hypokalemia and concomitant use of other AADs or QT-prolonging drugs. Careful dose titration and dose adjustment in renal insufficiency are essential to avoid risk of torsades de pointes. Typical adverse effects of beta-blockers such as bronchospasm, masking of hypoglycemia and rebound tachycardia, and hypertension on drug withdrawal may also be seen with sotalol. Concomitant use of sotalol with other QT-prolonging drugs increases the risk of torsades de pointes.

## Dofetilide

Dofetilide is a potent and selective IKr blocker that prolongs action potential duration and refractoriness, more so in the atrium than in the ventricle.[58] Dofetilide does not exhibit any negative inotropic properties and has no effect on conduction velocity.

Oral bioavailability of dofetilide exceeds 90% and peak plasma concentrations are attained in 2–3 hours. The drug is partially metabolized by CYP3A4 to inactive metabolites and excreted predominantly (80%) in the urine with an elimination half-life of 8–10 hours. Drug elimination is reduced and accumulation results in renal failure, necessitating dosage adjustment and/or drug discontinuation. Medications that can induce or inhibit CYP3A4 metabolism can affect dofetilide concentrations and can potentially lead to adverse effects.[59]

Dofetilide is primarily used in the restoration and maintenance of sinus rhythm in AF, especially in patients with structural heart disease. The Danish Investigators of Arrhythmia and Mortality on Dofetilide trial (DIAMOND), which evaluated dofetilide versus placebo on all-cause mortality in 1,518 patients with symptomatic congestive heart failure and severe left ventricular dysfunction, showed no difference in all-cause mortality between the two arms. A significant decrease in the risk of heart failure hospitalization was observed in the dofetilide group. In patients with AF, dofetilide resulted in a 12% conversion rate to sinus rhythm compared to 1% in the placebo group ($p < 0.05$) and once sinus rhythm was restored, dofetilide was significantly more effective in maintaining sinus rhythm than placebo (HR 0.35; 95% confidence interval 0.22–0.57; P < 0.001). Twenty-five cases of torsades de pointes were reported in the dofetilide group (3.3%) as compared with none in the placebo group.[60]

The recommended dosage of dofetilide is 500 µg twice daily, but the dose varies based on renal function. Given the risk of torsades de pointes, physicians are required to receive special training prior to prescribing dofetilide. The drug has to be initiated in the hospital with continuous electrocardiographic monitoring for either 3 days or 12 hours after conversion to sinus rhythm, whichever is greater. Creatinine clearance needs to be measured (using the Cockcroft-Gault formula) prior to initiation. A 500 mcg twice daily dosing is initiated only in patients with creatinine clearance more than 60 mL/min. The renal dosing algorithm for dofetilide is shown in **Table 4B**. Once initiated, if the QTc at 2–3 hours following the first dose is more than 15% from baseline or more than 500 msec (>550 msec for bundle branch block or intraventricular conduction delay), then the dose needs to be reduced. If the QTc is more than 500 msec (>550 msec for bundle branch block or intraventricular conduction delay) at any time during doses 2–6, dofetilide needs to be discontinued and an alternative drug sought. Despite initial enthusiasm with regard to the use of the drug,[61] its use has been tempered by strict regulations regarding its use.

The major adverse effect of dofetilide is torsades de pointes. The incidence is dose-dependent and is also influenced by structural heart disease and concomitant usage of QT-prolonging medications.[60,62] The overall incidence, during maintenance therapy on 500 µg twice daily, is around 1.7%.[63] Verapamil, trimethoprim, thiazides, azole antifungals and cimetidine should be discontinued prior to dofetilide initiation as concomitant administration results in markedly elevated plasma concentrations of dofetilide and increases risk of torsades de pointes.[59] Inducers of CYP3A4, such as phenobarbital and rifampin, can enhance dofetilide metabolism and decrease its efficacy. Dofetilide does not interact with digoxin or warfarin.

### TABLE 4B

**Dofetilide renal dosing algorithm**

| Creatinine clearance (measured by Cockcroft-Gault method) mL/min | Dosing frequency |
|---|---|
| >60 | 500 mcg twice daily |
| 40–60 | 250 mcg twice daily |
| 20–39 | 125 mcg twice daily |
| <20 | Contraindicated |
| Hemodialysis | Contraindicated |

## Ibutilide

Ibutilide is a methane sulfonamide analog of sotalol that is a potent blocker of IKr, resulting in prolongation of action potential duration and refractoriness. In addition, ibutilide also activates the slow inward sodium current.[64] Ibutilide is only available for intravenous use and is currently approved for rapid conversion of recent-onset AF and atrial flutter.

The Ibutilide Repeat Dose Study was a multicenter trial that randomly assigned 266 patients with AF of atrial flutter of recent-onset (3–45 days) to ibutilide or matching placebo. Ibutilide was administered as two 10-minute infusions of 1 mg, separated by 10 minutes. The overall conversion rate was 47% with ibutilide versus 2% with placebo ($p < 0.0001$), with the drug being more efficacious in atrial flutter than AF (63% vs 31%; $p < 0.0001$). The mean time to conversion was 27 minutes postinfusion. Among patients who received ibutilide, 8.3% developed torsades de pointes during infusion.[65] Ibutilide has also been shown to be efficacious in conversion of AF in patients with Wolff-Parkinson-White syndrome.

Ibutilide is given as an intravenous infusion over 10 minutes. Recommended dose is 1 mg given over 10 minutes. A second 1 mg dose, separated from the first dose by 10 minutes, can be given if the atrial arrhythmia persists. The drug has a half-life of ~ 6 hours and is primarily metabolized by the liver. No dosage adjustments are recommended for hepatic or renal dysfunction.

The major side effect of ibutilide is QTc prolongation and torsades de pointes, which developed in 8.3% of patients in the Ibutilide Repeat Dose Study.[65] Due to this, it is essential that patients receiving ibutilide have continuous electrocardiographic monitoring for 4–6 hours following treatment, with skilled personnel and resuscitation equipment available and ready. Ibutilide should be avoided in patients with baseline QTc prolongation (>440 msec), advanced structural heart disease, and electrolyte abnormalities such as hypokalemia or hypomagnesemia, given the higher risk of torsades de pointes in these situations.

The use of ibutilide for pharmacological conversion of AF or atrial flutter was never popular given modest efficacy, high risk of polymorphic VT, and the need for close monitoring following drug administration. Several studies have shown that concurrent administration of intravenous magnesium improves efficacy of ibutilide.[66-70] A better method to improve the safety and efficacy of ibutilide was addressed in a recent randomized trial. Fragakis et al. randomly assigned patients with recent-onset AF with rapid ventricular rate to receive ibutilide alone or a combination of ibutilide and esmolol and showed that intravenous beta-blockade resulted in a significant improvement in conversion rate (67% for the combination vs 46% for ibutilide alone) with marked reduction in immediate recurrence of AF. The combination of ibutilide plus esmolol proved to be safer also (no cases of polymorphic VT in the combination group vs 6.5% in the ibutilide group).[71] This combination of ibutilide and esmolol, along with newer drugs like vernakalant, may result in an expanded role for pharmacological agents in the restoration of sinus rhythm in AF.[72]

## Amiodarone

Amiodarone, a synthesized, iodinated benzofuran derivative, structurally similar to thyroxine, was identified with initial work with the ammi visnaga plant.[73] Although classified as a class III AAD, it is a complex and unique drug with properties spanning all four Vaughan-Williams classes. The exact mechanism responsible for its antiarrhythmic actions remains unclear. In animal studies, amiodarone has been shown to prolong action potential duration and refractoriness in the atria and the ventricles, the AV node, and Purkinje fibers.[74] Amiodarone also blocks inactivated sodium channels, slows phase 4 depolarization in sinus node, and delays AV nodal conduction.[75] Electrophysiological properties of amiodarone differ between intravenous and oral use. During intravenous use, amiodarone exhibits sodium and calcium channel blocking properties, has greater effect at higher heart rates and in depolarized tissue. This property makes it useful in treatment of ventricular arrhythmias in the setting of myocardial ischemia. Chronic oral therapy with amiodarone prolongs the PR and QT intervals on the surface electrocardiogram.

Amiodarone is highly lipid-soluble and has a large volume of distribution (20–200 l/kg).[76] Oral bioavailability is highly variable and it usually takes weeks before a steady state is reached, as it accumulates slowly in the adipose tissue. A dose of more than 10 g is usually needed to saturate the fat stores. Amiodarone is mostly metabolized to desethylamiodarone. Plasma half-life after intravenous administration ranges from 4.8 hours to 68.2 hours.[77] Elimination is slow and extremely variable with a half-life ranging from 13 days to 103 days. Dosage adjustment is not required in renal disease. Neither hemodialysis nor peritoneal dialysis removes amiodarone.

US Food and Drug Administration has currently approved amiodarone only for refractory, life-threatening ventricular arrhythmias, although the drug is widely used in the treatment of a variety of atrial and ventricular arrhythmias. Clinical data supporting the use of the amiodarone in ventricular arrhythmias is summarized below.

The European Myocardial Infarction Amiodarone Trial (EMIAT)[78] and Canadian Amiodarone Myocardial Infarction Arrhythmia Trial (CAMIAT)[79] were large randomized trials that evaluated the impact of amiodarone after myocardial infarction. In EMIAT, 1,486 postmyocardial infarction patients with a left ventricular ejection fraction less than 40% were randomly assigned to receive either amiodarone ($n = 743$; loading period followed by 200 mg/day), or matching placebo ($n = 743$). Presence of ventricular arrhythmia was not needed for inclusion. No difference in all-cause or cardiovascular death was seen after a median follow-up of 21 months. A 35% risk reduction ($p < 0.05$) in arrhythmic deaths was seen in the amiodarone group.[78]

In CAMIAT, 1,202 patients who were 6–45 days post-myocardial infarction and had a mean of at least 10 PVCs/hour were randomly assigned to amiodarone ($n = 606$) or placebo ($n = 596$) and followed for a mean of 1.8 years. Patients in the amiodarone group had a 48.5% reduction ($p = 0.016$) in the combined endpoint of resuscitation from ventricular fibrillation or arrhythmic death (3.3% in the amiodarone group vs 6.6% in the placebo group). There was no significant difference in all-cause mortality ($p = 0.13$) between the two groups.[79]

The EMIAT and CAMIAT showed that amiodarone given post-myocardial infarction can reduce arrhythmic death but did not improve total mortality. In a pooled post-hoc analysis of EMIAT and CAMIAT, the combination of amiodarone with a beta-blocker resulted in significant improvements in arrhythmic death or resuscitated cardiac arrest when compared to beta-blockers alone, amiodarone alone, or placebo. Nonsignificant reductions in total mortality were noted with the combination compared to those not receiving beta-blockers.[80]

Estudio Piloto Argentino de Muerte Sfibita y Amiodarone (EPAMSA), Grupo de Estudio de la Sobrevida en la Insuficiencia Cardiaca en Argentina (GESICA) and Congestive Heart Failure Survival Trial of Antiarrhythmic Therapy (CHF-STAT) were trials that evaluated the role of amiodarone in patients with congestive heart failure.[81-83] The EPAMSA randomized patients with a left ventricular ejection fraction less than or equal to 35% and asymptomatic ventricular arrhythmias to receive either amiodarone ($n = 66$) or no drug ($n = 61$). During a 12-month follow-up period, total mortality (10.6% vs 28.8%, $p = 0.02$) and sudden death (7% vs 20.4%, $p = 0.04$) were reduced in patients receiving amiodarone compared to placebo.[81]

The GESICA was a multicenter, randomized trial of 516 patients in Argentina with congestive heart failure and left ventricular systolic function less than or equal to 35% (39% with ischemic cardiomyopathy), but no history of symptomatic ventricular arrhythmias. The trial showed that patients receiving amiodarone had a 28% reduced risk of death and a 31% reduced risk of heart failure hospitalizations, when compared to placebo.[82]

The CHF-STAT, on the other hand, randomized 674 patients with congestive heart failure, a left ventricular ejection fraction less than or equal to 40%, and at least 10 PVCs/hour, to amiodarone ($n = 336$) or matching placebo ($n = 338$). Over a median follow-up of 45 months, amiodarone was associated with PVC suppression and improved left ventricular function. No difference in total mortality ($p = 0.6$) or sudden death ($p = 0.43$) was found between the two groups.[83] The reason for the difference in outcomes between GESICA and EPAMSA versus CHF-STAT has been attributed to the presence of a higher percentage of patients with ischemic cardiomyopathy in CHF-STAT. Drug discontinuation rate of amiodarone in these studies ranged from 20% to 40%.

A meta-analysis of 15 randomized controlled trials ($n = 8,522$) of amiodarone versus placebo for prevention of sudden cardiac death showed that amiodarone was associated with a 29% reduced risk of sudden cardiac death (7.1% vs 9.7%; OR 0.72, $p < 0.001$) and an 18% reduced risk of cardiovascular death (14.0% vs 16.3%; OR 0.82, $p = 0.004$). No significant difference in all-cause mortality was demonstrated. Patients who received amiodarone were more likely to have thyroid problems (OR 5.68; $p < 0.0001$), pulmonary toxicity (OR 1.97; $p = 0.002$), hepatotoxicity (OR 2.1; $p = 0.015$), or bradyarrhythmias (OR 1.78; $p = 0.008$) when compared to the control group.[84] The literature thus suggests that amiodarone is beneficial in treatment of ventricular arrhythmias in patients with cardiomyopathy and congestive heart failure. These findings also suggest that amiodarone is a reasonable option, albeit with risk for long-term side effects and no all-cause mortality benefit, for prevention of sudden cardiac death in patients who are not ICD candidates.

For most patients, however, the reason to use amiodarone to treat ventricular arrhythmias in patients with implantable devices is to suppress recurrent episodes of VT and ventricular fibrillation leading to ICD shocks. It is important to recognize that amiodarone can increase the threshold of energy necessary to defibrillate the patient and can slow the VT rates.[40]

Amiodarone is by far the most effective AAD to maintain sinus rhythm in patients with AF. The Canadian Trial of AF (CTAF) was a prospective, multicenter, randomized trial that randomly assigned 403 patients with at least one episode of AF in the past 6 months to receive amiodarone or either sotalol or propafenone. After a mean follow-up of 16 months, AF recurrence was noted in 35% of patients in the amiodarone group versus 63% in the sotalol or propafenone groups ($p < 0.001$). Adverse effects resulting in drug discontinuation was higher in the amiodarone group (18% vs 11% in the sotalol/propafenone group) but was not statistically significant ($p = 0.06$).[85]

The SAFE-T trial, which compared amiodarone against sotalol in the restoration and maintenance of sinus rhythm in patients with persistent AF, showed that amiodarone was equally efficacious as sotalol in restoring sinus rhythm but was vastly superior to sotalol in maintaining sinus rhythm. Major adverse events in the amiodarone group were comparable to placebo.[57] The relative safety of amiodarone when used in treatment of AF was illustrated in a Cochrane database review of 45 randomized controlled studies ($n = 12,559$) that evaluated the different AADs used for maintenance of sinus rhythm in AF. The effect on these drugs on mortality, thromboembolic events, and proarrhythmia were noted. The study found that class IA, class IC and class III drugs showed a significant reduction in AF recurrence (odds ratio 0.19–0.60, number needed to treat: 2–9) compared to placebo, but none improved mortality. Class IA drugs were associated with increased mortality and all drugs, except propafenone and amiodarone, increased the risk of proarrhythmia.[7]

Given huge volume of distribution, a loading dose regimen is essential to ensure onset of therapeutic action within a reasonable time frame. Loading can be done using intravenous or oral dosing. For outpatient initiation, we routinely employ a loading regimen (400 mg three to four times a day) that ensures a 10–15 g load within 7–10 days after initiation. Once the 10 g load is complete, the patient is switched to a maintenance dose of 200–400 mg a day. The loading dose is generally higher in those patients who have ventricular tachyarrhythmias and the long-term maintenance dose is higher as well. For patients with AF or other atrial arrhythmias, the maintenance dose can be as low as 100–200 mg a day with a load less than 10 g orally.

The manufacturer recommended and routinely used intravenous infusion regimen follows three phases over 24 hours: 150 mg over 10 minutes (with an additional bolus dose of 150 mg for patients with recurrent VT), followed by 1 mg/min over the next 6 hours, followed by 0.5 mg/min over next 18 hours. Infusion should preferably be through a central line to avoid risk of phlebitis. Intravenous amiodarone can result in hypotension and negative inotropy.

Amiodarone is well-tolerated in the long-term if close attention is paid to screen for and recognize adverse events.[86] Side effects are common and can range from 15% in the first year to 50% with long-term use. The majority of the side effects are extracardiac, with the most serious one being interstitial pneumonitis leading to pulmonary fibrosis.[87] This can be difficult to predict and challenging to diagnose.[87,88]

Amiodarone frequently affects thyroid function, but it can also cause hypersensitivity to the sun, cause skin color changes, have neurological effects (weakness, difficulty in walking especially in the elderly), effects on hepatic function and potentially optic neuritis. Corneal microdeposits are common but of little importance. Amiodarone can also cause sinus bradycardia and AV block.[86]

Amiodarone can increase serum levels of digoxin, quinidine, procainamide, flecainide, cyclosporine and warfarin. Although amiodarone can prolong QTc interval, risk of torsades de pointes is extremely rare, perhaps secondary to its multichannel blocking properties or to the uniformity by which it prolongs repolarization.

A comprehensive list of adverse reactions to amiodarone and their management is shown in **Table 5**. Fortunately, the majority of the adverse reactions can be easily managed and do not necessitate discontinuation of the drug. Adverse reactions to amiodarone depend, in part, on the dose and the duration of therapy. If long-term administration is considered, the lowest effective dose should be selected to minimize toxicity. Even then, regular and careful monitoring is essential to ensure patient safety. All patients at initiation of therapy should have a 12-lead electrocardiogram, chest X-ray, pulmonary function test (including DLCO), and laboratory evaluation for electrolytes and renal function, liver function, and thyroid function. An ophthalmological evaluation is recommended at baseline if there is visual impairment, and a follow-up evaluation should be done for new eye-related symptoms. Liver function and thyroid function tests are assessed every 6 months. An electrocardiogram and a chest X-ray should be repeated yearly. Follow-up pulmonary function tests should be done for new or unexplained dyspnea or if there are abnormalities in the chest X-ray compared to baseline.[86]

Amiodarone interferes with the clearance of many drugs, especially those that are highly protein bound. The major drug interactions of amiodarone are listed in **Table 3**. Of particular importance is the inhibition of warfarin and digoxin clearance by amiodarone, resulting in higher plasma levels of these drugs and necessitating dosage reduction or discontinuation. Warfarin dose should be reduced to half and digoxin should be discontinued if that particular patient is started on amiodarone.

## Dronedarone

Dronedarone is structurally similar to amiodarone but lacks the iodine moiety. It has multichannel blocking properties similar to amiodarone but it is not as potent. Dronedarone was initially developed with the aim to reducing or eliminating

**TABLE 5**

Incidence, diagnosis, and management of major adverse reactions to amiodarone

| Reaction | Incidence (%) | Diagnosis | Management |
|---|---|---|---|
| Pulmonary | 2 | Couth and/or dyspnea, especially with local or diffuse opacities on high-resolution CT scan and decrease in DLCO from baseline | Usually discontinue drug; corticosteroids may be considered in more severe cases; occasionally, can continue drug if levels high and abnormalities resolve; rarely, continue amiodarone with corticosteroid if no other option |
| Gastrointestinal tract | 30 | Nausea, anorexia and constipation | Symptoms may decrease with decrease in dose |
| | 15–30 | AST or ALT level greater than 2 times normal | If hepatitis considered, exclude other causes |
| | <3 | Hepatitis and cirrhosis | Consider discontinuation, biopsy or both to determine whether cirrhosis is present |
| Thyroid | 4–22 | Hypothyroidism | L-thyroxine |
| | 2–12 | Hyperthyroidism | Corticosteroids, propylthiouracil or methimazole; may need thyroidectomy |
| Skin | <10 | Blue discoloration | Reassurance; decrease in dose |
| | 25–75 | Photosensitivity | Avoidance of prolonged sun exposure; sunblock; decrease in dose |
| Central nervous system | 3–30 | Ataxia, paresthesias, peripheral polyneuropathy, sleep disturbance, impaired memory and tremor | Often dose dependent, and may improve or resolve with dose adjustment |
| Ocular | <5 | Halo vision, especially at night | Corneal deposits the norm; if optic neuropathy occurs, discontinue |
| | ≤1 | Optic neuropathy | Discontinue drug and consult an ophthalmologist |

*Contd...*

*Contd...*

| Reaction | Incidence (%) | Diagnosis | Management |
|---|---|---|---|
| | >90 | Photophobia, visual blurring and microdeposits | |
| Heart | 5 | Bradycardia and AV block | May need permanent cardiac pacing |
| | <1 | Proarrhythmia | May need to discontinue the drug |
| Genitourinary | <1 | Epididymitis and erectile dysfunction | Pain may resolve spontaneously |

(Abbreviations: ALT: Alanine aminotransferase; AST: Aspartate aminotransferase; DLCO: Diffusion capacity of carbone monoxide). [*Source*: Goldschlager N, Epstein AE, Naccarelli GV, et al. A practical guide for clinicians who treat patients with amiodarone: 2007. Heart Rhythm. 2007;4:1250-9 (Reference 86)]

amiodarone-induced toxicity while maintaining efficacy. For clinical purposes, dronedarone is classified as a Vaughan-Williams Class III AAD. Electrophysiological properties of dronedarone include inhibitory effects on the rapid delayed-rectifier, slow delayed-rectifier, acetylcholine-activated, and inward-rectifier potassium channels, inward sodium current, T-type and L-type calcium channels, and alpha-adrenoceptors and beta-adrenoceptors.[89,90] Dronedarone slows down sinus rate by suppression of sinus node automaticity and by changing the slope of phase 4 depolarization in the sinus node.[91] The drug also slows AV conduction, increase SAV nodal and ventricular effective refractory period, and has been shown to reduce VT and PVCs in ischemic animal models.[89,92]

Dronedarone has negligible proarrhythmic effect but has been shown to increase mortality in patients with acute heart failure.[93] Dronedarone is devoid of the many adverse effects and drug interactions associated with amiodarone. Pulmonary toxicity has not been reported with dronedarone. When compared to placebo, there was no significant difference in hyperthyroidism, hypothyroidism, neurological abnormalities, gastrointestinal and hepatic abnormalities with dronedarone. Similar to amiodarone, dronedarone causes mild increases in serum creatinine by inhibiting cation transport in the renal tubules. Glomerular filtration rate, however, is not affected.[94]

Dronedarone is metabolized by the hepatic CYP3A4 system and in turn is also a moderate inhibitor of the CYP3A4 system and weak CYP2D6 and P-glycoprotein inhibitor. These properties result in increased effects of drugs like cyclosporine, digoxin and some statins when coadministered with dronedarone.[94] Unlike amiodarone, dronedarone does not have any drug interactions with warfarin.

A synopsis of the randomized clinical trials that evaluated the impact of dronedarone in AF and heart failure are shown in **Table 6** and have been summarized in detail.[95] The European Trial in AF or Flutter Patients Receiving Dronedarone for the Maintenance of Sinus Rhythm (EURIDIS) and the American-Australian-African Trial with Dronedarone in AF or Flutter Patients for the Maintenance of Sinus Rhythm (ADONIS) compared dronedarone to placebo in maintaining sinus rhythm after conversion from atrial flutter or AF. The EURIDIS showed that at the end of 1 year of follow-up, 67% of patients on dronedarone had a recurrence of AF compared to 78% in the placebo group. In the ADONIS trial, 61% in the dronedarone group had recurrent AF compared to 73% in the placebo. Although significantly different from placebo, the high recurrence rate of AF with dronedarone cast doubts on its efficacy to maintain sinus rhythm.[96]

### TABLE 6

Summary of randomized clinical trials that assessed the efficacy and safety of dronedarone in patients with atrial fibrillation and heart failure

| Clinical trial | Patient profile | Number of patients | Intervention | Primary end point | Follow-up (months) | Results/Conclusions |
|---|---|---|---|---|---|---|
| DAFNE | Persistent AF post-cardioversion | 199 | Dronedarone (400–800 mg twice daily) versus placebo | Time to first recurrence of AF | 6 | Use of dronedarone was associated with a longer median time to AF recurrence (60 days vs 5.3 days for dronedarone and placebo respectively, $p = 0.026$; 55% relative risk reduction, $p = 0.001$); likewise, patients receiving dronedarone, 400 mg orally twice daily, were more likely to maintain sinus rhythm compared with patients receiving placebo |
| EURIDIS and ADONIS | Paroxysmal AF | 1,237 | Dronedarone 400 mg twice daily versus placebo | Time to first recurrence of AF | 12 | Dronedarone significantly lengthened the time to AF recurrence [41 days vs 96 days (EURIDIS) and 59 days vs 158 days (ADONIS) for dronedarone and placebo respectively], as well as symptoms associated with atrial fibrillation, compared with placebo. Ventricular rates during AF recurrence were significantly lower with dronedarone |
| DIONYSOS | Persistent AF for >3 days | 504 | Dronedarone (400 mg twice daily) versus amiodarone (600 mg and then 200 mg per day) | AF recurrence or drug intolerance resulting in discontinuation | 7 | More patients on dronedarone had AF recurrence or stopped the drug due to intolerance or lack of efficacy compared with patients receiving amiodarone (75.1% vs 58.8% for dronedarone and amiodarone respectively, HR 1.59). |

*Contd...*

*Contd...*

| Clinical trial | Patient profile | Number of patients | Intervention | Primary end point | Follow-up (months) | Results/Conclusions |
| --- | --- | --- | --- | --- | --- | --- |
| ERATO | Permanent AF with ventricular rates >80 bpm on rate-controlling agents | 630 | Dronedarone 400 mg twice daily versus placebo | Mean ventricular rate at 2 weeks | 1 | Dronedarone use was associated with decrease in ventricular rate, both at rest (12.3 bpm with dronedarone vs 0.2 bpm with placebo) and with exercise (25.6 bpm with dronedarone vs 2.2 bpm with placebo) |
| ATHENA | Paroxysmal or persistent AF or atrial flutter with one or more associated risk factors | 4,628 | Dronedarone (400 mg twice daily) versus placebo | Composite of all-cause mortality and cardiovascular hospitalization | 21±5 | The use of dronedarone was associated with decreased cardiovascular deaths and arrhythmic deaths compared with placebo (31.9% in dronedarone arm vs 39.8% in placebo arm, HR 0.76). There was also a decrease in hospitalizations for AF and acute coronary syndrome in patients receiving dronedarone compared with placebo |
| ANDROMEDA | Congestive heart failure (NYHA Class III–IV); left ventricular ejection fraction <35% | 617 | Dronedarone (400 mg twice daily) versus placebo | All-cause mortality or heart failure hospitalization | 2 | Trial was stopped early as dronedarone was associated with a significant increase in all-cause mortality (8.1% in the dronedarone arm vs 3.8% in placebo arm, HR 2.13) |

The DIONYSOS trial[97] [Randomized, Double-Blind TrIal to Evaluate the Efficacy and Safety of DrOnedarone (400 mg bid) Vs AmiodaroNe (600 mg qd for 28 daYS, then 200 mg qd Thereafter) for at least 6 months for the Maintenance of Sinus Rhythm in Patients with AF] directly compared dronedarone (400 mg twice daily) to amiodarone (600 mg every day for 28 days, then 200 mg every day thereafter) in restoration and maintenance of sinus rhythm in patients with AF. During a mean follow-up of 7 months, 64% of patients in the dronedarone arm had AF recurrence when compared to 42% in the amiodarone arm. Adverse event rates were high, but comparable between both drugs (39% with dronedarone vs 45% with amiodarone; HR 0.80, $p = 0.13$). There were fewer thyroid, neurological, skin, and eye-related adverse events with dronedarone except gastrointestinal side effects, which were higher in the dronedarone group. In summary, dronedarone was inferior to amiodarone in efficacy, but was more favorable than amiodarone in terms of safety.[97]

The Efficacy and Safety of Dronedarone for the Control of Ventricular Rate during AF (ERATO) study found dronedarone to be effective for ventricular rate control in patients with AF, both at rest and with exercise.[98]

The ATHENA (A Placebo-Controlled, Double-Blind, Parallel Arm Trial to Assess the Efficacy of Dronedarone 400 mg bid for the Prevention of Cardiovascular Hospitalization or Death from Any Cause in Patients with Atrial Fibrillation/Atrial Flutter) trial evaluated the effect of dronedarone in reducing a composite endpoint of death or cardiovascular hospitalizations in AF patients.[99] A total of 4,628 patients with paroxysmal or persistent AF and presence of risk factors for stroke and/or death were randomized to dronedarone or matching placebo and were followed for a median period of $21 \pm 5$ months. The study found that patients randomized to dronedarone had fewer cardiovascular deaths (HR = 0.71; 95% CI, 0.51–0.98; $P = 0.03$), as well as arrhythmic deaths (HR = 0.55; 95% CI, 0.34–0.88; $P = 0.01$), when compared to placebo. There were also fewer cardiovascular hospitalizations in the dronedarone arm. A post-hoc analysis of ATHENA showed that there were fewer strokes or transient ischemic attacks in the dronedarone group.[100] The ATHENA was the first trial to show mortality benefit with an AAD and was largely responsible for approval of dronedarone in the United States.

The Antiarrhythmic Trial with Dronedarone in Moderate to Severe Congestive Heart Failure Evaluating Morbidity Decrease (ANDROMEDA) compared dronedarone with placebo in patients with AF hospitalized with new or worsening heart failure and a left ventricular ejection fraction less than 35%.[93]

The study had to be terminated prematurely after a median follow-up of 2 months, as mortality was significantly increased in the dronedarone arm (8.1% vs 3.8% in the placebo arm). The increased mortality was predominantly attributed to deaths from worsening heart failure and treatment with dronedarone was the most powerful predictor of death. This study resulted in a black box warning for dronedarone that warns against its use in patients with New York Heart Association (NYHA) class IV heart failure or NYHA class II and III heart failure with recent decompensation requiring hospitalization or referral to a heart failure clinic.[94] Recent post-marketing data released by the manufacturer reports several cases of hepatocellular injury and at least two cases of acute hepatic failure requiring liver transplantation, which occurred at 4.5 months and 6 months following drug initiation. This has prompted a manufacturer recommendation to consider serial liver enzyme monitoring at least for the first 6 months while being on dronedarone.[101] PALLAS included patients at least 65 years old with at least 6-month history of permanent atrial fibrillation and risk factors for major vascular events. Patients received dronedarone or placebo. Of 3236 enrolled, the co-primary outcome of stroke, myocardial infarction, systemic embolism, or death from cardiovascular causes was higher with drenedarone (HR: 2.29; 95% confidence interval 1.34–3.94; P = 0.002). The death rate was higher with dronedarone (HR, 3.11; 95% confidence interval, 1.00–4.49; P = 0.046), including death from arrhythmia (HR, 3.26; 95% confidence interval 1.06–10.00; P = 0.03). There were more strokes (HR, 2.32; 95% confidence interval 1.11–4.88; P = 0.02) and more heart failure hospitalizations (HR, 1.81; 95% confidence interval 1.10 to 2.99; P = 0.02) with drenedarone.[101a]

Dronedarone, although not a very effective rhythm control drug by itself, remains the first AAD to show a reduction in cardiovascular mortality in AF patients with risk factors for stroke and/or death. Although mortality reduction is a significant finding, its clinical utility is unclear, as the primary goal in AF management is symptom reduction and improving quality of life. The favorable safety profile, as well as the fact that a loading dose is not needed, makes dronedarone an ideal drug to start as an outpatient. The fact that it improves mortality and reduces cardiovascular hospitalizations in patients with AF makes it attractive from a health care expenditure standpoint. On the other hand, it is our opinion that use of dronedarone should be avoided in patients with congestive heart failure and severe left ventricular dysfunction or for those patients with permanent AF. The 2011 ACCF/AHA/HRS focused update of the 2006 AF guidelines now include dronedarone.[31] It is fair to say that dronedarone has definitely expanded the horizon in terms of management options for AF.

## Azimilide

Azimilide dihydrochloride is a class III AAD that blocks both the rapid (IKr) and the slow (IKs) delayed-rectifier potassium channels.[102] It is different from the other class III drugs that only block IKr. Azimilide prolongs the action potential duration and refractoriness in atrial and ventricular myocardium and has been shown to cause dose-dependent QTc prolongation.[102] Unlike other class III drugs, azimilide does not exhibit reverse use-dependence, which is thought to be secondary to IKs blockade.

The AzimiLide post-Infarct surVival Evaluation (ALIVE) trial assessed the effect of azimilide on survival in patients who were 6–21 days post-myocardial infarction and had a left ventricular ejection fraction ranging from 15% to 35%. No survival advantage was seen with azimilide but the drug caused no excessive harm.[103] Azimilide is not approved for clinical use in the United States but is available in Europe, where it has been primarily used for suppressing ventricular arrhythmias in ICD patients.

The SHock Inhibition Evaluation with azimiLiDe (SHIELD) trial was a randomized, double-blind, placebo controlled, international trial of 633 patients that evaluated the effect of azimilide, either 75 mg ($n = 220$) or 125 mg ($n = 199$) daily, versus placebo ($n = 214$) on all-cause shocks plus symptomatic tachycardias terminated by antitachycardia pacing and appropriate ICD therapies.[104] All patients enrolled in the trial had an ICD implanted and had either a documented episode of cardiac arrest or spontaneous sustained VT with left ventricular ejection fraction less than or equal to 0.40 during 42 days prior to the first ICD implantation or an ICD shock for spontaneous VT or ventricular fibrillation within the previous 180 days. Over a median follow-up of 367 days, there was a significant 57% reduction in all-cause shocks plus antitachycardia pacing (ATP) therapies in the azimilide 75 mg per day group compared to placebo (HR = 0.43; CI 0.26–0.69, $P = 0.0006$). A 47% reduction was seen in the azimilide 125 mg per day group (HR = 0.53; CI 0.34–0.83, $P = 0.0053$). Both doses of azimilide decreased all-cause shocks but this was not statistically significant. When compared to placebo, azimilide 75 mg and 125 mg per day reduced appropriate ICD shocks and ATP by 48% ($p = 0.017$) and 62% ($p = 0.0004$) respectively. High (35–40%) but comparable rates of drug discontinuation was seen in both azimilide and placebo groups. Four patients in the azimilide group and one in the placebo group had torsades de pointes. Thus, it appears that azimilide has beneficial effects in prevention of ventricular arrhythmias in ICD patients.

On the other hand, azimilide was disappointing as a rhythm control drug for restoration and maintenance of sinus rhythm in AF. The North American Azimilide Cardioversion Maintenance Trial (ACOMET II) study compared azimilide (125 mg daily) with sotalol (160 mg twice daily) or placebo for maintaining sinus rhythm in 658 patients with persistent AF undergoing electrical cardioversion.[105] The primary endpoint was recurrence of AF. Azimilide was found to be superior to placebo, but was significantly inferior to sotalol in maintaining sinus rhythm.

The Azimilide Supraventricular Tachyarrhythmia Reduction (A-STAR) trial evaluated the effect of azimilide in maintaining sinus rhythm in patients with structural heart disease.[106] The trial randomized 220 patients to azimilide (125 mg daily) versus matching placebo, and patients were followed for time to first symptomatic AF recurrence. There was no significant difference between the azimilide and the placebo groups with respect to the primary endpoint. In terms of adverse effects, neutropenia was seen in 1% of patients who were on azimilide. A dose-dependent increase in torsades de pointes was noted with the incidence rates ranging from 0.3% for the 75 mg dose to 1.2% for the 100 mg dose.[107] Thus, in terms of AF rhythm control, the risk-benefit ratio was definitely not in favor of azimilide, and it is doubtful that this drug will be available for use in AF.

## Class IV Antiarrhythmic Drugs: Calcium Channel Antagonists

Verapamil blocks the L-type calcium channel and can be used to slow AV nodal conduction to control the ventricular response rate atrial flutter and AF, but it could also be used to prevent recurrence of AV nodal reentry and AV reentry supraventricular tachycardia. Furthermore, verapamil can prevent triggered activity and inhibit idiopathic right ventricular outflow tract tachycardias by this mechanism. Additionally, verapamil can affect reentrant mechanisms responsible for idiopathic left VT. The dose of verapamil is 120–480 mg a day in single or divided doses. Diltiazem, another calcium channel blocker that can be used to control the ventricular response rate in AF, is available in both intravenous and oral formulations.

## MISCELLANEOUS DRUGS

### Adenosine

Adenosine is an ultrashort acting purinergic agonist; it is vagotonic. It binds to the adenosine A1 receptor. Adenosine activates the $IK_{ACH,ADO}$ channels present in the atrium, sinus node and the AV node. This results in increased outward potassium current which leads to shortening of atrial action potential and membrane hyperpolarization and transient AV nodal block and sinus node depression.[108] These $IK_{ADO}$ channels are not present in the ventricular myocytes and, therefore adenosine has not much of an effect in the ventricular myocardium. Indirectly, adenosine has an antiadrenergic action due to a decrease in cyclic AMP. This property might be responsible for its suppressive effect on outflow tract ventricular arrhythmias as well as a subgroup of focal atrial tachycardias, which probably are delayed after depolarization-mediated triggered rhythms resulting from catecholamine-mediated calcium overload.

Adenosine has a rapid onset of action and intravenous administration of 6–12 mg adenosine results in sinus node slowing and transient AV block. This property is most often used to terminate AV node dependent paroxysmal AV nodal reentry and orthodromic AV reentry supraventricular tachycardias. Adenosine can stop idiopathic VTs, especially those that originate from the right ventricular outflow tract.[109] It can also terminate some atrial tachycardias.[110] Adenosine is commonly used during an electrophysiology study to determine the presence of a concealed accessory pathway. The vasodilatory properties of adenosine make it useful as a chemical alternative to exercise in the diagnosis of myocardial ischemia.

Adverse effects with adenosine typically include dyspnea, chest tightness, flushing and exacerbation of bronchospasm. These are typically short-lasting and resolve quickly. Adenosine should be used with caution in patients with severe reactive airway disease. Use of adenosine can result in AF in 10–15% of patients due to shortening of atrial refractory periods. Transplanted hearts are exquisitely sensitive to adenosine and significant dose reduction is required.[111] Methylxanthines, such as caffeine and theophylline, block adenosine receptors and counteract the effects of adenosine. Dipyridamole reduces the reuptake of adenosine, thereby prolonging the effect of adenosine. Due to this, those who are on oral dipyridamole undergoing a stress test should receive intravenous dipyridamole and not adenosine.

## NEWER DRUGS

### Tedisamil

Tedisamil is a class III AAD that blocks multiple potassium channels, including IKr, IKs, IKur, Ito and IKATP.[112] Tedisamil prolong atrial and ventricular action potentials, but its effects are more pronounced in the atrial tissue. It also suppresses sinus node function and has antianginal properties.[107] Tedisamil has a half-life of 8–13 hours, is not metabolized, and is renally excreted.

A randomized, placebo-controlled dose-response study ($n = 175$) showed that tedisamil at doses of 0.4 mg/kg and 0.6 mg/kg was superior to placebo in converting new-onset AF to sinus rhythm. Efficacy was modest with a 41% conversion rate for 0.4 mg/kg and 51% conversion rate for the 0.6 mg/kg tedisamil group. There were two cases of ventricular arrhythmias (one case of torsades de pointes and one case of monomorphic VT) in the tedisamil group.[112] Clearly, more studies are needed to evaluate the safety and efficacy of tedisamil in AF.

## Vernakalant

Vernakalant is the first in a class of AADs that are "atrial-selective". Atrial-selective drugs are being developed to target the ion channels or currents that are present in the atria and not in the ventricles. These include the ultrarapid potassium current IKur and the acetylcholine-mediated potassium channel $IK_{ACH}$. The goal for developing these drugs is to restore and maintain sinus rhythm in AF while avoiding the adverse ventricular events such as QTc prolongation and torsades de pointes.[113]

Vernakalant acts selectively in the atrium, targeting the following ion channels: IKur, $IK_{ACH}$, Ito and late INA.[114] The efficacy and safety of intravenous vernakalant (administered as a 10-minute intravenous infusion at a dose of 3 mg/kg; if AF had not been terminated within 15 minutes, a second 10-minute infusion be followed at a dose of 2 mg/kg) for the treatment of AF was assessed in the randomized, placebo-controlled, double-blind Atrial Arrhythmia Conversion Trials (ACT) I–III.[115-117] The ACT I and ACT III trials investigated vernakalant in the treatment of patients with sustained AF (duration > 3 hours, but not more than 45 days). A total of 336 patients were enrolled in ACT I and 276 patients in ACT III. The primary endpoint was conversion to sinus rhythm for at least 1 minute within 90 minutes of drug infusion. In both these trials, vernakalant was significantly better than placebo in converting AF to sinus rhythm.

In ACT I, sinus rhythm was achieved in 62% of patients receiving vernakalant compared with 4.9% of patients receiving placebo for AF of 3–48 hour duration.[115] In ACT III, 51.2% of patients receiving vernakalant converted to sinus rhythm compared with 3.6% of patients receiving placebo for AF of 3 hours to 7 days.[117] The median time to conversion was 10 minutes from the start of infusion and sinus rhythm was maintained for more than 24 hours in 97% of patients. Data from the ACT II trial, which investigated the efficacy of intravenous vernakalant in 150 patients with sustained AF (3–72 hours duration) that occurred between 24 hours and 7 days after coronary artery bypass graft and/or valvular surgery, showed a 47% conversion rate compared to 14% for placebo.[116]

In the AVRO (A Phase III Superiority Study of Vernakalant versus Amiodarone in Subjects With Recent Onset Atrial Fibrillation) trial, 254 patients with recent-onset AF (3–48 hours duration) were randomized to receive either intravenous vernakalant or intravenous amiodarone. Treatment with vernakalant converted 51.7% of patients to sinus rhythm at 90 minutes compared with 5.2% of patients treated with amiodarone. Both drugs were well tolerated.[118]

Vernakalant does not appear to be effective in AF of longer duration (>7 days) or in atrial flutter.[117] Preliminary studies have shown that oral vernakalant (5 mg/kg) is rapidly absorbed and well-tolerated. Studies are ongoing to determine efficacy and safety of the oral formulation. Vernakalant appears to have a good safety profile but concerns still exist. Most common side effects are nausea, dysgeusia, paresthesias and hypotension.[117] No episodes of drug-induced torsades de pointes were reported in the ACT trials.

Currently, vernakalant is approved in Europe for rapid conversion of recent-onset AF (≤7 days duration for non-surgery patients, and less than or equal to 3 days duration for post-cardiac surgery patients) to sinus rhythm in adults. The atrial selectivity, modest efficacy rate and excellent safety profile makes vernakalant an important addition to the armamentarium for pharmacological conversion of AF.

## Ivabradine

High resting sinus heart rates have been independently associated with mortality and adverse cardiovascular outcomes.[119,120] Ivabradine is a selective cardiac pacemaker ($I_f$) current blocker that slows sinus rates.[121] When compared to placebo in the BEAUTIFUL (morBidity mortality EvAlUaTion of the $I_f$ inhibitor ivabradine in patients with coronary disease and left ventricULar dysfunction) trial, ivabradine did not improve the composite outcome of cardiovascular death, hospitalizations for heart failure and/or acute myocardial infarction in patients with coronary artery and left ventricular dysfunction.[122] However, it did reduce fatal- and non-fatal myocardial infarction and coronary revascularization[122] and so may be useful as an antianginal drug, especially in combination with a beta-blocker.[123] Recent data would suggest that slowing heart rate may, in fact, improve outcomes in select patients with congestive heart failure[124] and with inappropriate sinus tachycardia.[125] Usual dosing range is 5–7.5 mg twice daily.

### Ranolazine

Ranolazine is an antianginal drug that also can affect the late and the peak inward sodium current, the late calcium current and the IKr and IKs currents, as well as the sodium/calcium exchanger.[126] In the Efficiency with Ranolazine for Less Ischemia in Non-ST elevation acute coronary syndromes (MERLIN)-TIMI 36 trial, ranolazine significantly lowered nonsustained VT and supraventricular tachyarrhythmias in patients with non-ST elevation myocardial infarction when compared to placebo.[127] Recent data from a canine model suggests that ranolazine, in combination with dronedarone, may be a potent combination to reduce AF.[128]

## EMERGING ANTIARRHYTHMIC DRUGS

Various novel AADs are presently being tested but they are nowhere near being considered valuable and/or valid therapies for arrhythmia suppression. Research continues with nifekalant, an IKr blocker for ventricular arrhythmias, celivarone, an amiodarone analogue, several IKur (and multichannel) blockers (AVE0118, AZD7009, NIP-141/142), sodium current blockers (pilsicainide and ranolazine), other amiodarone analogues (celivarone, ATI-2042 and PM101), selective IKs blockers (HMR1556) and Kv1.5 blockers (XEN-D0101).[129] Additionally, drugs are being tested that work by novel mechanisms including those that inhibit the atrial acetylcholine regulated potassium current, $IK_{ACH}$ (tertiapin-Q), those that target abnormal calcium handling via the ryanodine receptor RyR2 (calstabin-2), those that act as sodium/calcium exchange inhibitors (KB-R7943), those that block the stretch activated channels, those that are gap junction modifiers (rotigaptide, GAP-134), those that antagonize the serotonin 5-hydroxytryptamine receptors (RS-100-302) and those that are long-acting adenosine A1 receptors (tecadenoson and selodenoson).[129] Likely, new drugs will be developed that will focus on other approaches rather than simply blocking specific cardiac channels.

## ANTIARRHYTHMIC DRUG SELECTION IN ATRIAL FIBRILLATION

The 2011 ACC/AHA/ESC Guidelines for Management of AF provides recommendations regarding AAD selection if rhythm control is planned for AF **(Flowchart 1)**.[31] The recommendations are primarily based on AAD safety than on drug efficacy. For patients with no evidence of structural heart disease or have hypertension without substantial left ventricular hypertrophy, flecainide, propafenone, sotalol or dronedarone is first-line therapy, followed by amiodarone,

**FLOWCHART 1:** Antiarrhythmic drug selection, based on underlying structural heart disease, for maintenance of sinus rhythm in atrial fibrillation

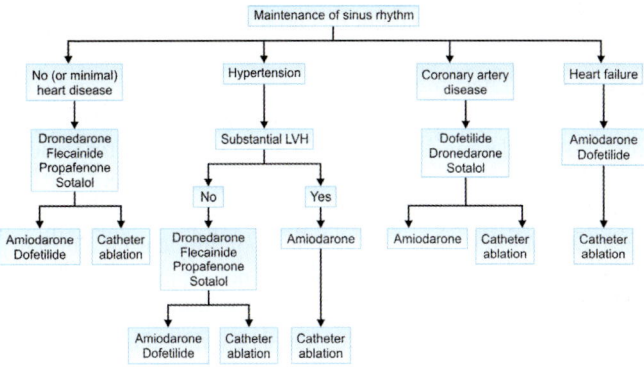

[*Source:* Modified from Wann LS, Curtis AB, January CT, et al. 2011 ACCF/AHA/HRS focused update on the management of patients with atrial fibrillation (updating the 2006 guideline): a report of the American College of Cardiology Foundation/American Heart Association Task Force on Practice Guidelines. Circulation. 2011;123:104-23 (Reference 31)]

dofetilide, or catheter ablation. For patients with hypertension and substantial left ventricular hypertrophy, amiodarone is the first choice drug, with catheter ablation as the second-line choice. In patients with coronary artery disease, dofetilide or sotalol is first-line, followed by amiodarone or catheter ablation. For heart failure patients, amiodarone or dofetilide is first-line therapy, followed by catheter ablation. Most recently, dronedarone has been included in the guidelines and has a role in the treatment of AF as stated in the package insert as "an AAD indicated to reduce the risk of cardiovascular hospitalization in patients with paroxysmal or persistent AF or atrial flutter, with a recent episode of AF or atrial flutter and associated cardiovascular risk factors (i.e. age >70, hypertension, diabetes, prior cerebrovascular accident, left atrial diameter ≥50 mm or left ventricular ejection fraction <40%), who are in sinus rhythm or who will be cardioverted".[94]

## OUTPATIENT VERSUS IN-HOSPITAL INITIATION FOR ANTIARRHYTHMIC DRUG THERAPY

The location of initiation of an AAD depends on the severity of the arrhythmia and the risk of starting the AAD. It is recommended that all class IA AADs be initiated in the hospital due to risk of torsades de pointes, which at times can be idiosyncratic and non-dose dependent. Class IB AADs, specifically mexiletine, can be started and titrated as an outpatient because the risk of proarrhythmia is small but, in most cases, this drug is started in the hospital due to the fact that most patients whom this drug is initiated have unstable ventricular arrhythmias. Class IC AADs can generally be started outside the hospital for AF as the early risk of proarrhythmia is low as long as the patient has no underlying structural heart disease and no evidence for cardiac ischemia. There is a small risk of rapid rates in AF with one-to-one conduction and atrial flutter but, with proper AV nodal blocking drugs, this risk can be offset. Amiodarone can be started as an outpatient for patients who have AF and atrial flutter as the proarrhythmic risk is low. On the other hand, most patients with VT are considered unstable and, therefore, the initiation of amiodarone normally begins in the hospital. A patient may not be fully loaded with amiodarone, but nevertheless the drug should be started in the hospital. Sotalol and dofetilide should be initiated in the hospital due to the risk of developing QT prolongation and torsades de pointes. Dofetilide must be started in the hospital based on strict guidelines about how the drug should be initiated and titrated. Dronedarone is generally not proarrhythmic and can be started outside the hospital.

## ANTIARRHYTHMIC DRUGS IN PREGNANCY AND LACTATION

An overview of the effect of various AADs in pregnancy and lactation is presented in **Table 7**. Sotalol is the only pregnancy category B drug [either animal-reproduction studies have not demonstrated a fetal risk but there are no controlled studies in pregnant women, or animal-reproduction studies have shown an adverse effect (other than a decrease in fertility) that was not confirmed in controlled studies in women in the first trimester (and there is no evidence of a risk in later trimesters)], while amiodarone is classified as pregnancy category D drug [there is positive evidence of human fetal risk, but the benefits from

**TABLE 7**

**Antiarrhythmic drugs in pregnancy and lactation**

| Drug | Pregnancy | Lactation |
|---|---|---|
| Quinidine | C | Excreted |
| Procainamide | C | Excreted |
| Disopyramide | C | Excreted |
| Mexiletine | C | Excreted |
| Flecainide | C | Excreted |
| Propafenone | C | ? |
| Sotalol | B | Excreted |
| Dofetilide | C | ? |
| Dronedarone | X | ? |
| Amiodarone | D | Excreted |

use in pregnant women may be acceptable despite the risk (e.g. if the drug is needed in a life-threatening situation or for a serious disease for which safer drugs cannot be used or are ineffective)]. Dronedarone is a pregnancy category X drug (studies in animals or human beings have demonstrated fetal abnormalities, or there is evidence of fetal risk based on human experience or both, and the risk of the use of the drug in pregnant women clearly outweighs any possible benefit. The drug is contraindicated in women who are or may become pregnant) and so is contraindicated in women who are or may become pregnant. The rest of the AADs are considered pregnancy category C drug [either studies in animals have revealed adverse effects on the fetus (teratogenic or embryocidal or other) and there are no controlled studies in women, or studies in women and animals are not available. Drugs should be given only if the potential benefit justifies the potential risk to the fetus]. The use of beta-blockers during pregnancy is relatively safe. The only exception is atenolol, which is a pregnancy category D drug.

## COMPARING ANTIARRHYTHMIC DRUGS TO IMPLANTABLE CARDIOVERTER DEFIBRILLATORS IN PATIENTS AT RISK OF ARRHYTHMIC DEATH

Several large, randomized, prospective, multicenter, controlled clinical trials compared ICDs versus AADs.[130,131] The Antiarrhythmics Versus Implantable Defibrillators (AVID) trial randomized patients resuscitated from a cardiac arrest to an ICD, empiric amiodarone (mean dose of 300 mg; 90% of patients) or sotalol (mean dose 250 mg) guided by electrophysiology study or Holter monitoring. The group runnings ICDs had a significant 39% (one year), 27% (two year) and 31% (three year) mortality reduction when compared to AADs. Only those patients with left ventricular ejection fraction between 20% and 34% showed a survival benefit with ICD (83%) when compared to amiodarone (72%).[130] The multicenter, prospective Sudden Cardiac Death Heart Failure Trial (SCD-HeFT) randomly assigned 2,521 medically managed ischemic and nonischemic cardiomyopathy patients with a left ventricular ejection fraction less than or equal to 35%, and NYHA functional class II–III heart failure to an ICD, amiodarone and placebo. Patients were followed for a median of 46 months. The primary endpoint was total mortality. The study showed that the ICD resulted in a 7.2% absolute and a 22% relative reduction in mortality, when compared to placebo and amiodarone. Amiodarone was no better than placebo in improving mortality. Patients with NYHA class III heart failure symptoms fared worse with amiodarone when compared to placebo (HR1.44, confidence interval 1.05–1.97, $P = 0.01$).[131]

In summary, ICDs are superior to AADs for both primary and secondary prevention of mortality, presumably due to sudden cardiac death. The AADs should be reserved only for patients who are not candidates for an ICD, who refuse ICD therapy, and for select patients with genetic disorders that respond well to a specific AAD.

## ANTIARRHYTHMIC DRUG-DEVICE INTERACTIONS

The ICD has emerged as the primary therapeutic modality for prevention of sudden cardiac death. Concomitant AAD therapy may be required in select ICD patients to suppress recurrent atrial and ventricular arrhythmias and to reduce the incidence and frequency of both appropriate and inappropriate shocks.[40] When used in this setting, AADs can affect device functioning in several ways:

- AADs can increase defibrillation and pacing thresholds
- AADs can slow VT rate to below the programmed ICD detection rate
- AADs can cause sinus and AV node dysfunction, resulting in bradycardia and AV block
- AADs can be proarrhythmic.

It is important to be aware of these potential interactions when selecting an appropriate AAD and also during device programming. Amiodarone and sotalol are the two most common AADs used in an ICD population. **Table 8** lists the effect of various AADs on pacing and defibrillation thresholds.

Class I AADs and amiodarone have been shown to raise the defibrillation threshold, whereas class III drugs such as sotalol and dofetilide tend to decrease the energy needed to defibrillate.[40] The results of the prospective, randomized, Optimal Pharmacologic Therapy in Cardioverter Defibrillator

### TABLE 8

**Effect of antiarrhythmic drugs on defibrillation and pacing thresholds**

| Drug | Pacing threshold | Defibrillation threshold |
|---|---|---|
| Quinidine | Increase | Increase |
| Procainamide | Increase | No change/increase |
| Lidocaine | No change | Increase |
| Flecainide | Increase | Increase |
| Beta-blockers | Increase | Decreases |
| Digoxin | Decrease | Decrease/no change |
| Ibutilide | Not known | Decrease |
| Sotalol | No effect | Decrease |
| Amiodarone | No effect | Increase |
| Dofetilide | No change | Decrease |
| Verapamil | Increase | Not known |

Patients (OPTIC) trial casts doubts regarding the clinical significance of these above-mentioned effects in the current era of high energy ICD devices. The OPTIC trial compared the effects of beta-blockers, beta-blocker plus amiodarone and sotalol on defibrillation energy requirements in 94 patients. The study showed that changes in defibrillation threshold with amiodarone and sotalol are at best modest and argues against repeat defibrillation threshold testing after initiating therapy with either drug.[132] The study also showed that amiodarone plus a beta-blocker was most effective in preventing ICD shocks at 1 year and was more effective than sotalol (10.3% vs 24.3% for sotalol; HR, 0.43; $P = 0.02$).[49]

## CONCLUSION

Antiarrhythmic drug therapy continues to play a critical role in the management of atrial and ventricular arrhythmias. The role of AADs has evolved in the face of advances in curative therapy for specific arrhythmias, as well as for underlying diseases. It is fair to say that the history of AAD therapy has come full circle: from the days of the CAST and SWORD trials showing increased mortality to demonstration of mortality benefit with dronedarone in the recent ATHENA trial. Irrespective of effects on survival, AADs are an integral part of the pharmacological armamentarium to combat AF, to treat ventricular arrhythmias in the structurally normal heart and in those with channelopathies, to suppress sustained ventricular arrhythmias in patients with structural heart disease who either have an ICD or are not candidates for one. The field of AAD therapy continues to evolve as newer drugs that target novel mechanisms are being actively developed and, with currently available drugs finding new indications for their use.

## REFERENCES

1. Gopinathannair R, Sullivan RM, Olshansky B. Update on medical management of atrial fibrillation in the modern era. Heart Rhythm. 2009;6:S17-22.
2. Wyse DG, Waldo AL, DiMarco JP, et al. A comparison of rate control and rhythm control in patients with atrial fibrillation. N Engl J Med. 2002;347:1825-33.
3. Vaughan Williams EM. A classification of antiarrhythmic actions reassessed after a decade of new drugs. J Clin Pharmacol. 1984;24: 129-47.
4. The Sicilian gambit. A new approach to the classification of antiarrhythmic drugs based on their actions on arrhythmogenic mechanisms. Task Force of the Working Group on Arrhythmias of the European Society of Cardiology. Circulation. 1991;84:1831-51.
5. Harrison DC. Antiarrhythmic drug classification: new science and practical applications. Am J Cardiol. 1985;56:185-7.

6. Hodgkin AL, Huxley AF. A quantitative description of membrane current and its application to conduction and excitation in nerve. J Physiol. 1952;117:500-44.
7. Lafuente-Lafuente C, Mouly S, Longas-Tejero MA, et al. Antiarrhythmic drugs for maintaining sinus rhythm after cardioversion of atrial fibrillation: a systematic review of randomized controlled trials. Arch Intern Med. 2006;166:719-28.
8. Belhassen B, Glick A, Viskin S. Efficacy of quinidine in high-risk patients with Brugada syndrome. Circulation. 2004;110:1731-7.
9. Mok NS, Chan NY, Chiu AC. Successful use of quinidine in treatment of electrical storm in Brugada syndrome. Pacing Clin Electrophysiol. 2004;27:821-3.
10. Mizusawa Y, Sakurada H, Nishizaki M, et al. Effects of low-dose quinidine on ventricular tachyarrhythmias in patients with Brugada syndrome: low-dose quinidine therapy as an adjunctive treatment. J Cardiovasc Pharmacol. 2006;47:359-64.
11. Olshansky B, Okumura K, Hess PG, et al. Use of procainamide with rapid atrial pacing for successful conversion of atrial flutter to sinus rhythm. J Am Coll Cardiol. 1988;11:359-64.
12. Olshansky B, Martins J, Hunt S. N-acetyl procainamide causing torsades de pointes. Am J Cardiol. 1982;50:1439-41.
13. Sherrid MV, Barac I, McKenna WJ, et al. Multicenter study of the efficacy and safety of disopyramide in obstructive hypertrophic cardiomyopathy. J Am Coll Cardiol. 2005;45:1251-8.
14. Lie KI, Wellens HJ, van Capelle FJ, et al. Lidocaine in the prevention of primary ventricular fibrillation. A double-blind, randomized study of 212 consecutive patients. N Engl J Med. 1974;291:1324-6.
15. Alexander JH, Granger CB, Sadowski Z, et al. Prophylactic lidocaine use in acute myocardial infarction: incidence and outcomes from two international trials. The GUSTO-I and GUSTO-IIb Investigators. Am Heart J. 1999;137:799-805.
16. Singh BN. Routine prophylactic lidocaine administration in acute myocardial infarction. An idea whose time is all but gone? Circulation. 1992;86:1033-5.
17. Hine LK, Laird N, Hewitt P, et al. Meta-analytic evidence against prophylactic use of lidocaine in acute myocardial infarction. Arch Intern Med. 1989;149:2694-8.
18. Stenson RE, Constantino RT, Harrison DC. Interrelationships of hepatic blood flow, cardiac output, and blood levels of lidocaine in man. Circulation. 1971;43:205-11.
19. Josephson ME, Kastor JA, Kitchen JG 3rd. Lidocaine in Wolff-Parkinson-White syndrome with atrial fibrillation. Ann Intern Med. 1976;84:44-5.
20. Wyman MG, Slaughter RL, Farolino DA, et al. Multiple bolus technique for lidocaine administration in acute ischemic heart disease. II. Treatment of refractory ventricular arrhythmias and the pharmacokinetic significance of severe left ventricular failure. J Am Coll Cardiol. 1983;2:764-9.
21. Ochs HR, Carstens G, Greenblatt DJ. Reduction in lidocaine clearance during continuous infusion and by coadministration of propranolol. N Engl J Med. 1980;303:373-7.
22. Feely J, Wilkinson GR, McAllister CB, et al. Increased toxicity and reduced clearance of lidocaine by cimetidine. Ann Intern Med. 1982;96:592-4.

23. Stein J, Podrid P, Lown B. Effects of oral mexiletine on left and right ventricular function. Am J Cardiol. 1984;54:575-8.
24. Campbell RW. Mexiletine. N Engl J Med. 1987;316:29-34.
25. International mexiletine and placebo antiarrhythmic coronary trial: I. Report on arrhythmia and other findings. Impact Research Group. J Am Coll Cardiol. 1984;4:1148-63.
26. Shimizu W, Aiba T, Antzelevitch C. Specific therapy based on the genotype and cellular mechanism in inherited cardiac arrhythmias. Long QT syndrome and Brugada syndrome. Curr Pharm Des. 2005;11:1561-72.
27. Fasola GP, D'Osualdo F, de Pangher V, et al. Thrombocytopenia and mexiletine. Ann Intern Med. 1984;100:162.
28. Bigger JT Jr. The interaction of mexiletine with other cardiovascular drugs. Am Heart J. 1984;107:1079-85.
29. Preliminary report: effect of encainide and flecainide on mortality in a randomized trial of arrhythmia suppression after myocardial infarction. The Cardiac Arrhythmia Suppression Trial (CAST) Investigators. N Engl J Med. 1989;321:406-12.
30. Roden DM, Woosley RL. Drug therapy. Flecainide. N Engl J Med. 1986;315:36-41.
31. Wann LS, Curtis AB, January CT, et al. 2011 ACCF/AHA/HRS focused update on the management of patients with atrial fibrillation (updating the 2006 guideline): a report of the American College of Cardiology Foundation/American Heart Association Task Force on Practice Guidelines. Circulation. 2011;123:104-23.
32. Konety SH, Olshansky B. The "pill-in-the-pocket" approach to atrial fibrillation. N Engl J Med. 2005;352:1150-1.
33. Buxton AE, Waxman HL, Marchlinski FE, et al. Right ventricular tachycardia: clinical and electrophysiologic characteristics. Circulation. 1983;68:917-27.
34. Watanabe H, Chopra N, Laver D, et al. Flecainide prevents catecholaminergic polymorphic ventricular tachycardia in mice and humans. Nat Med. 2009;15:380-3.
35. Benhorin J, Taub R, Goldmit M, et al. Effects of flecainide in patients with new SCN5A mutation: mutation-specific therapy for long-QT syndrome? Circulation. 2000;101:1698-706.
36. Effect of the antiarrhythmic agent moricizine on survival after myocardial infarction. The Cardiac Arrhythmia Suppression Trial II Investigators. N Engl J Med. 1992;327:227-33.
37. Vik-Mo H, Ohm OJ, Lund-Johansen P. Electrophysiologic effects of flecainide acetate in patients with sinus nodal dysfunction. Am J Cardiol. 1982;50:1090-4.
38. Alboni P, Botto GL, Baldi N, et al. Outpatient treatment of recent-onset atrial fibrillation with the "pill-in-the-pocket" approach. N Engl J Med. 2004;351:2384-91.
39. Muhiddin KA, Turner P, Blackett A. Effect of flecainide on cardiac output. Clin Pharmacol Ther. 1985;37:260-3.
40. Rajawat YS, Dias D, Gerstenfeld EP, et al. Interactions of anti-arrhythmic drugs and implantable devices in controlling ventricular tachycardia and fibrillation. Curr Cardiol Rep. 2002;4:434-40.
41. Hellestrand KJ, Burnett PJ, Milne JR, et al. Effect of the antiarrhythmic agent flecainide acetate on acute and chronic pacing thresholds. Pacing Clin Electrophysiol. 1983;6:892-9.

42. McLeod AA, Stiles GL, Shand DG. Demonstration of beta adrenoceptor blockade by propafenone hydrochloride: clinical pharmacologic, radioligand binding and adenylate cyclase activation studies. J Pharmacol Exp Ther. 1984;228:461-6.
43. Siddoway LA, Thompson KA, McAllister CB, et al. Polymorphism of propafenone metabolism and disposition in man: clinical and pharmacokinetic consequences. Circulation. 1987;75:785-91.
44. Lee JT, Kroemer HK, Silberstein DJ, et al. The role of genetically determined polymorphic drug metabolism in the beta-blockade produced by propafenone. N Engl J Med. 1990;322:1764-8.
45. Wagner F, Kalusche D, Trenk D, et al. Drug interaction between propafenone and metoprolol. Br J Clin Pharmacol. 1987;24:213-20.
46. Olshansky B, Rosenfeld LE, Warner AL, et al. The Atrial Fibrillation Follow-up Investigation of Rhythm Management (AFFIRM) study: approaches to control rate in atrial fibrillation. J Am Coll Cardiol. 2004;43:1201-8.
47. Olshansky B, Martins JB. Usefulness of isoproterenol facilitation of ventricular tachycardia induction during extrastimulus testing in predicting effective chronic therapy with beta-adrenergic blockade. Am J Cardiol. 1987;59:573-7.
48. Zicha S, Tsuji Y, Shiroshita-Takeshita A, et al. Beta-blockers as antiarrhythmic agents. Handb Exp Pharmacol. 2006;171:235-66.
49. Connolly SJ, Dorian P, Roberts RS, et al. Comparison of beta-blockers, amiodarone plus beta-blockers, or sotalol for prevention of shocks from implantable cardioverter defibrillators: the OPTIC Study: a randomized trial. JAMA. 2006;295:165-71.
50. Cheng J, Niwa R, Kamiya K, et al. Carvedilol blocks the repolarizing K+ currents and the L-type Ca2+ current in rabbit ventricular myocytes. Eur J Pharmacol. 1999;376:189-201.
51. Olshansky B. Management of atrial fibrillation after coronary artery bypass graft. Am J Cardiol. 1996;78:27-34.
52. Hohnloser SH, Woosley RL. Sotalol. N Engl J Med. 1994;331:31-8.
53. Waldo AL, Camm AJ, deRuyter H, et al. Effect of d-sotalol on mortality in patients with left ventricular dysfunction after recent and remote myocardial infarction. The SWORD Investigators. Survival With Oral d-Sotalol. Lancet. 1996;348:7-12.
54. Julian DG, Prescott RJ, Jackson FS, Szekely P. Controlled trial of sotalol for one year after myocardial infarction. Lancet. 1982;1:1142-7.
55. Mason JW. A comparison of seven antiarrhythmic drugs in patients with ventricular tachyarrhythmias. Electrophysiologic Study versus Electrocardiographic Monitoring Investigators. N Engl J Med. 1993;329:452-8.
56. Pacifico A, Hohnloser SH, Williams JH, et al. Prevention of implantable-defibrillator shocks by treatment with sotalol. d,l-Sotalol Implantable Cardioverter-Defibrillator Study Group. New Engl J Med. 1999;340:1855-62.
57. Singh BN, Singh SN, Reda DJ, et al. Amiodarone versus sotalol for atrial fibrillation. N Engl J Med. 2005;352:1861-72.
58. Baskin EP, Lynch JJ Jr. Differential atrial versus ventricular activities of class III potassium channel blockers. J Pharmacol Exp Ther. 1998;285:135-42.
59. Abel S, Nichols DJ, Brearley CJ, et al. Effect of cimetidine and ranitidine on pharmacokinetics and pharmacodynamics of a single dose of dofetilide. Br J Clin Pharmacol. 2000;49:64-71.

60. Torp-Pedersen C, Møller M, Bloch-Thomsen PE, et al. Dofetilide in patients with congestive heart failure and left ventricular dysfunction. Danish Investigations of Arrhythmia and Mortality on Dofetilide Study Group. N Engl J Med. 1999;341:857-65.
61. Olshansky B. Dofetilide versus quinidine for atrial flutter: viva la difference!? J Cardiovasc Electrophysiol. 1996;7:828-32.
62. Yap YG, Camm AJ. Drug induced QT prolongation and torsades de pointes. Heart. 2003;89:1363-72.
63. Elming H, Brendorp B, Pedersen OD, et al. Dofetilide: a new drug to control cardiac arrhythmia. Expert Opin Pharmacother. 2003;4: 973-85.
64. Murray KT. Ibutilide. Circulation. 1998;97:493-7.
65. Stambler BS, Wood MA, Ellenbogen KA, et al. Efficacy and safety of repeated intravenous doses of ibutilide for rapid conversion of atrial flutter or fibrillation. Ibutilide Repeat Dose Study Investigators. Circulation. 1996;94:1613-21.
66. Tercius AJ, Kluger J, Coleman CI, et al. Intravenous magnesium sulfate enhances the ability of intravenous ibutilide to successfully convert atrial fibrillation or flutter. Pacing Clin Electrophysiol. 2007;30:1331-5.
67. Patsilinakos S, Christou A, Kafkas N, et al. Effect of high doses of magnesium on converting ibutilide to a safe and more effective agent. Am J Cardiol. 2010;106:673-6.
68. Coleman CI, Sood N, Chawla D, et al. Intravenous magnesium sulfate enhances the ability of dofetilide to successfully cardiovert atrial fibrillation or flutter: results of the Dofetilide and Intravenous Magnesium Evaluation. Europace. 2009;11:892-5.
69. Steinwender C, Honig S, Kypta A, et al. Pre-injection of magnesium sulfate enhances the efficacy of ibutilide for the conversion of typical but not of atypical persistent atrial flutter. Int J Cardiol. 2010;141: 260-5.
70. Coleman CI, Kalus JS, Caron MF, et al. Model of effect of magnesium prophylaxis on frequency of torsades de pointes in ibutilide-treated patients. Am J Health Syst Pharm. 2004;61:685-8.
71. Fragakis N, Bikias A, Delithanasis I, et al. Acute beta-adrenoceptor blockade improves efficacy of ibutilide in conversion of atrial fibrillation with a rapid ventricular rate. Europace. 2009;11:70-4.
72. Gopinathannair R, Olshansky B. Ibutilide revisited: stronger and safer than ever. Europace. 2009;11:9-10.
73. Anrep GV, Barsoum GS, Kenawy MR, et al. Ammi visnaga in the treatment of the anginal syndrome. Br Heart J. 1946;8:171-7.
74. Mason JW. Amiodarone. N Engl J Med. 1987;316:455-66.
75. Mason JW, Hondeghem LM, Katzung BG. Amiodarone blocks inactivated cardiac sodium channels. Pflugers Arch. 1983;396:79-81.
76. Holt DW, Tucker GT, Jackson PR, et al. Amiodarone pharmacokinetics. Br J Clin Pract Suppl. 1986;44:109-14.
77. Plomp TA, van Rossum JM, Robles de Medina EO, et al. Pharmacokinetics and body distribution of amiodarone in man. Arzneimittelforschung. 1984;34:513-20.
78. Julian DG, Camm AJ, Frangin G, et al. Randomised trial of effect of amiodarone on mortality in patients with left-ventricular dysfunction after recent myocardial infarction: EMIAT. European Myocardial Infarct Amiodarone Trial Investigators. Lancet. 1997;349:667-74.
79. Cairns JA, Connolly SJ, Roberts R, et al. Randomised trial of outcome after myocardial infarction in patients with frequent or

repetitive ventricular premature depolarisations: CAMIAT. Canadian Amiodarone Myocardial Infarction Arrhythmia Trial Investigators. Lancet. 1997;349:675-82.
80. Boutitie F, Boissel JP, Connolly SJ, et al. Amiodarone interaction with beta-blockers: analysis of the merged EMIAT (European Myocardial Infarct Amiodarone Trial) and CAMIAT (Canadian Amiodarone Myocardial Infarction Trial) databases. The EMIAT and CAMIAT Investigators. Circulation. 1999;99:2268-75.
81. Garguichevich JJ, Ramos JL, Gambarte A, et al. Effect of amiodarone therapy on mortality in patients with left ventricular dysfunction and asymptomatic complex ventricular arrhythmias: Argentine Pilot Study of Sudden Death and Amiodarone (EPAMSA). Am Heart J. 1995;130:494-500.
82. Doval HC, Nul DR, Grancelli HO, et al. Randomised trial of low-dose amiodarone in severe congestive heart failure. Grupo de Estudio de la Sobrevida en la Insuficiencia Cardiaca en Argentina (GESICA). Lancet. 1994;344:493-8.
83. Singh SN, Fletcher RD, Fisher SG, et al. Amiodarone in patients with congestive heart failure and asymptomatic ventricular arrhythmia. Survival Trial of Antiarrhythmic Therapy in Congestive Heart Failure. N Engl J Med. 1995;333:77-82.
84. Piccini JP, Berger JS, O'Connor CM. Amiodarone for the prevention of sudden cardiac death: a meta-analysis of randomized controlled trials. Eur Heart J. 2009;30:1245-53.
85. Roy D, Talajic M, Dorian P, et al. Amiodarone to prevent recurrence of atrial fibrillation. Canadian Trial of Atrial Fibrillation Investigators. N Engl J Med. 2000;342:913-20.
86. Goldschlager N, Epstein AE, Naccarelli GV, et al. A practical guide for clinicians who treat patients with amiodarone: 2007. Heart Rhythm. 2007;4:1250-9.
87. Olshansky B, Sami M, Rubin A, et al. Use of amiodarone for atrial fibrillation in patients with preexisting pulmonary disease in the AFFIRM study. Am J Cardiol. 2005;95:404-5.
88. Olshansky B. Images in clinical medicine. Amiodarone-induced pulmonary toxicity. N Engl J Med. 1997;337:1814.
89. Manning AS, Bruyninckx C, Ramboux J, et al. SR 33589, a new amiodarone-like agent: effect on ischemia- and reperfusion-induced arrhythmias in anesthetized rats. J Cardiovasc Pharmacol. 1995;26: 453-61.
90. Djandjighian L, Planchenault J, Finance O, et al. Hemodynamic and antiadrenergic effects of dronedarone and amiodarone in animals with a healed myocardial infarction. J Cardiovasc Pharmacol. 2000;36: 376-83.
91. Sun W, Sarma JS, Singh BN. Electrophysiological effects of dronedarone (SR33589), a noniodinated benzofuran derivative, in the rabbit heart: comparison with amiodarone. Circulation. 1999;100:2276-81.
92. Finance O, Manning A, Chatelain P. Effects of a new amiodarone-like agent, SR 33589, in comparison to amiodarone, D,L-sotalol, and lignocaine, on ischemia-induced ventricular arrhythmias in anesthetized pigs. J Cardiovasc Pharmacol. 1995;26:570-6.
93. Kober L, Torp-Pedersen C, McMurray JJ, et al. Increased mortality after dronedarone therapy for severe heart failure. N Engl J Med. 2008;358:2678-87.

94. Dronedarone prescribing information. [online] MULTAQ website. Available from http://www.multaq.com/docs/consumer_pdf/pi.aspx [Accessed February 2011]
95. Sullivan RM, Olshansky B. Dronedarone: evidence supporting its therapeutic use in the treatment of atrial fibrillation. Core Evid. 2010;5:49-59.
96. Singh BN, Connolly SJ, Crijns HJ, et al. Dronedarone for maintenance of sinus rhythm in atrial fibrillation or flutter. N Engl J Med. 2007;357:987-99.
97. Le Heuzey JY, De Ferrari GM, Radzik D, et al. A short-term, randomized, double-blind, parallel-group study to evaluate the efficacy and safety of dronedarone versus amiodarone in patients with persistent atrial fibrillation: the DIONYSOS study. J Cardiovasc Electrophysiol. 2010;21:597-605.
98. Davy JM, Herold M, Hoglund C, et al. Dronedarone for the control of ventricular rate in permanent atrial fibrillation: the Efficacy and safety of dRonedArone for the cOntrol of ventricular rate during atrial fibrillation (ERATO) study. Am Heart J. 2008;156:527,e1-9.
99. Hohnloser SH, Crijns HJ, van Eickels M, et al. Effect of dronedarone on cardiovascular events in atrial fibrillation. N Engl J Med. 2009;360:668-78.
100. Connolly SJ, Crijns HJ, Torp-Pedersen C, et al. Analysis of stroke in ATHENA: a placebo-controlled, double-blind, parallel-arm trial to assess the efficacy of dronedarone 400 mg BID for the prevention of cardiovascular hospitalization or death from any cause in patients with atrial fibrillation/atrial flutter. Circulation. 2009;120:1174-80.
101. Sanofi Aventis. Important Drug Warning on Multaq: Letter to Healthcare Provider - Jan 14, 2011
101a. Connolly SJ, CammAJ, Halperin JL, et al. Dronedarone in high risk permanent atrial fibrillation. The New England Journal of Medicine. 2011;365:2258-76.
102. Lombardi F, Terranova P. Pharmacological treatment of atrial fibrillation: mechanisms of action and efficacy of class III drugs. Curr Med Chem. 2006;13:1635-53.
103. Camm AJ, Pratt CM, Schwartz PJ, et al. Mortality in patients after a recent myocardial infarction: a randomized, placebo-controlled trial of azimilide using heart rate variability for risk stratification. Circulation. 2004;109:990-6.
104. Dorian P, Borggrefe M, Al-Khalidi HR, et al. Placebo-controlled, randomized clinical trial of azimilide for prevention of ventricular tachyarrhythmias in patients with an implantable cardioverter defibrillator. Circulation. 2004;110:3646-54.
105. Lombardi F, Borggrefe M, Ruzyllo W, et al. Azimilide vs. placebo and sotalol for persistent atrial fibrillation: the A-COMET-II (Azimilide-CardiOversion MaintEnance Trial-II) trial. Eur Heart J. 2006;27:2224-31.
106. Kerr CR, Connolly SJ, Kowey P, et al. Efficacy of azimilide for the maintenance of sinus rhythm in patients with paroxysmal atrial fibrillation in the presence and absence of structural heart disease. Am J Cardiol. 2006;98:215-8.
107. Conway E, Musco S, Kowey PR. New horizons in antiarrhythmic therapy: will novel agents overcome current deficits? Am J Cardiol. 2008;102:12H-9H.

108. Lerman BB, Belardinelli L. Cardiac electrophysiology of adenosine. Basic and clinical concepts. Circulation. 1991;83:1499-509.
109. Wilber DJ, Baerman J, Olshansky B, et al. Adenosine-sensitive ventricular tachycardia. Clinical characteristics and response to catheter ablation. Circulation. 1993;87:126-34.
110. Kall JG, Kopp D, Olshansky B, et al. Adenosine-sensitive atrial tachycardia. Pacing Clin Electrophysiol. 1995;18:300-6.
111. Ellenbogen KA, Thames MD, DiMarco JP, et al. Electrophysiological effects of adenosine in the transplanted human heart. Evidence of supersensitivity. Circulation. 1990;81:821-8.
112. Hohnloser SH, Dorian P, Straub M, et al. Safety and efficacy of intravenously administered tedisamil for rapid conversion of recent-onset atrial fibrillation or atrial flutter. J Am Coll Cardiol. 2004;44:99-104.
113. Wijffels MC, Crijns HJ. Recent advances in drug therapy for atrial fibrillation. J Cardiovasc Electrophysiol. 2003;14:S40-7.
114. Naccarelli GV, Wolbrette DL, Samii S, et al. Vernakalant—a promising therapy for conversion of recent-onset atrial fibrillation. Expert Opin Investig Drugs. 2008;17:805-10.
115. Roy D, Pratt CM, Torp-Pedersen C, et al. Vernakalant hydrochloride for rapid conversion of atrial fibrillation: a phase 3, randomized, placebo-controlled trial. Circulation. 2008;117:1518-25.
116. Kowey PR, Dorian P, Mitchell LB, et al. Vernakalant hydrochloride for the rapid conversion of atrial fibrillation after cardiac surgery: a randomized, double-blind, placebo-controlled trial. Circ Arrhythm Electrophysiol. 2009;2:652-9.
117. Pratt CM, Roy D, Torp-Pedersen C, et al. Usefulness of vernakalant hydrochloride injection for rapid conversion of atrial fibrillation. Am J Cardiol. 2010;106:1277-83.
118. Camm AJ, Capucci A, Hohnloser SH, et al. A randomized active-controlled study comparing the efficacy and safety of vernakalant to amiodarone in recent-onset atrial fibrillation. J Am Coll Cardiol. 2011;57:313-21.
119. Fox K, Borer JS, Camm AJ, et al. Resting heart rate in cardiovascular disease. J Am Coll Cardiol. 2007;50:823-30.
120. Gopinathannair R, Sullivan RM, Olshansky B. Slower heart rates for healthy hearts: time to redefine tachycardia? Circ Arrhythm Electrophysiol. 2008;1:321-3.
121. DiFrancesco D, Camm JA. Heart rate lowering by specific and selective I(f) current inhibition with ivabradine: a new therapeutic perspective in cardiovascular disease. Drugs. 2004;64:1757-65.
122. Fox K, Ford I, Steg PG, et al. Ivabradine for patients with stable coronary artery disease and left-ventricular systolic dysfunction (BEAUTIFUL): a randomised, double-blind, placebo-controlled trial. Lancet. 2008;372:807-16.
123. Tardif JC, Ponikowski P, Kahan T. Efficacy of the I(f) current inhibitor ivabradine in patients with chronic stable angina receiving beta-blocker therapy: a 4-month, randomized, placebo-controlled trial. Eur Heart J. 2009;30:540-8.
124. Swedberg K, Komajda M, Bohm M, et al. Ivabradine and outcomes in chronic heart failure (SHIFT): a randomised placebo-controlled study. Lancet. 2010;376:875-85.
125. Calo L, Rebecchi M, Sette A, et al. Efficacy of ivabradine administration in patients affected by inappropriate sinus tachycardia. Heart Rhythm. 2010;7:1318-23.

126. Antzelevitch C, Belardinelli L, Zygmunt AC, et al. Electrophysiological effects of ranolazine, a novel antianginal agent with antiarrhythmic properties. Circulation. 2004;110:904-10.
127. Scirica BM, Morrow DA, Hod H, et al. Effect of ranolazine, an antianginal agent with novel electrophysiological properties, on the incidence of arrhythmias in patients with non ST-segment elevation acute coronary syndrome: results from the Metabolic Efficiency With Ranolazine for Less Ischemia in Non ST-Elevation Acute Coronary Syndrome Thrombolysis in Myocardial Infarction 36 (MERLIN-TIMI 36) randomized controlled trial. Circulation. 2007;116:1647-52.
128. Burashnikov A, Sicouri S, Di Diego JM, Belardinelli L, Antzelevitch C. Synergistic effect of the combination of ranolazine and dronedarone to suppress atrial fibrillation. J Am Coll Cardiol. 2010;56:1216-24.
129. Savelievap I, Camm J. Anti-arrhythmic drug therapy for atrial fibrillation: current anti-arrhythmic drugs, investigational agents, and innovative approaches. Europace. 2008;10:647-65.
130. A comparison of antiarrhythmic-drug therapy with implantable defibrillators in patients resuscitated from near-fatal ventricular arrhythmias. The Antiarrhythmics versus Implantable Defibrillators (AVID) Investigators. N Engl J Med. 1997;337:1576-83.
131. Bardy GH, Lee KL, Mark DB, et al. Amiodarone or an implantable cardioverter-defibrillator for congestive heart failure. N Engl J Med. 2005;352:225-37.
132. Hohnloser SH, Dorian P, Roberts R, et al. Effect of amiodarone and sotalol on ventricular defibrillation threshold: the optimal pharmacological therapy in cardioverter defibrillator patients (OPTIC) trial. Circulation. 2006;114:104-9.

CHAPTER 3

# Electrophysiology Studies

Indrajit Choudhuri, Masood Akhtar

## Chapter Outline

- Cardiac Electrophysiology Study: Philosophy, Requirements, and Basic Techniques
  - Cardiac Access and Catheterization
  - Signals and Filtering
- Fundamentals of the Cardiac Electrophysiology Study
  - Conventions
  - Normal Cardiac Electrophysiology
- Programmed Electrical Stimulation and Associated Electrophysiology
  - Continuous Pacing
  - Intermittent or Interrupted Pacing with Extrastimuli
  - Significance of "Short-Long-Short" Pacing Cycles
  - Clinical Application of "Routine" Electrophysiology Study and Anticipated Responses to Programmed Stimulation
  - Survivors of Sudden Cardiac Arrest
- Cardiac Electrophysiology Study for Evaluation of Drug Therapy
- Electrophysiology Study to Guide Ablative Therapy
  - Role of Three-dimensional Mapping Systems
- Complications

## INTRODUCTION

Clinical cardiac electrophysiology (EP) is a relatively new and continually evolving investigative field. Its modern underpinnings date back to the first description of the cardiac Purkinje fibers in 1839 by Czech neuroscientist Jan Evangelista Purkyně,[1] and description of the atrioventricular (AV) bundle by Wilhelm His Jr.[2] and accessory "AV bundles" by Albert Kent in 1893.[3] Such anatomic discoveries of the cardiac conduction system were the first murmurings of what would spawn an entirely independent arena of cardiac investigation that continued in this vein into the early 20th century with description of the AV node by Sunao Tawara[4] in 1906 and, finally, the sinoatrial node[5] in 1907 by Arthur Keith and his student Martin Flack. Einthoven's 1908 description of the modern electrocardiograph[6] heralded a new phase of cardiac electrophysiologic discovery, permitting arrhythmia description and electrocardiogram (ECG) correlation with clinical presentation. During his Nobel speech, Einthoven foretold that "a new chapter has been opened in the study of heart diseases ...." This was the primary mode of "electrophysiology study" during the first half of the 20th century, during which time the first case of idiopathic ventricular fibrillation (VF) was described, in 1929,[7] and the first long QT case reports were described by Jervell and Lange-Nielsen.[8]

Percutaneous and open-chest techniques for arrhythmia mapping were first described in the 1950s, but it was not until the late 1960s and 1970s that a reproducible technique for recording the His bundle potential[9] was demonstrated and utilized. Its role, as well as that of programmed electrical stimulation, for identification of site of origin and arrhythmia mechanism established the invasive cardiac EP study as a mainstay in diagnostic cardiology, expanding the frontiers of cardiac EP and cardiovascular disease. This breakthrough was accompanied by an increase in open-chest surgical therapy and ablation, providing further insight into the mechanisms of arrhythmias. However, demonstration of a closed-chest catheter technique for destruction of cardiac tissue[10] truly revolutionized the field. It provided a percutaneous option for patients to undergo diagnosis and treatment in the same setting.

Since those early and formative years, the comprehensive EP study has gradually evolved from a prolonged undertaking to a streamlined diagnostic process that attempts to identify clinically relevant mechanisms of arrhythmogenesis, and correlate these with symptomatology to guide therapy. This chapter focuses on fundamental aspects of the contemporary intracardiac EP study and should serve as a foundation for all cardiovascular disease practitioners seeking further insight into the electrophysiologic mechanisms of the human heart.

## CARDIAC ELECTROPHYSIOLOGY STUDY: PHILOSOPHY, REQUIREMENTS, AND BASIC TECHNIQUES

The contemporary EP study has been condensed, abbreviated and streamlined to capitalize on the basic science, and clinical foundations established since the 1800s to efficiently evaluate tendency toward arrhythmia and its underlying mechanisms. The EP studies are performed to investigate clinically documented rhythm disturbances or evaluate symptoms compatible with arrhythmic etiology, such as palpitations or syncope, for risk stratification of sudden death and evaluation of pharmacologic therapy.[11] Alternatively, not all arrhythmias and arrhythmic mechanisms may be evaluated by or necessitate study. For instance, an EP study is not generally indicated for evaluation of symptomatic bradycardia. Whether the mechanism is sinus node dysfunction or conduction disease in the AV node, His bundle or Purkinje system, permanent pacing is usually required, and demonstrating mechanism may be of more "academic" concern. In specific situations, however, demonstration of the level and mechanism of conduction block may be instrumental in guiding therapy, such as in apparent conduction block attributable to junctional extrasystoles in which beat suppression is required, whether medical or ablative, rather than pacing. Indications for EP study are shown in **Tables 1 to 13**.[12]

The cardiac EP study itself is a systematic evaluation of clinically relevant aspects of myocardial electrical stimulation

### TABLE 1
**Indications for electrophysiology study to evaluate sinus node function**

*Class I*
- Symptomatic patients in whom sinus node dysfunction is suspected as the cause of symptoms, but a causal relation between an arrhythmia and the symptoms has not been established after appropriate evaluation

*Class II*
- Patients with documented sinus node dysfunction in whom evaluation of AV or VA conduction or susceptibility to arrhythmias may aid in selection of the most appropriate pacing modality
- Patients with electrocardiographically documented sinus bradyarrhythmias to determine if abnormalities are due to intrinsic disease, autonomic nervous system dysfunction or the effects of drugs so as to help select therapeutic options
- Symptomatic patients with known sinus bradyarrhythmias to evaluate potential for other arrhythmias as the cause of symptoms

*Class III*
- Symptomatic patients in whom an association between symptoms and a documented bradyarrhythmia has been established and choice of therapy would not be affected by results of an electrophysiology study
- Asymptomatic patients with sinus bradyarrhythmias or sinus pauses observed only during sleep, including sleep apnea

Abbreviations: AV: Atrioventricular; VA: Ventriculoatrial

### TABLE 2
**Electrophysiology study indications for acquired AV block**

*Class I*
- Symptomatic patients in whom His-Purkinje block, suspected as a cause of symptoms, has not been established
- Patients with second-degree or third-degree AV block treated with a pacemaker who remain symptomatic and in whom another arrhythmia is suspected as a cause of symptoms

*Class II*
- Patients with second-degree or third-degree AV block in whom knowledge of the site of block or its mechanism or response to pharmacological or other temporary intervention may help direct therapy or assess prognosis
- Patients with premature, concealed junctional depolarizations suspected as a cause of second-degree or third-degree AV block pattern (i.e. pseudo-AV block)

*Class III*
- Symptomatic patients in whom the symptoms and presence of AV block are correlated by ECG findings
- Asymptomatic patients with transient AV block associated with sinus slowing (e.g. nocturnal type I second-degree AV block)

Abbreviations: AV: Atrioventricular; ECG: Electrocardiogram

### TABLE 3

**Electrophysiology study indications for chronic intraventricular conduction delay**

*Class I*
- Symptomatic patients in whom the cause of symptoms is not known

*Class II*
- Asymptomatic patients with bundle branch block in whom pharmacological therapy that could increase conduction delay or produce heart block is contemplated

*Class III*
- Asymptomatic patients with intraventricular conduction delay
- Symptomatic patients whose symptoms can be correlated with or excluded by ECG events

Abbreviations: ECG: Electrocardiogram

### TABLE 4

**Electrophysiology study indications for narrow QRS complex tachycardias**

*Class I*
- Patients with frequent or poorly tolerated episodes of tachycardia that do not adequately respond to drug therapy and for whom information about site of origin, mechanism and electrophysiological properties of the pathways of the tachycardia is essential for choosing appropriate therapy (drugs, catheter ablation, pacing or surgery)
- Patients who prefer ablative therapy to pharmacological treatment

*Class II*
- Patients with frequent episodes of tachycardia requiring drug treatment for whom there is concern about proarrhythmia or the effects of the antiarrhythmic drug on sinus node or AV conduction

*Class III*
- Patients with tachycardias easily controlled by vagal maneuvers and/or well-tolerated drug therapy who are not candidates for nonpharmacological therapy

Abbreviations: AV: Atrioventricular

### TABLE 5

**Electrophysiology study indications for wide QRS complex tachycardias**

*Class I*
- Patients with wide QRS complex tachycardia in whom correct diagnosis is unclear after analysis of available ECG tracings, and for whom knowledge of the correct diagnosis is necessary for patient care

*Class II*
- None

*Class III*
- Patients with ventricular or supraventricular tachycardia with aberrant conduction or preexcitation syndromes diagnosed with certainty by ECG criteria, and for whom invasive electrophysiological data would not influence therapy. However, data obtained at baseline EP study in these patients might be appropriate as a guide for subsequent therapy

Abbreviations: ECG: Electrocardiogram; EP: Electrophysiology

### TABLE 6
**Electrophysiology study indications for prolonged QT intervals**

*Class I*
- None

*Class II*
- Identification of a proarrhythmic effect of a drug in patients experiencing sustained ventricular tachycardia or cardiac arrest while receiving the drug
- Patients who have equivocal abnormalities of QT interval duration or TU wave configuration, with syncope or symptomatic arrhythmias, in whom catecholamine effects may unmask a distinct QT abnormality

*Class III*
- Patients with clinically manifest congenital QT prolongation, with or without symptomatic arrhythmias
- Patients with acquired prolonged QT syndrome with symptoms closely related to an identifiable cause or mechanism

### TABLE 7
**Electrophysiology study indications for Wolff-Parkinson-White syndrome**

*Class I*
- Patients being evaluated for catheter ablation or surgical ablation of an accessory pathway
- Patients with ventricular preexcitation who have survived cardiac arrest, or who have unexplained syncope
- Symptomatic patients in whom determination of the mechanism of arrhythmia, or knowledge of the electrophysiological properties of the accessory pathway and normal conduction system would help in determining appropriate therapy

*Class II*
- Asymptomatic patients with a family history of sudden cardiac death or with ventricular preexcitation, but no spontaneous arrhythmia, who engage in high-risk occupations or activities, and in whom knowledge of the electrophysiological properties of the accessory pathway or inducible tachycardia may help determine recommendations for further activities or therapy
- Patients with ventricular preexcitation who are undergoing cardiac surgery for other reasons

*Class III*
- Asymptomatic patients with ventricular preexcitation, except those in Class II above

and propagation and arrhythmic potential. It is conducted in a diagnostic cardiac EP or angiography suite, with minimum requirements of single-plane fluoroscopy and patient table/gantry; electrocardiac stimulator, signal filtering and recording system; and diagnostic electrode catheters through which cardiac stimulation and intracardiac signals/impulses may be sensed and delivered. Patients are generally studied in the postabsorptive state so as to minimize risk of aspiration while sedated or during

### TABLE 8

**Electrophysiology study indications for premature ventricular complexes, couplets, and nonsustained ventricular tachycardia**

*Class I*
- None

*Class II*
- Patients with other risk factors for future arrhythmic events, such as a low ejection fraction, positive signal-averaged ECG, and nonsustained VT on ambulatory ECG recordings in whom electrophysiology studies will be used for further risk assessment and for guiding therapy in patients with inducible VT
- Patients with highly symptomatic, uniform morphology premature ventricular complexes, couplets and nonsustained VT, who are considered as potential candidates for catheter ablation

*Class III*
- Asymptomatic or mildly symptomatic patients with premature ventricular complexes, couplets and nonsustained VT without other risk factors for sustained arrhythmias

Abbreviations: ECG: Electrocardiogram; VT: Ventricular tachycardia

### TABLE 9

**Electrophysiology study indications for unexplained syncope**

*Class I*
- Patients with suspected structural heart disease and syncope that remains unexplained after appropriate evaluation

*Class II*
- Patients with recurrent unexplained syncope without structural heart disease and a negative head-up tilt test

*Class III*
- Patients with a known cause of syncope for whom treatment will not be guided by electrophysiological testing

### TABLE 10

**Electrophysiology study indications for survivors of cardiac arrest**

*Class I*
- Patients surviving cardiac arrest without evidence of an acute Q-wave MI
- Patients surviving cardiac arrest occurring more than 48 hours after the acute phase of MI in the absence of a recurrent ischemic event

*Class II*
- Patients surviving cardiac arrest caused by bradyarrhythmia
- Patients surviving cardiac arrest thought to be associated with a congenital repolarization abnormality (long-QT syndrome) in whom the results of noninvasive diagnostic testing are equivocal

*Class III*
- Patients surviving a cardiac arrest that occurred during the acute phase (<48 hours) of MI
- Patients with cardiac arrest resulting from clearly definable specific causes, such as reversible ischemia, severe valvular aortic stenosis or noninvasively defined congenital or acquired long-QT syndrome

Abbreviation: MI: Myocardial infarction

arrhythmia induction that, at times, may provoke hemodynamic instability necessitating rapid arrhythmia termination, including external cardioversion/defibrillation. Patients undergo sterile skin preparation using iodine and other alcohol-based and nonalcohol-based scrubs, followed by draping to prevent cross-contamination from nonsterilized areas and to maintain patient dignity while permitting access to anticipated sites of vascular entry. Local anesthesia, as well as mild conscious sedation, is warranted to facilitate painless percutaneous vascular access, particularly in apprehensive patients. After diagnostic catheters are introduced, sedation may be lightened so as not to hinder arrhythmia induction, as some sedative drugs may alter properties of the cardiac conduction system. The awake

### TABLE 11
**Electrophysiology study indications for unexplained palpitations**

*Class I*
- Patients with palpitations who have a pulse rate documented by medical personnel as inappropriately rapid and in whom ECG recordings fail to document the cause of the palpitations
- Patients with palpitations preceding a syncopal episode

*Class II*
- Patients with clinically significant palpitations suspected to be of cardiac origin in whom symptoms are sporadic and cannot be documented. Studies are performed to determine the mechanisms of arrhythmias, direct or provide therapy, or assess prognosis

*Class III*
- Patients with palpitations documented to be due to extracardiac causes (e.g. hyperthyroidism)

Abbreviation: ECG: Electrocardiogram

### TABLE 12
**Electrophysiology study indications for guiding drug therapy**

*Class I*
- Patients with sustained ventricular tachycardia or cardiac arrest, especially those with prior MI
- Patients with AVNRT, AV reentrant tachycardia using an accessory pathway, or atrial fibrillation associated with an accessory pathway, for whom chronic drug therapy is planned

*Class II*
- Patients with sinus node reentrant tachycardia, atrial tachycardia, atrial fibrillation or atrial flutter without ventricular preexcitation syndrome, for whom chronic drug therapy is planned
- Patients with arrhythmias not inducible during control EPS, for whom drug therapy is planned

*Class III*
- Patients with isolated atrial or ventricular premature complexes
- Patients with ventricular fibrillation with a clearly identified reversible cause

Abbreviations: AVNRT: Atrioventricular nodal reentrant tachycardia; EPS: Electrophysiology study; MI: Myocardial infarction

> **TABLE 13**
>
> **Electrophysiology study indications for candidates and recipients of implantable electrical devices**
>
> *Class I*
> - Patients with tachyarrhythmias, before and during implantation, and final (predischarge) programming of an electrical device to confirm its ability to perform as anticipated
> - Patients with an implanted electrical antitachyarrhythmia device in whom changes in status or therapy may have influenced the continued safety and efficacy of the device
> - Patients who have a pacemaker to treat a bradyarrhythmia and receive a cardioverter-defibrillator, to test for device interactions
>
> *Class II*
> - Patients with previously documented indications for pacemaker implantation to test for the most appropriate long-term pacing mode and sites to optimize symptomatic improvement and hemodynamics
>
> *Class III*
> - Patients who are not candidates for device therapy

patient should be continuously reassured to promote relaxation and to prevent sudden movements that may result in catheter dislodgement and vascular injury, not to mention intracardiac trauma. Antiarrhythmic drugs are usually withheld prior to the study, although in select cases, they may be continued if clinical events occurred on specific agents or in an effort to promote tolerability of arrhythmias that may otherwise provoke severe symptoms or hemodynamic collapse.

A complete study evaluates sinus node automaticity and impulse propagation, atrial myocardial conduction properties, anterograde and retrograde conduction patterns, ventricular myocardial conduction properties, and associated arrhythmic tendency, both spontaneous and stimulated.[13] Not all assessments may be possible in every patient. Pharmacologic agents also are administered to modify intrinsic automaticity and conduction properties to expose occult arrhythmic potential.

## Cardiac Access and Catheterization

The recording of local activation signals during EP study is obtained through stationary electrode catheters **(Fig. 1)**, usually varying in size from 4 to 8 French. Standard diagnostic multipolar electrode catheters are introduced percutaneously through peripheral veins, such as the antecubital, femoral, subclavian or internal jugular veins, and then guided fluoroscopically to their intended intracardiac position. For safety and convenience, sites accessible transvenously via right cardiac chambers are chosen, usually the high-lateral right atrium or right atrial appendage, approximating the site of sinus endocardial breakthrough; His bundle region, approximating the site of atrioventricular nodal (AVN) conduction; within the coronary sinus (CS) that

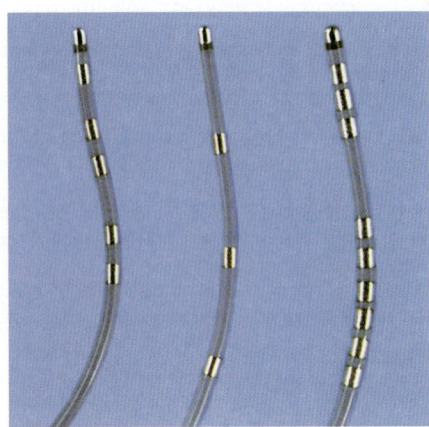

**FIGURE 1:** Examples of diagnostic multielectrode catheters. Three diagnostic catheters with different interelectrode spacing and electrode distribution are shown. Closer-spaced electrodes permit detection of high-frequency signals, such as His or accessory pathway potentials, with high degree of localization though at the expense of signal amplitude, whereas wider-spaced electrodes yield larger-amplitude signals, at the expense of localization accuracy

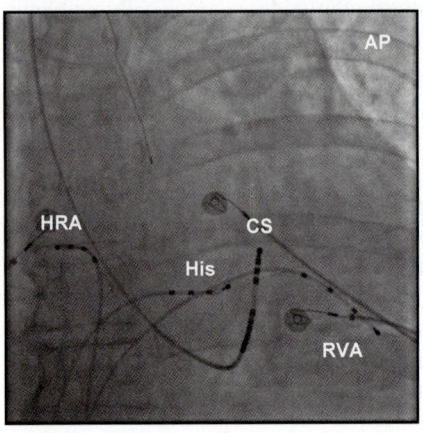

**FIGURE 2:** Anteroposterior (AP) fluoroscopic projection of "standard" intracardiac catheter locations. The high right atrial (HRA) catheter is positioned laterally in the right atrial appendage. The His catheter is positioned across the atrioventricular junction at the mid-to-superior septal aspect of the tricuspid valve. The right ventricular catheter is seated in the apex (RVA). The coronary sinus (CS) catheter is positioned with the proximal electrode approximately 1 cm from the CS ostium. (*Source:* Reproduced from Choudhuri et al. Principles and techniques of cardiac catheter mapping. In: Camm AJ, Saksena S (Eds). Electrophysiologic Disorders of the Heart, 2nd edition. St. Louis: Churchill-Livingstone (Elsevier) Inc; 2010. With permission from Elsevier.)

is posteriorly located, approximating a septal-lateral axis of activation of both the left atrium and ventricle; and the right ventricle (RV) **(Fig. 2)**. This "standard" catheter positioning approximates the normal conduction system axis, creating a skeleton of recording sites that define the sequence and timing of activation from all four cardiac chambers. Other recording sites, such as the right bundle-branch region just across the tricuspid valve and right ventricular septum or outflow tract, and even the pulmonary veins, may be sampled using traditional and specially designed catheters to further augment and enhance the diagnostic framework based on the suspected arrhythmia.

Transseptal catheterization (see below) via the right atrium to access left atrium is invaluable, particularly to approach pulmonary veins in atrial fibrillation (AF) ablation and for ventricular tachycardia (VT) ablation in patients with mechanical aortic valve in whom the left ventricle (LV) would be otherwise inaccessible through the aortic retrograde approach. Continuous heparinization is desirable for left-heart catheterization to avoid thromboembolic complications. Catheterization into the pericardial space using a subxiphoid or subcostal approach permits access for mapping and ablation of arrhythmias of epicardial origin.

## Signals and Filtering

Once intracardiac diagnostic catheters are appropriately positioned and connectivity established, the recorded signals, intracardiac electrograms (EGMs), are displayed simultaneously on a multichannel digital recording system along with several unfiltered surface ECG leads. Signal filtering between 30–40 Hz and 500 Hz is best suited for sharp intracardiac signals such as those from the His bundle and accessory pathways **(Figs 3A to F)**. A high-pass filter setting of more than 50–100 Hz reduces undesirable low-frequency signals. In addition, 60-cycle interference and its harmonics can be eliminated with a notch filter tuned to 60 Hz,[11] although potentially at the expense of other low-amplitude physiologic signals in that frequency range.

## FUNDAMENTALS OF THE CARDIAC ELECTROPHYSIOLOGY STUDY

Identifying pathology during the cardiac EP study requires a keen awareness of normal electrophysiologic characteristics, an understanding of the principles guiding intracardiac EGM interpretation and knowledge of expected responses to programmed stimulation.[14] Further, these assessments must be universally understood and communicable to other practitioners at various levels of training and expertise, including electrophysiologists, cardiologists, physician extenders, lab technicians, nurses and even health care providers not directly practicing in the field of cardiovascular diseases.

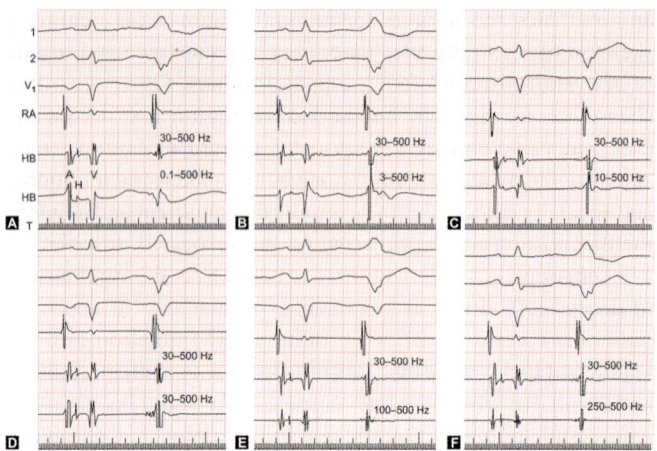

**FIGURES 3A TO F:** Effects of various filtering frequencies on the morphologic appearance of intracardiac electrograms. The tracings from top to bottom are electrocardiographic leads 1, 2, V1, right atrial (RA), two His bundle (HB) electrograms and timeline (T). In each Panel, the first beat is of sinus origin and is followed by a spontaneous ventricular premature beat. The top HB, RA and right ventricle are filtered at 30–500 Hz (i.e. the usual filtering frequencies). The bottom HB tracing shows the effect of various filtering frequencies on the appearance. The low-frequency signals are mostly eliminated at high-band-pass filter frequency settings above 10 Hz (Panel C). The low-band-pass filter settings above 500 Hz generally do not have a significant effect on the intracardiac electrogram appearance. It should be pointed out that the high-band-pass setting reduces the overall magnitude of the electrogram, necessitating an increase in amplification. It should also be noted that, at all frequencies depicted, the HB deflection can be clearly identified. (*Source:* Akhtar M. Invasive cardiac electrophysiologic studies: An introduction. In: Parmley WW, Chatterjee K (Eds). Cardiology: Physiology Pharmacology Diagnosis. Philadelphia: Lippincott; 1991. With permission from Lippincott Williams and Wilkins)

## Conventions

Several conventions are used to describe electrical events in a standardized manner, which are briefly described in this chapter. The most fundamental of these is that of timing and intervals. Measurements of most EP intervals are made in milliseconds, similar to usual intervals on the standard ECG. However, rates of atrial and ventricular events are determined in beats per minute and are converted to milliseconds by dividing the rate into 60,000 (the number of milliseconds per minute), yielding the rate of that particular event (e.g. heart rate) in milliseconds (per beat or occurrence). Hence sinus rhythm, usually 60–100 beats per minute, corresponds to cycle length 1,000–600 ms, respectively.

Diagnostic EP catheters have multiple electrodes, which create various recording unipoles and bipoles. By convention, the distal or tip electrode is designated electrode "1" and more proximal electrodes are numbered sequentially with increasing distance from the tip electrode. Further, whereas intracardiac EGMs are typically bipolar signals—and hence require two electrodes (usually adjacent) to generate the signal—a quadripolar catheter can generate up to three bipolar EGMs ("1–2", "2–3", "3–4"), and a decapolar catheter can display up to nine bipoles from adjacent electrodes. Other pairs can be configured as well. Often, all bipoles are not displayed; it depends on the clinical utility, particular catheter and/or its location and mapping resolution required.

## Normal Cardiac Electrophysiology

### Normal Propagation Patterns

An awareness of two principles aids in proper interpretation of intracardiac EGMs. The first is that propagation within myocardium is generally radial, although with some directionality, due to anisotropic conduction and presence of structural barriers. This pattern of propagation during a stable rhythm results in activation at various myocardial areas simultaneously. Consider an atrial tachycardia arising from the low crista terminalis **(Fig. 4A)**. Repetitive depolarization at this site is itself not detected, as this location is not standard for diagnostic catheters. Only after radial propagation through the right atrium, away from the tachycardia origin, and arrival at sites where catheters are located, e.g. the high right atrium (HRA) and His, are EGMs first recorded. In this case conduction times to these locations are fairly similar, resulting in near-simultaneous activation at both sites. The EGMs should neither be interpreted as representing a tachycardia arising simultaneously from those sites nor rapidly propagating from one site to the other so as to "appear" near-simultaneous, but rather more accurately explained by radial spread of a wave of activation arising from a location that has relatively similar conduction times to those catheter locations. This principle is inherently not specific, and other endocardial sites may also be equidistant from the HRA and His catheters **(Fig. 4B)**. Hence, radial spread permits tachycardias arising from various intracardiac sites to produce similar EGM patterns **(Fig. 4C)**. Definitive identification of involved sites requires more detailed cardiac mapping.

With respect to the AV conduction system, such tissues, in fact, should be considered to exhibit at least bidirectional propagation unless absence of this capability is demonstrated, even if unidirectional propagation predominates. For instance, in sinus rhythm or when pacing the atrium, it is expected that the impulse will conduct to the ventricles in an otherwise

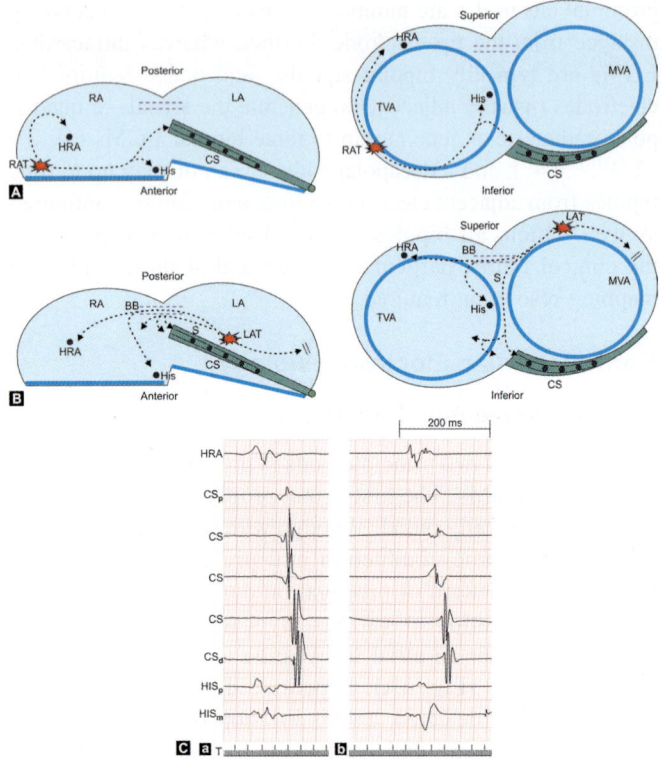

**FIGURES 4A TO C:** Radial propagation from two different focal tachycardia origins resulting in similar electrogram (EGM) activation sequences. (A) Cranial (upper) and left anterior oblique caudal (lower) schematic depictions of the atria with focal tachycardia site of origin in the low crista terminalis and (B) anterior left atrium (LA) above the mitral valve annulus (MVA). (A) Low crista terminalis (CT) tachycardia focus: wavefronts propagate superiorly along the CT into the right atrial (RA) appendage as well as simultaneously along the floor of the RA and then anteriorly and superiorly to arrive at the His region. Local conduction properties and similarity in distance between the tachycardia focus and the high right atrium (HRA) and His results in similar activation times to these two sites. The wavefront traveling along the RA floor also penetrates the septum posteriorly and activates the coronary sinus (CS) from proximal to distal. (B) Anterior LA tachycardia focus: wavefronts are shown to propagate along Bachmann's bundle (BB) and the interatrial septum (S). The rapidly conducted BB wavefront then propagates radially within the RA to arrive at the HRA and His electrodes with relatively similar activation times. Conduction block to the lateral LA (from scar or ablation) prevents a wavefront from activating the CS electrodes from distal to proximal. The S-wavefront travels inferiorly along the interatrial septum and then activates the CS from proximal to distal, possibly through direct penetration or after crossing the septum and entering the RA. The more rapid conduction across BB may explain earlier activation at the HRA and His as compared to the CS. (C) EGM patterns from a (left) and b (right). Surface ECG leads and intracardiac EGM channels are shown with timeline (T). In both tachycardias, activation is earliest at HRA and His, followed by proximal to distal activation in the coronary sinus. Red—arrhythmia focus; lime green—valve annuli; teal—coronary sinus; violet dashed lines—Bachmann's bundle; dotted lines with arrowhead—activation wavefronts. HRA, His and CS electrodes are shown. Gray signifies "behind" other structures. TVA: Tricuspid valve annulus

healthy heart. However, it should not be discounted that retrograde, i.e. ventriculoatrial (VA), conduction may also be responsible for cardiac rhythm events. Consider the situation of aberrant conduction induced by cycle length variation. Often one finds aberrancy, i.e. "bundle branch block," persists for several beats. This phenomenon develops due to anterograde conduction of a supraventricular impulse solely along the unblocked bundle and then spread of activation across the interventricular septum to invade and travel retrogradely along the previously blocked bundle, rendering it refractory to anterograde conduction by the next arriving supraventricular impulse and thereby maintaining aberrancy **(Figs 5A to C)**.

In addition to these technical and physiologic aspects, it should be recognized that microscopic, molecular and cellular properties underlying myocardial membrane ion channel function, excitation-contraction, cellular automaticity, conduction velocity and tissue refractoriness to name a few, directly impact EGMs observed during EP study. These aspects

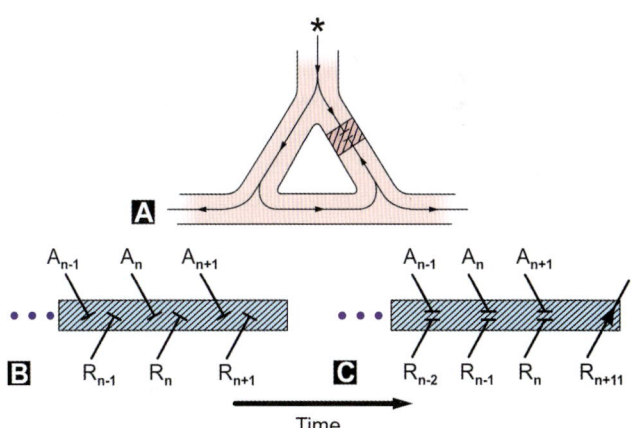

**FIGURES 5A TO C:** Schematic representation of a generalized linking phenomenon. (A) Depiction of a hypothetical macroreentry circuit into which successive impulses (asterisk) enter and preferentially traverse one limb as a result of persistent functional block (shaded region) in the contralateral limb. (B) and (C) Two distinct mechanisms whereby the functional block can be dynamically maintained. Each of the two Panels is a "blow-up" of the region of block as it is invaded by successive (i.e. $n-2, n-1, n, ...$) anterograde (A) and retrograde (R) impulses over time. (B) Shows impulse interference, whereas (C) depicts impulse collision. (*Source:* Lehmann MH, Denker S, Mahmud R, et al. Linking: a dynamic electrophysiologic phenomenon in macroreentry circuits. Circulation. 1985;71:254-65)

are critical to a fundamental foundation on which to develop an understanding of cardiac EP in all its manifestations. As an initial consideration of the final consequence of these processes, we will introduce here normal conduction patterns, and a perspective on the most normal of electrophysiologic manifestations—sinus rhythm.

## Sinus Rhythm and Normal Atrioventricular Conduction Parameters

An almost tacit assumption in the interpretation of intracardiac EGMs and, more importantly, in the understanding of cardiology as well as EP is the recognition of sinus rhythm. It remains the most important and common cardiac rhythm, yet its study and comprehension are typically limited to the identification of regular cardiac activity with ECG demonstration of a P-wave preceding a QRS complex. Understanding sinus rhythm through intracardiac EGMs conveys a wealth of fundamental EP, and cardiovascular physiology that is critical to understanding normal phenomena and pathology. It is also important to recognize that the depiction of sinus rhythm through intracardiac EGMs, and all rhythms for that matter, is in large part dependent upon the established construct that anticipates catheters positioned in standard locations, such as HRA, CS, His bundle region and RV, as described above.

Sinus rhythm originates from the sinoatrial node, located epicardially at the junction of the right atrium with the anterolateral aspect of the superior vena cava. Its important property of automaticity generates an electrical depolarization in a regular manner that is the sinus rhythm. Depolarizations occur usually every 1,000–600 ms, termed the sinus cycle length, corresponding to a heart rate of 60–100 bpm. Once the sinus node depolarizes, resulting phenomena, including propagation to and within the atrium, atrial contraction, conduction along the normal AV pathway and activation of the ventricles, are all secondary and need not necessarily occur while in sinus rhythm. However, these secondary phenomena signify that sinus rhythm is associated with the other normal electromechanical cardiac events necessary for maintenance of circulation.

After exiting the sinoatrial node, the sinus impulse propagates both epicardially and endocardially. The endocardial breakthrough is in the posterior lateral right atrium, typically somewhat below but still in close proximity to the actual sinus node. Hence, a catheter in this region will be the first to detect an EGM during sinus rhythm. By the time endocardial breakthrough has occurred, the impulse has also stimulated enough atrial myocardium to initiate inscription of the ECG

P-wave. The impulse spreads rapidly due to presence of sodium channels, with some degree of preferential conduction anteriorly toward the septum with breakthrough into the left atrium anteromedially, and inferiorly along the lateral and posterior walls and a second left atrial breakthrough in close proximity to the CS ostium.[15] In general, the resulting wavefront propagates radially away from the sinus node and toward the AV node.

The proximal electrodes of a correctly positioned His catheter lie intra-atrially, mid to anteriorly, and are the poles that typically identify the next atrial deflection of sinus rhythm as the impulse propagates toward the AV node. The time interval between the onset of the P-wave and the arrival of the atrial impulse at the His bundle catheter is a measure of intra-atrial conduction time (IACT) and typically is less than 30 ms in adults with healthy atrial myocardium **(Fig. 6)**. The impulse also has traveled posteroseptally into and along the CS musculature that is activated like the surrounding myocardium, through sodium channels, producing high-frequency EGMs. The septal-to-left-lateral activation along the CS results in a proximal-to-distal coronary sinus EGM activation pattern. While biatrial activation is occurring, the impulse also encounters the AVN region, where absence of sodium channels rendering conduction dependent primarily upon slow calcium channels, as well as other anatomic, histologic, and electrical phenomena results in slow conduction within the AV node. The resultant low-amplitude and slowly propagating electrical wavefronts do not generate a discernable wave or deflection on surface ECG or intracardiac electrodes using standard catheters and filtering. The delay permits completion of passive ventricular filling, allowing the eventual atrial contraction to prime the ventricles, thereby augmenting stroke volume. After the impulse leaves the AV node, it encounters the His bundle where cell membranes do once again incorporate sodium channels, and rapid conduction resumes, generating a high-frequency deflection. The location of the His bundle, anatomically within several millimeters to 1 cm anterior and superior to the AV node, provides a surrogate marker to measure AVN conduction time, which is measured between the atrial EGM on the bipole identifying the largest His EGM and the first rapid deflection of the His EGM (A-H interval), and is typically less than 125 ms **(Fig. 6)**. Propagation along the His-Purkinje system (HPS) results in myocardial breakout at various points along the LV septum and soon after at the mid-to-distal RV septum. Standard catheter positioning does not typically employ LV catheters but does incorporate an

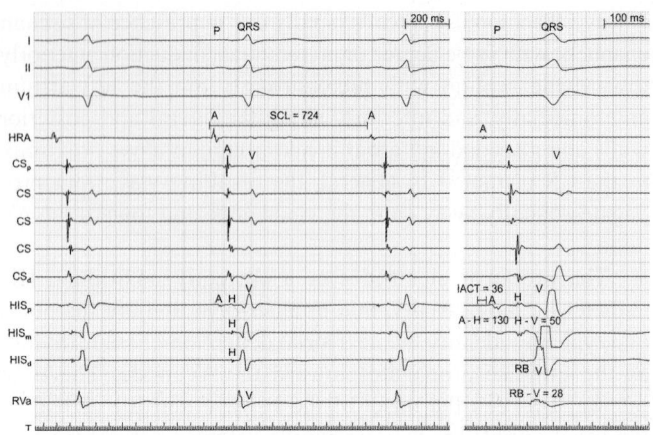

**FIGURE 6:** Baseline conduction parameters. Surface electrocardiographic (ECG) leads and intracardiac channels are shown during sinus rhythm at 100 mm/s (left) and 200 mm/s (right) sweep. Atrial electrograms (AEGMs) span the duration of the P-wave while ventricular electrograms (VEGMs) are aligned with and span the QRS duration. Sinus cycle length is measured from onset of AEGM on the HRA channel to the onset of the next HRA EGM (calipers, Left panel). The His bundle EGM (H) is the largest high-frequency signal between the AEGM and ventricular electrogram (VEGM) on the His recording channels. Intra-atrial conduction time (IACT) estimates the conduction time from the sinus to the atrioventricular (AV) node and is measured from onset of the P-wave to onset of the AEGM on the His catheter (calipers, right Panel). A-H interval, analogous to AV nodal conduction, is measured from onset of AEGM on His catheter to first high-frequency component of the His bundle deflection. The H-V interval assesses His-Purkinje conduction and is measured from the first high-frequency deflection of the His EGM to the onset of the ventricular depolarization whether VEGM or surface QRS. Right panel: A low-amplitude high-frequency EGM is seen on the distal His channel just preceding the VEGM. There is no discernable AEGM on this channel, signifying the electrode is far enough distal across the tricuspid valve so as not to be able to detect atrial activity. This EGM is generated by the right bundle branch. The RB–V interval is typically less than 30 ms. (Abbreviations: CS: Coronary sinus; HRA: High right atrium; RB: Right bundle; SCL: Sinus cycle length)

RV catheter, usually at the apex, and it is this catheter that displays the first ventricular EGM. The time for normal HPS activation to reach the ventricles, measured from the His bundle recording to the earliest ventricular activation (H-V interval), whether on surface ECG or intracardiac EGMs, is 35–55 ms **(Fig. 6)**. The IACT, A-H and H-V intervals together comprise the ECG P-R interval.

Simultaneously or very soon after RV activation, RV septal activation is detected as ventricular EGMs on the His bundle electrodes across the tricuspid valve and along the basal ventricular septum, resulting from both transseptal impulse

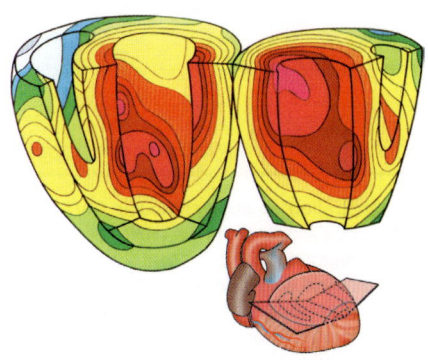

**FIGURE 7:** Three-dimensional isochronic representation of activation of the heart. Inset shows section levels. (*Source:* Durrer et al. Total excitation of the isolated human heart. Circulation. 1970;41: 899-912. Wolters Kluwer Health, with permission)

spread and radial propagation of the impulse from the RV septal breakout site **(Figs 6 and 7)**.[16] The CS catheter also detects ventricular signals; however, these are not the signals of earliest LV activation. Rather, the latest area to be activated in the LV is the base, and it is this basal posterior LV activation that is detected on the CS catheter. Unlike the ventricular EGMs seen on the RV and His catheter, the CS ventricular EGMs are of lower amplitude and lower frequency because the CS electrodes are not in contact with LV myocardium and may even reside more than 10 mm away, due to the location of the CS catheter within the coronary sinus and anatomic variations in the relationship between the coronary sinus and the mitral valve annulus.[17]

## PROGRAMMED ELECTRICAL STIMULATION AND ASSOCIATED ELECTROPHYSIOLOGY

Two distinct patterns of pacing are applied during an EP study: (i) continuous pacing and (ii) interrupted pacing. These techniques permit comprehensive assessment of all relevant aspects of myocardial conduction and clinical arrhythmia tendency. If arrhythmia is induced, variations of these techniques may be employed during tachycardia to diagnose underlying mechanism(s).

### Continuous Pacing

Continuous pacing, in which each stimulus is referred to as "S1," is performed at fixed cycle lengths or with gradual decrementation in cycle length (i.e. gradual increase in heart rate). Continuous fixed-cycle-length pacing is used for study of sinus node function and integrity of subsidiary pacemakers,

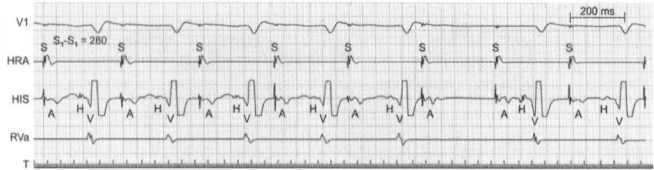

**FIGURE 8:** Demonstration of atrioventricular (AV) conduction block. Surface lead ($V_1$), intracardiac channels (HRA, His, RVa), and timeline (T) are shown. Continuous atrial pacing at 280 ms ($S_1$–$S_1$) results in conduction to the ventricle with progressive prolongation in A–H interval until conduction block. Notice that each pacing stimulus is associated with an atrial electrogram (A) and the A–H interval prolongs with each successive paced stimulus until no H is seen after the sixth paced A. Conducted beats are associated with a fixed H–V interval, confirming conduction delay above the His, i.e. in the AV node. With continued pacing, a new Wenckebach cycle ensues

arrhythmia induction and overdrive pacing of a tachycardia ("entrainment," not discussed here). Continuous pacing with gradual decrementation in cycle length is performed to evaluate myocardial stimulation limits, and usually delivered until the occurrence of a desired event such as induction of tachycardia, conduction block **(Fig. 8)** or failure to achieve 1:1 myocardial capture. As such, this technique provides an overall measure of ability of a particular tissue, chamber or region to respond to sequential stimuli; is a measure of absolute limits of myocardial responsiveness; and may provide insight into tachycardia mechanism.

With fixed-cycle-length pacing at cycle lengths that are not excessively short, but shorter than sinus so as to avoid competition and interference, healthy atrial and ventricular tissue will respond in a 1:1 manner with minimal conduction delay (latency) between the stimulus artifact and the onset of the associated EGM and P-wave or QRS down to cycle lengths shorter than 250 ms.[14] As cycle length is decremented, a point is reached beyond which latency increases. This observation suggests the pacing cycle length is encroaching on the ability of local tissue capture and propagation to the surrounding myocardium in a 1:1 manner, and further pacing acceleration can result in failure of capture or breakup and fractionation of the propagating wavefront, provoking fibrillation, whether atrial or ventricular. The observation of increased latency should signal the operator to discontinue the pacing drive as induction of fibrillation may, at the least, prevent complete electrophysiologic evaluation in the case of AF, unless fibrillation is the desired result, and may require emergent electrical conversion if VF is induced.

## Intermittent or Interrupted Pacing with Extrastimuli

The second pacing format is intermittent pacing with delivery of premature (or extra) stimuli. This pacing format is advantageous for studying myocardial and conduction system refractory periods and for creation of local conduction block to facilitate induction of reentrant arrhythmias. With this approach, a series of six to ten paced beats are delivered at a constant cycle length (drive train of S1s) and are followed by at least one extrastimulus (S2) coupled to the last beat of the basic drive at a cycle length shorter than the S1 drive cycle. The S2 is initiated late during electrical diastole and the coupling interval is progressively decreased with successive drives, thereby "scanning diastole" until myocardial capture and/or conduction cannot be achieved, i.e. the effective refractory period of myocardial **(Fig. 9)** or AV conduction **(Figs 10A to C)**[11,18] or conduction delay promotes arrhythmia induction **(Fig. 11)**. When latency or frank tissue refractoriness is encountered, the refractory period of downstream tissues may not be determinable. In such cases, the S2 coupling interval may be slightly increased so as to permit myocardial capture or avoid latency, then additional extrastimuli may be introduced (S3, S4, S5, etc.). Since antegrade AVN conduction is

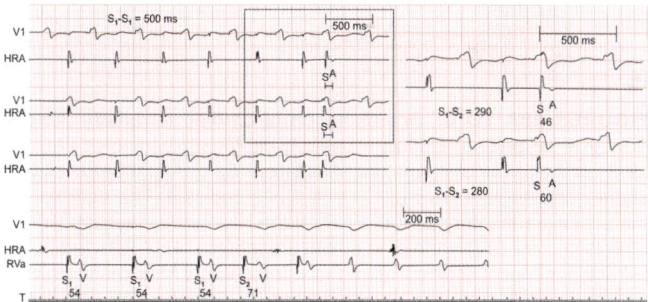

**FIGURE 9:** Latency and atrial effective refractory period. Surface lead (V1) and intracardiac channel (HRA) are shown with three successive atrial pacing drives at 500-ms cycle length ($S_1$–$S_1$) and progressively abbreviated extrastimuli ($S_2$) (upper left Panel). In the first drive, atrial extrastimulation coupled at 290 ms results in a stim-A (latency) time of 46 ms (inset; expanded view to the right). In the second drive, the extrastimulation at 280 ms is associated with increased latency (60 ms) (inset; expanded view to the right). In the third drive, further decrementation in the extrastimulus coupling interval to 270 ms results in loss of atrial capture with no atrial electrogram associated with the extrastimulus, thereby establishing the atrial effective refractory period as 600:270 ms. Surface lead (V1) and intracardiac channels (HRA, RVa) are shown (lower Panel). Ventricular pacing at 350 ms ($S_1$–$S_1$) is associated with a $S_1$–V time (latency) of 54 ms. The first ventricular extrastimulus is associated with increased latency, $S_2$–V time of 71 ms

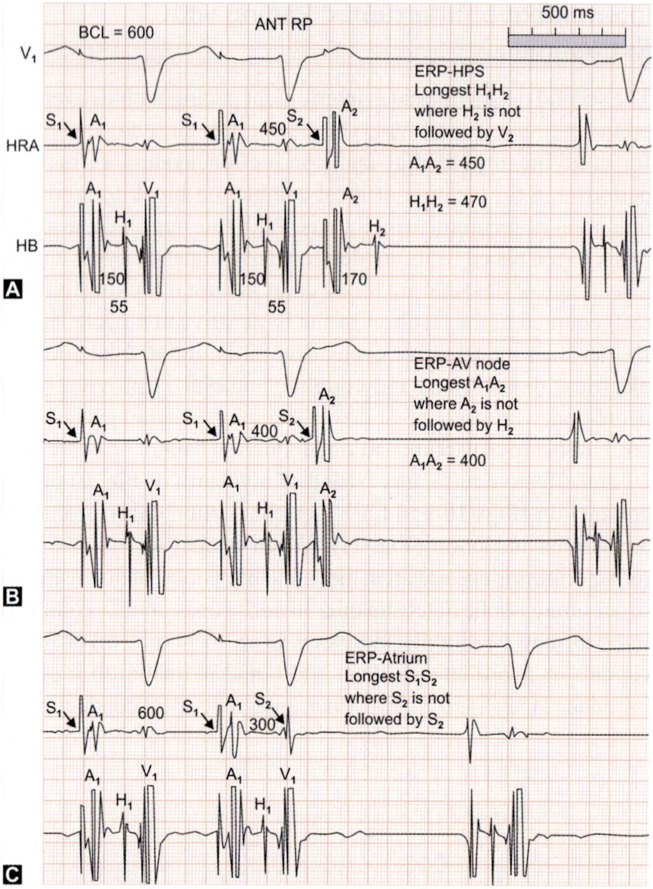

**FIGURES 10A TO C:** Determination of cardiac refractory periods during atrial pacing. During a basic cycle-length pacing at 600 ms ($S_1S_1$ or $A_1A_1$), atrial premature stimulation ($S_2$ or $A_2$) at progressively shorter coupling intervals ($S_1S_2$ or $A_1A_2$) is depicted. The definition of the effective refractory period (ERP) of the His-Purkinje system (HPS), atrioventricular (AV) node, and atrium are labeled. ANT RP: Antegrade refractory period. (*Source:* Reproduced from Akhtar M. Invasive cardiac electrophysiologic studies: an introduction. In: Parmley WW, Chatterjee K (Eds). Cardiology: Physiology, Pharmacology, Diagnosis. Philadelphia: Lippincott; 1991. With permission from Lippincott Williams and Wilkins)

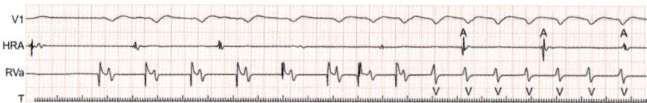

**FIGURE 11:** Ventricular tachycardia (VT) induction with extrastimuli. Surface lead (V1) and intracardiac channels (HRA, RVa) are shown. A six-beat ventricular drive at 350 ms is followed by two premature extrastimuli, inducing a wide complex tachycardia (V) with left bundle branch block morphology at 230 ms and atrioventricular dissociation, i.e. VT. A, atrial electrogram

generally poorer than HPS conduction, evaluating the refractory period of the HPS is limited by AVN conduction and may not always be determinable. Various agents can be administered to shorten refractoriness and improve conduction, thereby permitting evaluation of HPS conduction if AVN conduction improves adequately. Refractory periods of myocardial tissue are dependent upon and vary directly with drive cycle length. Therefore multiple drive cycles are used to assess dynamicity of refractoriness. Whereas a 600 ms cycle length drive may yield a tissue-effective refractory period of 300 ms, a 400 ms cycle length drive would be expected to shorten tissue refractoriness.[19]

## Significance of "Short-long-short" Pacing Cycles

At times, the aforementioned pacing maneuvers are unsuccessful in inducing a suspected arrhythmia. For documented or suspected arrhythmias that rely on conduction delay for arrhythmia initiation (i.e. reentrant arrhythmias), "short-long-short" pacing—a variation on the extrastimulus technique—may prove useful in promoting conduction delay that initiates tachyarrhythmias when other maneuvers do not. This is applicable to all forms of reentrant tachyarrhythmias, whether in the atrium, ventricle or involving the HPS. A drive of 6–8 beats is delivered at a specific cycle length, followed by the last beat of the drive coupled at a cycle length greater than that of the preceding beats in the drive. This long-coupled beat has the effect of shortening myocardial refractoriness **(Figs 12A to F)**.[19,20] This allows an extrastimulus to be coupled at shorter cycle lengths ("short-long-short") than with fixed-cycle-length pacing, which may then create adequate myocardial propagation slowing to support reentry. This pacing technique has a divergent effect in the HPS, that is, HPS refractoriness increases with "short-long-short" sequences, which may directly enhance conduction delay and promote reentry within the HPS **(Figs 13A to F)**.

For induction of reentrant supraventricular tachycardias (SVTs), single, double or more extrastimuli may be delivered **(Fig. 14)**. For induction of VT, up to three ventricular extrastimuli are typically employed. The sensitivity of pacing protocols seems to be directly related to the number of extrastimuli utilized.[16] However, this occurs at the expense of specificity as polymorphic VT/VF can be induced at very short coupling intervals by using multiple ventricular extrastimuli in patients otherwise without arrhythmic risk. Regardless of pacing protocol, induction of sustained monomorphic VT constitutes a specific response and is seldom induced in patients who are not

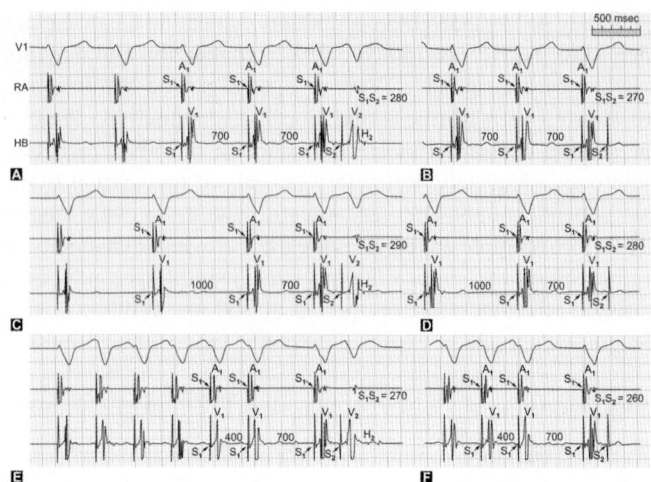

**FIGURES 12A TO F:** Effect of abrupt cycle length change on refractoriness of the ventricular muscle. The effective refractory period (ERP) of the ventricular muscle during constant cycle length (method I) is 270 ms (A and B). A change of $CL_P \rightarrow CL_R$ from 1,000 → 700 ms (method IIA, C and D) lengthens ERP of the ventricular muscle to 280 ms, whereas a change of $CL_P \rightarrow CL_R$ from 400 → 700 ms (method IIB, E and F) shortens the ERP of the ventricular muscle to 260 ms. Note that for the same $CL_R$, ERP of the ventricular muscle varies directly with $CL_P$. (*Source:* Modified from Denker S, Lehmann MH, Mahmud R, et al. Divergence between refractoriness of His-Purkinje system and ventricular muscle with abrupt changes in cycle length. Circulation. 1983;68;1212-21)

clinically prone to such arrhythmias. In contrast, the induction of polymorphic VT/VF with three extrastimuli at short coupling intervals can be nonspecific.[11]

## Relation of Pacing Technique to Anticipated Arrhythmia Mechanism and Inducibility

Reentrant arrhythmias lend themselves best to study because the reentrant nature of arrhythmias creates a reproducible and regular activation sequence that can be evaluated when the arrhythmia is sustained. Tissue refractoriness and conduction slowing are necessary factors in the initiation and maintenance of reentrant arrhythmias, which can be achieved readily through continuous pacing with progressive cycle length decrementation and premature extrastimulation, with or without pharmacologic facilitation.

Unifocal triggered rhythms also can be evaluated by EP study. However, their tendency toward arrhythmia induction can be more challenging, being somewhat more sensitive to hormonal changes and limited by sedation and attendant varying catecholamine levels. Rapid, long pacing drives, through

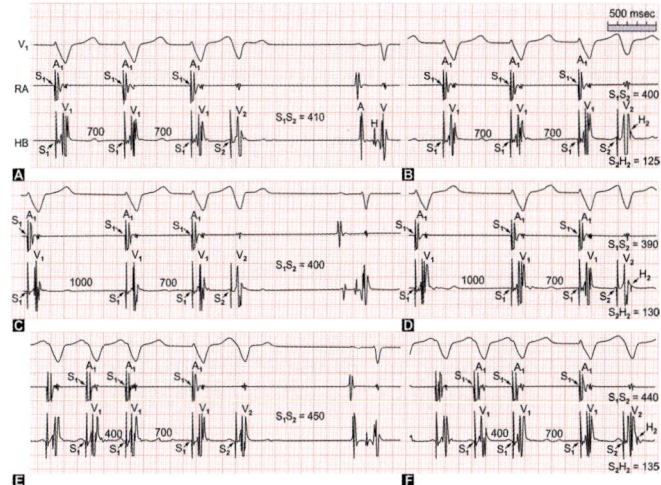

**FIGURES 13A TO F:** Effect of abrupt cycle length change on refractoriness of the His-Purkinje system (HPS). The tracings show the retrograde relative refractory period (RRP) of the HPS during methods I, IIA and IIB. With method I (A and B), $H_2$ emergence from the local ventricular electrogram is noted at an $S_1S_2$, interval of 400 ms. A change of $CL_p \to CL_R$ from 1,000 → 700 ms (method IIA, C and D) shortens the RRP of the HPS to 390 ms, whereas a change of $CL_p \to CL_R$ from 400 → 700 ms (method IIB, E and F) lengthens the RRP of the HPS to 440 ms. Note that for the same $CL_R$, the RRP of the HPS is 50 ms longer during method IIB compared with method IIA, but remarkably $CL_p$ is 600 ms shorter during method IIB compared with method IIA. (Abbreviations: $V_1$: Surface electrocardiographic lead; RA: Right atrial electrogram; HB: His bundle electrogram. All measurements are in milliseconds). (*Source:* Modified from Denker S, Lehmann MH, Mahmud R, et al. Divergence between refractoriness of His-Purkinje system and ventricular muscle with abrupt changes in cycle length. Circulation. 1983;68;1212-21)

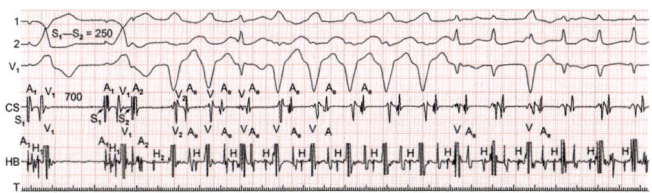

**FIGURE 14:** Induction of supraventricular tachycardia (SVT) in Wolff-Parkinson-White syndrome. The tracings are labeled. Atrial pacing from coronary sinus (CS) is done at a 700-ms basic cycle. During the basic drive pacing, left free-wall accessory pathway conduction to the ventricle produces ventricular preexcitation. A single premature beat ($S_2$) blocks in the accessory pathway and conducts over the normal pathway with a left bundle branch block morphology, and the SVT is initiated. Note the intermittent normalization of the QRS complex during this SVT. (*Source:* Modified from Jazayeri MR, Caceres J, Tchou P, et al. Electrophysiologic characteristics of sudden QRS axis deviation during orthodromic tachycardia. J Clin Invest. 1989;83:952-9. With permission from American Society for Clinical Investigation)

promotion of myocyte calcium overload and associated delayed afterdepolarizations, may permit induction of arrhythmias known to occur related to this mechanism, such as focal atrial tachycardias and certain idiopathic VT. Further, catecholamine administration (e.g. isoproterenol) enhancing cAMP-mediated adrenergic stimulation and associated diastolic calcium overload may facilitate induction with or without pacing. Interestingly, often it is not a specific level of catecholamine but rather the flux in serum concentration that permits induction (e.g. wash-in or wash-out phases).

## Clinical Application of "Routine" Electrophysiology Study and Anticipated Responses to Programmed Stimulation

While few would consider the findings of even a normal comprehensive EP study "routine," it is important to utilize a systematic approach that permits complete evaluation of the myocardium and conduction system, including sinus node automaticity and impulse propagation, atrial myocardial conduction properties, anterograde and retrograde AV conduction patterns, ventricular myocardial conduction properties and associated arrhythmic tendency, including attempts at arrhythmia induction in the baseline state and during pharmacologic facilitation. The comprehensive EP study is performed in the following stages:
- Atrial continuous pacing with and without cycle length decrementation
- Atrial premature stimulation with extrastimuli
- Ventricular continuous pacing with and without cycle length decrementation
- Ventricular premature stimulation with extrastimuli
- Short bursts of rapid atrial or ventricular pacing.[14]

While not all aspects may be assessed in each patient and the order may vary according to patients' needs and tolerances, this framework provides a method to perform a comprehensive electrophysiologic evaluation of intrinsic conduction properties and arrhythmogenic tendency. Burst pacing is rarely employed for the study of normal cardiac EP and is generally used in arrhythmia induction or termination and will not be elaborated upon here.[14]

## Baseline Observations

Irrespective of the presenting rhythm or potential rhythm of interest, the surface ECG provides valuable information against which the intracardiac patterns of activation can be compared.

It is negligent to disregard the baseline surface ECG, and basic observations of P-R, R-P, QRS and QT intervals should be made at the initiation of and throughout every study. Once baseline observations have been made, including intrinsic cycle length, activation sequence and parameters of AV conduction **(Fig. 6)**, the active study can be performed with particular attention to appropriate stimulation protocols, myocardial capture and conduction, and activation of other chambers in response to each stimulus and nonstimulated responses. The large majority of patients, undergoing diagnostic EP study, are present in sinus rhythm; hence, the study usually starts with atrial stimulation. Alternately, as induction of AF presents a barrier to study completion, it may be pragmatic to perform atrial stimulation last in patients with a history of AF or sick sinus.

## Atrial Stimulation for Evaluation of Sinus Node Function; and Atrial and Atrioventricular Nodal Conduction Properties

The assessment of sinus node automaticity should be included in all comprehensive EP studies, but particularly in patients presenting with dizziness, dyspnea on exertion, presyncope, syncope or other manifestations of sinus node dysfunction in whom diagnosis cannot be made noninvasively.[14] Right atrial pacing for 30 sec to 1 min at a fixed cycle length shorter than sinus causes sinus suppression. Abrupt discontinuation of pacing permits determination of the time for the first automatic intrinsic/escape sinus beat to return, this interval is termed the sinus node recovery time (SNRT). The SNRT is evaluated at various drive cycles ranging between the sinus cycle length and usually 400 ms. Sinus node recovery times less than 1,525 ms are generally considered normal. By deducting the predominant sinus cycle length from this interval, one can obtain the so-called corrected SNRT **(Fig. 15)**. In one series[21] the value for corrected SNRT was less than 525 ms in normal individuals but exceeded this in patients with overt sinus node dysfunction.

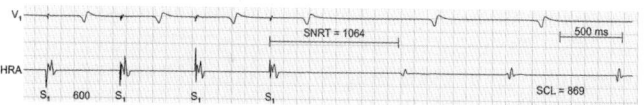

**FIGURE 15:** Measurement of sinus node recovery time (SNRT). A train of $S_1$'s at 600 ms is delivered over 30 seconds to achieve sinus suppression. With discontinuation of pacing, the time for first sinus return cycle is 1,064 ms, which is within normal limits. Correcting for the predominant sinus cycle length (SCL), 869 ms, yields a corrected SNRT of 195 ms, also within normal limits

In the vast majority of patients with true sinus node disease, sinoatrial conduction abnormalities are the predominant reason for sinus node dysfunction. Sinoatrial conduction time (SACT) in the absence of obvious sinus node disease is less than 100 ms. The SACT is evaluated in similar fashion to the SNRT by atrial pacing, in this case, just faster than sinus, hence avoiding significant sinus node suppression. The return cycle then represents the time for the last paced beat to enter the sinus node, reset it and propagate back to the pacing catheter. Again, deducting the predominant sinus cycle from the return cycle should approximate the propagation time to and from the sinus node, and the SACT is thus one-half of this value. SACTs in excess of 125 ms[14] are felt to represent important sinoatrial conduction disease. This interval is most accurate when the HRA catheter is positioned in close proximity to the sinus node, in the posterolateral aspect of the right atrium, and not in the right atrial appendage, which may be associated with prolonged conduction intervals between it and the sinus node.

The sensitivity of SNRT for the detection of sinus node dysfunction is 54%, whereas that of SACT is 51%, with a combined sensitivity of approximately 64%. Poor sensitivity of such testing relates in part to the possibility that in previous studies, documented episodes of sinus bradycardia or sinus arrest due to neurocardiogenic mechanisms may not have been excluded.[22] The specificity of both tests combined is approximately 88%. In patients with bradycardia/tachycardia syndrome, EP testing may also be necessary for the proper diagnosis and therapy of the concomitant tachyarrhythmia or bradyarrhythmia, as AV conduction is frequently abnormal in patients with sinus node dysfunction.[14]

After evaluation of sinus node automaticity, AV conduction capabilities are assessed in patients presenting in sinus rhythm. With successively faster pacing or shorter cycle length, AVN conduction time as measured by the A-H interval initially accommodates to the pacing drive, then prolongs until a pacing cycle length at which the Wenckebach pattern of conduction block is observed **(Fig. 8)**. In most, this AV block cycle length is reached between 600 ms and 300 ms, although in some patients with "enhanced AVN conduction" 1:1 conduction may be maintained at cycle lengths less than 300 ms, often with electrocardiographically abbreviated P-R intervals. The clinical significance of this is unclear but has been seen in patients with rapid ventricular response in AF and atrial flutter. At the other extreme, observing a pattern of Wenckebach at cycle lengths greater than 600 ms is unusual but may be seen in healthy young adults with elevated vagal tone and should not be considered abnormal unless it is not reversible by vagolytic or sympathomimetic agents or occurs during exercise.[14]

Next, atrial premature extrastimuli are introduced for evaluation of atrial myocardial refractoriness and effective refractory period of the AV node. As mentioned previously, a fixed drive cycle coupled with a single premature extrastimulus is repeatedly delivered to scan diastole until absence of a particular event, whether that be conduction to the ventricles, arrival at the His, or myocardial capture, thereby establishing effective refractory periods at that cycle length **(Figs 10A to C)**. If the atrial effective refractory period is reached before demonstrating AV nodal refractoriness, then prolongation of the extrastimulus coupling interval and introduction of additional extrastimuli may permit such demonstration **(Figs 16A and B)**.

In addition, with atrial extrastimulus testing, "dual AVN physiology" may be observed. With gradual decrementation in the extrastimulus coupling interval, there is an abrupt prolongation in the H1-H2 interval, signifying conduction block in the AVN "fast" pathway but resulting in the atrial paced impulse arriving at the AV node by an alternate pathway that is associated with longer H1-H2 intervals ("slow-pathway"). Dual AVN pathways are felt to underlie the mechanism of

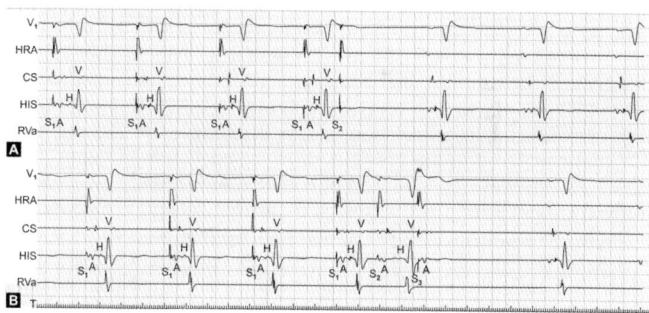

**FIGURES 16A AND B:** Atrial effective refractory period (AERP) and atrioventricular nodal effective refractory period (AVNERP). Surface lead ($V_1$) and intracardiac channels (HRA, CS, HIS, RVa) are shown. (A) Atrial pacing results in 1:1 atrial capture and conduction through AVN to the ventricle during the basic drive ($S_1$). The single premature atrial extrastimulus ($S_2$) fails to capture the atrial myocardium (AERP)—observe no atrial electrogram follows the $S_2$, unlike the $S_1$'s—so no impulse propagates through the AVN to activate the ventricle. Sinus rhythm ensues in the absence of atrial pacing. (B) Having reached AERP, the $S_2$ coupling is increased by 30 ms to permit atrial capture and evaluation of AVN refractoriness. In this case the $S_2$ captures atrial myocardium as evidenced by presence of an A electrogram and conducts through the AVN to the ventricle, and an $S_3$ that also captures the atrium fails to conduct to the ventricles and blocks above the His bundle as no His deflection is seen, i.e. in the AVN. In this case, the AVNERP was established at 600:300:360 ms

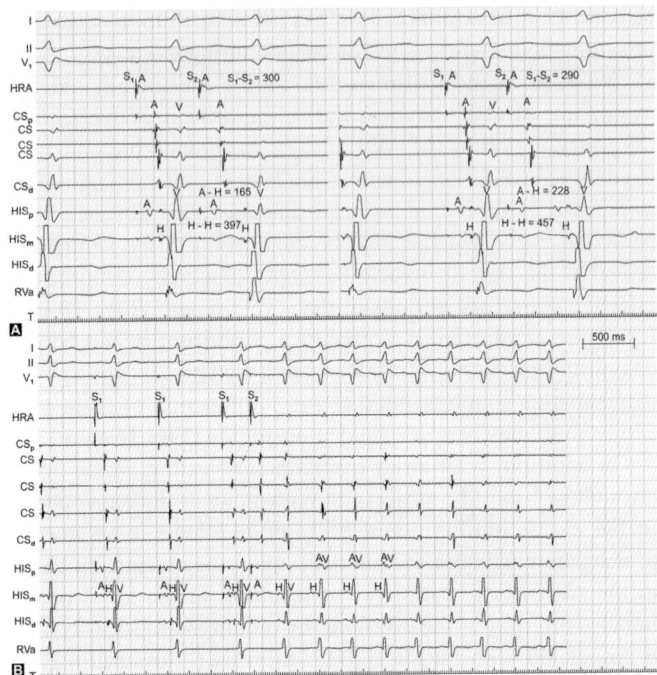

**FIGURES 17A AND B:** Dual atrioventricular node (AVN) physiology and induction of AVNRT. (A) On left side: the last paced beat of a 600-ms basic drive ($S_1$) is shown, with a single atrial extrastimulus coupled at 300 ms ($S_2$) that conducts through the AVN to the ventricles with A-H 165 ms and H-H 397 ms. On right side: with the same basic drive, the atrial extrastimulus coupling interval is decremented by 10 ms to 290 ms, resulting in conduction through the AVN to the ventricle with marked conduction delay in excess of 50 ms (A-H 228 ms, H-H 457 ms) compared to the previous drive, compatible with "jump" to a slow AVN pathway. (B) Induction of common AVN reentry tachycardia (AVNRT). Atrial pacing at 600 ms (in a different patient) is followed by a single extrastimulus coupled at 360 ms ($S_2$) that is associated with A-H delay and initiation of a tachycardia with narrow QRS identical to sinus, and electrogram (EGM) pattern with nearly simultaneous atrial and ventricular activation preceded by a His deflection. A narrow QRS tachycardia preceded by His EGM must be, in general, supraventricular as each impulse activates the His before activating the ventricles and hence must be arising above the His, i.e. atrium or AVN. In AVNRT, an atrial premature complex or premature extrastimulus provokes AVN fast pathway block and results in conduction over a slow AVN pathway manifested by a prolonged P-R and A-H. After conducting over the slow AVN pathway, the wavefront reaches a lower turnaround/branch point and propagates along the AVN "fast pathway" retrogradely to the atria while continuing along the His-Purkinje system (HPS) toward the ventricles. This results in nearly simultaneous atrial and ventricular activation, and the distinctive pattern of complete alignment of all atrial and ventricular EGM and superimposed P's and QRS's

AVN reentry tachycardia, and both continuous atrial pacing and atrial premature extrastimuli will often induce AVN reentry **(Figs 17A and B)**, although other SVTs may be initiated by this mechanism as well.

## Evaluation of Atrioventricular Conduction Disease

In appropriate patients, EP study is warranted to evaluate the site and mechanism of AV block. A discernible His bundle recording enables one to determine the exact site of AV conduction abnormality, i.e. proximal, within or distal to the His bundle region. This, in combination with surface ECG morphology of conducted beats, enables one to identify precisely the location of conduction abnormality. The finding of a prolonged H-V interval, greater than 60 ms, is evidence of HPS conduction disease **(Figs 18A and B)**. If 1:1 AV conduction is present in patients suspected of intermittent AV block, atrial pacing with cycle length decrementation should be performed to evaluate reproducibility of AV block. AV block in the HPS is abnormal during continuous atrial pacing **(Fig. 19)** but may be a physiologic response during atrial extrastimulation or with sudden rate change related to asynchronous initiation of atrial pacing.[14]

In asymptomatic patients with first-degree AV block (prolonged P-R interval), electrophysiologic assessment is unnecessary regardless of the QRS morphology of the conducted beats, although in asymptomatic individuals with second-degree AV block electrophysiologic assessment is used to identify site of block **(Figs 20A to C)**. Patients with intra-Hisian or infra-Hisian block tend to have a more unpredictable course, and permanent pacing is desirable.[23] Even though intranodal block usually presents as Wenckebach's phenomenon or Mobitz type I, it is not uncommon to see Wenckebach

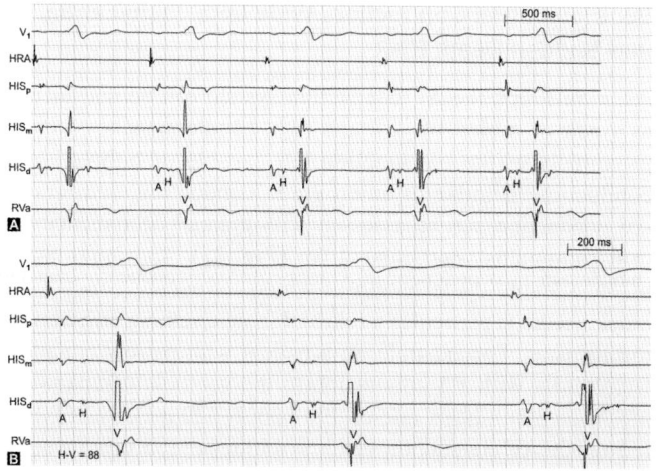

**FIGURES 18A AND B:** His-Purkinje disease. Sinus rhythm is shown in a patient with recurrent syncope and right bundle branch block at 50 mm/s (A) and 100 mm/s (B) sweep. The A-H interval is 138 ms and H-V interval is 88 ms at baseline heart rate of 69 bpm

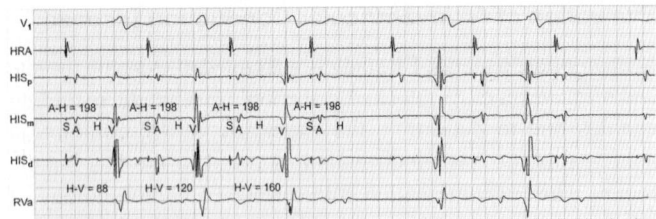

**FIGURE 19:** Infra-Hisian Wenckebach. With atrial pacing at cycle length 700 ms (heart rate 86 bpm), 1:1 atrial capture results in A-H stabilization at 198 ms and progressive prolongation in the H-V until the fourth paced complex fails to conduct beyond the His to reach the ventricles. The prolonged but stable A-H interval of 198 ms signifies atrioventricular conduction disease, but the Wenckebach pattern of conduction and block below the His implies significant distal conduction system disease as well

within the His or in the HPS distal to the His bundle. There is no difference in prognosis regardless of how the intra-Hisian or infra-Hisian second-degree block manifests itself, i.e. type I versus type II. In symptomatic patients with second-degree AV block, the role of EP study is limited because permanent pacing is the appropriate intervention. On the other hand, if the patient's symptoms cannot be explained on the basis of AV block and may be related to another arrhythmia, such as VT, EP study should be considered. In patients with third-degree or complete AV block, EP studies are seldom required; permanent pacing is the obvious option in symptomatic patients.[11]

## Ventricular Stimulation and Assessment of Ventriculoatrial Conduction, Wide QRS Tachycardia, and Sudden Death Risk

As a construct, the AV conduction system can be considered as a pair of cables, the left and right bundle branches, joined proximally at the AV junction; all capable of bidirectional conduction. Similar to anterograde conduction, retrograde AV conduction can be assessed with gradual decrementation in a continuous ventricular drive **(Fig. 21)**. The importance of the specific cycle length at which retrograde conduction block occurs is particularly relevant in determining arrhythmia mechanisms. For example, in a patient with narrow complex tachycardia with 1:1 conduction to the ventricle of unknown mechanism at a heart rate of 190 bpm (~320 ms), identifying retrograde AV conduction block at 450 ms (~135 bpm) would suggest that the tachycardia cannot be one in which the ventricles would be required to conduct to the atrium to maintain the

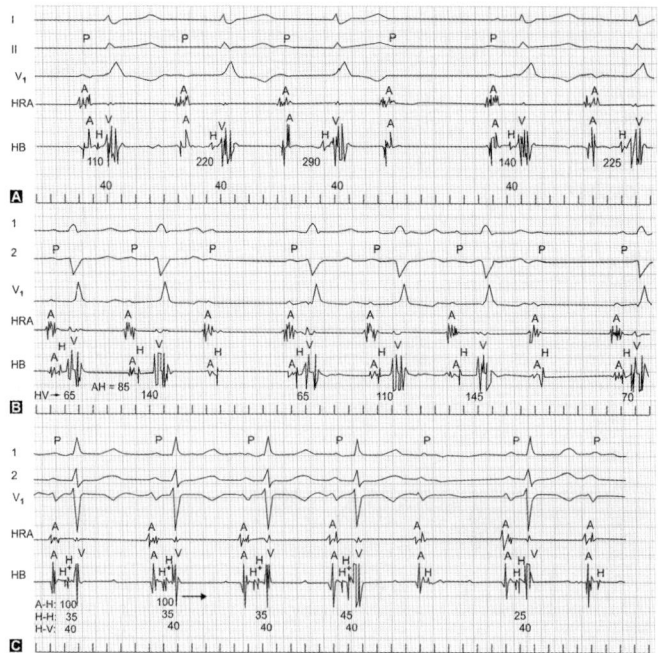

**FIGURES 20A TO C:** His bundle (HB) electrograms in atrioventricular (AV) block. The tracings are from three different patients with second-degree AV block. In Figures A and B, the conducted QRS complexes are wide and associated with bundle branch block. (A) The block is within the AV node (i.e. the A-wave on the HB is not followed by an HB deflection). (B) It can be appreciated that the block is distal to the HB even though the surface electrocardiogram (ECG) demonstrates a Wenckebach phenomenon. The latter can obviously occur in the His-Purkinje system as well, as depicted in this figure. (C) The site of the block is within the HB. This is suggested by split HB potentials (labeled H and H⁺), and the block is distal to the H but proximal to the H⁺. Intra-His block is difficult to diagnose from the surface ECG but can be suspected when a Mobitz type II occurs in association with a normal P-R interval and a narrow QRS complex. (*Source:* Modified from Akhtar M. Invasive cardiac electrophysiologic studies: an introduction. In: Parmley WW, Chatterjee K (Eds). Cardiology: Physiology, Pharmacology, Diagnosis. Philadelphia: Lippincott; 1991. With permission from Lippincott, Williams and Wilkins)

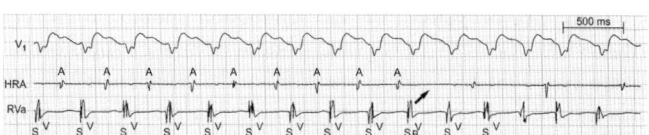

**FIGURE 21:** Retrograde conduction block. Continuous ventricular pacing (S) with gradual decrementation in cycle length results in 1:1 ventricular capture (V) and conduction to the atria (A) with progressive prolongation in V-A time until the 10th paced complex (S$_B$V) fails to conduct to the atrium and a 2:1 retrograde conduction pattern is established

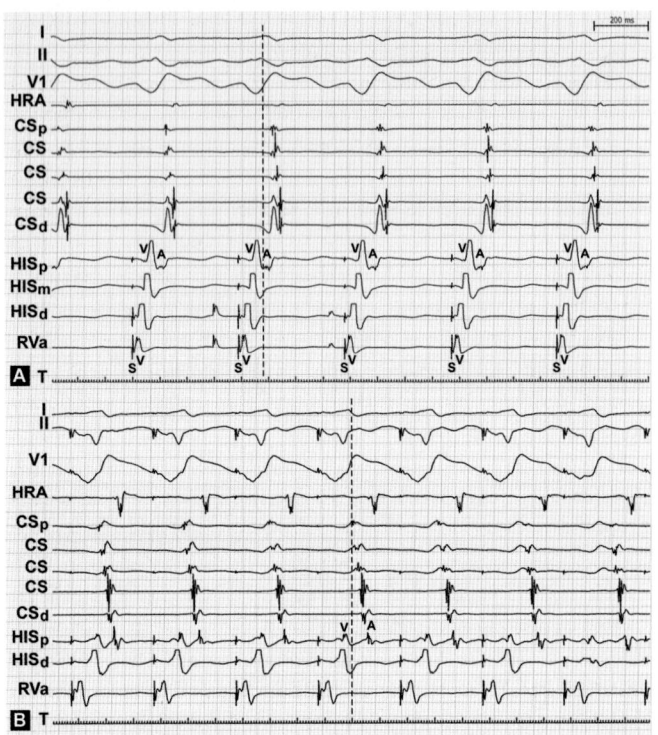

**FIGURES 22A AND B:** Retrograde conduction patterns. Surface electrocardiogram and intracardiac channels show two examples of ventricular pacing resulting in 1:1 myocardial capture and conduction to the atria over different pathways. (A) Retrograde atrial activation is earliest at the proximal His (A) followed by coronary sinus (CS) activation from proximal to distal, suggesting retrograde conduction over an anteriorly located pathway, i.e. the atrioventricular node (AVN) fast pathway. (B) Retrograde atrial activation is earliest in the proximal CS and later in the His (A), suggesting retrograde conduction over a posteriorly located pathway, i.e. AVN slow pathway or posteroseptal accessory pathway. Dashed lines identify earliest atrial activation. (Abbreviations: V: Ventricular electrogram; S: Stimulus artifact)

tachycardia. Instead, the tachycardia mechanism is more likely entirely independent of VA conduction, i.e. atrial tachycardia.

The specific pattern of VA conduction should be closely observed. Whereas retrograde conduction over the normal pathway would be expected to activate the atria earliest at the His catheter, which is in closest proximity to the AV node **(Fig. 22A)**, retrograde AV conduction over other pathways would result in altered activation sequences (aside from pathways very close to the AV node, such as anteroseptal accessory pathways). For instance, retrograde activation that is earliest in the proximal CS is suggestive of a posteriorly located midline pathway **(Fig. 22B)**, such as an AVN slow pathway or a posteroseptal accessory pathway; retrograde activation earliest at the distal CS would suggest presence of a left free-wall accessory pathway or left-sided AVN;[24] and retrograde activation that is earliest at the HRA would suggest presence of a right free-wall accessory pathway.

Retrograde conduction refractory periods of HPS, AV node and accessory pathways may be determined through the use of ventricular premature extrastimuli. It is once again identified as the longest ventricular S1-S2 that fails to conduct beyond His or to the atria and is analogous to the anterograde refractory periods. Ventricular extrastimuli are also of use in assessing ventricular refractoriness and for arrhythmia induction, whether supraventricular or ventricular. Multiple extrastimuli are important for induction protocols when assessing tendency toward arrhythmia, particularly wide QRS tachycardia, which may occur due to a variety of electrophysiologic mechanisms, both from supraventricular and ventricular origins **(Figs 23A to D)**.[25] The underlying nature of the wide QRS tachycardia is critical for both prognosis and therapy, and EP studies have proven invaluable for distinguishing the various etiologies. With few exceptions, when the nature of the arrhythmic problem is not known and the direction of therapy is not clear, patients with wide QRS tachycardia should undergo EP study. This is particularly true in situations where nonpharmacologic therapy is the desired goal.[11]

In patients with features suggesting high risk of sudden death, such as structural heart disease and LV dysfunction as well as evidence of ventricular ectopy or arrhythmia in this setting, utilization of multiple extrastimuli (maximum 3), including short-long-short sequences, from multiple ventricular sites in the baseline state and under pharmacologic stress or stimulation is necessary. The induction of monomorphic VT is a specific response reflecting tendency of the myocardial substrate to support this type of arrhythmia and clinically appropriate therapy should be rendered accordingly. Alternately, the induction of polymorphic VT can be a nonspecific response in patients without structural heart disease, when triple extrastimuli at short coupling intervals are utilized; hence, the delivery of additional ventricular extrastimuli should be weighed carefully against the risk of a nonspecific finding.[11]

## Role of Electrophysiology Study in Evaluation of Unexplained Syncope

While neurocardiogenic mechanisms constitute the most common causes of syncope in patients with ostensibly normal hearts and should be evaluated through tilt testing **(Figs 24A to G)**,[22,23,25,26] EP study is integral in evaluating patients

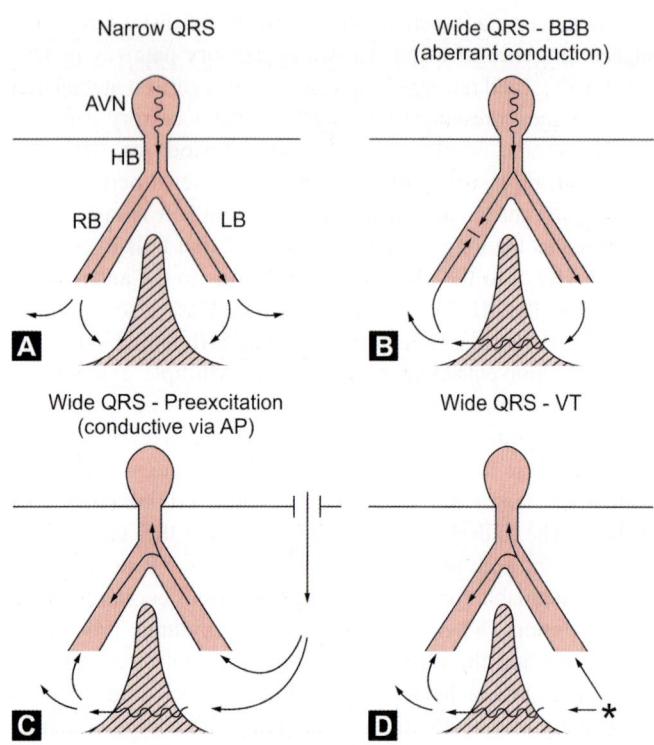

**FIGURES 23A TO D:** Wide QRS tachycardia mechanisms. Routes of impulse propagation during a wide QRS tachycardia in various settings are depicted. It should be noted that only in A and B, His bundle activation expected to precede ventricular activation. This helps the delineation from other causes of wide QRS tachycardia, shown in C and D. (Abbreviations: AP: Accessory pathway; BBB: Bundle branch block; VT: Ventricular tachycardia). (*Source:* Modified with permission from Akhtar M. Techniques of electrophysiologic evaluation. In: Fuster V, O'Rourke RA, Walsh RA, Poole-Wilson P (Eds). Hurst's The Heart, 12th edition. New York: The McGraw-Hill Companies, Inc.; 2007. With permission from The McGraw-Hill Companies, Inc.)

with syncope that remains unexplained, particularly those with heart disease.[27] During such studies, all arrhythmic possibilities, such as sinus node dysfunction, AV conduction abnormalities, SVT and VT, should be excluded. Patients with underlying structural heart disease, such as old myocardial infarction, primary myocardial disease or poor LV function, generally have underlying VT to explain syncope **(Figs 25A and B)**. When arrhythmias occur in patients without overt structural heart disease, sinus node dysfunction, AV block (particularly intra-Hisian block) or SVT may be more likely. Less frequently, VT can occur in the absence of an overt structural heart disease.[11]

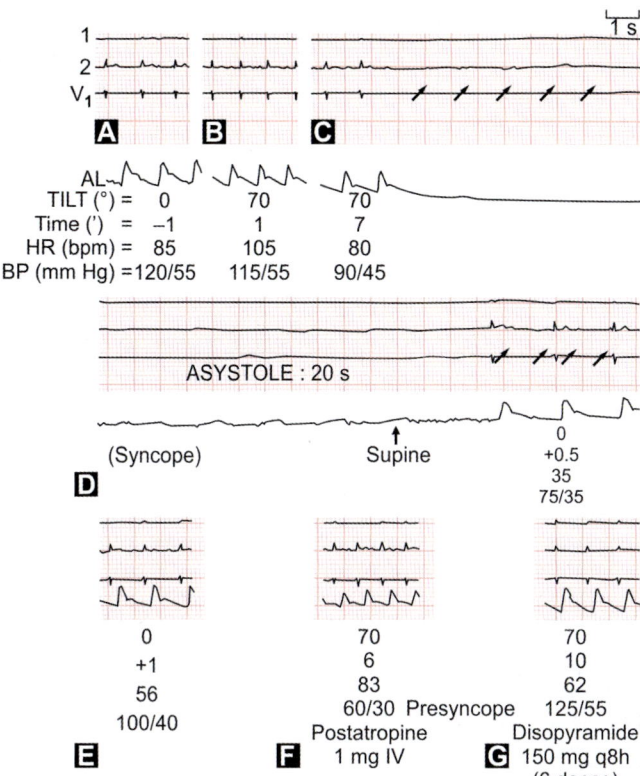

**FIGURES 24A TO G:** Asystole in neurocardiogenic syncope. Note the normal heart rate (HR) and blood pressure (BP) in supine position. At the beginning of head-up tilt at 70 degrees (B), some degree of tachycardia is noted. Seven minutes after the onset of tilt (C), an episode of atrioventricular block occurs and is followed by sinus arrest and a total asystole of 20 seconds. Syncopal episodes follow. Presyncope is still present when asystole is prevented by atropine (F). Findings in C might tempt one to prescribe permanent pacing, an inappropriate choice of therapy. In this patient with neurocardiogenic syncope, disopyramide (G) prevented hypotension and syncope without the need for a permanent pacemaker. (*Source:* Modified from Sra JS, Jazayeri MR, Avitall B, et al. Comparison of cardiac pacing with drug therapy in the treatment of neurocardiogenic (vasovagal) syncope with bradycardia or asystole. N Engl J Med. 1993;328:1085-90. With permission from Massachusetts Medical Society)

## Survivors of Sudden Cardiac Arrest

In many patients with documented episodes of cardiac arrest from the onset, VF can be documented as the initial cause. Patients dying suddenly often have underlying structural heart disease (usually coronary artery disease or primary myocardial disease) and are prone to VT/VF due to electrical instability.

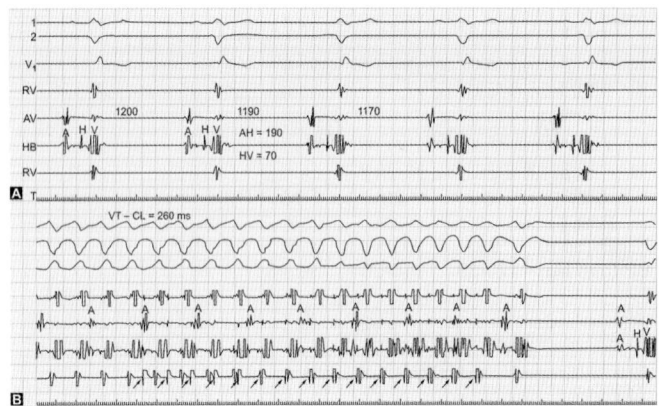

**FIGURES 25A AND B:** Arrhythmic causes of syncope. (A) Sinus rhythm in a patient with unexplained syncope. Sinus bradycardia, bifascicular block, and a long P-R interval from surface electrocardiogram suggest possible conduction system disease etiology. (B) In this patient, however, ventricular tachycardia was inducible with ventricular extrastimulation and was the actual cause of syncope. Control of ventricular tachycardia (VT) without a pacemaker was sufficient to prevent syncope in this patient. Termination of tachycardia and restoration of sinus rhythm are shown in Figure B. (Abbreviation: CL: Cycle length). (Source: Modified from Akhtar M. Techniques of electrophysiology evaluation. In: Fuster V, O'Rourke RA, Walsh RA, Poole-Wilson P (Eds). Hurst's The Heart, 12th edition. New York: The McGraw-Hill Companies Inc.; 2007. With permission from The McGraw-Hill Companies Inc.)

It seems prudent to investigate both the nature and extent of organic heart disease and also to assess vulnerability to recurrent VT/VF. At present, EP study is considered a routine part of the overall patient assessment in this group of individuals.[11,28,29]
In survivors of VT/VF, EP study is desirable for a variety of reasons:
- In our experience, almost 40% of patients with monomorphic VT in association with idiopathic dilated cardiomyopathy and valvular heart disease have bundle branch reentry (BBR) as the underlying mechanism **(Fig. 26)**. We feel this arrhythmia is preferably managed with bundle branch ablation, which is curative, rather than with an implantable cardioverter-defibrillator (ICD) alone.
- Several VT morphologies or other arrhythmias may be identified in addition to the presenting/clinical VT. Lack of awareness of such arrhythmias may complicate patient management. For example, rapid SVT may require separate attention to prevent unnecessary ICD shocks, either through antiarrhythmic therapy or by ablation. The coexistence of sick sinus or conduction system disease may be aggravated by antiarrhythmic therapy and necessitate

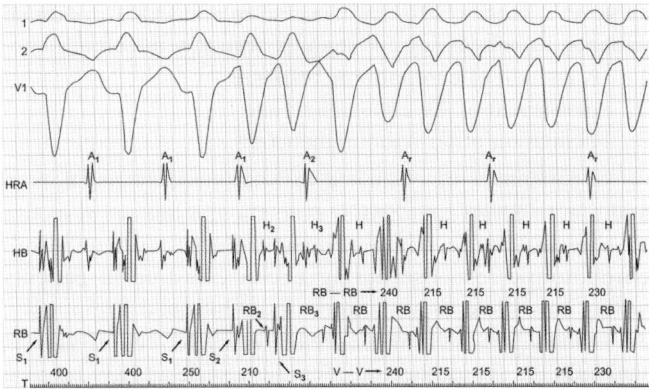

**FIGURE 26:** Induction of sustained ventricular tachycardia due to bundle branch reentry (BBR). The surface electrocardiogram and intracardiac tracings are labeled. Basic cycle length ($S_1S_1$) is 400 ms during ventricular pacing. Sustained BBR is induced with two extrastimuli ($S_2S_3$). Note that the His bundle and right bundle (RB) deflections precede the QRS, suggesting supraventricular tachycardia with aberrant conduction. However, there is 2:1 ventriculoatrial (VA) block, indicating the ventricular nature of this tachycardia. Without His bundle/right bundle (HB/RB) recordings, the diagnosis can be difficult and, consequently, the likelihood of inappropriate therapy will be high. RB-RB and V-V (ventricular) intervals are labeled. (*Source:* Modified from Caceres J, Jazayeri M, McKinnie J, et al. Sustained bundle branch reentry as a mechanism of clinical tachycardia. Circulation. 1989:79:256-70. With permission from Wolters Kluwer Health)

pacing. Identification preoperatively could contribute to appropriate device selection.
- Rarely, supraventricular arrhythmia may trigger VT/VF. This may happen in patients with severe coronary artery disease, congestive heart failure or Wolff-Parkinson-White syndrome to name a few scenarios. Elimination of the underlying triggers should be the primary therapeutic approach with the need for an ICD, a secondary concern.

## Cardioactive Agents

The invasive EP study often incorporates a phase in which observation and programmed stimulation is conducted under pharmacologic influence in an attempt to facilitate arrhythmia induction. Two agents are commonly employed, isoproterenol and procainamide.

Isoproterenol is a sympathomimetic amine with primary activity on beta-adrenergic receptors type 1 and type 2. Its overall effects are to increase inotropic and chronotropic response as well as improve conduction system and myocardial propagation. Usually a dose of 1–3 mcg/min is adequate to

achieve at least a 20–25% increase in sinus rate in normal patients. Its role during EP study is primarily threefold:
1. Evaluation of chronotropic response
2. Assessment of AV conduction; and
3. Facilitation of arrhythmia induction.

In patients being evaluated for sinus node dysfunction, in addition to SNRT and SACT evaluation, a blunted response to catecholamine stimulation correlates with impaired chronotropic response during exercise testing in patients with sinus node dysfunction. As mentioned previously, AV conduction with Wenckebach at cycle lengths greater than 600 ms should be considered unusual but may be seen in patients with high vagal tone. However, this finding should not be considered abnormal in isolation, and only if a lack of improvement with adrenergic stimulation is demonstrated should pathology be suspected. Isoproterenol infusion aids such evaluation in that it can improve both AVN and HPS conduction. In addition to assessment of sinus node automaticity and AV conduction, isoproterenol facilitates arrhythmia induction and sustainability both through modulation of AVN and accessory pathway conduction; and by provoking myocardial refractory period shortening that permits shorter coupling during atrial and ventricular premature extrastimulation to enhance tissue conduction delay, thereby promoting reentry. A pronounced benefit of isoproterenol is seen in patients with structurally normal hearts and idiopathic VT, in whom rapid atrial pacing and isoproterenol facilitates induction of triggered activity.[30] Finally, isoproterenol can reverse antiarrhythmic effects, primarily through its actions on ion channels. Therefore, efficacy of antiarrhythmic therapy is also an important contribution of isoproterenol to the EP study.

Procainamide is a class IA antiarrhythmic drug with primary effects in blocking sodium channels. It increases tissue refractoriness and slows conduction in the atria, HPS and ventricles, with variable effects on the AV node.[31,32] Its primary role in the clinical EP study is to assess propensity toward AV conduction block and VT induction. In patients with normal HPS conduction, procainamide may introduce mild conduction delay. However, in patients with moderate conduction disease, manifested by H-V intervals greater than 65 ms, a dose of 10 mg/kg can unmask a profound tendency toward conduction block. If the H-V prolongs beyond 100 ms, this is an indication for permanent pacing. As stated, conduction slowing is an anticipated effect of procainamide therapy and hence is the primary mechanism of its antiarrhythmic qualities in acute VT suppression.

However, conduction slowing can also promote reentry, particularly in patients with structural heart disease. Hence, it is employed during EP study to evaluate VT inducibility but also explains why class I antiarrhythmics are contraindicated for long-term maintenance therapy in patients with structural heart disease.

## CARDIAC ELECTROPHYSIOLOGY STUDY FOR EVALUATION OF DRUG THERAPY

In patients with relatively benign cardiac arrhythmias, EP testing to assess efficacy of pharmacologic therapy is unnecessary in most situations, and clinical course can be observed to determine whether control has been achieved. Also, in patients with ICDs, antiarrhythmic therapy can be assessed clinically as the device will record and treat events according to its specific programming. However, for patients with potentially life-threatening tachycardias like VT or with severe manifestations of cardiac arrhythmias, such as syncope or presyncope, in whom device therapy is not present, it is desirable to assess efficacy of pharmacologic intervention. A technique of drug testing has been developed whereby the elimination of inducibility of a given tachycardia is assessed following a drug administration. If drug therapy eliminates induction, addition of isoproterenol may demonstrate reversal of therapeutic drug effect.[33] This is helpful in considering additional or alternative therapy. Failure of serial drug testing is associated with a significant recurrence rate and is a strong indication for nonpharmacologic intervention.[11]

## ELECTROPHYSIOLOGY STUDY TO GUIDE ABLATIVE THERAPY

### Role of Three-dimensional Mapping Systems

Treatment of cardiac arrhythmias is dictated by various factors impacting risk related to therapy including potential adverse effects, tolerance of the patient to specific treatment and anticipated likelihood of long-term success. One modality employed for arrhythmia diagnosis and treatment is catheter ablation, through which sites involved in arrhythmia genesis and maintenance are mapped and targeted for local tissue destruction via a percutaneous catheter-based procedure. Identification of specific location(s) that initiate or maintain arrhythmia is challenging for a variety of reasons, but particularly so due to difficulty in returning the catheter to a particular location with any measure of precision and accuracy, given the various cardiac and respiratory motions that impact how catheters and the heart interface, as well as the ambiguity of depth perception on two-dimensional (2D) fluoroscopy.

Three-dimensional (3D) electroanatomic mapping systems are an integral tool in interventional EP as they provide a manner to visualize inside of the heart and reliably guide catheters to specific locations that would otherwise prove challenging with 2D fluoroscopy alone.[34,35] These systems have two important capabilities: (i) creating and visualizing endocardial geometry as a 3D model and (ii) superimposing timing and voltage information to create "maps" that visually indicate whether a particular region is activated earlier or later than others, and if a particular region is healthy or scarred according to the amplitude of the local EGM. These color-coded maps are displayed with a virtual rendition of the mapping catheter, so as to convey visually the relationship of the catheter to the surrounding myocardial chamber—a relationship that is often ambiguous on fluoroscopy. These systems are able to achieve a fairly high level of precision and can annotate catheter tip locations in the 3D model to provide targets for navigation and provide a "history" of previous sites mapped. However, as the true 3D anatomical position of the virtual catheter is not known to the system—only the relative position of the catheter within the mapping system's 3D coordinate space is known—accuracy is compromised. To address this, these systems can import an actual 3D anatomic volume, such as a computed tomography (CT) scan or magnetic resonance image (MRI), of a particular cardiac chamber to define the anatomic coordinate system.[36] The anatomic volume can be aligned and rotated to best approximate the orientation of 3D map, which is then superimposed **(Fig. 27)**. This process of "registering" the volume and map permits better anatomical localization by comparing the created 3D map to the known anatomy to verify that all areas have been accounted for and to identify true location of the catheter and its relationship to sometimes highly important and sensitive structures, such as pulmonary veins for AF ablation.[37] Whether a mapping system is used with or without a 3D volume, the images are displayed on a separate view from live fluoroscopy, requiring the operator to incorporate information from both image sources and perform a mental real-time registration of these images. The latest commercial fluoroscopy systems have the capability to register a 3D volume directly with live fluoroscopy so true catheter location can be visualized within a registered anatomic model[38] **(Fig. 28)**. The registered volumetric and fluoroscopic image can then be compared to the image of the

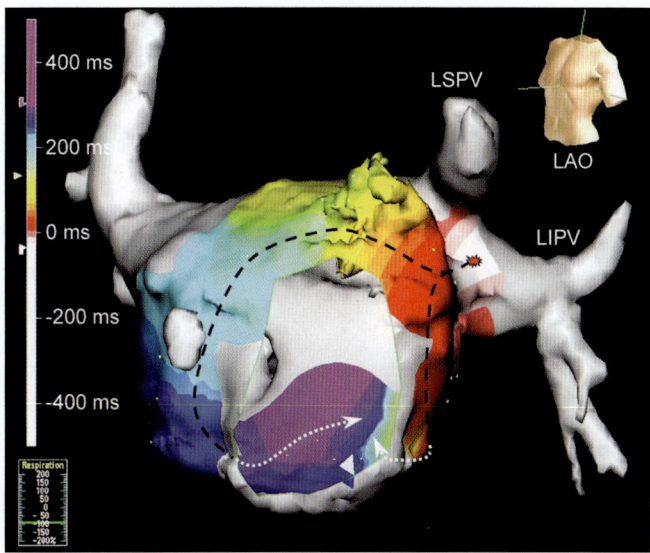

**FIGURE 27:** Left atrial (LA) computed tomography (CT) with 3D electroanatomic map (LAO projection). Electroanatomic mapping data is superimposed on a 3D LA geometry and then registered with a CT of the LA and pulmonary veins (PVs) to convey activation timing. Activation with respect to a timing reference is depicted according to the color scale (left). The mitral valve annulus is demarcated by a green perimeter. Earliest activation, displayed in white, is seen arising from between the left superior and inferior PVs, and then propagates (red to orange to yellow, etc. according to color scale) counterclockwise and clockwise around the mitral valve annulus (black dashed lines). The two wavefronts pass posteriorly (white dashed arrow) and meet (seen by looking through the mitral valve annulus into the LA) where crowding of isochrones suggests conduction block, in this case, between the mitral valve annulus and the inferior aspect of the inferior pulmonary vein, i.e. across the posterior mitral isthmus

3D map, which may also be registered to the 3D volume. In an effort to achieve both accuracy and precision, this method still requires operator to assimilate information from multiple image sources. The systems that register various imaging modalities with live fluoroscopy must also overcome the challenges posed by cardiac and respiratory motion evident on live fluoroscopy but not accounted for by the static anatomic and virtual models. Techniques to compensate for this motion, cardiac and respiratory "gating,"[39] permit a method of continuous and real-time registration between the live fluoroscopy and the 3D model. Incorporating voltage and timing maps directly into the CT or MRI volumes that are registered to fluoroscopy would provide a means to unite all these imaging modalities and technologies into a single system.

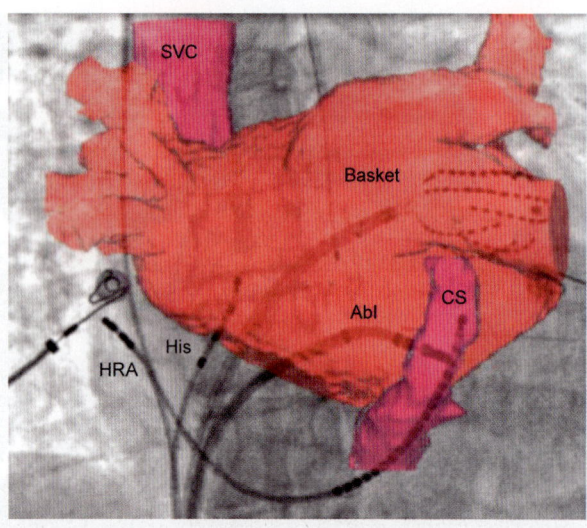

**FIGURE 28:** Live fluoroscopy with overlay of registered left atrial (LA) computed tomography (CT). An anteroposterior projection of live fluoroscopy shows multiple catheters including a multielectrode basket catheter and ablation catheter positioned through transseptal sheaths into the LA. The LA CT reconstruction is registered with fluoroscopy to demonstrate specific anatomy and catheter locations. The basket catheter is positioned in the left inferior pulmonary vein and the ablation catheter is positioned at the mitral valve annulus and, hence, appears in close proximity to the duodecapolar catheter within the coronary sinus (CS). The CT is depicted in posteroanterior projection in order for anatomical alignment between the two modalities. (Abbreviations: Abl: Ablation catheter; HRA: High right atrium; SVC: Superior vena cava)

## COMPLICATIONS

The contemporary EP study is safe when performed in appropriate facilities by trained physicians and personnel. Although patient factors, such as age, anatomy and associated comorbidities, must be considered in individualizing procedural risk, the major complication rate is approximately 1% and for death is 1:1,000. The complications are the same as those anticipated with other forms of cardiac catheterization as well some more particular to the EP study. These include inadequate hemostasis with or without vascular injury including local bleeding from access sites with adjacent extension, hematoma, AV fistula, pseudoaneurysm and major vessel perforation; vascular and intracardiac thrombosis with or without pulmonary or systemic embolism including stroke; cardiac injury, including myocardial infarction, coronary artery and CS dissection, and myocardial perforation and associated pericardial effusion with or without cardiac tamponade; tachyarrhythmias and bradyarrhythmias and injury to the conduction system rarely necessitating permanent pacing; esophageal injury and pulmonary vein stenosis in left atrial ablation; skin injuries associated with direct cardioversion and radiation/fluoroscopy exposure; phrenic nerve injury and paralysis; decompensated heart failure; infection; and allergic reactions primarily to administered agents, such as iodine, anesthetics, antibiotics, blood products and protamine.

## CONCLUSION

It remains a truism that the past offers insight into the present and future. The historical discoveries that have coalesced into the field of clinical cardiac EP are truly awesome and compel one to take pause and reflect on the origins of this burgeoning field. Cardiologists of the 21st century must still see in the electrocardiogram all the electrical processes and associated physical manifestations underlying cardiac EP described over the past 100 years to fully appreciate normal processes and cardiovascular pathology. These fundamental principles, outlined here, form a clinical foundation for every student of the medical arts who is interested and invested in cardiovascular diseases to begin to comprehend the most fundamental of physiologic questions, "Why does the heart beat?"

## ACKNOWLEDGMENTS

The authors gratefully acknowledge the assistance of Brian Miller and Brian Schurrer in the preparation of illustrations and Barbara Danek, Joe Grundle and Katie Klein in editing the manuscript.

## REFERENCES

1. John HJ. Jan Evangelista Purkynê: Czech Scientist and Patriot, 1787-1869. Philadelphia: American Philosophical Society; 1959.
2. His W Jr. Die Thätigkeit des embryonalen Herzens und deren Bedeutung für die Lehre von der Herzbewegung bein Erwachsenen. Arb Med Klinik Leipzig. 1893;1:14-50.
3. Kent AF. Researches on the structure and function of the mammalian heart. J Physiol. 1893;14:i2-254.
4. Tawara S. Eine anatomisch-histologische studie über das atrio-ventrikular bündel und die Purkinjeschen fäden. Das Reizleitungssystem des Säugetierherzens. Jena, Germany: Verlag von Gustav Fischer; 1906. p. 200.
5. Keith A, Flack M. The form and nature of the muscular connections between the primary divisions of the vertebrate heart. J Anat Physiol. 1907;41:172-89.
6. Einthoven W. Weiteres über das elektrokardiogram. Pflüger Arch ges Physiol. 1908;122:517-48.

7. Dock W. Transitory ventricular fibrillation as a cause of syncope and its prevention by quinidine sulfate. Am Heart J. 1929;4:709-14.
8. Jervell A, Lange-Nielsen F. Congenital deaf-mutism, functional heart disease with prolongation of the Q-T interval and sudden death. Am Heart J. 1957;54:59-68.
9. Scherlag BJ, Lau SH, Helfant RH, et al. Catheter technique for recording His bundle activity in man. Circulation. 1969;39:13-8.
10. Scheinman MM, Morady F, Hess DS, et al. Catheter-induced ablation of the atrioventricular junction to control refractory supraventricular arrhythmias. JAMA. 1982;248:851-5.
11. Akhtar M. Techniques of electrophysiologic evaluation. In: Fuster V, O'Rourke R, Walsh R, Poole-Wilson P (Eds). Hurst's The Heart, 12th edition. New York: The McGraw-Hill Companies; 2008. pp. 1064-76.
12. Guidelines for Clinical Intracardiac Electrophysiological and Catheter Ablation Procedures. A report of the American College of Cardiology/American Heart Association Task Force on practice guidelines. (Committee on Clinical Intracardiac Electrophysiologic and Catheter Ablation Procedures), Developed in collaboration with the North American Society of Pacing and Electrophysiology. Circulation. 1995;92:673-91.
13. Buxton AE, Calkins H, Callans DJ, et al. ACC/AHA/HRS 2006 key data elements and definitions for electrophysiological studies and procedures: a report of the American College of Cardiology/American Heart Association Task Force on Clinical Data Standards (ACC/AHA/HRS Writing Committee to Develop Data Standards on Electrophysiology). J Am Coll Cardiol. 2006;48:2360-96.
14. Akhtar M, Mahmud R, Tchou P, et al. Normal electrophysiologic responses of the human heart. Cardiol Clin. 1986;4(3):365-86.
15. Lemery R, Birnie D, Tang AS, et al. Normal atrial activation and voltage during sinus rhythm in the human heart: an endocardial and epicardial mapping study in patients with a history of atrial fibrillation. J Cardiovasc Electrophysiol. 2007;18:402-8.
16. Durrer D, van Dam RT, Freud GE, et al. Total excitation of the isolated human heart. Circulation 1970;41:899-912.
17. Becker AE. Left atrial isthmus: anatomic aspects relevant for linear catheter ablation procedures in human. J Cardiovasc Electrophysiol. 2004;15:809-12.
18. Josephson ME. Clinical Cardiac Electrophysiology: Techniques and Interpretations, 4th edition. Philadelphia: Lippincott Williams and Wilkins; 2008. pp. 39-47.
19. Denker S, Lehmann MH, Mahmud R, et al. Divergence between refractoriness of His-Purkinje system and ventricular muscle with abrupt changes in cycle length. Circulation. 1983;68:1212-21.
20. Denes P. The effect of cycle length on the atrial refractory period. Pacing Clin Electrophysiol. 1984;7:1108-14.
21. Narula OS, Scherlag BJ, Samet P, et al. Atrioventricular block. Localization and classification by His bundle recordings. Am J Med. 1971;50:146-65.
22. Sra JS, Jazayeri MR, Avitall B, et al. Comparison of cardiac pacing with drug therapy in the treatment of neurocardiogenic (vasovagal) syncope with bradycardia or asystole. N Engl J Med. 1993;328:1085-90.

23. Dhingra RC, Wyndham C, Bauernfeind R, et al. Significance of block distal to the His bundle induced by atrial pacing in patients with chronic bifascicular block. Circulation. 1979;60:1455-64.
24. Nakagawa H, Jackman WM. Catheter ablation of paroxysmal supraventricular tachycardia. Circulation. 2007;116:2465-78.
25. Akhtar M, Jazayeri M, Avitall B, et al. Electrophysiologic spectrum of wide QRS complex tachycardia. In: Zipes DP, Jalife J (Eds). Cardiac Electrophysiology: From Cell to Bedside. Orlando: WB Saunders; 1990. p. 635.
26. Sra JS, Anderson AJ, Sheikh SH, et al. Unexplained syncope evaluated by electrophysiologic studies and head-up tilt testing. Ann Intern Med. 1991;114:1013-9.
27. Strickberger SA, Benson DW, Biaggioni I, et al. AHA/ACCF scientific statement on the evaluation of syncope: from the American Heart Association Councils on Clinical Cardiology, Cardiovascular Nursing, Cardiovascular Disease in the Young, and Stroke, and the Quality of Care and Outcomes Research Interdisciplinary Working Group; and the American College of Cardiology Foundation: in collaboration with the Heart Rhythm Society: endorsed by the American Autonomic Society. Circulation. 2006;113:316-27.
28. Akhtar M, Garan H, Lehmann MH, et al. Sudden cardiac death: management of high-risk patients. Ann Intern Med. 1991;114:499-512.
29. Morady F, Scheinman MM, Hess DS, et al. Electrophysiologic testing in the management of survivors of out-of-hospital cardiac arrest. Am J Cardiol. 1983;51:85-9.
30. Lerman BB, Stein K, Engelstein ED, et al. Mechanism of repetitive monomorphic ventricular tachycardia. Circulation. 1995;92:421-9.
31. Pronestyl injection package insert (Princeton Pharmaceutical—US), Rev 8/91, Rec 2/93.
32. Coyle JD, Lima JJ. Procainamide. In: Evans WE, Schentag JJ, Jusko WJ (Eds). Applied Pharmacokinetics: Principles of Therapeutic Drug Monitoring, 3rd edition. Vancouver: Applied Therapeutics; 1992; pp. 1-33.
33. Jazayeri MR, Van Wyhe G, Avitall B, et al. Isoproterenol reversal of antiarrhythmic effects in patients with inducible sustained ventricular tachyarrhythmias. J Am Coll Cardiol. 1989;14:705-11.
34. Gepstein L, Hayam G, Ben-Haim SA. A novel method for nonfluoroscopic catheter-based electroanatomical mapping of the heart. Circulation. 1997;95:1611-22.
35. Sra J, Thomas JM. New techniques for mapping cardiac arrhythmias. Indian Heart J. 2001;53:423-44.
36. Dong J, Calkins H, Solomon SB, et al. Integrated electroanatomic mapping with three-dimensional computed tomographic images for real-time guided ablations. Circulation. 2006;113:186-94.
37. Sra J. Cardiac image registration. J Atr Fibrillation. 2008;1:145-60.
38. Sra J, Krum D, Malloy A, et al. Registration of three-dimensional left atrial computed tomographic images with projection images obtained using fluoroscopy. Circulation. 2005;112:3763-8.
39. Sra J, Ratnakumar S. Cardiac image registration of the left atrium and pulmonary veins. Heart Rhythm. 2008;5:609-17.

CHAPTER

# Syncope

4

*Vijay Ramu, Fred Kusumoto, Nora Goldschlager*

## Chapter Outline

- Epidemiology
  - Incidence and Prevalence of Syncope
  - Economic Burden of Syncope
  - Causes and Classification of Syncope
- Diagnostic Tests
  - History and Physical Examination
  - Blood Tests
  - Electrocardiogram
  - Echocardiography
  - Exercise Testing
  - Continuous ECG Monitoring
  - Signal Averaged ECG
  - Upright Tilt-table Testing
  - Electrophysiology Study
  - Cardiac Catheterization
  - Neurologic Tests
- Approach to the Evaluation of Syncope
- Specific Patient Groups
  - Vasovagal (Neurocardiogenic) Syncope
  - Hypertrophic Cardiomyopathy
  - Nonischemic Cardiomyopathy
  - Congenital Heart Disease
  - Elderly Patients
- Syncope and Driving
- Guidelines

## INTRODUCTION

Syncope is a sudden and transient loss of consciousness associated with loss of postural tone, followed by complete and spontaneous recovery. The term "syncope" originates from the Greek word "Synkoptein" which means—cutting short (koptein- "to cut"). In the first six centuries, many Greek philosophers and physicians speculated on the causes of syncope. Claudius Galen, a famous Greek physician, suggested that syncope was a problem of both the stomach and the heart.[1] The mechanism for transient loss of consciousness associated with syncope is cerebral hypoperfusion with reduced blood flow to the reticular activating system. A common phrase used in clinical medicine is presyncope which is considered to represent a warning or prodrome for frank syncope. In the case of presyncope, symptoms, such as dizziness and graying out, are not followed by frank loss of consciousness. Many physicians evaluate and treat presyncope in a similar manner to syncope; although a reasonable approach, there is no strong clinical data to support similar etiologies and outcomes.

It is important to acknowledge that the definition of syncope (and thus etiologic classification) varies even among experts. Using a more general definition of transient loss of consciousness, some experts include neurologic (e.g. seizure and concussion), metabolic (e.g. hypoxia) and psychiatric conditions as forms of syncope, while others who emphasize cerebral hypoperfusion consider syncope as one of the several causes of transient loss of consciousness and classify some

**FLOWCHART 1:** Classification of mechanistic causes for transient loss of consciousness and syncope

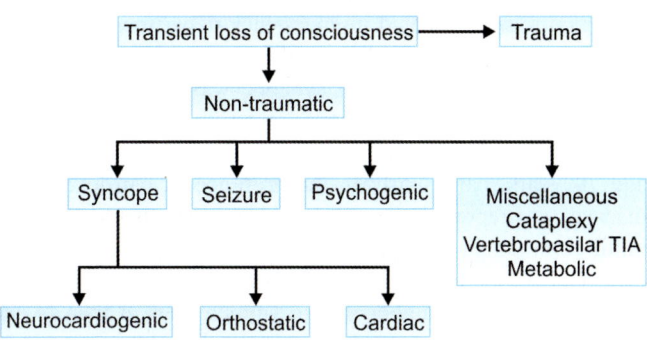

neurologic, metabolic and psychiatric mechanisms as separate entities **(Flowchart 1)**.[2,3] Using this more restrictive definition of syncope, the three most important causes of transient loss of consciousness are: (i) syncope; (ii) seizure, and (iii) psychogenic blackouts. Other rare causes include metabolic disorders, such as hypoglycemia or hypoxia, intoxication, and psychiatric problems such as cataplexy or pseudosyncope. Determining the correct cause of syncope is the key to approaching therapy, if the initial working clinical diagnosis is erroneous, subsequent investigations and even the final diagnosis and treatment may also be incorrect.[4] Regardless of definition, syncope may represent a harbinger for sudden death, and often the diagnostic evaluation focuses on identifying or ruling out potential life-threatening causes of syncope, such as ventricular arrhythmias or aortic stenosis. For this reason, the diagnostic workup for the patient with syncope revolves around two different but related issues: (i) identification of the specific mechanism for syncope and (ii) risk stratification to estimate short-term and long-term risk of adverse outcomes. It is important for the clinician to remember that the diagnostic workup of syncope requires a patient-specific approach and diagnostic tests must carefully be chosen. Several medical societies, including the European Society of Cardiology (ESC) and the American Heart Association/American College of Cardiology Foundation (AHA/ACCF) have provided comprehensive guidelines or scientific statements for the diagnosis and management of syncope, although some aspects are not without controversy.[2-5]

## EPIDEMIOLOGY

### Incidence and Prevalence of Syncope

Several studies have attempted to evaluate the incidence and prevalence of syncope. Obtaining accurate figures is difficult, since it has been estimated that only 25–50% of syncopal episodes are reported to medical professionals and only 2–5% of episodes are evaluated in emergency department settings.[2] In the most recent report from the Framingham study, the incidence of a first report of syncope was 6.2/1000 person-years follow-up and a ten-year cumulative incidence of 6%.[6] In addition, a sharp increase in the incidence of syncope after age 70 years, particularly in the presence of cardiovascular disease, was reported. Several studies suggest that syncope is quite common in younger populations.[7–9] For example, in a cohort of 62 medical students, 32% reported a prior episode of syncope, with a higher rate in women than in men (42% and 31% respectively).[9] Collectively, the data suggest a bimodal distribution of a first episode of syncope, with a first peak between the first and the second decade (with a female predominance) and another peak that begins after the age of 60 years and is gender-independent.[2] To summarize, up to approximately 30–40% of people will have an episode of syncope during their lifetime, and of these 30–40% will have a recurrent episode within three years.[1–9]

Although syncope may be the first symptom for a patient at high risk for sudden death and adverse clinical outcomes, examination of cohort studies suggests that the mortality rate ranges from 1–2% at 30 days to 7–8% at one year.[6,10] Examination of large hospital databases suggests that in-hospital mortality for syncope is also low (0.28%) with almost all deaths occurring in patients over 60 years of age.[8] However, in the Framingham report, patients who were thought to have a cardiac cause of syncope had a six month mortality rate of 10%.[6] The additional risk conferred simply by the presence of syncope is controversial and has varied from no additional risk to a 30% increase in risk in cohort studies using matched controls.[6,11] For example, Kapoor et al. found that mortality was the same in patients with or without syncope and was instead dependent on the presence and type of underlying cardiac disease and other comorbidities.[11] In contrast, in a population based study from the Framingham data, syncope was associated with a 30% increase in mortality compared to patients without syncope **(Fig. 1)**.[6]

## Economic Burden of Syncope

It has been estimated that syncope accounts for 1–3% of emergency department evaluations and 1–6% of hospital admissions.[12,13] Several studies have estimated that the annual cost of management and treatment of patients with syncope ranges 1.7–2.4 billion dollars in the United States.[8,14] The cost for the management of patients with syncope varies widely,

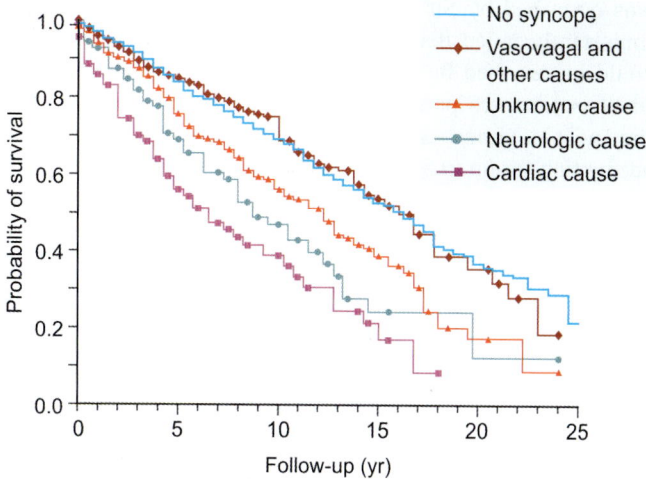

**FIGURE 1:** Survival curves from Framingham data for patients with different types of syncope. (*Source:* Soteriades ES, Evans JC, Larson MG, et al. Incidence and prognosis of syncope. N Engl J Med. 2002;347: 878-85)

dependent on the diagnostic tests ordered and whether or not an implantable cardiac rhythm device (pacemaker or ICD) is used.[8] In addition to the cost of diagnostic evaluation and specific therapies, syncope can be associated with injuries and significant psychological disability that can increase cost and have a significant impact on quality-of-life.[15–17] Major injuries, such as fractures and motor vehicle accidents were reported in about 6% of patients.[15] Minor injuries, including bruises and lacerations were reported in 27–29% of patients with syncope.[15] Elderly patients have a higher incidence of injuries when compared to younger patients, with a dramatic increase after age of 70 years, the incidence almost doubles between the sixth and the seventh decades of life.[15–17] Syncope is associated with a significant reduction in quality-of-life indices, particularly in the presence of recurrent episodes, associated comorbidities and in women.[15–17]

## Causes and Classification of Syncope

As discussed earlier, syncope is but one cause of transient loss of consciousness. Syncope can further be classified into three general causes: (i) reflex or neurally mediated syncope; (ii) syncope due to orthostatic hypotension, and (iii) cardiac syncope (**Flowchart 1**).

Reflex or neurally mediated syncope is the most common cause of syncope. One form, often called vasovagal syncope or the "common faint", is the single most common cause of syncope. The pathophysiological basis for vasovagal syncope was described by Sir Thomas Lewis in the early 1900s and importantly noted that although the bradycardia component could be reversed by atropine, hypotension and altered state of consciousness persisted. Although vasovagal syncope is the most common cause of syncope in younger patients, accounting for up to 50% of cases, it is important to recognize that even in patients more than 65 years old vasovagal syncope accounts for approximately 30% of cases. In all forms of reflex syncope, triggering of the afferent limb of a reflex arc leads to hypotension due to vasodilation (vasodepressor effect) and decreased heart rate (vagal effect). In vasovagal syncope the afferent limb can be triggered by a variety of conditions such as heat, hypovolemia, pain, or fear and anxiety while in conditions often called—situational syncope, triggering occurs from specific actions such as micturition, cough or swallow. Finally, particularly in older patients, the afferent limb can be triggered by carotid sinus stimulation.

A second cause of syncope is orthostatic hypotension. Normally with standing, accumulation of fluid in the legs results in the initiation of a complex neurologic reflex response that maintains systemic blood pressure. In orthostatic hypotension this response is insufficient, leading to a decrease in systemic blood pressure. An abnormal response is usually defined as a 20–30 mm Hg drop in systolic blood pressure on standing. Reduction in blood pressure and the increase in heart rate (which may be attenuated in older patients and in those with diabetes) are usually observed immediately after standing but can be delayed for a short period of time and thus blood pressure measurements should continue for several minutes after standing. Orthostatic hypotension can be due to a primary abnormality of the autonomic nervous system (either isolated or affecting multiple systems) or secondary (e.g. diabetes, Parkinson's disease and uremia). Diabetes is the most common cause of autonomic neuropathy in the United States and can be associated with relatively high mortality rates (25–50% mortality at 5–8 years).[18] Another form of autonomic dysfunction that can be associated with syncope is the postural orthostatic tachycardia syndrome (POTS); up to 20–30% of patients with POTS will report a prior history of syncope or presyncope.[19] Investigators believe that in POTS, hypotension on standing does not occur but maintenance of upright blood pressure requires an abnormal increase in heart rate, leading to sustained sinus tachycardia (usually >110 bpm). Although patients with POTS and orthostatic hypotension can present with symptoms of syncope they more commonly complain of symptoms such as fatigue, palpitations, rapid and pounding heart rates, and dizziness.

The third cause of syncope is a cardiac abnormality. Etiologies of cardiac syncope can broadly be divided into abnormal heart rhythms and obstruction to flow. Both rapid heart rates and slow heart rates can cause cerebral hypoperfusion and syncope. Obstruction to blood flow can be due to aortic stenosis, pulmonary valve stenosis or dynamic left ventricular outflow tract obstruction in some patients with hypertrophic cardiomyopathy (HCM). Patients with anomalous coronary arteries can sometimes present with exertional syncope, particularly in those whose right coronary artery originates from the left system with the right coronary artery passing between the aorta and pulmonary artery.[20] Large population studies have shown that of the multiple causes of syncope, a cardiac etiology carries the worst prognosis **(Fig. 1)**.[6]

## DIAGNOSTIC TESTS

The evaluation of syncope is often challenging and in up to 40% of cases no specific cause can be identified; this is specially true in older patients.[2,3,6] The history and physical examination play an essential role in evaluating patients with syncope. Several studies have attempted to evaluate the diagnostic yield of different tests in clinical practice.[2,3,21] It is important to acknowledge that the highest diagnostic yield of any test in unselected populations of patients with syncope is at best probably 25–30% with no test providing a "gold standard", underscoring the importance of careful initial evaluation by the history and the physical examination and choosing subsequent tests based on this initial assessment.[2–4]

### History and Physical Examination

The history and physical examination play an important role in establishing cause of transient loss of consciousness and, in particular, for differentiating between syncope and seizure as this distinction can be difficult **(Flowchart 1)**. For example, up to 40% of patients with syncope will have generalized seizure-like activity and myoclonic jerking due to cerebral hypoperfusion.[22] Symptoms, such as prolonged confusion after the episode (postictal state) and tongue biting, are suggestive of seizure.[23,24] An altered sense of smell, taste or an aura, such as a sense of déjà vu prior to the event are suggestive of a temporal lobe seizure.[24] Focal neurologic signs and symptoms during or after the event (Todd's palsy), also make seizure more likely than syncope. Urinary incontinence, although more commonly observed with seizures, does not completely rule out syncope. In syncopal patients, recovery is complete and often, but not always, rapid, as contrasted with seizures in which recovery is slow due to the postictal state and associated confusion. Petit mal or absence seizures are occasionally misdiagnosed as syncope. A key feature that favors the diagnosis of absence seizures is preserved postural tone despite unresponsiveness. Temporal lobe seizures can have a long duration with varied levels of consciousness. Several investigators have evaluated specific characteristics of the history for differentiating among the various causes of syncope and between syncope and seizure.[24,25] In a cohort of 539 patients a simple point score system correctly differentiated seizure from syncope in 94% of patients with a sensitivity of 94% and a specificity of 94%.[24] Tongue biting, postictal confusion, head turning to one side and prodromal déjà vu or jamais vu were more suggestive of seizures while presyncope, diaphoresis prior to the episode, or loss of consciousness associated with standing made seizure less likely.

Once the clinician has decided that an episode of transient loss of consciousness is most likely due to syncope, the history and physical examination can provide further clues as to its specific cause.[26,27] Pertinent questions should include a history prior of cardiac disease and diagnosis, if known; family history of arrhythmias, syncope and sudden death; knowledge of an abnormal electrocardiogram, medications, positional changes that occurred prior to the syncopal spell (including head-turning), prodromal symptoms, and history of prior syncopal events. Features of the clinical history that are more commonly associated with significant arrhythmias such as ventricular tachycardia or bradycardia due to advanced or complete heart block include male sex, age older than 50 years, fewer than three episodes of syncope and duration of warning prior to syncope of less than 6 seconds.[26-28] Conversely, symptoms, such as blurred vision, nausea, vomiting, warmth, diaphoresis and prolonged fatigue after syncope, have been associated with neurally mediated syncope rather than ventricular tachycardia or complete heart block. In a cohort of 341 patients, the presence of suspected cardiac disease was the strongest predictor of cardiac cause of syncope and absence of cardiac disease had a negative predictive value for a cardiac cause of 97%.[26] However, a more recent study found that the value of clinical history for distinguishing between cardiac and neurally mediated syncope was significantly reduced in older patients.[21] Finally, syncope associated with exertion has traditionally been identified as "high-risk" due to its association with valvular aortic stenosis, HCM, congenital coronary anomalies and channelopathies such as Long QT syndrome, but no large studies have been performed in these groups.[29] Using a similar approach for distinguishing between syncope and seizure, Sheldon and his coworkers identified several historical features that can help

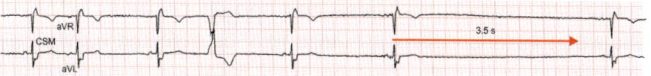

**FIGURE 2:** Simultaneously recorded leads aVR and aVF rhythm strips obtained during carotid sinus massage (CSM). Initially sinus slowing is observed and ultimately a 3.5 second sinus pause is noted. A pause with ventricular asystole more than 3 seconds, due either to sinus pause/arrest or AV block, defines carotid sinus hypersensitivity

to differentiate between vasovagal syncope and cardiac causes of syncope **(Fig. 2)**.[30,31] Symptoms such as lightheadedness associated with pain or medical settings or with prolonged sitting or standing, and a sensation of warmth or sweatiness prior to the episode made vasovagal syncope more likely. Conversely, a history of diabetes, prior arrhythmia, no recollection of the episode and palpitations preceding syncope more likely made a cardiac cause. Similarly, a first event over 35 years of age or bystanders describing the patient "turning blue" were also associated with a cardiac cause of syncope.

The physical examination has a central role in evaluation of syncope. A complete cardiac examination should be performed to assess the presence of structural cardiac disease. Orthostatic vital signs are an essential part of the physical examination in a patient. A drop of systolic blood pressure exceeding 20 mm Hg or a decrease in diastolic blood pressure of more than 10 mm Hg is considered to be diagnostic for orthostatic hypotension. The POTS is defined as an increase in heart rate more than 25–30 beats per minute within 5 minutes of standing with symptoms. Cardiac auscultatory findings, such as murmurs or gallops, are important for identifying the presence and severity of structural cardiac disease such as aortic stenosis and HCM. On palpation, findings, such as a left ventricular heave or a sustained left ventricular impulse, can alert the clinician to the presence of structural heart disease, such as left ventricular hypertrophy or ventricular dilation, due to cardiomyopathy or past myocardial infarction that will make more likely a cardiac cause of syncope.

Finally, performing carotid sinus massage (CSM) is important in patients with syncope over 40 years of age.[2] Neurologic complications during CSM occur rarely (0.17–0.45% of patients) and it is important to confirm the absence of carotid bruits or neurologic symptoms suggestive of a stroke or transient ischemic attack and before performing the maneuver.[2] Continuous electrocardiographic monitoring is essential and continuous blood pressure monitoring is highly desirable. To perform CSM, firm continuous pressure for 5–10 seconds should be applied to the right carotid artery at the level of the cricoid cartilage and after several minutes the same maneuver should be repeated on the left side. Ideally, CSM should be performed in both the supine and upright positions since an abnormal vasodepressor response may be detected only when the patient is upright. An abnormal response (carotid sinus hypersensitivity) is usually defined as ventricular asystole more than 3 seconds (due either to sinus pause or AV block) or a fall in systemic blood pressure more than 50 mm Hg **(Fig. 2)**. Formal CSM is often included as a part of the tilt-table test in some institutions.

It is important to remember that the history and physical examination play a critical role for risk stratification of the patient with syncope.[2,3] As outlined previously, patients with a cardiac cause of syncope have a far worse short-term and long-term prognosis compared to syncope due to other causes. For this reason, the history and physical examination help the clinician decide whether structural cardiac disease is present and provide an initial estimate for the likelihood of a cardiac cause for syncope.

## Blood Tests

Routine blood tests including electrolytes, tests for anemia (hematocrit or hemoglobin) and glucose, although commonly performed, generally have a low diagnostic yield in evaluation of syncope.[2,3] The exception may be the presence of anemia; anemia has been incorporated into several prognostic algorithms used for risk stratification of patients with syncope.[32,33] Frequently, patients with syncope are evaluated for myocardial infarction, even in the absence of an infarction pattern on the electrocardiogram. Despite this approach, the yield of diagnostic evaluation for myocardial infarction for patients with syncope is less than 1%.[34] In a small study investigators found that higher brain natriuretic peptide (BNP) levels could be used to identify patients with syncope due to cardiac causes or worse outcomes, and a recently published risk stratification schema for patients with syncope used a BNP more than 300 pg/mL as one criteria for identifying high-risk patients.[33,35] It is important to note that the most recent guidelines do not specifically recommend any blood tests for the evaluation of syncope.[2,3]

## Electrocardiogram

A 12 lead ECG is a basic part of the workup in all patients with syncope.[2,3] Although the diagnostic yield of a baseline ECG is low (5–10%), it is an inexpensive and widely available test that can be used to quickly risk stratify patients, particularly if it is abnormal.[2,3,29] Baseline sinus bradycardia, atrioventricular block and intraventricular conduction block (left or right bundle branch block) suggest the possibility of bradycardia as a cause for syncope. Presence of Q waves that suggest the possibility of a prior myocardial infarction or other findings, such as left ventricular hypertrophy make structural heart disease, and thus a cardiac cause of syncope, more likely. The presence of premature ventricular depolarizations may have some prognostic information in patients with syncope. In an older study of 235 patients with syncope, the presence of frequent or paired premature ventricular contractions was associated with higher mortality and risk of sudden death.[36]

Finally, there are some ECG patterns that can be used to identify potential causes for syncope: Wolff-Parkinson-White syndrome, Long QT syndrome, Brugada syndrome, arrhythmogenic right ventricular cardiomyopathy and HCM **(Figs 3 to 7A and B)**.

## Echocardiography

Both the AHA/ACCF and ESC guidelines state that the echocardiogram plays a central role in syncopal patients with suspected cardiac disease.[2,3] In a review of over 2,000 elderly patients admitted for syncope at a single center, echocardiograms were obtained in 40% of patients and abnormalities were identified in almost 70%.[21] However, results from the echocardiogram affected management in less than 5% of the cases.[21]

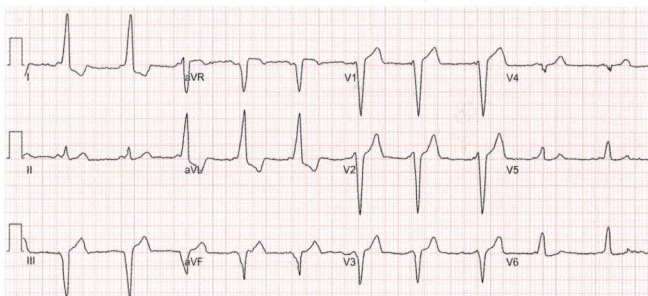

**FIGURE 3:** ECG from a patient with Wolff-Parkinson-White syndrome due to the presence of a right freewall accessory pathway. Preexcitation of the right ventricle leads to a short PR interval and a QRS complex with a left bundle branch block morphology and a negative delta wave in aVR [*Source:* Kusumoto FM. ECG Interpretation. Pathophysiology to Clinical Application. New York: Blackwell-Springer; 2009 (with permission)]

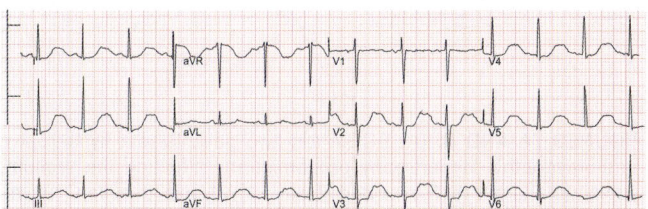

**FIGURE 4:** ECG from a patient with syncope and long QT syndrome (*Source:* Kusumoto FM. ECG Interpretation. Pathophysiology to Clinical Application. New York: Blackwell-Springer; 2009)

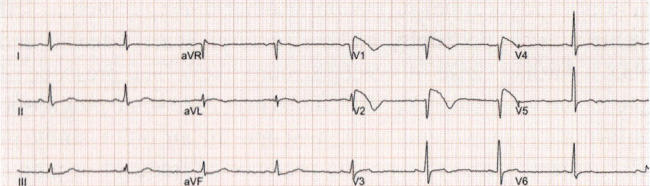

**FIGURE 5:** ECG from a patient with Brugada syndrome showing the characteristic type I right bundle branch block pattern and downsloping ST segment elevation in $V_1$ and $V_2$. Type I Brugada syndrome is more specific than the "saddle back" ST segment contour in the type II pattern. ECG patterns in Brugada syndrome can be quite variable, even over short periods of time; questionable diagnoses can be clarified during the intravenous infusion of a sodium-channel blocking drug (*Source:* Kusumoto FM. ECG Interpretation. Pathophysiology to Clinical Application. New York: Blackwell-Springer; 2009)

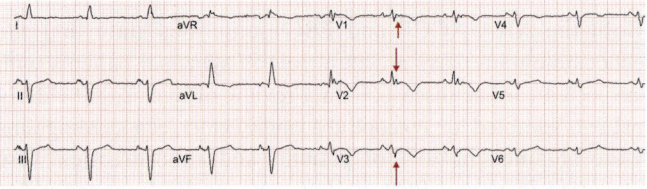

**FIGURE 6:** ECG from a patient with arrhythmogenic right ventricular cardiomyopathy. There are prominent anterior forces and precordial T wave inversion is present in addition to the highly specific epsilon waves (arrows) that represent delayed conduction in the right ventricle. (*Source:* Kusumoto FM. ECG Interpretation. Pathophysiology to Clinical Application. New York: Blackwell-Springer; 2009)

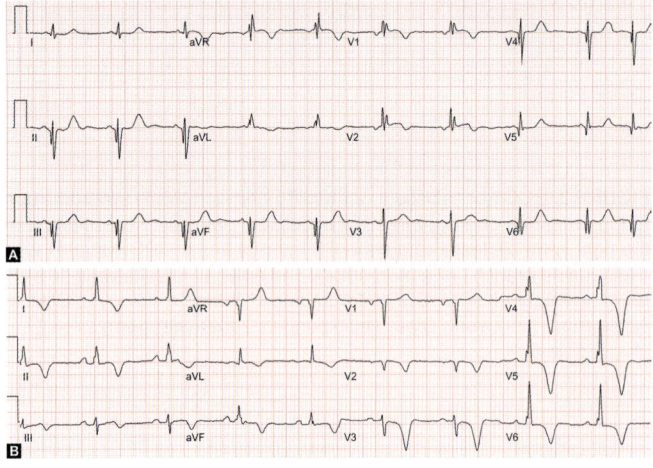

**FIGURES 7A AND B:** (A) ECGs from two patients with hypertrophic cardiomyopathy. Hypertrophy predominantly affecting the septum, leading to a larger than expected R wave in V1 and "pseudo Q waves in the inferolateral leads. (B) Hypertrophy affecting the cardiac apex, leading to deep lateral T wave inversions (*Source:* Kusumoto FM. ECG Interpretation. Pathophysiology to Clinical Application. New york: Blackwell-Springer; 2009)

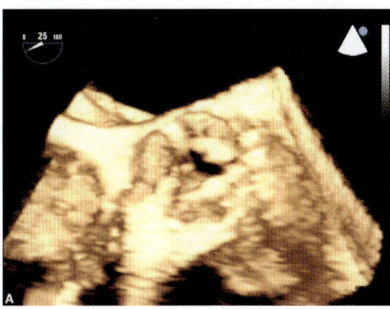
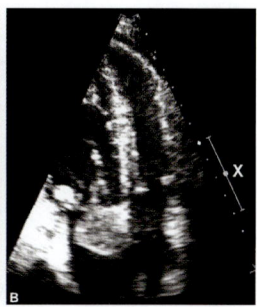

**FIGURES 8A AND B:** (A) Diagnostic echocardiographic images in patients that presented with syncope. Three-dimensional transesophageal echocardiographic image of a patient with severe aortic stenosis. (B) Four chamber transthoracic echocardiographic image of a sessile left atrial myxoma attached to the interatrial septum. (*Source:* Emery Kapples, Jeannine Hiers and Carolyn Landolfo)

Echocardiography has a low diagnostic yield in patients with a normal physical examination and normal ECG and need not necessarily be obtained in all patients with syncope. A structural abnormality noted during echocardiography does not per se establish a diagnostic cause for syncope. However, the presence of severe aortic stenosis or rarer conditions, such as atrial myxoma, is usually diagnostic of etiology **(Figs 8A and B)**.

## Exercise Testing

Exercise testing has a low diagnostic yield in the evaluation of syncope (<5%).[29] However, it may particularly be useful in those patients with exertional syncope. Published guidelines are not uniform in their recommendations; however, the AHA/ACCF scientific statement (but not the ESC guidelines) suggest that exercise testing should be more widely applied to any patient with unexplained syncope, particularly those with coronary artery disease or those at risk for coronary artery disease.

Perhaps even more useful than the identification of ischemia in the patient with syncope is the evaluation of hemodynamic and heart rhythm responses to exercise testing. Development of atrioventricular block with exercise is always abnormal and suggests bradycardia as the mechanism for symptoms. An abnormal decrease in blood pressure during or after exercise may be an important clue for the mechanism of exertional syncope in a patient with HCM.

## Continuous ECG Monitoring

### External Devices (24 Hours Ambulatory ECG Recorders, Event Recorders)

Since intermittent bradycardia or tachycardia are the most common cardiac etiologies for syncope, an ECG obtained

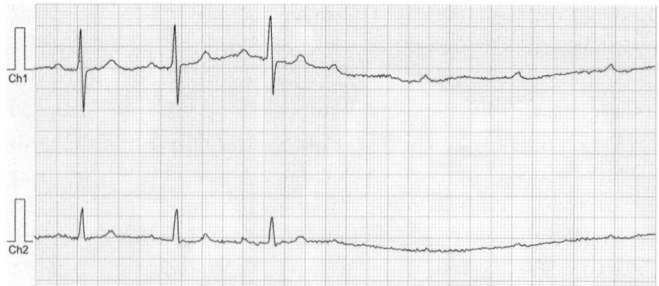

**FIGURE 9:** Simultaneously recorded rhythm strips from an external ECG event monitor with looping memory documenting intermittent high grade AV block associated with symptoms. The patient had a prior positive tilt-table test for a diagnosis of vasovagal syncope, but due to continued symptoms despite medical treatment underwent further diagnostic testing for arrhythmias. A permanent pacemaker was placed with resolution of symptoms

when the patient is having symptoms is critical for determining whether an arrhythmia is the cause of symptoms. Although asymptomatic arrhythmias can be helpful in suggesting a possible mechanism in some settings, the diagnostic yield of short periods of rhythm monitoring, whether telemetry monitoring during a hospital admission or traditional 24–48 hours ambulatory ECG (Holter) monitoring, is very low.[29] Since most episodes of syncope are usually separated by long periods of time the yield of monitoring less than 48 hours is at best 1–2%.[29]

External recorders that have a loop memory that continuously acquires and deletes ECG information can provide longer periods of rhythm monitoring but studies have provided conflicting reports on the utility of these devices due to the sporadic nature of syncope events. An external event recorder is an attractive diagnostic test for patients with near syncope that occurs frequently (e.g. weekly) **(Fig. 9)**. Event monitors that do not provide continuous ECG monitoring and instead are applied by the patient to the chest when symptoms occur may be useful in the patient with palpitations or dizziness, but are of little use in the diagnostic evaluation of syncope due to patient incapacitation during the episodes.

## Implantable Loop Recorders

More recently implantable loop recorders (ILRs) that are placed subcutaneously in the left upper chest and that have larger memories and the ability to continuously monitor the ECG for more than 1 year have been developed by several manufacturers. These small devices have a battery life that lasts for 18–24 months. Some clinicians use a program system analyzer to optimize placement of the device to obtain good cardiac signals. The two electrodes used for recording cardiac electrical activity are usually placed on either end of a rectangular shaped device.

Once implanted the ILR will record tachycardia or bradycardia using rate parameters defined and programmed by the clinician. The ILR can also be manually activated with a hand-held activator by a bystander or by the patient after the episode. The ILR is usually programmed to save data for a prespecifed time (e.g. 5 minutes) prior to manual activation. Several studies have documented the usefulness of the ILR for evaluating patients with syncope.[2,37–39] In the largest study to date, of 392 patients with syncope who underwent placement of an ILR, 103 (26%) had recurrent syncope and of these, 53 received specific therapy based on the findings recorded on the ILR (usually bradycardia requiring implantation of a permanent pacemaker).[39] Patients that received specific therapy based on the ILR results had a significant reduction in the incidence of recurrent syncope (10% vs 41%).[39] Importantly, approximately 30–40% of patients will have a recurrent episode of syncope not associated with an arrhythmia, and while this finding does not allow definitive therapy it does essentially rule out arrhythmia as a cause for the patient's symptoms and can be reassuring to the patient.[40]

Current guidelines recommend an ILR for patients with recurrent but infrequent episodes of syncope in whom there is a high index of suspicion for an arrhythmogenic cause after a negative initial workup.[2,3] The ILR has gradually supplanted invasive electrophysiology studies and tilt-table testing as the diagnostic test of choice for patients with syncope. The ILR is ideal for obtaining a heart rhythm correlation for a patient with intermittent symptoms. As discussed above, the most common abnormal finding recorded by the ILR is transient bradycardia, although supraventricular and ventricular tachycardias **(Fig. 10)** can sometimes be observed. In addition, identifying normal heart rates during an episode of syncope is extremely useful as this finding essentially rules out a primary arrhythmia mechanism for syncope.

## Signal Averaged ECG

In some patients with structural heart disease (cardiomyopathy or prior myocardial infarction), low amplitude signals in the terminal portion of the QRS complexes can sometimes be observed using special recording techniques that obtain a number of QRS complexes (allowing random noise to cancel out) and use special filtering algorithms. Late potentials are thought to arise from delayed depolarization of the abnormal myocardium and thus reflect nonhomogeneous depolarization. The signal averaged ECG (SAECG) may be useful in rare

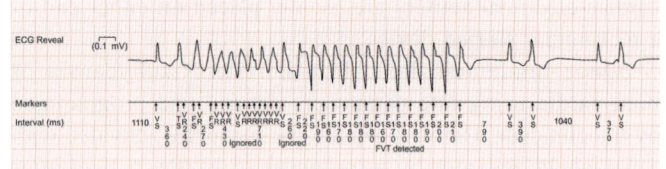
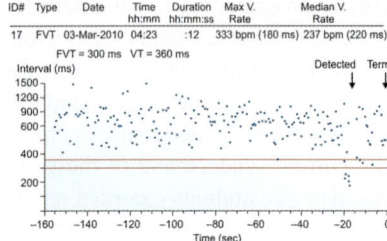

**FIGURE 10:** Tracings from an ILR showing nonsustained polymorphic ventricular tachycardia that was recorded during sleep (04:23). The device was set to automatically capture rapid heart rates (FVT: Fast ventricular tachycardia, in this case defined as a rhythm with a cycle length <300 ms). Actual electrograms (Top) and histograms (Bottom) from the event can be obtained. The histogram shows sudden onset of rapid ventricular activity (FS: fibrillation sense) separated by short cycle lengths. The rhythm spontaneously terminates with normal ventricular signal and bigeminy (VS: ventricular sense). Although this abnormal rhythm was not recorded during symptoms results from the ILR suggest ventricular arrhythmias as the cause for the patient's symptoms

circumstances, for example in some cases the SAECG can help to identify patients with arrhythmogenic right ventricular cardiomyopathy.[41] In these patients, fatty infiltration of the right ventricle leads to an abnormal SAECG recording that appears to correlate with myocardial fibrosis obtained by biopsy. Other preliminary data suggest that the SAECG may also be useful in identifying patients with Brugada syndrome who are at higher risk for ventricular arrhythmias.[42] In general, however, the SAECG provides little additional diagnostic information and is not routinely used.

## Upright Tilt-table Testing

Upright tilt-table testing is commonly obtained in the diagnostic workup of syncope.[2,43] The physiology of standing is complex, but when the patient is moved from the supine to the upright position, approximately 300–600 mL of blood pools in the lower extremities and lower portion of the abdomen, which in turn leads to a 25–50% decrease in intravascular volume.[43] In response to the decrease in stroke volume a complex interplay of various cardioregulatory systems normally results in maintenance of blood pressure despite the redistribution of blood.

Upright tilt-table testing was first applied in clinical medicine 25 years ago as a method for evaluating a patient's hemodynamic response to orthostatic stress and identifying patients likely to develop vasovagal syncope.[44] Several protocols have been developed and there is no uniformity of approach, but generally the patient is positioned on a table at an angle of 60–70 degrees for 30–45 minutes, with only foot support. Some protocols use isoproterenol infusion or nitroglycerin to increase the likelihood of eliciting a vasovagal response. The tilt-table test is used to quantify orthostatic hypotension and to attempt to induce a vasovagal response. In those patients who have a vasovagal response, hemodynamic monitoring during the test can quantify the relative and absolute changes in blood pressure and heart rate. Several different hemodynamic responses to tilt-table testing can be observed in **(Fig. 11)**:

1. The *normal response* consists of an increase in heart rate of approximately 10–15 beats per minute, an elevation of diastolic blood pressure of about 10 mm Hg, and little change in systolic blood pressure.
2. *Orthostatic/POTS response*: Orthostatic hypotension is defined as a reduction in systolic blood pressure of at least 20 mm Hg or a reduction in diastolic blood pressure of at least 10 mm Hg. The POTS pattern consists of a sustained increase in heart rate of at least 30 beats/min or a sustained pulse rate of 120 beats per minute with no profound hypotension. Both of these responses are usually observed within the first 5–15 minutes of tilting; however, in older patients, the orthostatic response may be delayed.[45]
3. *Neurocardiogenic response*: Initially, blood pressure and heart rate remain stable. However, after 10–20 minutes a sudden decrease in blood pressure and heart rate will be observed. Some investigators further divide this response into primary vagal, primary vasodepressor or a mixed

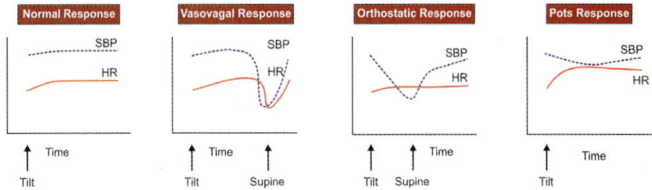

**FIGURE 11:** Tilt-table hemodynamic responses in different conditions. Normally in tilt-table testing, with the upright position normal baroceptor function maintains a relatively constant heart rate (HR) and systolic blood pressure (SBP). In the vasovagal response after a relatively long period of time a drop in SBP rapidly followed by a drop in HR is observed. In the orthostatic response, with standing an almost immediate but gradual drop in SBP is observed and HR remains unchanged. In the postural orthostatic tachycardia syndrome (POTS) response SBP is maintained by a significant increase in HR

response, depending on the relative magnitude of blood pressure and heart rate changes.

4. Some investigators also classify as a response of *psychogenic reaction* in which patients develop symptoms with no changes in heart rate or blood pressure.

In patients referred for evaluation for syncope, the most commonly observed abnormal finding during tilt-table testing is a mixed form of the neurocardiogenic response (35–45%) in all age groups. The second most common response is age dependent: a bradycardia response observed in patients less than 35 years of age and a pure vasodepressor response in older patients.[46]

Although tilt-table testing has been useful as an experimental test for providing physiologic data in patients with vasovagal syncope and orthostatic hypotension and is widely performed, its clinical application has not been well defined. First, it is important to note that the tilt-table test has poor reproducibility. When patients with a positive tilt-table test are subsequently reevaluated, approximately 50% will have a negative test regardless of whether they were treated or not.[47,48] Second, several studies have found that the likelihood of recurrent episodes of syncope was similar in patients with a positive response and a negative response. Third, abnormalities identified by tilt-table testing do not predict the likelihood of bradycardia events that are documented by stored ILR data.[49] Despite these shortcomings tilt-table testing may be useful for evaluating some patients with syncope, particularly those with orthostatic hypotension. The ESC has developed detailed guidelines for the methodology, indications and diagnostic criteria for tilt-table testing.[2] In general, tilt-table testing is recommended when it is important to identify whether the patient is susceptible to vasovagal syncope (e.g. a patient with a structurally normal heart that has a single episode of syncope associated with significant injury) or to help differentiate between reflex syncope from orthostatic hypotension. The tilt-table test has been shown to have little use for guiding therapy or as a follow-up tool.[50]

## Electrophysiology Study

The electrophysiology study (EPS) is an invasive test that may be useful for workup of syncope in selected patients.[29] In EPS, using specialized electrode catheters placed in the heart, the clinician can define cardiac electrophysiologic properties, such as sinus node and AV node function, and evaluate the mechanism for any inducible ventricular tachycardia or supraventricular tachycardia under controlled conditions.

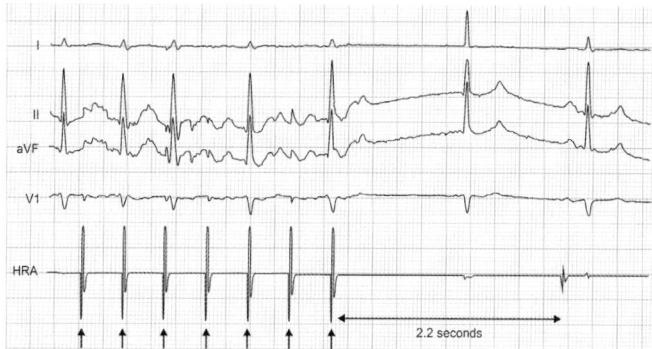

**FIGURE 12:** Abnormal sinus node recovery time (SNRT). The atria are paced (arrows) at 150 bpm for 30 seconds. Notice the patient has AV block during atrial pacing (this would be expected due to normal decremental conduction properties of the AV node). With cessation of pacing a junctional beat occurs but it takes 2.2 seconds for sinus node activity to return. (*Source:* Kusumoto FM, Goldschlager N. Cardiac Pacing for the Clinician, 2nd edition. New York: Springer; 2008. pp 647-94)

Bradycardia can be due to siunus node dysfunction and/or atrioventricular conduction abnormalities. There are several parameters used in EPS for evaluation of the sinus node. The most commonly used parameter is the sinus node recovery time (SNRT). To measure the SNRT, atrial pacing is performed for 30 seconds and the sinus node response on cessation of pacing is evaluated. An abnormal SNRT is shown in **Figure 12**. In this case, sinus node activation is observed 2.2 seconds after cessation of pacing. Unfortunately, parameters for sinus node dysfunction have highly variable sensitivity (25–70%) and specificity (45–100%) for the clinical diagnosis. The EPS is more useful for evaluation of atrioventricular function and can be used to determine the site of AV conduction block **(Fig. 13)**. Atrioventricular block that develops at or below the level of the His bundle (infra-Hisian block) portends a poor prognosis since intrinsic pacemaker activity of ventricular tissue is not only slow and unresponsive to autonomic influences, but is also notoriously unreliable, even in the short term. A baseline His-to-ventricular (HV) interval more than 100 ms or prolongation of the HV interval to more than 100 ms with procainamide stress has also been shown to be useful for identifying patients at high risk for the development of syncope due to bradycardia. It is important to acknowledge that EPS is useful for defining the site of block in the presence of fixed block but is not useful for evaluating the patient that develops intermittent atrioventricular block at the level of the AV node.

Tachycardia accounts for approximately 15–25% of patients with syncope.[51,52] The EPS may be useful for identifying supraventricular tachycardia but usually this diagnosis is made

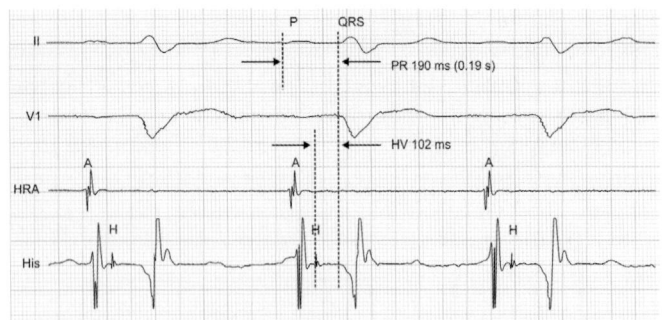

**FIGURE 13:** Electrograms demonstrating significant prolongation of the HV interval (102 ms) in the setting of a PR interval at the upper limit of normal (190 ms) that suggests infra-Hisian disease. (Abbreviations: H: His bundle electrogram; A: Atrial electrogram). (*Source:* Kusumoto FM. Understanding Intracardiac EGMs and ECGs, 1st edition. Hoboken New Jerse: Wiley-Blackwell; 2010)

with continuous ECG monitoring since any arrhythmia induced at EPS may not represent a clinically relevant arrhythmia or the mechanism for syncope. Traditionally, the utility of EPS in the patient with syncope has focused on induction of ventricular tachycardia. In patients with myocardial scars (due to past myocardial infarction or any process that produces myocardial fibrosis) and ventricular tachycardia due to reentry utilizing slowly conducting channels within the scar, programmed stimulation of the ventricle during EPS can be used to assess risk for future ventricular arrhythmias. Although protocols vary among institutions, generally pacing is performed from two sites in the right ventricle delivering one, two or three extrastimuli after a basic pacing train of eight beats that ensures uniform capture of ventricular tissue. Data from older studies suggest that EPS can be useful for evaluating risk of sudden death in patients with syncope and a prior myocardial infarction.[53,54]

A decade ago, EPS and ventricular stimulation protocols for induction of ventricular tachycardia played a central role in the diagnostic evaluation of syncope in patients with structural heart disease. Since then several important trends have relegated EPS to only occasional use in certain patient groups with syncope. First, several landmark studies have shown that many patients with structural heart disease will receive a mortality benefit from an empiric implantable cardiac defibrillator (ICD) irrespective of the presence or absence of syncope. For example, many patients with an ejection fraction less than 30% due to prior myocardial infarction or an ejection fraction less than 35% in the presence of heart failure symptoms are candidates for ICD placement whether or not they have syncope.[55,56] Second, the ILR appears to be an excellent option for many patients with structural heart disease and syncope who are not candidates for empiric ICD implantation based on ejection fraction and heart failure symptoms alone.[51,52,57] Currently, EPS is reasonable for evaluating patients with syncope and coronary artery disease with prior myocardial infarction that do not meet criteria for an ICD implant or those patients that meet criteria for ICD implant, but where further risk stratification information might change a clinical decision (usually whether or not to implant an ICD). The EPS is also reasonable for the patient with syncope and evidence for abnormal atrioventricular conduction where defining the site of block will impact clinical decision-making. For example, if infra-Hisian block was found at EPS a permanent pacemaker would be implanted.

## Cardiac Catheterization

Cardiac catheterization is generally not indicated for the workup of syncope unless accompanied by symptoms suggestive of significant coronary artery disease.

## Neurologic Tests

Computed tomography (CT) scans, electroencephalography (EEG) and carotid duplex scans are often obtained for the evaluation of patients with transient loss of consciousness. Multiple studies have shown that the diagnostic yield of these tests is extremely low (1–3%) in unselected patient populations.[2,3,29] It is recommended that these tests should be ordered only if indicated by clinical findings that specifically suggest a neurologic process.

## APPROACH TO THE EVALUATION OF SYNCOPE

As outlined in the preceding sections, there are many tests available for the assessment of syncope and indiscriminate use of diagnostic tests can lead to an expensive evaluation that provides little insight into the management of the patient. Although it is difficult to provide a "one size fits all" algorithm for managing patients with syncope some general guidelines are useful. Recently published guidelines emphasize the importance of the history and a comprehensive physical examination in the initial evaluation of syncope and also recommend a baseline ECG. Since future risk is largely dependent on whether the patient has a cardiac cause of syncope and whether the patient has structural heart disease most diagnostic and risk-stratification algorithms use this issue as the first decision point **(Flowchart 2)**.

Patients with no history, physical examination or ECG findings suggestive of cardiac disease have a fairly low likelihood of a significant cardiac cause for syncope. If the history is suggestive of vasovagal syncope no further evaluation will

**FLOWCHART 2:** Diagnostic evaluation of a patient with syncope

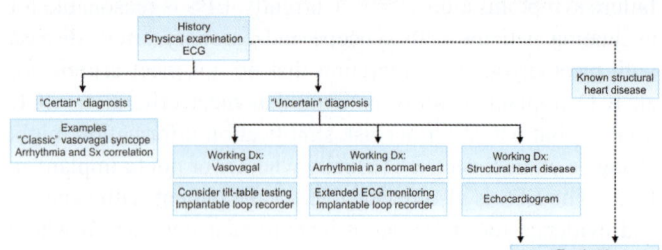

be required in many patients. Similarly if the patient presents to the emergency department with symptoms that correlate with an arrhythmia then specific treatment can be initiated. However, even after a comprehensive initial evaluation, the clinician may be unsure of the mechanism of syncope. If the clinician is confident that the patient has no cardiac disease, but is uncertain of whether an arrhythmia is present extended ECG monitoring will be useful. Particularly, in patients with syncope associated with injury, an external event recorder or an ILR may be appropriate depending on the frequency of symptoms. If the patient does not have structural heart disease and the clinician is unsure of the patient's hemodynamic response to orthostatic stress, tilt-table testing may be a reasonable next test. Tilt-table testing may be helpful particularly for identifying patients with orthostatic hypotension or POTS.

When the clinician is uncertain as to whether structural heart disease is present an echocardiogram can be extremely useful for obtaining information on cardiac anatomy and function. An exercise test may be useful in selected patients with exertional syncope. Cardiac tests, such as 24-hour ambulatory ECG monitoring (duration of evaluation is too short), cardiac enzymes, electrophysiologic tests and cardiac catheterization, and neurologic tests, such as carotid ultrasound, CT scan and EEG, have very little utility.

In the patient with structural heart disease identified by history, physical examination and ECG the appropriate workup will depend on the type of disease present **(Flowchart 3)**. Issues with specific cardiac conditions are described in the following section, but several general comments can be made. Patients with structural heart disease can often have vasovagal syncope but the physician should have a low threshold for further cardiac evaluation. Patients at high risk for ventricular arrhythmias (prior myocardial infarction and EF < 30%, EF < 30–35% with Class II or III heart failure symptoms) can be referred directly for an ICD. For patients with syncope who have coronary artery disease and prior myocardial infarction associated with wall motion abnormalities, EPS may be useful for determining

**FLOWCHART 3:** Diagnostic and therapeutic considerations in a patient with syncope and structural heart disease

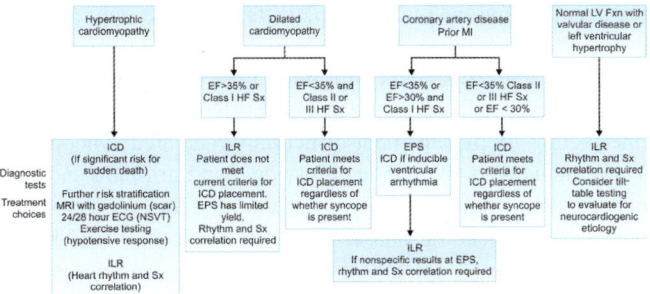

*Abbreviations:* ICD: Implantable cardiac defibrillator; ILR: Implantable loop recorder; EF: Ejection fraction)

whether ventricular arrhythmias can be induced with premature ventricular stimulation. The EPS may also be useful for patients with atrioventricular conduction disease identified on ECG to assess the site of block and therefore prognosis and management strategies. Often, an ILR to evaluate the cardiac rhythm during a subsequent episode of syncope will be the most useful diagnostic test for determining the cause of syncope.

## SPECIFIC PATIENT GROUPS

### Vasovagal (Neurocardiogenic) Syncope

In vasovagal syncope there are widely spaced episodes of a temporary loss of consciousness associated with a fall in arterial blood pressure followed by an almost instantaneous profound slowing of the heart rate. Neurocardiogenic fainting usually occurs in a standing position and is triggered by stressful conditions or pain. The onset may be abrupt or associated with warning symptoms, such as fatigue, weakness, nausea, sweating, pallor, visual disturbances, abdominal discomfort, headache, pins-and-needles sensations and feelings of depersonalization, lightheadedness or vertigo. Vasovagal syncope is the most common etiology of syncope regardless of the population studied. The diagnosis of vasovagal syncope is generally made by the history and physical examination although tilt table testing may be necessary in some cases to provide confirmatory evidence in some patients; this may especially be the case in patients who are amnesic for the syncopal spell and who cannot therefore provide a sufficiently detailed history.

Treatment of vasovagal syncope, particularly in patients with frequent symptoms, can be challenging due to the sporadic nature of symptoms and the absence of therapeutic options that have been validated by large clinical trials. As emphasized by the 2009 European Society Guidelines, explanation of the diagnosis and counseling and reassurance of the patient remain the cornerstone of treatment for patients with neurally mediated syncope.[2] About one-third of patients will have recurrent symptoms but many patients will have only a single event. Recurrent episodes of vasovagal syncope are more likely in women and in patients with more than 3 prior episodes.[58]

Physical counter pressure maneuvers have emerged as a first line therapy in management of neurally mediated syncope.[2,59] These maneuvers include leg crossing, hand gripping and arm or buttock muscle tensing in an effort to raise the blood pressure during the impending phase. A recent multicenter trial found that training in counter pressure maneuvers was associated with a 40% decrease in the likelihood of recurrent syncope. Similarly, "tilt training" or "standing training" is a management option in a patient who is educated and highly motivated.[60] Patients are asked to stand approximately 10–15 cm from a wall (to reduce the likelihood of significant injury in case of a fall) for gradually longer periods, usually 3–5 minutes twice daily initially, increasing to 30 minutes twice daily over time. This form of training improves tolerance to standing although long-term compliance with the training regimen can sometimes be a limiting factor in successful treatment.

Many drugs, including beta blockers, selective serotonin receptor inhibitors (SSRI), disopyramide, theophylline, scopolamine, ephedrine, midodrine, clonidine and other medications have been tried in treatment of this condition.[47,61–64] Although small studies and trials have been published that suggest benefit from therapy, the intermittent and inconsistent occurrence of symptoms make evaluation of treatment extremely difficult. In addition, the pathophysiology-triggers, relative degrees/importance of bradycardia and hypotension are probably extremely heterogeneous and thus it is not surprising that individual responses vary markedly. Beta blockade, although traditionally popular, has recently been shown in a randomized multicenter trial to have no clinical benefit.[61] In the prevention of syncope trial (POST), 208 patients with vasovagal syncope were randomized to metoprolol or placebo, and after one year follow-up there were no differences in symptoms or quality-of-life detected between the two groups.[61] At least in part due to these results, beta blockade is no longer considered a frontline therapy for neurocardiogenic syncope. Vasoconstrictors, such as the alpha-agonist midodrine, have been used for treatment of neurocardiogenic syncope in an effort to treat the hypotension associated with the episodes. In one placebo controlled study, 80% of patients randomized to midodrine did not have recurrent symptoms at one year follow-up compared to 13% in the placebo group.[62] In general, midodrine was well tolerated although it should be noted that older patients who may develop hypertension while taking midodrine were not evaluated. Since patients may identify periods when they are more likely to develop symptoms, midodrine has also been administered as a "pill in pocket" strategy in certain patients who are educated and motivated. Not all studies using vasoconstriction for treating vasovagal syncope have been successful. In the vasovagal syncope international study (VASIS), etilefrine, another alpha agonist, was studied in patients with vasovagal syncope.[47] One hundred twenty six patients were randomized to oral etilefrine or placebo and after one year follow-up syncope occurred in 22–24% of patients, without a difference between the two treatment arms and no change in the time to first occurrence of syncope.

Almost two decades ago, permanent cardiac pacing was proposed as a potential treatment option for patients with neurally mediated syncope associated with significant bradycardia.[65,66] Initial nonrandomized trials suggested an important effect of pacing for reducing episodes of syncope.[65,66] Subsequent placebo controlled studies, however, suggested that there was a significant placebo effect associated with pacing therapy and that therefore pacing therapy per se could not be shown to have a beneficial effect.[67,68] The most recent study, the International Study of Syncope of Uncertain Etiology (ISSUE)-2 evaluated whether significant bradycardia identified by ILRs could be used to better identify patients that could benefit from pacing therapy.[69] Interestingly, 53 patients that received pacemakers due to bradycardia identified by ILR reported a statistically significant 41% decrease in recurrent syncope compared to those patients that did not receive an ILR (and consequently did not receive a permanent pacemaker). A large randomized trial (ISSUE-3) has been initiated to validate this diagnostic and therapeutic strategy. At this time, pacing therapy plays a small role in management and only in selected patients with neurocardiogenic syncope who have frequent recurrent symptoms primarily associated with bradycardia or asystolic pauses in rhythm.

## Hypertrophic Cardiomyopathy

Hypertrophic cardiomyopathy (HCM) is a diverse genetic disorder that often affects proteins in the sarcomere and that is associated with left ventricular hypertrophy. In a small percentage of patients the interventricular septum is preferentially affected and during systole a dynamic gradient in the left ventricular outflow tract can be observed (hypertrophic obstructive cardiomyopathy or HOCM). Several cohort studies have found that the occurrence of syncope is a major risk factor for sudden death with a 1.7–5-fold increase in risk.[70–72] Although ventricular arrhythmias are the most concerning possibility for the cause of syncope in these patients, they can have syncope from many other mechanisms, including supraventricular arrhythmias (particularly atrial fibrillation with loss of atrial contraction and rapid ventricular rates), bradycardia, left ventricular outflow tract obstruction and abnormal reflex peripheral blood pressure responses (hypotension) due to stimulation of pressure receptors in the body of the left ventricle. In the largest study to date, in 1,511 patients with HCM followed for more than 5 years, syncope occurred in 205 (14%), of these, 52 had symptoms suggestive of a neurally mediated episode (episode associated with a trigger such as coughing, micturition or change in position) and 153 patients had "unexplained" syncope.[73] Risk of sudden death was 5-fold higher in patients with syncope within 6 months of their evaluation; conversely, older patients (>40 years old) and an episode of syncope more than 5 years before the initial evaluation were not found to be at increased risk of sudden cardiac death.

Collectively, the data from published studies suggest that while syncope is an important symptom that may herald an increased risk of sudden death, it requires thoughtful clinical evaluation of the patient. Family history of sudden cardiac death (first degree relative with sudden death before age 50), documented nonsustained ventricular tachycardia and degree of left ventricular hypertrophy (>3 cm) have been identified as risk factors for sudden death. Emerging risk factors include a hypotensive response after exercise and late gadolinium enhancement on magnetic resonance imaging. Although some have advocated ICD placement in HCM patients with multiple risk factors, one cohort study found similar rates of appropriate ICD therapy in patients with 1, 2 or 3 risk factors.[72] For patients who do not receive an ICD, an ILR may be a reasonable diagnostic option.[53,71,74]

## Nonischemic Cardiomyopathy

For many years the presence of syncope in a patient with nonischemic cardiomyopathy has been considered an ominous sign, associated with increased risk of sudden cardiac death due to ventricular arrhythmias.[75,76] In the largest cohort to date, 26% of 108 patients with nonischemic cardiomyopathy and syncope had significant ventricular arrhythmias during follow-up, a rate that was not statistically different from a comparison group of patients that presented with sustained ventricular arrhythmias.[76] Post-hoc analysis from two of the large ICD trials suggests that syncope in patients with nonischemic cardiomyopathy can have multiple mechanisms other than ventricular arrhythmias. In the defibrillators in nonischemic cardiomyopathy treatment evaluation (DEFINITE) Trial, 458 patients with nonischemic cardiomyopathy were randomized to receive standard medical therapy or standard medical therapy and an ICD.[77] After randomization, there was no significant difference for the development of syncope between the two groups (standard therapy: 34% vs standard therapy + ICD: 39%). Of the patients in the ICD arm that had syncope, two-thirds were not associated with delivery of ICD therapy, suggesting a mechanism other than ventricular arrhythmias for the symptoms. The sudden cardiac death heart failure trial (SCD-HeFT) randomized patients with heart failure symptoms and reduced ejection fraction less than 35% (approximately 50% were nonischemic) to placebo, ICD or amiodarone.[78] Regardless of the treatment arm, syncope occurred in approximately 14% of patients after randomization. Although syncope was associated with appropriate ICD therapy in patients randomized to the ICD arm, total mortality was increased by 40% equally in all three arms. Ventricular arrhythmias were thought to be the presumptive cause of syncope in less than 15% of cases. Syncope may be an important symptom for increased risk of sudden death in patients with nonischemic cardiomyopathy but may be due to mechanisms other than ventricular arrhythmias and may identify a group of patients at higher risk for total mortality.

## Congenital Heart Disease

Approximately one million adults in the United States have congenital heart disease. Depending on the abnormality, patients with congenital heart disease are at higher risk for different arrhythmias. For example, significant arrhythmias will develop in approximately 80% of patients with D-transposition of the great arteries (dTGA) who have undergone an atrial switch repair (Mustard or Senning) **(Fig. 14)**.[79–85] Sinus node dysfunction (20–40%) that often requires permanent pacing

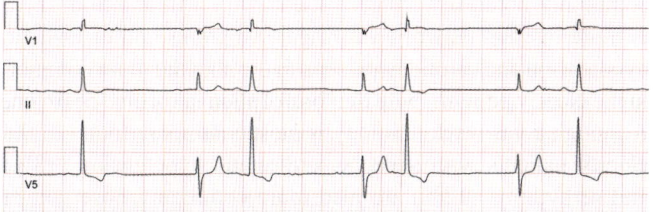

**FIGURE 14:** Three lead ECG rhythm strip from a syncopal patient with d-transposition of the great arteries who had undergone a Mustard procedure. Severe sinus node dysfunction is present with a slow sinus rate. In addition a prominent R wave diagnostic for right ventricular hypertrophy is observed in lead $V_1$. Cardiac rhythm causes for syncope in this patient include bradycardia due to sinus node dysfunction, ventricular tachycardia due to right ventricular dilation and, less likely, atrial flutter

is very common as are development of atrial tachycardias (4–20%).[79–81] Atrioventricular block and ventricular tachycardia (due to right ventricular dysfunction and fibrosis that develop due to long-term contraction against systemic pressures) have all been described as causes of syncope. In one long-term follow-up study, syncope developed in 6% of patients. In addition, sudden death occurred in approximately 7% of patients between 6–19 years after repair.[82] In a recently published multicenter study of a 149 patient cohort, sudden death and sustained ventricular arrhythmias occurred in approximately 9% of patients.[83] A QRS duration more than 140 ms was associated with a 14-fold increase in risk of ventricular arrhythmias. Atrial tachycardias were present in 44% of patients but was not a significant predictor for risk of ventricular arrhythmias or sudden death. Finally, in a cohort study of 37 patients with d-TGA who received ICDs, the annual rate of appropriate shocks was 0.5% for primary prevention and 6.0% for secondary prevention.[84] Taken together the data illustrate the complexity of managing adults with repaired congenital heart disease who present with syncope.

## Elderly Patients

Syncope in the elderly patients can particularly be difficult from both a diagnostic and therapeutic viewpoint. First, there is significant overlap between syncope and "unexplained falls" and many older patients have limited recall of events surrounding the symptoms.[84] In studies that have specifically evaluated syncope in the elderly patients, in approximately 50% the mechanism cannot be determined, although neurally mediated syncope was, somewhat surprisingly, found to be the most common cause (22%).[19,21,45,46] The natural history of vasovagal syncope may be different in the elderly patients with some reports suggesting an association with malignancy and other terminal conditions.[86] Elderly patients often do not have the typical prodrome of vasovagal syncope that is commonly noted in younger patients.[19,21,46] In addition, there is more overlap between vasovagal symptoms and orthostatic hypotension and carotid hypersensitivity, both are more common in the elderly patients, with the former often iatrogenic in origin. Although a vasovagal etiology is most commonly found in the elderly patients with syncope, orthostatic hypotension (13%) and arrhythmia (12%) are also more commonly observed in the elderly patients than in younger patients.[19]

Treatment of syncope in the elderly patients can be more challenging due to the multifactorial etiology and difficulty in determining a precise cause despite multiple diagnostic tests. In particular, neurologic tests, such as CT scans and EEGs, have very low diagnostic yield (<5%), as is true for younger patients.[19] Some treatment strategies are limited in the elderly patients, as, for example, the older patient with vasovagal syncope, in whom medications, such as midodrine and fludrocortisone, are not tolerated due to accompanying hypertension.

## SYNCOPE AND DRIVING

The clinician is often faced with questions about the safety of driving in patients with syncope. In the largest study to date, 3,877 consecutive patients with syncope were identified from a large database.[87] Within this group, 381 patients (9.8%) had an episode of syncope while driving (driving group). When compared to the non-driving group, the syncope while driving group was younger, more likely to be male, and more likely to have cardiovascular disease. Syncope during driving was most commonly due to neurocardiogenic causes (37%) with no cause determined in 23% and arrhythmia in 12%. Patients in the driving group had a slightly higher prevalence of accompanying injury (driving: 29% vs non-driving: 24%) but no difference in hospitalization rates (driving: 17% vs non-driving: 15%). Recurrent syncope in the driving group occurred in 72 patients, out of which 35 of whom had recurrent syncope more than 6 months after the initial evaluation.

Driving has become almost an essential component for functioning in today's society for many people. Driver incapacitation for medical reasons (e.g. seizure or syncope) has important public safety ramifications.[88-90] Laws for mandatory physician reporting of medical conditions that can impact driving exist in some states (e.g. New Jersey, Pennsylvania), but not in other states (e.g. Florida, New Mexico). What constitutes an important reportable medical condition varies significantly, particularly in the case of syncope where etiology and prognosis vary widely. In the United Kingdom, for a noncommercial license, a simple faint with definite provocative factors does not lead to any driving restrictions, while unexplained syncope with a high risk of recurrence leads to a mandatory 6 month period during which driving is not permitted. The overall impact of mandatory physician reporting of patients with cardiac conditions probably has a negligible impact on motor vehicle accident-related mortality and morbidity in a larger population. As emphasized in a recent editorial, the risk of driving accidents related to recurrent syncope is significantly lower than the risk of severe accidents from high risk groups such as young drivers, the elderly patients, or distracted drivers.[88] The most recent published guidelines recommend a minimum of 6 months of abstinence from driving after a syncopal event, with resumption of driving permitted if no further episodes have occurred.[90]

## GUIDELINES FOR EMERGENCY DEPARTMENT EVALUATION

One of the most significant sources for high cost in patients with syncope is hospital admission that ranges 26–60%.[2,3,37,91] For this reason a number of investigators have evaluated the utility of algorithms for identifying patients at higher risk for significant events and have developed syncope management units (similar to the concept of chest pain units) that allow expedited and more efficient management of patients with syncope. The components of the risk stratification rules vary and some have debated whether the rules are effective for reducing cost **(Table 1)**.

The Osservatorio Epidemiologico sulla Sincope nel Lazio (OESIL) risk score was derived from a patient cohort of 270 patients that presented with syncope in 6 community hospitals in the Lazio region of Italy.[91,92] Multivariate analysis identified four independent predictors that predicted risk: History of cardiovascular disease, age more than 65 years, syncope without a prodrome and an abnormal ECG. The 12 month mortality increased linearly from 0% (no risk factors present) to 57% (all four risk factors present). Several subsequent studies have validated the OESIL risk score for predicting one year risk in other patient cohorts.[91–93]

The San Francisco Syncope Rule was developed to predict short-term outcomes (7 days after the index event).[32] The investigators evaluated multiple variables but simplified their rule to include five elements: (1) abnormal ECG; (2) shortness of breath; (3) hematocrit less than 30%; (4) systolic blood pressure less than 90 mm Hg and (5) a history heart failure. Often the mnemonic chess is used to more easily remember the components: C: Congestive heart failure; H: Hematocrit; E: ECG; S: Systolic blood pressure; S—shortness of breath. Similar to the OESIL risk score subsequent studies have generally validated the utility of the San Francisco Syncope Rule.[93]

In the short-term prognosis of syncope (STePS) study 676 patients with syncope were evaluated at both 10 days and 1 year.[94] Severe outcomes (death, major therapeutic procedures, readmission to the hospital within 10 days) were observed in 6.1% of patients (mainly rehospitalization) at 10 day follow-up. Severe outcomes were observed mainly in patients who were admitted (14.7%) compared to those who were discharged from the emergency department (2.0%). The main mechanistic cause for severe outcomes was arrhythmia-related (25/41 patients, most often due to implantation of a permanent pacemaker), although five patients died had a variety of causes, none specifically

## TABLE 1
### Comparison of the components and primary outcomes for four different risk stratification schemes

| Algorithm | Age | Components: Sx and Hx | PE | ECG | Anemia | $O_2$ | BNP | Endpoint |
|---|---|---|---|---|---|---|---|---|
| OESIL | ≥65 years | Sudden onset CVD | | Q waves, ST Δ's, LVH, BBB, Arrhythmia | | | | 1 year Mortality |
| SFSR* | | SOB, HF Hx | SBP <90 mm Hg | Δ's, from a prior ECG, Arrhythmia | Hct ≤30% | | | 7 day Mortality and serious outcomes |
| STePS† | >65 years | Trauma, Sudden onset Hx CVA or CVD, Male | | Q waves LVH, LBBB, Arrhythmia | | | | 10 day and 1 year Mortality and serious outcomes |
| ROSE** | | Chest pain | | Q waves, LBBB, HR ≤50 bpm | Hb ≤ 9 g/dL; fecal blood | ≤94% | >300 pg/mL | 30 day serious outcomes |

(**ABBREVIATIONS:** OESIL: Osservatorio Epidemiologico sulla Sincope nel Lazio; SFSR: *San Francisco Syncope Rule. Serious Outcomes: Myocardial infarction, arrhythmia, pulmonary embolism, stroke, subarachnoid hemorrhage, significant hemorrhage or any condition causing or likely to cause a return ED visit and hospitalization for a related event; †STePS: Short Term Prognosis of Syncope. Serious Outcomes: Need for major therapeutic procedures and early (within 10 days) readmission to hospital; **ROSE: Risk stratification of Syncope in the Emergency Department. Serious Outcomes: Acute myocardial infarction, serious arrhythmia, hemorrhage, pulmonary embolus; Hx: History; PE: Physical examination; $O_2$: Oxygen saturation; BNP: Brain natriuretic peptide; Time frame: Time of endpoint evaluation; HR: Heart rate; PVC: Premature ventricular contractions; SOB: Shortness of breath; CVD: Cardiovascular disease; CVA: Cerebrovascular accident; LVH: Left ventricular hypertrophy; LBBB: Left bundle branch block) arrhythmia related (disseminated intravascular coagulation, pulmonary edema, aortic dissection, pulmonary embolism and stroke). Predictors of short-term risk included an abnormal ECG, concomitant trauma and absence of preceding symptoms. Interestingly, factors associated with long-term adverse outcomes were different from the short-term risk predictors and included age more than 65 years and history of neoplasm, cerebrovascular disease and heart disease (structural heart disease or ventricular arrhythmias).

In the risk stratification of syncope in the emergency department (ROSE) study a cohort of 550 patients with syncope was evaluated.[33] One-month serious outcome (acute myocardial infarction, life-threatening arrhythmia, requirement for ICD or permanent pacemaker implant, hemorrhage requiring transfusion, pulmonary embolus or significant neurologic event) or all-cause death occurred in 40 (7.3%) patients in the derivation cohort. Independent predictors were B-type BNP concentration ≥300 pg/ml [odds ratio (OR): 7.3], positive fecal occult blood (OR: 13.2), hemoglobin ≤9.0 g/dL (OR: 6.7), oxygen saturation ≤94% (OR: 3.0) and Q-wave on the presenting ECG (OR: 2.8). One-month serious outcome or all-cause death occurred in 39 (7.1%) patients in the validation cohort. The ROSE rule (the presence of any of the independent predictors) had a sensitivity and specificity of 87.2% and 65.5% respectively, and a negative predictive value of 98.5% for a serious outcome or death at one month. An elevated BNP concentration alone was a major predictor of serious cardiovascular outcomes (8 of 22 events, 36%) and all-cause deaths (8 of 9 deaths, 89%).

Another strategy for reducing the cost of syncope is streamlining the process of evaluation using syncope or transient loss of consciousness units. In the Syncope Evaluation in the Emergency Department Study (SEEDS), 103 patients were randomized to standard care or a specialized syncope unit that provided early evaluation and focused diagnostic testing.[95] Patients randomized to the syncope unit were less likely to require hospital admission (syncope unit: 43% vs standard care: 98%) and more likely to have a presumptive diagnosis (syncope unit: 67% vs standard care: 10%) on discharge from the emergency department or from the syncope unit.

Several groups have argued the obvious importance of developing a consistent method for risk stratification of patients with syncope.[2,3,37] Although the currently published rules vary some basic points can be made. First it is important to decide whether or not to assign a working diagnosis of a cardiac cause of syncope. All of the risk stratification schemes use two or more criteria for identifying a group of patients with a higher likelihood for cardiac syncope. An abnormal ECG (Q waves, bundle branch block or atrioventricular block) is a component in all of the risk stratification schemes. Criteria, such as age more than 65 (OESIL),[90] history of cardiovascular disease (OESIL)[91] or presence of congestive heart failure (San Francisco Syncope Rule)[32] and elevated BNP (ROSE)[33] are all parameters that increase the likelihood of identifying a patient with a cardiac cause for syncope. Second, criteria that evaluate for noncardiac causes of syncope focus on conditions associated with higher short-term risk such as pulmonary embolus ($O_2$ saturation <90% in the San Francisco Syncope Rule or <94% in ROSE),[32,33] significant anemia (hematocrit <30% in the San Francisco Syncope Rule and hemoglobin <9.0 g/dL in ROSE),[32,33] or shock from any cause (SBP <90 mm Hg in the San Francisco Syncope Rule)[32] or gastrointestinal bleeding (fecal occult blood in ROSE).[33]

## GUIDELINES/OFFICIAL RECOMMENDATIONS

There have been formal statements on syncope from two cardiology groups. The AHA/ACCF in collaboration with the Heart Rhythm Society published a scientific statement on the evaluation of syncope in 2006.[3] The writing group recommends a history, physical examination and ECG in all patients with syncope. If the cause of syncope remains unexplained (not neurally mediated or orthostatic) they recommend an echocardiogram, an exercise test and ischemia evaluation. If these tests are normal, no additional testing is required for an isolated benign event. However, if recurrent episodes of syncope or if the episode is associated with significant injury, the clinician should use tests that evaluate the cardiac rhythm during symptoms. The choice among 24–48 hour ambulatory ECG monitoring, external event recorder or an ILR will depend on the frequency of the episodes and the severity of symptoms (syncope vs presyncope). The scientific statement provides a concise practical framework on the evaluation of syncope but does not address emergency room evaluation or subsequent treatment.

More recently, the ESC has published a comprehensive guidelines document that discusses both diagnosis and management of syncope.[2] Similar to the scientific statement from the cardiology societies based in the United States, they recommend an initial evaluation that includes history,

physical examination (with orthostatic blood pressure measurements) and an ECG. They emphasize that the clinician should attempt to answer three specific questions:
1. Is it a syncopal episode or not?
2. Has an etiologic diagnosis been determined?
3. Is there evidence for a high risk of a cardiovascular event or death?

If these questions cannot be answered with the initial evaluation, additional tests, such as an echocardiography and other types of cardiac imaging, exercise testing, tilt-table testing, cardiac rhythm monitoring and other tests, can be chosen depending on the clinical situation. The European Guidelines do not provide a simple algorithmic approach but rather emphasize that the clinician must carefully choose tests in individual patients to risk stratify the patient and identify a specific etiology of syncope so that a diagnosis for specific treatment plan can be developed.

## REFERENCES

1. Papavramidou N, Tziakas D. Galen on "syncope". Int J Cardiol. 2010;142:242-4.
2. European Heart Rhythm Association (EHRA), Heart Failure Association (HFA), Heart Rhythm Society (HRS), et al. Guidelines for the diagnosis and management of syncope (version 2009): the Task Force for the Diagnosis and Management of Syncope of the European Society of Cardiology (ESC). Eur Heart J. 2009;30: 2631-71.
3. Strickberger SA, Benson DW, Biaggioni I, et al. AHA/ACCF scientific statement on the evaluation of syncope: from the American Heart Association Councils on Clinical Cardiology, Cardiovascular Nursing, Cardiovascular Disease in the Young, and Stroke, and the Quality of Care and Outcomes Research Interdisciplinary Working Group; and the American College of Cardiology Foundation In Collaboration With the Heart Rhythm Society. J Am Coll Cardiol. 2006;47:473-84.
4. Petkar S, Cooper P, Fitzpatrick AP. How to avoid a misdiagnosis in patients presenting with transient loss of consciousness. Postgrad Med J. 2006;82:630-41.
5. Benditt DG, Olshansky B, Wieling W. The ACCF/AHA scientific statement on syncope needs rethinking. J Am Coll Cardiol. Ad Hoc Syncope Consortium. 2006;48:2598-9; (author reply) Epub 2006;2599.
6. Soteriades ES, Evans JC, Larson MG, et al. Incidence and prognosis of syncope. N Engl J Med. 2002;347:878-85.
7. Dermaskin G, Lamb LE. Syncope in a population of healthy young adults; incidence, mechanisms, and significance. J Am Med Assoc. 1958;168:1200-7.
8. Alshekhlee A, Shen WK, Mackall J, et al. Incidence and mortality rates of syncope in the United States. Am J Med. 2009;122:181-8.

9. Serletis A, Rose S, Sheldon AG, et al. Vasovagal syncope in medical students and their first degree relatives. Eur Heart J. 2006;27:1965-70.
10. Quinn J, McDermott, Kramer N, et al. Death after emergency department visits for syncope: how common and can it be predicted? Ann Emerg Med. 2007;49:420-7.
11. Kapoor WN, Hanusa BH. Is syncope a risk factor for poor outcomes? Comparison of patients with and without syncope. Am J Med. 1996;100:646-55.
12. Day SC, Cook EF, Funkenstein H, et al. Evaluation and outcome of emergency room patients with transient loss of consciousness. Am J Med. 1982;73:15-23.
13. Silverstein MD, Singer DE, Mulley AG, et al. Patients with syncope admitted to medical intensive care units. JAMA. 1982;248:1185-9.
14. Sun BC, Emond JA, Camargo CA. Direct medical costs of syncope-related hospitalizations in the United States. Am J Cardiol. 2005;95:668-71.
15. van Dijk N, Sprangers MA, Boer KR, et al. Quality of life within one year following presentation after transient loss of consciousness. Am J Cardiol. 2007;100:672-6.
16. van Dijk N, Sprangers MA, Colman N, et al. Clinical factors associated with quality of life in patients with transient loss of consciousness. J Cardiovasc Electrophysiol. 2006;17:998-1003.
17. Bartoletti A, Fabiani P, Bagnoli L, et al. Physical injuries caused by a transient loss of consciousness: main clinical characteristics of patients and diagnostic contribution of carotid sinus massage. Eur Heart J. 2008;29:618-24.
18. Ewing DJ, Campbell IW, Clarke BF. The natural history of diabetic autonomic neuropathy. Q J Med. 1980;49:95-108.
19. Ojha A, McNeeley K, Heller E, et al. Orthostatic syndromes differ in syncope frequency. Am J Med. 2010;123:245-9.
20. Cheitlin MD, MacGregor J. Congenital anomalies of coronary arteries: role in the pathogenesis of sudden cardiac death. Herz. 2009;34:268-79.
21. Mendu ML, McAvay G, Lampert R, et al. Yield of diagnostic tests in evaluating syncopal episodes in older patients. Arch Intern Med. 2009;169:1299-305.
22. Grubb BP, Gerard G, Roush K, et al. Differentiation of convulsive syncope and epilepsy with head-up tilt testing. Ann Intern Med. 1991;115:871-6.
23. Sheldon R, Rose S, Ritchie D, et al. Historical criteria that identify Syncope from Seizures. J Am Coll Cardiol. 2002;40:142-8.
24. Benbadis SR, Wolgamuth BR, Goren H, et al. Value of tongue biting in the diagnosis of seizures. Arch Intern Med. 1995;155:2346-9.
25. Hoefnagels WA, Padberg GW, Overweg J, et al. Syncope or seizure? A matter of opinion. Clin Neurol Neurosurg. 1992;94:153-6.
26. Alboni P, Brignole M, Menozzi C, et al. Diagnostic value of history in patients with syncope with or without heart disease. J Am Coll Cardiol. 2001;37:1921-8.
27. Del Rosso A, Ungar A, Maggi R. Clinical predictors of cardiac syncope at initial evaluation in patients referred to general hospital EGSYS Score. Heart. 2008;94:1620-6.

28. Sheldon R, Hersi A, Ritchie D, et al. Syncope and structural heart disease: historical criteria for vasovagal syncope and ventricular tachycardia. J Cardiovasc Electrophysiol. 2010;21:1358-64.
29. Linzer M, Yang EH, Estes NA 3rd, et al. Diagnosing syncope. Part 1: value of history, physical examination, and electrocardiography. Clinical efficacy assessment project of the American College of Physicians. Ann Intern Med. 1997;126:989-96.
30. Sheldon R, Rose S, Connolly S, et al. Diagnostic criteria for vasovagal syncope based on a quantitative history. Eur Heart J. 2006;27:344-50.
31. Sheldon R, Hersi A, Ritchie D, et al. Syncope and structural heart disease: historical criteria for vasovagal syncope and ventricular tachycardia. J Cardiovasc Electrophysiol. 2010;21:1358-64.
32. Quinn JV, Stiell IG, McDermott DA, et al. Derivation of the San Francisco Syncope Rule to predict patients with short-term serious outcomes. Ann Emerg Med. 2004;43:224-32.
33. Reed MJ, Newby DE, Coull AJ, et al. The ROSE (risk stratification of syncope in the emergency department) study. J Am Coll Cardiol. 2010;55:713-21.
34. Link MS, Lauer EP, Homoud MK, et al. Low yield of rule-out myocardial infarction protocol in patients presenting with syncope. Am J Cardiol. 2001;88:706-7.
35. Reed MJ, Newby DE, Coull AJ, et al. Role of brain natriuretic peptide (BNP) in risk stratification of adult syncope. Emerg Med J. 2007;24:769-73.
36. Kapoor WN, Cha R, Peterson JR, et al. Prolonged electrocardiographic monitoring in patients with syncope: importance of frequent or repetitive ventricular ectopy. Am J Med. 1987;82:20-8.
37. Huff JS, Decker WW, Quinn JV, et al. American College of Emergency Physicians. Clinical policy: critical issues in the evaluation and management of adult patients presenting to the emergency department with syncope. Ann Emerg Med. 2007;49:431-44.
38. Krahn AD, Klein GJ, Yee R, et al. Randomized assessment of syncope trial: conventional diagnostic testing versus a prolonged monitoring strategy. Circulation. 2001;104:46-51.
39. Brignole M, Sutton R, Menozzi C, et al. International Study on Syncope of Uncertain Etiology 2 (ISSUE 2) Group. Early application of an implantable loop recorder allows effective specific therapy in patients with recurrent suspected neurally mediated syncope. Eur Heart J. 2006;27:1085-92.
40. Pierre B, Fauchier L, Breard G, et al. Implantable loop recorder for recurrent syncope: influence of cardiac conduction abnormalities showing up on resting electrocardiogram and of underlying cardiac disease on follow-up developments. Europace. 2008;10:477-81.
41. Marcus FI, Zareba W, Calkins H, et al. Arrhythmogenic right ventricular cardiomyopathy/dysplasia clinical presentation and diagnostic evaluation: results from the North American Multidisciplinary Study. Heart Rhythm. 2009;6:984-92.
42. Furushima H, Chinushi M, Hirono T, et al. Relationship between dominant prolongation of the filtered QRS duration in the right precordial leads and clinical characteristics in Brugada syndrome. J Cardiovasc Electrophysiol. 2005;16:1311-7.
43. Benditt DG, Ferguson DW, Grubb BP, et al. Tilt table testing for assessing syncope. J Am Coll Cardiol. 1996;28:263-75.

44. Kenny RA, Ingram A, Bayliss J, et al. Head-up tilt: a useful test for investigating unexplained syncope. Lancet. 1986;1:1352-5.
45. Podoleanu C, Maggi R, Brignole M, et al. Lower limb and abdominal compression bandages prevent progressive orthostatic hypotension in elderly persons: a randomized single-blind controlled study. J Am Coll Cardiol. 2006;48:1425-32.
46. Kurbaan AS, Bowker TJ, Wijesekera N, et al. Age and hemodynamic responses to tilt testing in those with syncope of unknown origin. J Am Coll Cardiol. 2003;41:1004-7.
47. Raviele A, Brignole M, Sutton R, et al. Effect of etilefrine in preventing syncopal recurrence in patients with vasovagal syncope: a double-blind, randomized, placebo-controlled trial. The Vasovagal Syncope International Study. Circulation. 1999;99:1452-7.
48. Moya A, Permanyer-Miralda G, Sagrista-Sauleda J, et al. Limitations of head-up tilt test for evaluating the efficacy of therapeutic interventions in patients with vasovagal syncope: results of a controlled study of etilefrine versus placebo. J Am Coll Cardiol. 1995;25:65-9.
49. Moya A, Brignole M, Menozzi C, et al. International Study on Syncope of Uncertain Etiology (ISSUE) Investigators. Mechanism of syncope in patients with isolated syncope and in patients with tilt-positive syncope. Circulation. 2001;104:1261-7.
50. Petkar S, Fitzpatrick A. Tilt table testing: transient loss of consciousness discriminator or epiphenomenon? Europace. 2008;10:747-50.
51. Sud S, Klein GJ, Skanes AC, et al. Predicting the cause of syncope from clinical history in patients undergoing prolonged monitoring. Heart Rhythm. 2009;6:238-43.
52. Entem FR, Enriquez SG, Cobo M, et al. Utility of implantable loop recorders for diagnosing unexplained syncope in clinical practice. Clin Cardiol. 2009;32:28-31.
53. Lacroix D, Dubuc M, Kus T, et al. Evaluation of arrhythmic causes of syncope: correlation between Holter monitoring, electrophysiologic testing, and body surface potential mapping. Am Heart J. 1991;122:1346-54.
54. Ruskin JN. Role of invasive electrophysiological testing in the evaluation and treatment of patients at high risk for sudden cardiac death. Circulation. 1992;85:I152-9.
55. Moss AJ, Zareba W, Hall WJ, et al. Multicenter Automatic Defibrillator Implantation Trial II Investigators. Prophylactic implantation of a defibrillator in patients with myocardial infarction and reduced ejection fraction. N Engl J Med. 2002;346:877-83.
56. Maron BJ, Shen WK, Link MS, et al. Efficacy of implantable cardioverter-defibrillators for the prevention of sudden death in patients with hypertrophic cardiomyopathy. Efficacy of implantable cardioverter-defibrillators for the prevention of sudden death in patients with hypertrophic cardiomyopathy. N Engl J Med. 2000;342:365-73.
57. Menozzi C, Brignole M, Garcia-Civera R, et al. International Study on Syncope of Uncertain Etiology (ISSUE) Investigators. Mechanism of syncope in patients with heart disease and negative electrophysiologic test. Circulation. 2002;105:2741-5.
58. Aydin MA, Maas R, Mortensen K, et al. Predicting recurrence of vasovagal syncope: a simple risk score for the clinical routine. J Cardiovasc Electrophysiol. 2009;20:416-21.

59. van Dijk N, Quartieri F, Blanc JJ, et al. PC-Trial Investigators. Effectiveness of physical counterpressure maneuvers in preventing vasovagal syncope: the Physical Counterpressure Manoeuvres Trial (PC-Trial). J Am Coll Cardiol. 2006;48:1652-7.
60. Duygu H, Zoghi M, Turk U, et al. The role of tilt training in preventing recurrent syncope in patients with vasovagal syncope: a prospective and randomized study. Pacing Clin Electrophysiol. 2008;31:592-6.
61. Sheldon RS, Amuah JE, Connolly SJ, et al. Prevention of syncope trial. Effect of metoprolol on quality of life in the prevention of syncope trial. J Cardiovasc Electrophysiol. 2009;20:1083-8.
62. Perez-Lugones A, Schweikert R, Pavia S, et al. Usefulness of midodrine in patients with severely symptomatic neurocardiogenic syncope: a randomized control study. J Cardiovasc Electrophysiol. 2001;12:935-8.
63. Grubb BP, Wolfe DA, Samoil D, et al. Usefulness of fluoxetine hydrochloride for prevention of resistant upright tilt induced syncope. Pacing Clin Electrophysiol. 1993;16:458-64.
64. Takata TS, Wasmund SL, Smith ML. Serotonin reuptake inhibitor (paxil) does not prevent the vasovagal reaction associated with carotid sinus massage and/or lower body negative pressure in healthy volunteers. Circulation. 2002;106:1500-4.
65. Connolly SJ, Sheldon R, Roberts RS, et al. The North American Vasovagal Pacemaker Study (VPS). A randomized trial of permanent cardiac pacing for the prevention of vasovagal syncope. J Am Coll Cardiol. 1999;33:16-20.
66. Sutton R, Brignole M, Menozzi C, et al. Dual-chamber pacing in the treatment of neurally mediated tilt-positive cardioinhibitory syncope: pacemaker versus no therapy: a multicenter randomized study. The Vasovagal Syncope International Study (VASIS) Investigators. Circulation. 2000;102:294-9.
67. Connolly SJ, Sheldon R, Thorpe KE, et al. VPS II Investigators. Pacemaker therapy for prevention of syncope in patients with recurrent severe vasovagal syncope: Second Vasovagal Pacemaker Study (VPS II): a randomized trial. JAMA. 2003;289:2224-9.
68. Sud S, Massel D, Klein GJ, et al. The expectation effect and cardiac pacing for refractory vasovagal syncope. Am J Med. 2007;120:54-62.
69. Brignole M, Sutton R, Menozzi C, et al. International Study on Syncope of Uncertain Etiology 2 (ISSUE 2) Group. Early application of an implantable loop recorder allows effective specific therapy in patients with recurrent suspected neurally mediated syncope. Eur Heart J. 2006;27:1085-92.
70. Dimitrow PP, Chojnowska L, Rudzinski T, et al. Sudden death in hypertrophic cardiomyopathy: old risk factors re-assessed in a new model of maximalized follow-up. Eur Heart J. 2010;31:842-8.
71. Haghjoo M, Faghfurian B, Taherpour M, et al. Predictors of syncope in patients with hypertrophic cardiomyopathy. Pacing Clin Electrophysiol. 2009;32:642-7.
72. Maron BJ, Spirito P, Shen WK, et al. Implantable cardioverter-defibrillators and prevention of sudden cardiac death in hypertrophic cardiomyopathy. JAMA. 2007;298:405-12. Erratum in: JAMA. 2007;298:1516.
73. Spirito P, Autore C, Rapezzi C, et al. Syncope and risk of sudden death in hypertrophic cardiomyopathy. Circulation. 2009;119:1703-10.

74. Pezawas T, Stix G, Kastner J, et al. Implantable loop recorder in unexplained syncope: classification, mechanism, transient loss of consciousness and role of major depressive disorder in patients with and without structural heart disease. Heart. 2008;94:e17.
75. Knight BP, Goyal R, Pelosi F, et al. Outcome of patients with nonischemic dilated cardiomyopathy and unexplained syncope treated with an implantable defibrillator. J Am Coll Cardiol. 1999;33: 1964-70.
76. Phang RS, Kang D, Tighiouart H, et al. High risk of ventricular arrhythmias in patients with nonischemic dilated cardiomyopathy presenting with syncope. Am J Cardiol. 2006;97:416-20.
77. Ellenbogen KA, Levine JH, Berger RD, et al. Defibrillators in Non-Ischemic Cardiomyopathy Treatment Evaluation (DEFINITE) Investigators. Are implantable cardioverter defibrillator shocks a surrogate for sudden cardiac death in patients with nonischemic cardiomyopathy? Circulation. 2006;113:776-82.
78. Olshansky B, Poole JE, Johnson G, et al. SCD-HeFT Investigators. Syncope predicts the outcome of cardiomyopathy patients: analysis of the SCD-HeFT study. J Am Coll Cardiol. 2008;51:1277-82.
79. Hayes CJ, Gersony WM. Arrhythmias after the Mustard operation for transposition of the great arteries: a long-term study. J Am Coll Cardiol. 1986;7:133-7.
80. Gillette PC, Wampler DG, Shannon C, et al. Use of cardiac pacing after the Mustard operation for transposition of the great arteries. J Am Coll Cardiol. 1986;7:138-41.
81. Gelatt M, Hamilton RM, McCrindle BW, et al. Arrhythmia and mortality after the Mustard procedure: a 30-year single-center experience. J Am Coll Cardiol. 1997;29:194-201.
82. Wilson NJ, Clarkson PM, Barratt-Boyes BG, et al. Long-term outcome after the mustard repair for simple transposition of the great arteries. 28-year follow-up. J Am Coll Cardiol. 1998;32:758-65.
83. Schwerzmann M, Salehian O, Harris L, et al. Ventricular arrhythmias and sudden death in adults after a Mustard operation for transposition of the great arteries. Eur Heart J. 2009;30:1873-9.
84. Khairy P, Harris L, Landzberg MJ, et al. Sudden death and defibrillators in transposition of the great arteries with intra-atrial baffles: a multicenter study. Circ Arrhythm Electrophysiol. 2008;1: 250-7.
85. Cummings SR, Nevitt MC, Kidd S. Forgetting falls. The limited accuracy of recall of falls in the elderly. J Am Geriatr Soc. 1988;36:613-6.
86. Venkatraman V, Lee L, Nagarajan DV. Lymphoma and malignant vasovagal syndrome. Br J Haematol. 2005;130:323.
87. Sorajja D, Nesbitt GC, Hodge DO, et al. Syncope while driving: clinical characteristics, causes, and prognosis. Circulation. 2009;120: 928-34.
88. Curtis AB, Epstein AE. Syncope while driving: How safe is safe? Circulation. 2009;120:921-3.
89. Simpson CS, Hoffmaster B, Mitchell LB, et al. Mandatory physician reporting of drivers with cardiac disease: Ethical and practical considerations. Can J Cardiol. 2004;20:1329-34.
90. Epstein AE, Baessler CA, Curtis AB, et al. Addendum to "personal and public safety issues related to arrhythmias that may affect consciousness: implications for regulation and physician recommen-

dations: a medical/scientific statement from the American Heart Association and the North American Society of Pacing and Electrophysiology": public safety issues in patients with implantable defibrillators: a scientific statement from the American Heart Association and the Heart Rhythm Society. Circulation. 2007;115: 1170-6.
91. Brignole M, Disertori M, Menozzi C, et al. Management of syncope referred urgently to general hospitals with and without syncope units evaluation of guidelines in syncope study group. Europace. 2003;5:293-8.
92. Colivicchi F, Ammirati F, Melina D, et al. OESIL (Osservatorio Epidemiologico sulla Sincope nel Lazio) Study Investigators Development and prospective validation of a risk stratification system for patients with syncope in the emergency department: the OESIL risk score. Eur Heart J. 2003;24:811-9.
93. Dipaola F, Costantino G, Perego F, et al. STePS investigators. San Francisco Syncope Rule, Osservatorio Epidemiologico sulla Sincope nel Lazio risk score, and clinical judgment in the assessment of short-term outcome of syncope. Am J Emerg Med. 2010;28:432-9.
94. Costantino G, Perego F, Dipaola F, et al. STePS Investigators. Short- and long-term prognosis of syncope, risk factors, and role of hospital admission: results from the STePS (Short-Term Prognosis of Syncope) study. J Am Coll Cardiol. 2008;51:276-83.
95. Shen WK, Decker WW, Smars PA, et al. Syncope Evaluation in the Emergency Department Study (SEEDS): a multidisciplinary approach to syncope management. Circulation. 2004;110:3636-45.

**CHAPTER 5**

# Atrial Fibrillation

*Vasanth Vedantham, Jeffrey E Olgin*

## Chapter Outline

- Definition and Classification
- Epidemiology
  - Incidence and Prevalence
  - Natural History
- Etiology and Pathogenesis
  - Structural Heart Disease
  - Electrophysiological Abnormalities
  - Noncardiac Causes
  - Lone Atrial Fibrillation
- Diagnosis
  - Presentation
  - Physical Examination
  - Electrocardiogram
  - Diagnostic Testing
- Management
  - New-onset Atrial Fibrillation
  - Rate Control versus Rhythm Control in Recurrent AF
  - Restoration of Sinus Rhythm
  - Maintenance of Sinus Rhythm— Pharmacological Approaches
  - Maintenance of Sinus Rhythm— Invasive Approaches
  - Strategies for Rate Control
  - Prevention of Thromboembolism
- Guidelines

## INTRODUCTION

Atrial fibrillation (AF) is the most common sustained arrhythmia in adults and is associated with substantial morbidity, mortality, and cost. AF is characterized by disorganized atrial electrical activity and irregular ventricular rates. AF can result in heart failure, thromboembolism, impaired quality of life and may increase mortality. While AF is frequently associated with structural heart disease, it can occur in isolation (lone AF) or in association with noncardiac diseases. In the coming years it is projected that AF will be seen with increasing frequency both by cardiologists and by non-specialists. The purpose of this chapter is to provide a broad overview of our current understanding of this complex arrhythmia, and to provide a framework for clinical decision making in patients with AF.

## DEFINITION AND CLASSIFICATION

Atrial fibrillation (AF) is easily recognized on the surface of ECG as an irregular supraventricular rhythm (irregular QRS complexes), with a loss of clear P-waves and/or the presence of fibrillatory waves **(Fig. 1)**. AF in response to a reversible cause (e.g. hyperthyroidism, pericarditis, hypoxia, pneumonia, surgery, pulmonary embolism) is called "secondary AF". AF can also occur in association with valve disease (typically mitral stenosis or regurgitation), in association with other structural heart disease (congestive heart failure, right ventricular dysfunction) or other known risks (e.g. hypertension, pulmonary disease, sleep apnea). AF that occurs without any overt heart disease, pulmonary disease or hypertension (HTN) is called "lone AF".

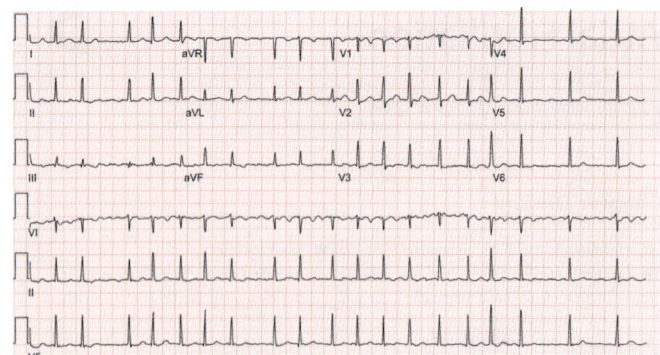

**FIGURE 1:** Typical ECG of atrial fibrillation showing an absence of P waves, the presence of fibrillatory waves (visible in lead V1), and a rapid, irregular ventricular response

While several classification schemes have been proposed for AF, the most widely used is based on the duration of AF episodes and whether intervention is required to terminate AF.[1] AF is called "paroxysmal" when episodes terminate spontaneously in less than seven days from onset. When two or more such episodes occur, paroxysmal AF is called "recurrent". When AF lasts longer than seven days or requires pharmacological or electrical conversion, it is called "persistent". AF that is resistant to drugs or cardioversion is called "permanent". It should be noted that these definitions are not necessarily always clean; for example, some patients with persistent AF may have periods where their AF is paroxysmal. Moreover, while there is evidence that there is a progression of AF from paroxysmal to persistent to permanent, this does not occur in every patient.

## EPIDEMIOLOGY

### Incidence and Prevalence

Atrial fibrillation is the most common clinical arrhythmia, both in the population at large and in hospitalized patients. According to a large population-based study (ATRIA), the prevalence of AF in the general population is roughly 1%, translating to about 3 million patients in the United States.[2] The prevalence of AF in ATRIA also rose steeply with age, with a prevalence of 0.1% in patients less than 50 years old and 9% in patients over 80 years old. Data from ATRIA also show higher incidence of AF among men than women (1.1% vs 0.9%), and among whites than blacks (2.2% vs 1.5% among patients older than 50 years). Other prospective longitudinal population-based studies[3,4] have explored the incidence of AF, and have shown a marked increase with age, from about 0.1% per year in patients between 55 years

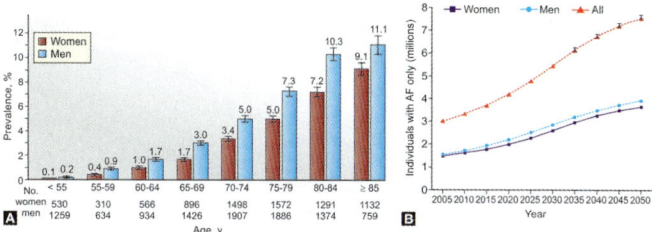

**FIGURES 2A AND B:** (A) Increasing prevalence of AF as a function of age in men and women in a large cross-sectional study. (B) Projected increase in the prevalence of AF to 2050. (*Source:* Reproduced with permission from reference 2 (A) and reference 5 (B))

and 60 years old to as high as 5% per year in patients older than 80 years, consistent with the studies of AF prevalence **(Fig. 2A)**. Since the age distribution of the population in the developed world is shifting toward older ages, the overall incidence of AF is rising. Available projections based on these longitudinal studies forecast about 5 million patients in the United States with AF by 2025 **(Fig. 2B)**.[5] In addition, since many episodes of paroxysmal AF are either asymptomatic, self-limited or occur in unmonitored patients, the true prevalence of AF is likely to be significantly higher than that of diagnosed AF.

## Natural History

Left untreated, the rapid ventricular rates associated with AF and the loss of atrial mechanical activity can lead to cardiomyopathy and heart failure, conditions which themselves can perpetuate AF through their effects on cardiac hemodynamics. While AF is often thought of as a progressive disease, with paroxysmal AF eventually progressing to persistent and permanent AF, they are patients in whom it does not progress. Although long-term data from large studies are not available, smaller studies have shown that about 25% of patients with paroxysmal AF will progress to permanent AF within 5 years.[6]

Not surprisingly, given its association with cardiovascular and systemic disease, the diagnosis of AF is associated with adverse long-term and short-term clinical outcomes. Although longitudinal studies in patients with lone AF have not revealed an adverse prognosis,[7] patients with AF in the context of cardiovascular disease had an approximate doubling of all-cause mortality in two large population-based studies, even after adjustment for the contributions of age, sex and comorbid conditions **(Figs 3A and B)**.[8,9]

The most common serious complication of AF is thromboembolic stroke. Compared to the general population, the relative risk of stroke in patients with AF is 2.4 for women and 3.0

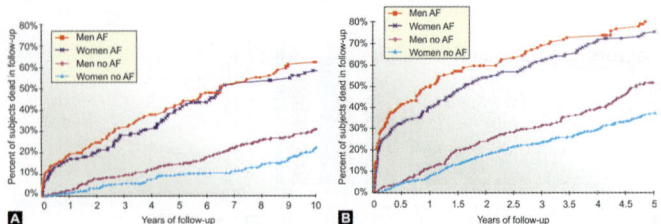

**FIGURES 3A AND B:** Kaplan-Meier curves for patients with AF and matched controls from the Framingham Study cohort. (A) Data for patients aged 55–74 (B) Data from patients aged 75–94. (*Source:* Modified from reference 9)

for men.[10] Additional factors, such as congestive heart failure, diabetes, HTN, prior stroke and age, increase the risk of stroke. The CHADS2 score is a useful risk assessment tool to calculate risk of stroke in patients with AF.[11] Compared to patients with carotid artery disease, strokes associated with AF are on average more severe (larger territory) and transient ischemic attacks are longer lasting, presumably because embolic particles are larger in AF patients.[12] As a result, long-term outcomes, both in terms of morbidity and mortality, are worse for patients who suffer strokes due to AF than for those whose strokes are due to carotid disease.[13–15] Recent epidemiological studies have also shown an increased risk of Alzheimer's disease and other forms of dementia in patients with AF, even in the absence of stroke.[16,17] This association appears to be independent of common risk factors for both conditions and confers increased mortality in the subset of AF patients who experience cognitive decline.[18] Conflicting data exist on the effect of AF on heart failure progression and mortality in heart failure patients. In the studies of left ventricular dysfunction (SOLVD) and Candesartan in heart failure—assessment of reduction in mortality (CHARM) trials, development of AF was associated with significantly worse outcomes than patients without AF.[19,20] However, no significant outcomes differences attributable to AF were observed in the vasodilator heart failure trial (V-HeFT) studies.[21]

## ETIOLOGY AND PATHOGENESIS

While short episodes of AF lasting a few seconds can be induced in normal atria, longer episodes require a vulnerable atrial substrate. This vulnerable substrate can be due to atrial enlargement, atrial fibrosis or other electrophysiological abnormalities of the atrial myocardium **(Table 1)**. For spontaneous AF to occur, there also needs to be a trigger. This is typically premature atrial depolarizations or short bursts of atrial tachycardia that interact with a vulnerable substrate to spontaneously induce AF. While the triggering activity may

**TABLE 1**
**Factors predisposing to atrial fibrillation**

| |
|---|
| *Electrophysiological abnormalities* <br> • Enhanced automaticity (focal AF) <br> • Conduction abnormality (reentry) |
| *Atrial pressure elevation* <br> • Mitral or tricuspid valve disease <br> • Myocardial disease (primary or secondary, leading to systolic or diastolic dysfunction) <br> • Semilunar valve abnormalities (causing ventricular hypertrophy) <br> • Systemic or pulmonary hypertension <br> • Intracardiac tumors or thrombi |
| *Atrial ischemia* <br> • Coronary artery disease |
| *Inflammatory or infiltrative atrial disease* <br> • Pericarditis <br> • Amyloidosis <br> • Myocarditis <br> • Age-induced atrial fibrotic change |
| *Drugs* <br> • Alcohol <br> • Caffeine |
| *Endocrine disorders* <br> • Hyperthyroidism <br> • Pheochromocytoma |
| *Changes in autonomic tone* <br> • Increased parasympathetic activity <br> • Increased sympathetic activity |
| *Primary or metastatic disease in or adjacent to the atrial wall* |
| *Postoperative* <br> • Cardiac, pulmonary or esophageal surgery |
| *Congenital heart disease* |
| *Neurogenic* <br> • Subarachnoid hemorrhage <br> • Nonhemorrhagic, major stroke |
| *Idiopathic (lone AF)* |
| *Familial AF* |
| (*Source:* Reference 1) |

arise from anywhere in the atria, evidence to date suggests that majority arise from the pulmonary veins.

## Structural Heart Disease

AF is most frequently associated with underlying structural heart disease. In the Framingham Study cohort, the major echocardiographic predictors of the development of AF, apart from valvular disease, were LV systolic dysfunction, LV hypertrophy, and atrial enlargement.[22] In addition, a variety of noncardiac conditions can result in AF, and AF can occur in the absence of any other discernable cardiac or noncardiac disease.

## Valvular Heart Disease

Diseases of the mitral and tricuspid valves result in pressure and/or volume overload of the atria, causing marked dilation and adverse atrial remodeling, predisposing to AF. Depending on the number of valves involved and the severity of the lesions, the prevalence of AF in patients with rheumatic heart disease ranges from 16% for isolated mitral regurgitation to as high as 70% for patients with a combination of mitral stenosis, mitral regurgitation and tricuspid regurgitation.[23]

## Heart Failure

Heart failure is a major cause of AF, due to the effects of chronically elevated left-sided pressures on left atrial structure and function, and the activation of neurohormonal cascades that lead to atrial remodeling. Between 10% and 50% of patients with LV dysfunction have also AF, depending on the severity of LV dysfunction and NYHA functional class.[24] Rapid rates associated with AF can also cause heart failure. In such cases, treatment of AF or rate control can result in significant improvement in heart failure symptoms and ejection fraction.[25,26]

## Ischemic Heart Disease

Cardiac ischemia is a common cause of AF likely due to a combination of elevated filling pressures in the left atrium, metabolic stress and inflammation. Approximately 5–10% of patients experiencing an acute MI will present in AF and this subset has a worse prognosis.[27,28]

## Hypertrophic Cardiomyopathy

The estimated incidence of AF among patients with hypertrophic cardiomyopathy varies from around 10% to 30%.[29] Because of poor LV compliance, these patients often depend on the atrial contribution to cardiac output and require relatively longer times for ventricular filling. As a result, they tolerate AF poorly, and can exhibit marked hemodynamic deterioration and severe symptoms associated with rapid AF.

In addition to the above-mentioned lesions, adult survivors of congenital heart disease often develop AF, either alone or in combination with other atrial tachyarrhythmias. Pericardial disease can also cause AF, both due to the effects of inflammation and hemodynamic sequelae of compromised ventricular filling. Related to this, patients undergoing cardiac surgery have an incidence of mostly self-limited postoperative AF as high as 30–40%.[30,31]

## Electrophysiological Abnormalities

AF is associated with several other electrophysiological disorders of the heart, which in some cases may trigger episodes of AF directly and in other cases may be indicators of a diseased atrial substrate that is prone to developing AF.

### Atrial Tachycardia and Pulmonary Venous Activity

The pulmonary venous myocardium has been identified as a major source of AF triggers, through a combination of abnormal automaticity, triggered activity and the proximity of autonomic ganglia.[32,33] Other thoracic venous structures (superior vena cava and coronary sinus) and embryological venous remnants (vein and ligament of Marshall) may also trigger AF.[34] The precise mechanisms that regulate electrical activity of the pulmonary venous myocardium are unknown. Intensive research is ongoing into the anatomy, embryology, and electrophysiological properties of the myocardial cells within the pulmonary veins.[35]

### Supraventricular Tachycardia

AV node reentry tachycardia and AV reentry tachycardia utilizing an accessory pathway can trigger AF. In such cases, elimination of the SVT with catheter ablation or treatment with medications may eliminate the trigger for AF.[36,37] Other reentrant arrhythmias, such as atrial flutter, can similarly degenerate into AF, and catheter ablation of the predisposing arrhythmia may reduce AF frequency. An increased incidence of AF has been documented in patients with ventricular preexcitation due to an accessory pathway, even in the absence of spontaneous or inducible tachycardia.[38,39]

### Conduction System Disease

Sinus node dysfunction and prolongation of the PR interval are both associated with AF, presumably due to common underlying atrial pathophysiology.[40,41] Some evidence supports the idea that bradycardia in patients with sinus node dysfunction may itself predispose to AF episodes, and that atrial pacing may reduce AF burden in such patients.

### Cardiac Nervous System Dysfunction

Imbalance between the sympathetic and parasympathetic arms of the cardiac autonomic nervous, in either direction, can lead to AF. Sympathetic activation can lead to enhanced activity of ectopic foci, which can trigger AF.[42] Prolonged sympathetic activation, as occurs in heart failure, can also lead to adverse structural remodeling of atrial tissue. On the other hand, enhanced vagal tone shortens the refractory period of atrial tissue, facilitating atrial reentry. The latter mechanism may account for a subset of patients with lone AF, such as highly trained athletes with high vagal tone, and those whose episodes of AF occur predominantly in sleep.[43]

## Noncardiac Causes

Hypertension is the most common noncardiac cause of AF, with a prevalence of 70% among AF patients enrolled in the atrial fibrillation follow-up investigation of rhythm management (AFFIRM) trial.[44] In population-based series, hypertension confers an adjusted relative risk of 1.5 for the development of AF.[45] While hypertension undoubtedly can lead to AF via an increased LV stiffness and left-sided filling pressures, the increased activation of the renin-angiotensin-aldosterone system (RAAS) may directly cause adverse electrical remodeling within the atria. Hyperthyroidism is common cause of AF. One percent of patients with new-onset AF have overt hyperthyroidism, while an additional 5–6% have subclinical hyperthyroidism.[46] In other studies, subclinical hyperthyroidism confers a relative risk of 3–5% for the development of AF.[47,48] Conversely, about 5–15% of patients with hyperthyroidism develop AF.[49] Conditions associated with systemic inflammation, metabolic stress or atrial enlargement, such as diabetes,[45] obesity,[50] postsurgical state[51] and sepsis[52] are all associated with the development of AF.

Chronic obstructive pulmonary disease (COPD) is associated with a relative risk of 1.3–1.8 for the development of AF, depending on the severity of lung disease.[53] The pathogenesis of AF in the setting of lung disease is likely to be a combination of direct effects of inflammation on the atria, metabolic stress of hypoxia and hemodynamic effects of chronically elevated right-sided pressures. Obstructive sleep apnea (OSA) is also associated with AF, and may be an under-recognized cause of the arrhythmia. The relative risk of AF in patients with OSA is as high as 2.8.[54] Treatment of OSA with continuous positive airway pressure can reduce the frequency of episodes of AF. With the increasing incidence of obesity, AF related to OSA is likely to occur with increasing frequency.

A number of substances are known causes of AF. While moderate alcohol consumption does not significantly increase the risk of AF, heavy alcohol use is strongly associated with AF.[55–57] In addition, a variety of medications can precipitate episodes of AF, including modulators nervous system function, diuretics and cardiac inotropic agents.[58] Although anecdotal evidence exists in support of caffeine as a precipitant of AF, an association between AF and caffeine consumption has not been proven in larger studies.[59]

## Lone atrial fibrillation

About 20–30% of patients with AF have no discernable cardiac or noncardiac cause.[60,61] Studies of tissue in patients with lone AF have identified numerous abnormalities, including subclinical cardiomyopathy as well as atrial fibrosis, but it is not clear whether these are a cause or a consequence of the arrhythmia.[62] Propensity to develop AF is highly heritable, even after adjustment for other risk factors.[63,64] Genetic mapping studies in patients with familial AF have identified rare monogenic causes for AF, including mutations in cardiac ion channels and accessory proteins, gap junctions, and other genes relevant to atrial biology **(Table 2)**.[65] However, the vast majority of lone AF appears to be non-Mendelian, implying polygenic or epigenetic etiology. Candidate gene association studies in AF have identified several alleles that confer increased risk of AF, including variants in ion channels and associated proteins as well as regulators of the RAAS system **(Table 2)**. More recent candidate gene data have uncovered genetic mosaicism in AF patients, in which atrial cardiomyocytes, but not peripheral blood lymphocytes, harbor disease-causing gap junction mutations.[66,67] Unbiased genome-wide association studies have also identified risk-conferring polymorphisms, including one at a non-coding locus on Chr 4q25, which confers a roughly fourfold lifetime relative risk for AF in carriers.[68] The polymorphism is in a non-transcribed area of the genome where it likely confers a regulatory function on nearby genes. PITX2c is located downstream of this polymorphism, and is important for establishing the identity of the pulmonary venous myocardium and in suppressing automaticity in left-sided remnants of embryonic cardiac pacemaking tissue.[69] Intensive research is ongoing to determine precisely how this allele might contribute to AF pathogenesis via an effect on PITX2C. Clinical implications of this research might include identifying novel drug targets, but such advances will require considerable progress in our understanding of the regulatory networks controlling atrial structure and function.

## DIAGNOSIS

## Presentation

As assessed by remote monitoring, most individual episodes of AF are asymptomatic and many patients are unaware of their arrhythmia.[70] Those who are symptomatic exhibit a broad

### TABLE 2
**Genetic causes of atrial fibrillation**

| *Mendelian AF, Candidate gene resequencing, and rare variants* | |
|---|---|
| **Gene symbol/ Locus** | **Gene name** |
| GJA5 | Connexin 40 |
| KCNQ1 | Potassium voltage-gated channel, KQT-like subfamily, member 1 |
| NPPA | Natriuretic peptide precursor A |
| LMNA | Lamin A/C |
| KCNA5 | Potassium voltage-gated channel, shaker-related subfamily, member 5 |
| KCNE2 | Potassium voltage-gated channel, Isk-related family, member 2 |
| KCNH2 | Potassium voltage-gated channel, subfamily H, member 2 |
| KCNJ2 | Potassium inwardly rectifying channel, subfamily J, member 2 |
| SCN5A | Sodium channel, voltage-gated, type V, alpha-subunit |
| Chr 5p13 | Unknown |
| Chr 6q14-q16 | Unknown |
| Chr 10q22-q24 | Unknown |
| Chr 10p11-q21 | Unknown |
| *Candidate gene association studies* | *Gene name* |
| ACE | Angiotensin-converting enzyme |
| AGT | Angiotensinogen |
| GJA5 | Connexin 40 |
| KCNE1 | Potassium voltage-gated channel, Isk-related family, member 1 |
| KCNH2 | Potassium voltage-gated channel, subfamily H, member 2 |
| *Genome-wide association studies* | *Candidate gene symbol and name* |
| Chr 4q25 | PITX2 Paired-like homeodomain 2 |
| Chr 16q22 | ZFHX3 Zinc finger homeodomain |
| Chr 1p21 | KCNN3 Calcium-activated potassium channel |

(*Source:* Reference 65)

spectrum of complaints, most commonly palpitations, dyspnea and symptoms of congestive heart failure.[71] The hemodynamic consequences and symptoms associated with AF are usually related to the ventricular rate and the presence or absence of underlying heart disease such as valvular disease, LV dysfunction or active coronary disease. Patients in any of the latter categories may tolerate AF poorly. Rarely, when a rapidly conducting accessory pathway is present (WPW syndrome), AF can lead to ventricular fibrillation and sudden cardiac death.[72] In otherwise healthy patients without accessory pathways, syncope is an unusual presentation for AF. Noncardiac symptoms associated with AF include polyuria (related to ANF release by distended atria) and thromboembolic events, most commonly acute embolic stroke.

## Physical Examination

The physical examination of patients with AF is also highly variable. Patients in AF can have heart rates ranging from bradycardia to extreme tachycardia depending on the integrity of AV nodal conduction, medications, autonomic tone and the presence of accessory pathways. Due to loss of atrial mechanical function, A-waves are absent from the jugular venous pulsation and a fourth heart sound is not audible in AF. Owing to the variable ventricular filling time in AF, the intensity of heart sounds can change beat to beat. It should also be recognized that commonly used diagnostic maneuvers used to evaluate heart murmurs, such as Valsalva, handgrip, and respiratory variation, are of limited utility in the patient with AF because the variable filling time will cause variation in murmur intensity independent of the effects of preload and afterload. Signs of poor perfusion or congestive heart failure can be seen when AF occurs in the setting of valvular disease, LV systolic or diastolic dysfunction, acute myocardial infarction, or when long-standing tachycardia leads to cardiomyopathy.

## Electrocardiogram

Although the hallmarks of AF on the surface ECG are loss of P-waves and an irregular ventricular response, AF must still be distinguished from other supraventricular tachycardias associated with an irregular ventricular response, including atrial flutter with variable block and multifocal atrial tachycardia **(Fig. 4)**. The latter is typically associated with lung disease, and is characterized by at least three distinct P-wave morphologies on the surface ECG, whereas the fibrillatory waves in AF are more rapid and variable in morphology. In certain patients, however, fibrillatory waves can be "coarse" and at first glance may be difficult to distinguish from flutter waves. Careful examination of the surface ECG in such patients usually unmasks subtle variation in the amplitudes and frequency of apparent flutter waves that reveal the true rhythm to be AF. In the setting of complete AV block with a junctional escape, accelerated junctional rhythms, ventricular tachycardia or

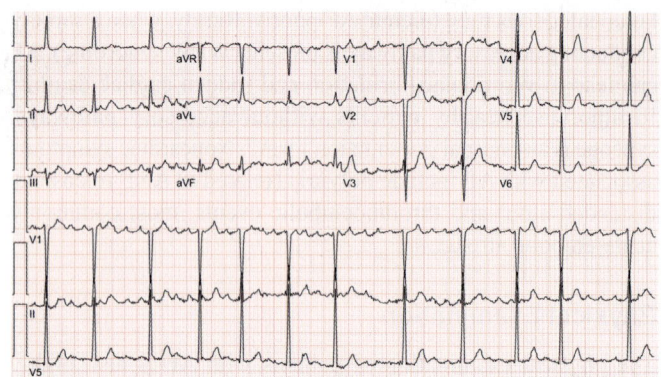

**FIGURE 4:** ECG shows coarse atrial fibrillation, sometimes mistaken for atrial flutter or erroneously called "fib-flutter" due to the apparent flutter waves in lead V1. However, the variable atrial cycle length and changing morphology of atrial depolarizations makes this ECG diagnostic of atrial fibrillation

ventricular pacing, the ventricular rate in AF can be regular. In such cases, it is important to focus on the nature of the atrial activity to make the diagnosis of AF. Sometimes, temporary inhibition of pacing function may be necessary to unmask AF. When QRS morphology is highly variable in AF and ventricular rate exceeds 200 beats per minute, ventricular preexcitation or ventricular tachycardia should be suspected.

## Diagnostic Testing

All patients presenting with new-onset AF should undergo appropriate testing for a reversible cause.[1] A careful history of medication use and substance use, particularly alcohol, should be obtained. AF can often be the only presenting sign of hyperthyroidism, so all patients with new-onset AF should receive an assessment of thyroid function. Additional laboratory testing should be guided by the history and physical examination. Although AF frequently accompanies acute coronary syndromes and stable coronary artery disease, it is rarely the only sign of active ischemia. Thus, in the absence of other symptoms suggestive of active ischemic heart disease, it is not necessary for patients with AF to undergo stress testing or coronary angiography. All patients with AF should receive an echocardiogram, since the most common cause of AF is structural heart disease.

Because episodes of AF can be brief, can occur in sleep and can be asymptomatic, the clinical history is often not reflective of a patient's overall burden of AF (the fraction of time the patient is in AF).[70,73] In patients with symptoms compatible with AF, but no clear evidence at presentation, cardiac monitoring for extended periods can be helpful. Unless episodes are very frequent, Holter monitoring is often of insufficient duration to capture the episodes. Event monitors with telephonic transmission and automatic triggering algorithms for detecting and recording AF can be very useful to confirm a diagnosis, determine whether symptoms are due to AF, determine the burden of AF and determine whether symptoms are due to poorly controlled ventricular rates.[74]

## MANAGEMENT

### New-onset Atrial Fibrillation

Newly diagnosed AF should prompt a diagnostic work up as outlined above for a reversible cause. In general, when a reversible cause is identified, initial treatment should be directed at the underlying precipitating factor rather than the AF. Patients with new-onset AF presenting to the emergency room can in most cases be managed safely without hospital admission.[75] Hospital admission may be warranted for patients with concurrent medical conditions requiring inpatient treatment, for the elderly, and for patients with significant structural heart disease, ischemia, hemodynamic instability or preexcited AF.

Immediate reversion to sinus rhythm is warranted in patients with hemodynamic instability, active ischemia, severe heart failure symptoms, or AF with ventricular preexcitation.[1] In such cases, while it is ideal to confirm absence of an intracardiac thrombus using a transesophageal echocardiogram, this may not be possible due to the urgency of the situation. At a minimum, systemic anticoagulation should be administered prior to cardioversion unless strongly contraindicated.

When urgent cardioversion is not indicated, it is acceptable to pursue an initial rate control strategy for patients who are mildly or moderately symptomatic. This approach allows time for a diagnostic workup, treatment of a potentially reversible cause of AF, and for an assessment of thromboembolic risk. About 70% of patients with new-onset AF of less than 72 hours duration will spontaneously convert to sinus rhythm without intervention.[76] For the remainder, reversion to sinus rhythm with cardioversion, either electrical or chemical, may be reasonable if the risk of short-term AF recurrence is relatively low or unknown. This is the case in younger patients (<65) with structurally normal hearts and in patients with reversible causes for AF once the underlying cause is addressed. Antiarrhythmic drug therapy is generally reserved for patients with recurrent AF and is not routinely administered to patients after cardioversion for new-onset AF. Appropriate thromboembolic prophylaxis is essential before and after cardioversion (discussed under heading "Prevention of Thromboembolism").

## Rate Control versus Rhythm Control in Recurrent AF

In approaching the patient with recurrent paroxysmal or persistent AF, the clinician must decide whether to attempt to maintain sinus rhythm. Theoretically, maintaining sinus rhythm would prevent symptoms associated with AF, normalize heart rate, maintain AV synchrony and the atrial contribution to cardiac output, and prevent deleterious atrial remodeling. Moreover, the epidemiological data on outcomes in AF raise hope that patients in whom sinus rhythm can be maintained might have improved quality of life and reduced mortality. On the other hand, the pharmacological tools available to maintain sinus rhythm are limited, and the attempt to maintain AF may be associated with side effects of these medications.

Two large clinical trials, AFFIRM and rate control versus electrical cardioversion (RACE), along with several smaller trials, have tested prospectively whether attempting to maintain sinus rhythm using antiarrhythmic drugs in patients with AF results is better clinical outcomes than simply controlling the ventricular rate in AF.[77,78] Two critical findings emerged from these studies: first, that there was no clear mortality benefit, cardiovascular benefit or clinically significant functional improvement associated with pursuing a rhythm control strategy in the patients enrolled in these studies; and second, that the incidence of thromboembolic events was similar regardless of the strategy chosen **(Figs 5A and B)**. The latter finding likely

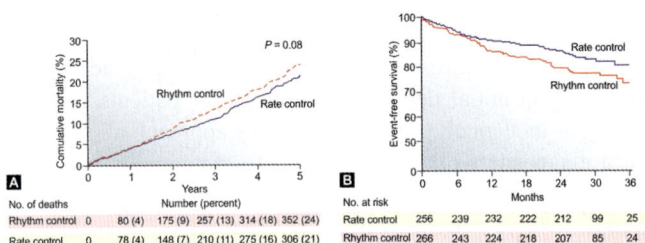

**FIGURES 5A AND B:** Comparison of outcomes in patients with AF pursuing a rhythm control and rate control strategy in two large clinical trials: (A) Data from AFFIRM shows no statistically significant difference in overall mortality, although there is a trend toward increased mortality with rhythm control. (*Source:* Wyse DG, Waldo AL, DiMarco JP, et al. A comparison of rate control and rhythm control in patients with atrial fibrillation. N Engl J Med. 2002;347:1825-33, with permission). (B) Similar findings were seen in the RACE trial for a composite endpoint including mortality and other adverse events. (*Source:* Van Gelder IC, Hagens VE, Bosker HA, et al. A comparison of rate control and rhythm control in patients with recurrent persistent atrial fibrillation. N Engl J Med. 2002;347:1834-40, with permission)

reflected the relatively poor efficacy of rhythm control in these patients: although 63% remained free of symptomatic AF in AFFIRM at 5 years after randomization, many of these patients had subclinical episodes of AF that contribute to thromboembolic risk.

It is important to recognize that the failure to demonstrate benefit to rhythm control is not due to a clinical equivalence between sinus rhythm and rate-controlled AF. Rather, patients in the rhythm control arms of these trials had antiarrhythmic drug-related side effects and an overall increase in noncardiovascular death,[79] along with frequent recurrences of AF despite the attempt at rhythm control. Indeed, analysis of mortality data from AFFIRM and RACE has shown that patients who were in sinus rhythm throughout the study, regardless of treatment arm chosen, had a hazard ratio for mortality of 0.53 compared to those in AF.[79,80] However, it is unclear whether the improved mortality was due to the fact that they were in sinus rhythm or whether those patients in whom sinus rhythm could be maintained with available therapy had fewer comorbidities or other factors that was associated with lower mortality risks regardless of what rhythm they were in. In these studies, any benefit was likely counterbalanced by drug toxicity and limited efficacy in the other patients, yielding a net equivalence of the two approaches. Thus, while maintaining sinus rhythm is in general a desirable outcome, the pharmacological tools employed in these studies lacked the safety and efficacy to do so in a way that provided net benefit in an unselected patient population. It should be noted that these studies were undertaken prior to the wide availability of catheter ablation for AF, which might significantly change the efficacy of sinus rhythm maintenance and the frequency of adverse events associated with rhythm control.

At a minimum, these studies legitimized rate control as a reasonable treatment option in asymptomatic patients. Currently, there is no algorithmic or guideline-driven approach that determines in whom rhythm control should be attempted. Relevant considerations are age, comorbidities, patient preference, risk of antiarrhythmic drugs, the likelihood of maintaining sinus rhythm, and whether the patient has symptoms due to AF even with adequate rate control. Thus, in older patients with structural heart disease and/or hypertension and no symptoms, rate control may be a reasonable first strategy; while in younger patients or in patients with lone AF with symptoms, rhythm control may be a reasonable initial choice.

## Restoration of Sinus Rhythm

Once a rhythm control strategy is selected, the initial step is to restore sinus rhythm. Reversion to sinus rhythm without early recurrence of AF is more likely when the duration of AF is less than one year, the left atrium is not markedly enlarged, and structural heart disease is minimal.[81] In unselected patients with AF, electrical cardioversion is associated with a 1–2% short-term risk of clinical thromboembolism in the absence of anticoagulation.[82] This risk can be reduced to an acceptable level if patients are therapeutic on warfarin for 1 month prior to the procedure, or if a transesophageal echo performed immediately prior to cardioversion reveals no intracardiac thrombus and intravenous heparin is administered prior to cardioversion.[83] It is also safe to cardiovert low-risk patients without a history of rheumatic heart disease or prior thromboembolism when the duration of the AF episode is less than 48 hours without a TEE. In these cases, intravenous heparin or equivalent should be administered before cardioversion.[84] After cardioversion, either electrical or chemical, a period of atrial stunning ensues, in which atrial function is reduced and the potential for thrombus formation remains high.[85] For that reason, patients should be therapeutically anticoagulated from the time of cardioversion for at least 1 month. After that, thromboembolic risk should be reassessed and addressed as indicated.

Electrical cardioversion is highly effective for restoration of sinus rhythm in patients with paroxysmal AF. Seventy to ninety percent of patients can be converted to SR using biphasic shocks.[86] Chemical cardioversion is also effective for reversion to sinus rhythm, but in general it is not as effective as electrical cardioversion. Class III agents, such as amiodarone, ibutilide and dofetilide, are the most effective drugs for cardioversion of long-standing AF, while class 1C agents flecainide and propafenone are also effective when the duration of AF is less than 7 days.[1] When using ibutilide or dofetilide, special attention should be paid to electrolytes and QT interval, as these medications confer a significant short-term risk of torsades-de-pointes. Facilitated electrical cardioversion, in which patients are loaded on an antiarrhythmic drug prior to cardioversion, can improve the success rate for patients who fail conventional electrical cardioversion.[87]

## Maintenance of Sinus Rhythm—Pharmacological Approaches

In general, once a decision to attempt rhythm control has been made, the choice of antiarrhythmic drugs is primarily dictated by risks and side-effect potential, which are largely determined by the presence or absence of structural heart disease. **Flowchart 1** shows a scheme recommended in the 2006 ACC/AHA/ESC Guidelines for management of AF. This section presents an overview of the Class 1 and Class 3 medications

**FLOWCHART 1:** 2006 American College of Cardiology (ACC)/American Heart Association (AHA)/European Society of Cardiology (ESC) algorithm for antiarrhythmic drug therapy to maintain sinus rhythm in patients with recurrent paroxysmal or persistent AF. Patients should first be categorized by severity of heart disease (left to right) and treatment selection should proceed from top to bottom. Within boxes, drugs are listed alphabetically and not by order of preference

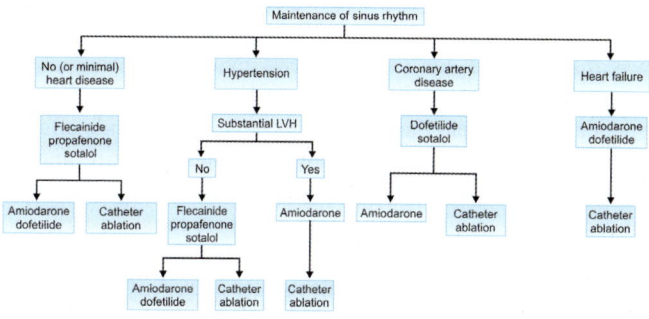

(*Source:* Reference 1)

used to maintain sinus rhythm and the evidence supporting their use in specific populations.

## Class 1: Antiarrhythmic Drugs

The class 1C medications flecainide and propafenone are effective for maintaining sinus rhythm in patients with paroxysmal AF and are widely used as first-line therapy in selected patients.[88–90] While these medications can be taken on a standing basis for rhythm maintenance, they can also be used on an as-needed basis ("pill-in-the-pocket" approach) for patients with symptomatic paroxysmal AF.[91] The main cardiac side effects of these agents are pro-arrhythmia and increased mortality in patients with ischemic or structural heart disease,[92] bradyarrhythmias in patients with infranodal conduction system disease, and worsening heart failure in patients with LV dysfunction. 1C agents are therefore not recommended for use in patients with any structural heart lesions. Since the ventricular pro-arrhythmia is thought to occur during ischemia, even asymptomatic patients on 1C agents should be screened for ischemic heart disease while receiving drug therapy. In addition, class 1C medications can convert AF to atrial flutter that can be conducted 1:1 by the AV node, resulting in extremely rapid ventricular response.[93] Patients on 1C medications should therefore also take an AV nodal blocking agent unless AV conduction is not present or is significantly impaired. In the past, the Class 1A antiarrhythmic drugs, such as quinidine or procainamide, have been used for sinus rhythm maintenance in patients with AF. These agents are no longer widely used for treatment of AF because they are often poorly tolerated.[94]

## Class 3: Antiarrhythmic Drugs

Although amiodarone is not FDA approved for the treatment of AF, it is widely used for this indication and is the most effective antiarrhythmic agent for sinus rhythm maintenance.[95] Although it has minimal proarrhythmic effects, its extracardiac side effect profile is significant, particularly at higher doses and with prolonged treatment.[96] In clinical trials, amiodarone was discontinued due to side effects more often than sotalol or propafenone. Nevertheless, amiodarone had significantly greater efficacy than the other drugs.[97] Despite its side effects, it is also safe to use in patients with heart failure and in patients with CAD.[98,99] For these reasons (efficacy and safety), amiodarone is frequently the drug of choice for short-term use and for patients with significant comorbidities. Nevertheless, because of extracardiac side effects, the 2006 ACC/AHA/ESC guidelines recommend that amiodarone be used as a second-line therapy except for patients with heart failure, moderate-to-severe systolic dysfunction, or hypertension and significant left ventricular hypertrophy. Amiodarone has an additional use in the prevention of postoperative AF after cardiac surgery, where a short perioperative course cuts the incidence of AF by about 50%.[100]

Sotalol is roughly equal in efficacy to propafenone in randomized trials and significantly less effective that amiodarone in maintaining sinus rhythm.[97] Its effects are highly dose-dependent, with greater rhythm maintenance effects at higher doses and predominantly beta-blocking effects at lower doses.[101] The presence of beta-blocking effects can be helpful in slowing ventricular rate during recurrences of AF.[102] Side effects can include bradycardia and QT interval prolongation. For that reason, it is essential to monitor the QT interval in patients taking sotalol and to avoid use of the drug in patients at risk for QT interval prolongation or with significant impairment in drug clearance due to kidney disease. In most cases, inpatient monitoring is required for initiation of sotalol due to the risk of torsades de pointes and bradyarrhythmias.

Dofetilide is at least as effective as sotalol in maintaining sinus rhythm in patients with recurrent AF.[103] Although dofetilide prolongs the QT interval in a dose-dependent fashion and can cause torsades de pointes, an increase in mortality has not been observed in clinical trials of dofetilide use for AF.[104] This is likely because dofetilide is administered during inpatient monitoring with careful assessment of QT interval and dose adjustment or discontinuation as needed. Dofetilide is also safe and effective in patients with severe heart failure, in whom it reduces AF recurrences and heart failure hospitalizations without affecting mortality.[105] However, because of the potential for proarrhythmia with this medication, dofetilide is as considered a second-line medication except for patients with heart failure or coronary artery disease.

Dronedarone is a non-iodinated chemical derivative of amiodarone that lacks extracardiac side effects, such as pulmonary, thyroid and ocular toxicity. Several large clinical trials have found that it can be effective in maintaining sinus rhythm, although less than amiodarone. However, unlike amiodarone, dronedarone appears to cause increased mortality in patients with severe heart failure.[106] In patients with normal cardiac function or mild-to-moderate heart failure, a large prospective randomized trial found a significant reduction in a composite primary outcome of hospitalization for cardiovascular causes and cardiovascular death, without a significant effect on overall mortality.[107]

## Modulators of the RAAS System

Due to the role of the RAAS system in the pathogenesis of AF, the use of angiotensin receptor blockers (ARB) and angiotensin converting enzyme inhibitors has been explored in AF. A recent meta-analysis of 23 randomized trials showed a roughly 33% reduction in AF episodes associated with RAAS inhibition.[108] Particularly for patients with hypertension, left ventricular dysfunction or diabetes, but possibly even for patients without many comorbidities, RAAS inhibition can be an important adjunctive therapy for AF.

## Maintenance of Sinus Rhythm—Invasive Approaches

Nonpharmacological approaches to maintenance of sinus rhythm include catheter ablation, surgical ablation, pacing, and atrial defibrillation. The finding that physiological pacing modes in bradycardic patients with AF can prevent recurrences has led to a variety of pacing strategies to maintain sinus rhythm, including multisite pacing, alternative site pacing, and overdrive suppression of AF. With the exception of small subpopulations these strategies have not been shown to be effective.[109] Patients with AF should therefore not receive permanent pacemakers for the purpose of AF suppression, although selection of physiological pacing modes and minimization of ventricular pacing can prevent AF episodes in patients receiving pacemakers for other reasons. Thus far, the safety, efficacy and tolerability of implantable atrial defibrillators have not been demonstrated in large trials, and therefore these devices remain largely investigational.

## Catheter Ablation

The development of catheter ablation for AF began with the observation that rapid firing originating in the pulmonary veins frequently triggered AF.[32] This finding led to the idea of electrically isolating the pulmonary venous myocardium from the left atrium using radiofrequency ablation.[110–112] Since the original description of this approach, intensive development and testing of different approaches and techniques has taken place **(Figs 6A and B)**.

At present, ablation techniques for AF vary widely among practitioners and the optimal approach has yet to be defined.[113] Most commonly, contiguous or nearly contiguous lesions are created around each of the four pulmonary veins, and electrical isolation is confirmed with a combination of recording and pacing within the pulmonary veins. Many ablationists also target nonpulmonary vein AF triggers such as autonomic ganglia, the ligament and vein of Marshall, the superior vena cava,[114] the posterior left atrial wall and the left atrial appendage.[115] There has also been interest in modifying the atrial substrate using catheter ablation, particularly for patients with permanent AF. Additional ablations are sometimes performed in areas displaying complex, fractionated electrograms that may play a role in AF maintenance,[116] although the added benefit of such an approach has yet to be established in large clinical trials.[117]

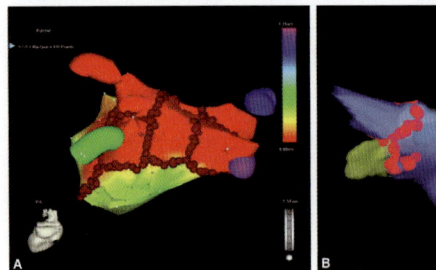
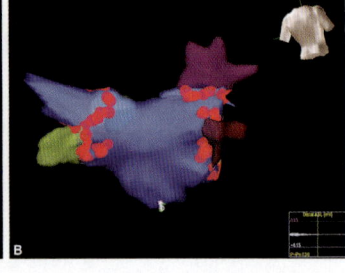

**FIGURES 6A AND B:** Two approaches to atrial fibrillation ablation: (A) Linear left atrial ablation. Linear lesions (red balls) are shown superimposed on a posterior-anterior view of the left atrium generated with the CARTO electroanatomical mapping system (Biosense). The following lines are represented: A wide area circumferential ablation around the left and right pulmonary veins, a line from the left inferior pulmonary vein to the mitral annulus, lines between the upper and the lower pulmonary veins, and two lines along the posterior left atrium connecting the circumferential ablations. Reproduced with permission from: Pappone C and Santinelli V. Heart Rhythm. 2006;3:1105-9. (B) Segmental pulmonary vein isolation. In this approach, lesions (red balls) are created surrounding the ostia of the four pulmonary veins in ordert to achieve electrical isolation. A PA view is shown, with lesions superimposed on a CT registered electroanatomic map created with the NavX system (St. Jude Medical). (*Source.* Dr Nitish Badhwar)

Because ablation causes atrial inflammation in the short run, early recurrence of AF is common after ablation and does not necessarily indicate long-term procedural non-success.[118] For this reason, success is usually defined as symptomatic improvement with no evidence of AF after a post-procedure blanking period. When AF recurs after the blanking period, the most common reason is reconnection of pulmonary venous myocardium to the atria, which may necessitate additional ablation procedures.[119] Catheter ablation is most effective in patients with paroxysmal AF, normal atrial size and minimal structural heart disease.[120] Success rates of 70–90% at 1 year have been reported for such patients.[113] When patients with structural heart disease, heart failure and increased atrial size are included in such studies, success rate declines. Success can be achieved in some patients by adding an antiarrhythmic medication after ablation, even when the medication was not effective prior to ablation.

Although AF ablation has become safer as techniques have improved, there is potential for serious complications. In addition to the usual risks associated with invasive cardiac procedures, such as thromboembolism, cardiac perforation, and vascular complications, AF ablation also carries the risks of iatrogenic left atrial flutter,[121] pulmonary vein stenosis,[122] extracardiac nerve injury[113] and atrial–esophageal fistula.[123] Although these complications are uncommon, they can be devastating when they occur. These concerns, along with limited data on long-term efficacy, are reflected in the 2006 ACC/AHA/ESC guidelines, in which ablation is second-line therapy for AF. While the results of initial trials of ablation versus antiarrhythmic drug therapy for paroxysmal AF have generally favored the invasive approach,[124–126] this comparison has not yet been made in large multicenter trials, nor has the question of whether anticoagulation can be discontinued after successful AF ablation. Trials currently underway will hopefully shed additional light on these issues and help to clarify the proper place of catheter ablation in a rhythm maintenance strategy.

## Surgical Procedures for AF Maintenance

Surgical procedures to maintain sinus rhythm in patients with AF predate the era of catheter ablation.[127] The Cox MAZE procedure, which may include a "cut and sew" approach or an ablation approach, involves making a patchwork of lesions in the atrium to create lines of scar. The lines of scar presumably prevent reentry circuits from sustaining and are believed to prevent vulnerable atrial substrate from maintaining AF. As with catheter ablation, many different lesion sets, ablative methods and strategies have been employed for surgical

AF ablation.[128–130] Observational studies suggest that these techniques are highly effective; however, large randomized trials have not been carried out and surgical technique is highly variable, so overall efficacy of surgical management of AF is not known.[131,132] In addition, sinus node injury or exit block requiring permanent pacing can complicate the procedures, and atrial function may be permanently impaired.[133,134] Finally, as with catheter ablation, lesion sets can be proarrhythmic by creating conduction barriers and slowing conduction velocity, thereby facilitating macroreentrant circuits.[135] Because these procedures require open cardiac surgery with cardioplegia and cross-clamping of the aorta, they are usually performed on patients who require cardiac surgery for a structural lesion associated with AF such as mitral valve disease.

## Strategies for Rate Control

The recommended target for rate control in AF is 80 bpm at rest and less than 120 bpm with moderate activity, although these numbers are not based on prospective trials.[1] Recent prospective data, in which patients randomized to a permissive rate control arm had no worse outcomes than patients in whom strict rate control was achieved have called these numbers into question.[136] Further research will be necessary to determine the optimal targets for rate control.

The mainstays for rate control in AF are beta-adrenergic blockers and the non-dihydropyridine calcium channel blockers diltiazem and verapamil. These medications slow AV nodal conduction and prolong AV nodal refractoriness, thereby reducing the frequency of fibrillatory waves that can be conducted to the ventricles. In retrospective studies of AFFIRM patients, beta-blockers alone or combined with digoxin were most effective for rate control.[137] In patients with reduced EF and symptoms of heart failure in AF, intravenous calcium channel blockers should be avoided due to the potential for causing symptomatic hypotension and in severe cases precipitating cardiogenic shock. In such patients, beta-blockers, amiodarone or digoxin are preferred. As oral therapy for ambulatory AF patients, digoxin can be a useful adjunctive agent for rate control, but is less effective as monotherapy for rate control in AF and should not be used in this way.[137] Dronederone is also an effective rate control agent and can be used for this purpose even if not effective at maintaining sinus rhythm. The side-effect profile and pharmacokinetics may make this more attractive than amiodarone for this purpose.

Not infrequently, patients with paroxysmal AF and rapid ventricular response requiring rate control medications will experience sinus pauses or symptomatic bradycardia when AF converts spontaneously to sinus rhythm. Since rate control medications are essential in these patients, permanent pacemaker implantation may become necessary to permit higher doses of nodal agents. Conversely, in patients in whom attempts at rate control have failed or are not tolerated and are not candidates for rhythm control, catheter ablation of the AV junction with pacemaker implantation is highly effective for rate control.[138,139] For patients whose symptoms are primarily due to elevated heart rates, this is a very effective therapy for symptomatic AF.

## Prevention of Thromboembolism

It is imperative that all patients with AF undergo risk stratification for thromboembolic events, regardless of AF type and regardless of treatment strategy (rhythm control vs rate control).[1] Anticoagulation with warfarin lowers the risk of stroke for nearly all patients with AF;[140] however, in the lowest risk patients, the risk of major bleeding due to warfarin therapy exceeds the value of this marginal risk reduction.[141] For such patients, aspirin can be an acceptable alternative. To facilitate the categorization of AF patients by stroke risk, prediction tools have been developed based on pooled data from several large stroke prevention trials and other smaller studies.

Patients with AF due to rheumatic heart disease represent the highest risk group because of the marked atrial enlargement and consequent stasis that typically accompanies mitral stenosis. These patients should be anticoagulated with warfarin unless a strong contraindication is present. For patients with nonvalvular AF, several studies have evaluated the clinical predictors for stroke. The Atrial Fibrillation Investigators (AFI), using pooled data from several trials, and the Stroke Prevention and Atrial Fibrillation (SPAF) investigators each used data from the non-treatment arms of primary prevention trials of stroke in AF patients.[142,143] Clinical factors that predicted risk of stroke were then integrated to form the CHADS2 score, and this tool was validated using results from the National Registry of Atrial Fibrillation.[11] The CHADS2 score assigns 1 point each for a history of congestive heart failure, hypertension, age greater than 74, diabetes mellitus and two points for a prior history of systemic embolic event. An overall risk score of 0 suggests low risk, 1 or 2 suggests intermediate risk and greater than 2 is considered high risk **(Table 3)**. Based on these data, current guidelines recommend aspirin for patients with a CHADS2 score of 0 and warfarin in patients with a CHADS2 score of 2 or greater. Patients with a CHADS2 score of 1 may use aspirin or warfarin depending on comorbidities and patient preference. Currently there are no other approved pharmacological treatments to prevent strokes; however, there are several direct

### TABLE 3

Event rates by risk factor, baseline CHADS2 score, and anticoagulation status in 11,526 adults with atrial fibrillation and no contraindications to warfarin therapy at baseline

| CHADS2 score (no. of patients) | Event rate (per 100 person-years) (95% confidence interval) | | Crude rate ratio (95% confidence interval) |
|---|---|---|---|
| | Taking warfarin | Not taking warfarin | |
| 0 (2557) | 0.25 (0.11–0.55) | 0.49 (0.30–0.78) | 0.50 (0.2–1.28) |
| 1 (3862) | 0.72 (0.50–1.03) | 1.52 (1.19–1.94) | 0.47 (0.30–0.73) |
| 2 (2955) | 1.27 (0.94–1.72) | 2.50 (1.98–3.15) | 0.51 (0.35–0.75) |
| 3 (1555) | 2.20 (1.61–3.01) | 5.27 (4.15–6.70) | 0.42 (0.28–0.62) |
| 4 (556) | 2.35 (1.44–3.83) | 6.02 (3.90–9.29) | 0.39 (0.20–0.75) |
| 5 or 6 (241) | 4.60 (2.72–7.76) | 6.88 (3.42–13.84) | 0.67 (0.28–1.60) |

(Source: Reference 148)

thrombin inhibitors and factor Xa inhibitors under various phases of study.[144] Nonpharmacological treatments to prevent strokes in AF have targeted the left atrial appendage, either by occlusion or removal. While this has traditionally been accomplished by surgery,[145] there are several transvenous devices under investigation that occlude or ligate the left atrial appendage.[146] These approaches may be useful in patients at high risk for stroke, but who are not capable of taking warfarin or in patients who have had a stroke from AF on therapeutic doses of warfarin. While small clinical trials have demonstrated the feasibility of percutaneous approaches,[147,148] large randomized clinical trials to test rigorously for reduction in stroke have not been carried out.

## CONCLUSION

Although major advances in the understanding and treatment of AF have occurred recently—such as the understanding of the role of the pulmonary veins in AF and the development of catheter ablation for AF—we still do not have an understanding of the etiology of the underlying substrate of AF and thus no targeted treatments to prevent or reverse AF exist. Current treatment strategies are aimed at preventing stroke, and either rate control or rhythm control to prevent rapid ventricular rates and symptoms. If the latter approach is chosen, the choice of rhythm control drugs is dictated by side-effect profile and risk, rather than efficacy, since efficacies are similar for long-term. Until rigorous multicenter randomized trials have been completed, ablation is reserved for symptomatic patients with paroxysmal AF who are intolerant or resistant to pharmacological rhythm control. Success rate for ablation of persistent and permanent AF is significantly lower than that for paroxysmal AF. In general, treatment options for persistent and permanent AF are limited to stroke prevention and rate control. Future research will hopefully define the substrate(s) that predispose to AF and thereby allow directed therapy to prevent or reverse AF.

## Guidelines

### 2006 ACC/AHA ESC guidelines: Pharmacological management of newly discovered AF

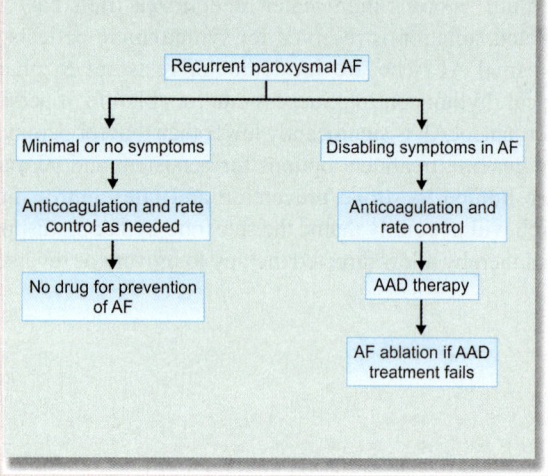

(Abbreviation: AAD: Antiarrhythmic drugs). (*Source:* Fuster V, et al. J Am Coll Cardiol. 2006;48(4):e149-e246)

### 2006 ACC/AHA/ESC guidelines: Pharmacological management of recurrent paroxysmal AF

(*Source:* Fuster V, et al. J Am Coll Cardiol. 2006;48(4):e149-e246)

## 2006 ACC/AHA ESC guidelines: Pharmacological management of recurrent persistent or permanent AF

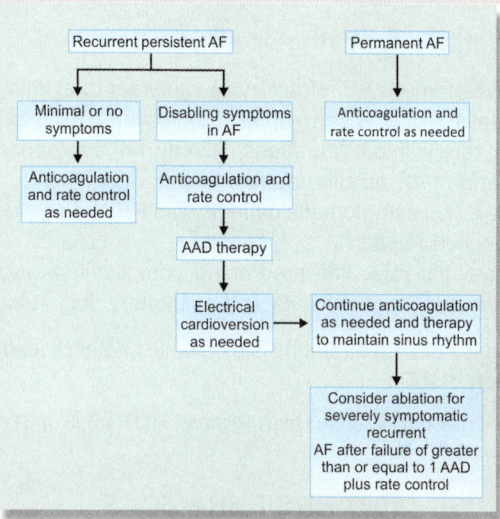

(*Source:* Fuster V, et al. J Am Coll Cardiol. 2006;48(4):e149-e246)

## 2006 ACC/AHA/ESC guidelines: Antiarrhythmic approaches to maintain sinus rhythm in patients with recurrent paroxysmal or persistent AF who require sinus rhythm*

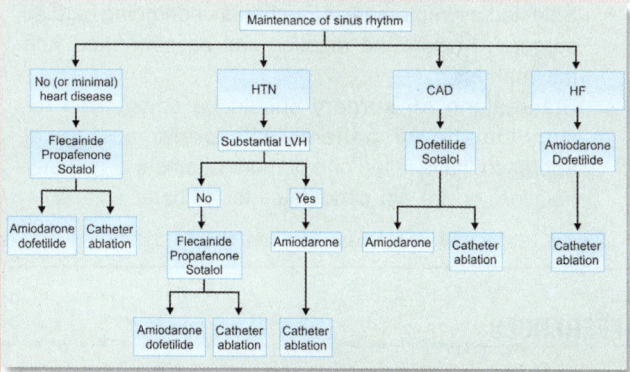

*Within each box, drugs are listed alphabetically and not in order of suggested use. The vertical flow indicates order of preference under each condition. The seriousness of heart disease proceeds from left to right, and selection of therapy in patientw with multiple conditions depends on the most serious condition present. (Abbreviation: LVH: Left ventricular hypertrophy). (*Source:* Fuster V, et al. J Am Coll Cardiol. 2006;48(4):e149-e246).

## 2007 HRS/EHRA/ECAS Expert Consensus Statement

### Indications for catheter AF ablation

- Symptomatic AF refractory or intolerant to at least one Class 1 or 3 antiarrhythmic medication
- In rare clinical situations, it may be appropriate to perform AF ablation as first-line therapy
- Selected symptomatic patients with HF and/or reduced ejection fraction
- Potential rare, life-threatening complications include atrio-esophageal fistula and pulmonary vein stenosis

[Presence of a LA thrombus is a contraindication to catheter ablation of AF]

(*Source:* Calkins H, et al. Heart Rhythm. 2007;4(6):816-61)

## 2007 HRS/EHRA/ECAS Expert Consensus Statement

### Indications for surgical ablation

- Symptomatic AF patients undergoing other cardiac surgery
- Selected asymptomatic AF patients undergoing cardiac surgery in whom the ablation can be performed with minimal risk
- Stand-alone AF surgery should be considered for symptomatic AF patients who prefer a surgical approach, have failed one or more attempts at catheter ablation, or are not candidates for catheter ablation

(*Source:* Calkins H, et al. Heart Rhythm. 2007;4(6):816-61)

## REFERENCES

1. Fuster V, Rydén LE, Cannom DS, et al. ACC/AHA/ESC 2006 Guidelines for the Management of Patients with Atrial Fibrillation: a report of the American College of Cardiology/American Heart Association Task Force on Practice Guidelines and the European Society of Cardiology Committee for Practice Guidelines (Writing Committee to Revise the 2001 Guidelines for the Management of Patients With Atrial Fibrillation): developed in collaboration with the European Heart Rhythm Association and the Heart Rhythm Society. Circulation. 2006;114:e257-e354.
2. Go AS, Hylek EM, Phillips KA, et al. Prevalence of diagnosed atrial fibrillation in adults: national implications for rhythm management and stroke prevention: the AnTicoagulation and Risk Factors in Atrial Fibrillation (ATRIA) study. JAMA. 2001;285:2370-5.

3. Miyasaka Y, Barnes ME, Gersh BJ, et al. Secular trends in incidence of atrial fibrillation in Olmsted County, Minnesota, 1980 to 2000, and implications on the projections for future prevalence. Circulation. 2006;114:119-25.
4. Heeringa J, van der Kuip DAM, Hofman A, et al. Prevalence, incidence and lifetime risk of atrial fibrillation: the Rotterdam study. European Heart Journal. 2006;27:949-53.
5. Naccarelli GV, Varker H, Lin J, et al. Increasing prevalence of atrial fibrillation and flutter in the United States. Am J Cardiol. 2009;104:1534-9.
6. Kerr CR, Humphries KH, Talajic M, et al. Progression to chronic atrial fibrillation after the initial diagnosis of paroxysmal atrial fibrillation: results from the Canadian Registry of Atrial Fibrillation. Am Heart J. 2005;149:489-96.
7. Kopecky SL, Gersh BJ, McGoon MD, et al. The natural history of lone atrial fibrillation. A population-based study over three decades. N Engl J Med. 1987;317:669-74.
8. Stewart S, Hart CL, Hole DJ, et al. A population-based study of the long-term risks associated with atrial fibrillation: 20-year follow-up of the Renfrew/Paisley study. Am J Med. 2002;113:359-64.
9. Benjamin EJ, Wolf PA, D'Agostino RB, et al. Impact of atrial fibrillation on the risk of death: the Framingham heart study. Circulation. 1998;98:946-52.
10. Frost L, Engholm G, Johnsen S, et al. Incident stroke after discharge from the hospital with a diagnosis of atrial fibrillation. Am J Med. 2000;108:36-40.
11. Gage BF, Waterman AD, Shannon W, et al. Validation of clinical classification schemes for predicting stroke: results from the National Registry of Atrial Fibrillation. JAMA. 2001;285:2864-70.
12. Anderson DC, Kappelle LJ, Eliasziw M, et al. Occurrence of hemispheric and retinal ischemia in atrial fibrillation compared with carotid stenosis. Stroke. 2002;33:1963-7.
13. Lin HJ, Wolf PA, Kelly-Hayes M, et al. Stroke severity in atrial fibrillation. The Framingham Study. Stroke. 1996;27:1760-4.
14. Jorgensen HS, Nakayama H, Reith J, et al. Acute stroke with atrial fibrillation. The Copenhagen Stroke Study. Stroke. 1996;27:1765-9.
15. Lamassa M, Di Carlo A, Pracucci G, et al. Characteristics, outcome, and care of stroke associated with atrial fibrillation in Europe: data from a multicenter multinational hospital-based registry (The European Community Stroke Project). Stroke. 2001;32:392-8.
16. Ott A, Breteler MM, de Bruyne MC, et al. Atrial fibrillation and dementia in a population-based study. The Rotterdam Study. Stroke. 1997;28:316-21.
17. Kilander L, Andren B, Nyman H, et al. Atrial fibrillation is an independent determinant of low cognitive function: a cross-sectional study in elderly men. Stroke. 1998;29:1816-20.
18. Miyasaka Y, Barnes ME, Petersen RC, et al. Risk of dementia in stroke-free patients diagnosed with atrial fibrillation: data from a community-based cohort. Eur Heart J. 2007;28:1962-7.
19. Dries DL, Exner DV, Gersh BJ, et al. Atrial fibrillation is associated with an increased risk for mortality and heart failure progression in patients with asymptomatic and symptomatic left ventricular systolic dysfunction: a retrospective analysis of the SOLVD trials. Studies of left ventricular dysfunction. J Am Coll Cardiol. 1998;32:695-703.

20. Olsson LG, Swedberg K, Ducharme A, et al. Atrial fibrillation and risk of clinical events in chronic heart failure with and without left ventricular systolic dysfunction: results from the Candesartan in Heart failure-Assessment of Reduction in Mortality and morbidity (CHARM) program. J Am Coll Cardiol. 2006;47:1997-2004.
21. Carson PE, Johnson GR, Dunkman WB, et al. The influence of atrial fibrillation on prognosis in mild to moderate heart failure. The V-HeFT Studies. The V-HeFT VA Cooperative Studies Group. Circulation. 1993;87:VI102-10.
22. Vaziri SM, Larson MG, Benjamin EJ, et al. Echocardiographic predictors of nonrheumatic atrial fibrillation. The Framingham Heart Study. Circulation. 1994;89:724-30.
23. Diker E, Aydogdu S, Ozdemir M, et al. Prevalence and predictors of atrial fibrillation in rheumatic valvular heart disease. Am J Cardiol. 1996;77:96-8.
24. Maisel WH, Stevenson LW. Atrial fibrillation in heart failure: epidemiology, pathophysiology, and rationale for therapy. Am J Cardiol. 2003;91:2D-8D.
25. Kieny JR, Sacrez A, Facello A, et al. Increase in radionuclide left ventricular ejection fraction after cardioversion of chronic atrial fibrillation in idiopathic dilated cardiomyopathy. Eur Heart J. 1992;13:1290-5.
26. Redfield MM, Kay GN, Jenkins LS, et al. Tachycardia-related cardiomyopathy: a common cause of ventricular dysfunction in patients with atrial fibrillation referred for atrioventricular ablation. Mayo Clin Proc. 2000;75:790-5.
27. Crenshaw BS, Ward SR, Granger CB, et al. Atrial fibrillation in the setting of acute myocardial infarction: the GUSTO-I experience. Global utilization of streptokinase and TPA for occluded coronary arteries. J Am Coll Cardiol. 1997;30:406-13.
28. Wong CK, White HD, Wilcox RG, et al. New atrial fibrillation after acute myocardial infarction independently predicts death: the GUSTO-III experience. Am Heart J. 2000;140:878-85.
29. Robinson K, Frenneaux MP, Stockins B, et al. Atrial fibrillation in hypertrophic cardiomyopathy: a longitudinal study. J Am Coll Cardiol. 1990;15:1279-85.
30. Pires LA, Wagshal AB, Lancey R, et al. Arrhythmias and conduction disturbances after coronary artery bypass graft surgery: epidemiology, management, and prognosis. Am Heart J. 1995;129:799-808.
31. Maisel WH, Rawn JD, Stevenson WG. Atrial fibrillation after cardiac surgery. Ann Intern Med. 2001;135:1061-73.
32. Haissaguerre M, Jais P, Shah DC, et al. Spontaneous initiation of atrial fibrillation by ectopic beats originating in the pulmonary veins. N Engl J Med. 1998;339:659-66.
33. Chen SA, Hsieh MH, Tai CT, et al. Initiation of atrial fibrillation by ectopic beats originating from the pulmonary veins: electrophysiological characteristics, pharmacological responses, and effects of radiofrequency ablation. Circulation. 1999;100:1879-86.
34. Chen SA, Tai CT, Hsieh MH, et al. Radiofrequency catheter ablation of atrial fibrillation initiated by spontaneous ectopic beats. Curr Cardiol Rep. 2000;2:322-8.
35. Wit AL, Boyden PA. Triggered activity and atrial fibrillation. Heart Rhythm. 2007;4:S17-S23.
36. Sauer WH, Alonso C, Zado E, et al. Atrioventricular nodal reentrant tachycardia in patients referred for atrial fibrillation ablation: response

to ablation that incorporates slow-pathway modification. Circulation. 2006;114:191-5.
37. Katritsis DG, Giazitzoglou E, Wood MA, et al. Inducible supraventricular tachycardias in patients referred for catheter ablation of atrial fibrillation. Europace. 2007;9:785-9.
38. Sharma AD, Klein GJ, Guiraudon GM, et al. Atrial fibrillation in patients with Wolff-Parkinson-White syndrome: incidence after surgical ablation of the accessory pathway. Circulation. 1985;72:161-9.
39. Hamada T, Hiraki T, Ikeda H, et al. Mechanisms for atrial fibrillation in patients with Wolff-Parkinson-White syndrome. J Cardiovasc Electrophysiol. 2002;13:223-9.
40. Cheng S, Keyes MJ, Larson MG, et al. Long-term outcomes in individuals with prolonged PR interval or first-degree atrioventricular block. JAMA. 2009;301:2571-7.
41. Sanders P, Morton JB, Kistler PM, et al. Electrophysiological and electroanatomic characterization of the atria in sinus node disease: evidence of diffuse atrial remodeling. Circulation. 2004;109:1514-22.
42. Coumel P. Autonomic influences in atrial tachyarrhythmias. J Cardiovasc Electrophysiol. 1996;7:999-1007.
43. Herweg B, Dalal P, Nagy B, et al. Power spectral analysis of heart period variability of preceding sinus rhythm before initiation of paroxysmal atrial fibrillation. Am J Cardiol. 1998;82:869-74.
44. Baseline characteristics of patients with atrial fibrillation: the AFFIRM study. Am Heart J. 2002;143:991-1001.
45. Benjamin EJ, Levy D, Vaziri SM, et al. Independent risk factors for atrial fibrillation in a population-based cohort. The Framingham Heart Study. JAMA. 1994;271:840-4.
46. Krahn AD, Klein GJ, Kerr CR, et al. How useful is thyroid function testing in patients with recent-onset atrial fibrillation? The Canadian Registry of Atrial Fibrillation Investigators. Arch Intern Med. 1996;156:2221-4.
47. Sawin CT, Geller A, Wolf PA, et al. Low serum thyrotropin concentrations as a risk factor for atrial fibrillation in older persons. N Engl J Med. 1994;331:1249-52.
48. Auer J, Scheibner P, Mische T, et al. Subclinical hyperthyroidism as a risk factor for atrial fibrillation. Am Heart J. 2001;142:838-42.
49. Frost L, Vestergaard P, Mosekilde L. Hyperthyroidism and risk of atrial fibrillation or flutter: a population-based study. Arch Intern Med. 2004;164:1675-8.
50. Wang TJ, Parise H, Levy D, et al. Obesity and the risk of new-onset atrial fibrillation. JAMA. 2004;292:2471-7.
51. Polanczyk CA, Goldman L, Marcantonio ER, et al. Supraventricular arrhythmia in patients having noncardiac surgery: clinical correlates and effect on length of stay. Ann Intern Med. 1998;129:279-85.
52. Salman S, Bajwa A, Gajic O, et al. Paroxysmal atrial fibrillation in critically ill patients with sepsis. J Intensive Care Med. 2008;23: 178-83.
53. Buch P, Friberg J, Scharling H, et al. Reduced lung function and risk of atrial fibrillation in the Copenhagen City Heart Study. Eur Respir J. 2003;21:1012-6.
54. Gami AS, Pressman G, Caples SM, et al. Association of atrial fibrillation and obstructive sleep apnea. Circulation. 2004;110:364-7.
55. Frost L, Vestergaard P. Alcohol and risk of atrial fibrillation or flutter: a cohort study. Arch Intern Med. 2004;164:1993-8.

56. Mukamal KJ, Tolstrup JS, Friberg J, et al. Alcohol consumption and risk of atrial fibrillation in men and women: the Copenhagen City Heart Study. Circulation. 2005;112:1736-42.
57. Djousse L, Levy D, Benjamin EJ, et al. Long-term alcohol consumption and the risk of atrial fibrillation in the Framingham study. Am J Cardiol. 2004;93:710-3.
58. van der Hooft CS, Heeringa J, van Herpen G, et al. Drug-induced atrial fibrillation. J Am Coll Cardiol. 2004;44:2117-24.
59. Frost L, Vestergaard P. Caffeine and risk of atrial fibrillation or flutter: the Danish Diet, Cancer, and Health Study. Am J Clin Nutr. 2005;81:578-82.
60. Levy S, Maarek M, Coumel P, et al. Characterization of different subsets of atrial fibrillation in general practice in France: the ALFA study. The College of French Cardiologists. Circulation. 1999;99: 3028-35.
61. Kannel WB, Abbott RD, Savage DD, et al. Epidemiologic features of chronic atrial fibrillation: the Framingham study. N Engl J Med. 1982;306:1018-22.
62. Frustaci A, Chimenti C, Bellocci F, et al. Histological substrate of atrial biopsies in patients with lone atrial fibrillation. Circulation. 1997;96:1180-4.
63. Fox CS, Parise H, D'Agostino RB, Sr, et al. Parental atrial fibrillation as a risk factor for atrial fibrillation in offspring. JAMA. 2004;291:2851-5.
64. Marcus GM, Smith LM, Vittinghoff E, et al. A first-degree family history in lone atrial fibrillation patients. Heart Rhythm. 2008;5: 826-30.
65. Lubitz SA, Ozcan C, Magnani JW, et al. Genetics of atrial fibrillation: implications for future research directions and personalized medicine. Circ Arrhythm Electrophysiol. 2010;3:291-9.
66. Gollob MH, Jones DL, Krahn AD, et al. Somatic mutations in the connexin 40 gene (GJA5) in atrial fibrillation. N Engl J Med. 2006;354:2677-88.
67. Thibodeau IL, Xu J, Li Q, et al. Paradigm of genetic mosaicism and lone atrial fibrillation: physiological characterization of a connexin 43-deletion mutant identified from atrial tissue. Circulation. 2010;122:236-44.
68. Gudbjartsson DF, Arnar DO, Helgadottir A, et al. Variants conferring risk of atrial fibrillation on chromosome 4q25. Nature. 2007;448: 353-7.
69. Mommersteeg MT, Brown NA, Prall OW, et al. Pitx2c and Nkx2-5 are required for the formation and identity of the pulmonary myocardium. Circ Res. 2007;101:902-9.
70. Israel CW, Gronefeld G, Ehrlich JR, et al. Long-term risk of recurrent atrial fibrillation as documented by an implantable monitoring device: implications for optimal patient care. J Am Coll Cardiol. 2004;43:47-52.
71. Lok NS, Lau CP. Presentation and management of patients admitted with atrial fibrillation: a review of 291 cases in a regional hospital. Int J Cardiol. 1995;48:271-8.
72. Timmermans C, Smeets JL, Rodriguez LM, et al. Aborted sudden death in the Wolff-Parkinson-White syndrome. Am J Cardiol. 1995;76:492-4.

73. Page RL, Wilkinson WE, Clair WK, et al. Asymptomatic arrhythmias in patients with symptomatic paroxysmal atrial fibrillation and paroxysmal supraventricular tachycardia. Circulation. 1994;89:224-7.
74. Kinlay S, Leitch JW, Neil A, et al. Cardiac event recorders yield more diagnoses and are more cost-effective than 48-hour Holter monitoring in patients with palpitations. A controlled clinical trial. Ann Intern Med 1996;124:16-20.
75. Michael JA, Stiell IG, Agarwal S, et al. Cardioversion of paroxysmal atrial fibrillation in the emergency department. Ann Emerg Med. 1999;33:379-87.
76. Danias PG, Caulfield TA, Weigner MJ, et al. Likelihood of spontaneous conversion of atrial fibrillation to sinus rhythm. J Am Coll Cardiol. 1998;31:588-92.
77. Wyse DG, Waldo AL, DiMarco JP, et al. A comparison of rate control and rhythm control in patients with atrial fibrillation. N Engl J Med. 2002;347:1825-33.
78. Van Gelder IC, Hagens VE, Bosker HA, et al. A comparison of rate control and rhythm control in patients with recurrent persistent atrial fibrillation. N Engl J Med. 2002;347:1834-40.
79. Corley SD, Epstein AE, DiMarco JP, et al. Relationships between sinus rhythm, treatment, and survival in the Atrial Fibrillation Follow-Up Investigation of Rhythm Management (AFFIRM) study. Circulation. 2004;109:1509-13.
80. Steinberg JS, Sadaniantz A, Kron J, et al. Analysis of cause-specific mortality in the Atrial Fibrillation Follow-up Investigation of Rhythm Management (AFFIRM) study. Circulation. 2004;109:1973-80.
81. Dittrich HC, Erickson JS, Schneiderman T, et al. Echocardiographic and clinical predictors for outcome of elective cardioversion of atrial fibrillation. Am J Cardiol. 1989;63:193-7.
82. Gentile F, Elhendy A, Khandheria BK, et al. Safety of electrical cardioversion in patients with atrial fibrillation. Mayo Clin Proc. 2002;77:897-904.
83. Klein AL, Grimm RA, Murray RD, et al. Use of transesophageal echocardiography to guide cardioversion in patients with atrial fibrillation. N Engl J Med. 2001;344:1411-20.
84. Weigner MJ, Caulfield TA, Danias PG, et al. Risk for clinical thromboembolism associated with conversion to sinus rhythm in patients with atrial fibrillation lasting less than 48 hours. Ann Intern Med. 1997;126:615-20.
85. Black IW, Fatkin D, Sagar KB, et al. Exclusion of atrial thrombus by transesophageal echocardiography does not preclude embolism after cardioversion of atrial fibrillation. A multicenter study. Circulation. 1994;89:2509-13.
86. Wozakowska-Kaplon B, Janion M, Sielski J, et al. Efficacy of biphasic shock for transthoracic cardioversion of persistent atrial fibrillation: can we predict energy requirements? Pacing Clin Electrophysiol. 2004;27:764-8.
87. Oral H, Souza JJ, Michaud GF, et al. Facilitating transthoracic cardioversion of atrial fibrillation with ibutilide pretreatment. N Engl J Med. 1999;340:1849-54.
88. Meinertz T, Lip GY, Lombardi F, et al. Efficacy and safety of propafenone sustained release in the prophylaxis of symptomatic paroxysmal atrial fibrillation (The European Rythmol/Rytmonorm

Atrial Fibrillation Trial [ERAFT] study). Am J Cardiol. 2002;90: 1300-6.
89. Pritchett EL, Page RL, Carlson M, et al. Efficacy and safety of sustained-release propafenone (propafenone SR) for patients with atrial fibrillation. Am J Cardiol. 2003;92:941-6.
90. Aliot E, Denjoy I. Comparison of the safety and efficacy of flecainide versus propafenone in hospital out-patients with symptomatic paroxysmal atrial fibrillation/flutter. The Flecainide AF French Study Group. Am J Cardiol. 1996;77:66A-71A.
91. Alboni P, Botto GL, Baldi N, et al. Outpatient treatment of recent-onset atrial fibrillation with the "pill-in-the-pocket" approach. N Engl J Med. 2004;351:2384-91.
92. Echt DS, Liebson PR, Mitchell LB, et al. Mortality and morbidity in patients receiving encainide, flecainide, or placebo. The cardiac arrhythmia suppression trial. N Engl J Med. 1991;324:781-8.
93. Feld GK, Chen PS, Nicod P, et al. Possible atrial proarrhythmic effects of class 1C antiarrhythmic drugs. Am J Cardiol. 1990;66:378-83.
94. Stafford RS, Robson DC, Misra B, et al. Rate control and sinus rhythm maintenance in atrial fibrillation: national trends in medication use, 1980-1996. Arch Intern Med. 1998;158:2144-8.
95. Zimetbaum P. Amiodarone for atrial fibrillation. N Engl J Med. 2007;356:935-41.
96. Goldschlager N, Epstein AE, Naccarelli GV, et al. A practical guide for clinicians who treat patients with amiodarone: 2007. Heart Rhythm. 2007;4:1250-9.
97. Roy D, Talajic M, Dorian P, et al. Amiodarone to prevent recurrence of atrial fibrillation. Canadian Trial of Atrial Fibrillation Investigators. N Engl J Med. 2000;342:913-20.
98. Effect of prophylactic amiodarone on mortality after acute myocardial infarction and in congestive heart failure: meta-analysis of individual data from 6500 patients in randomised trials. Amiodarone Trials Meta-Analysis Investigators. Lancet. 1997;350:1417-24.
99. Bardy GH, Lee KL, Mark DB, et al. Amiodarone or an implantable cardioverter-defibrillator for congestive heart failure. N Engl J Med. 2005;352:225-37.
100. Daoud EG, Strickberger SA, Man KC, et al. Preoperative amiodarone as prophylaxis against atrial fibrillation after heart surgery. N Engl J Med. 1997;337:1785-91.
101. Benditt DG, Williams JH, Jin J, et al. Maintenance of sinus rhythm with oral d,l-sotalol therapy in patients with symptomatic atrial fibrillation and/or atrial flutter. d,l-Sotalol Atrial Fibrillation/Flutter Study Group. Am J Cardiol. 1999;84:270-7.
102. Juul-Moller S, Edvardsson N, Rehnqvist-Ahlberg N. Sotalol versus quinidine for the maintenance of sinus rhythm after direct current conversion of atrial fibrillation. Circulation. 1990;82:1932-9.
103. Singh S, Zoble RG, Yellen L, et al. Efficacy and safety of oral dofetilide in converting to and maintaining sinus rhythm in patients with chronic atrial fibrillation or atrial flutter: the symptomatic atrial fibrillation investigative research on dofetilide (SAFIRE-D) study. Circulation. 2000;102:2385-90.
104. Pritchett EL, Wilkinson WE. Effect of dofetilide on survival in patients with supraventricular arrhythmias. Am Heart J. 1999;138: 994-7.
105. Torp-Pedersen C, Moller M, Bloch-Thomsen PE, et al. Dofetilide in patients with congestive heart failure and left ventricular dysfunction.

Danish Investigations of Arrhythmia and Mortality on Dofetilide Study Group. N Engl J Med. 1999;341:857-65.
106. Kober L, Torp-Pedersen C, McMurray JJ, et al. Increased mortality after dronedarone therapy for severe heart failure. N Engl J Med. 2008;358:2678-87.
107. Hohnloser SH, Crijns HJ, van Eickels M, et al. Effect of dronedarone on cardiovascular events in atrial fibrillation. N Engl J Med. 2009;360:668-78.
108. Schneider MP, Hua TA, Bohm M, et al. Prevention of atrial fibrillation by Renin-Angiotensin system inhibition a meta-analysis. J Am Coll Cardiol. 2010;55:2299-307.
109. Knight BP, Gersh BJ, Carlson MD, et al. Role of permanent pacing to prevent atrial fibrillation: science advisory from the American Heart Association Council on Clinical Cardiology (Subcommittee on Electrocardiography and Arrhythmias) and the Quality of Care and Outcomes Research Interdisciplinary Working Group, in collaboration with the Heart Rhythm Society. Circulation. 2005;111:240-3.
110. Oral H, Knight BP, Tada H, et al. Pulmonary vein isolation for paroxysmal and persistent atrial fibrillation. Circulation. 2002;105: 1077-81.
111. Pappone C, Rosanio S, Oreto G, et al. Circumferential radiofrequency ablation of pulmonary vein ostia: a new anatomic approach for curing atrial fibrillation. Circulation. 2000;102:2619-28.
112. Haissaguerre M, Shah DC, Jais P, et al. Electrophysiological breakthroughs from the left atrium to the pulmonary veins. Circulation. 2000;102:2463-5.
113. Cappato R, Calkins H, Chen SA, et al. Updated worldwide survey on the methods, efficacy, and safety of catheter ablation for human atrial fibrillation. Circ Arrhythm Electrophysiol. 2010;3:32-8.
114. Tsai CF, Tai CT, Hsieh MH, et al. Initiation of atrial fibrillation by ectopic beats originating from the superior vena cava: electrophysiological characteristics and results of radiofrequency ablation. Circulation. 2000;102:67-74.
115. Di Biase L, Burkhardt JD, Mohanty P, et al. Left atrial appendage: an underrecognized trigger site of atrial fibrillation. Circulation. 2010;122:109-18.
116. Nademanee K, McKenzie J, Kosar E, et al. A new approach for catheter ablation of atrial fibrillation: mapping of the electrophysiologic substrate. J Am Coll Cardiol. 2004;43:2044-53.
117. Oral H, Chugh A, Yoshida K, et al. A randomized assessment of the incremental role of ablation of complex fractionated atrial electrograms after antral pulmonary vein isolation for long-lasting persistent atrial fibrillation. J Am Coll Cardiol. 2009;53:782-9.
118. Oral H, Knight BP, Ozaydin M, et al. Clinical significance of early recurrences of atrial fibrillation after pulmonary vein isolation. J Am Coll Cardiol. 2002;40:100-4.
119. Verma A, Kilicaslan F, Pisano E, et al. Response of atrial fibrillation to pulmonary vein antrum isolation is directly related to resumption and delay of pulmonary vein conduction. Circulation. 2005;112:627-35.
120. Berruezo A, Tamborero D, Mont L, et al. Pre-procedural predictors of atrial fibrillation recurrence after circumferential pulmonary vein ablation. Eur Heart J. 2007;28:836-41.
121. Oral H, Scharf C, Chugh A, et al. Catheter ablation for paroxysmal atrial fibrillation: segmental pulmonary vein ostial ablation versus left atrial ablation. Circulation. 2003;108:2355-60.

122. Saad EB, Marrouche NF, Saad CP, et al. Pulmonary vein stenosis after catheter ablation of atrial fibrillation: emergence of a new clinical syndrome. Ann Intern Med. 2003;138:634-8.
123. Pappone C, Oral H, Santinelli V, et al. Atrio-esophageal fistula as a complication of percutaneous transcatheter ablation of atrial fibrillation. Circulation. 2004;109:2724-6.
124. Pappone C, Augello G, Sala S, et al. A randomized trial of circumferential pulmonary vein ablation versus antiarrhythmic drug therapy in paroxysmal atrial fibrillation: the APAF study. J Am Coll Cardiol. 2006;48:2340-7.
125. Piccini JP, Lopes RD, Kong MH, et al. Pulmonary vein isolation for the maintenance of sinus rhythm in patients with atrial fibrillation: a meta-analysis of randomized, controlled trials. Circ Arrhythm Electrophysiol. 2009;2:626-33.
126. Wilber DJ, Pappone C, Neuzil P, et al. Comparison of antiarrhythmic drug therapy and radiofrequency catheter ablation in patients with paroxysmal atrial fibrillation: a randomized controlled trial. JAMA. 2010;303:333-40.
127. Cox JL, Boineau JP, Schuessler RB, et al. Electrophysiologic basis, surgical development, and clinical results of the maze procedure for atrial flutter and atrial fibrillation. Adv Card Surg. 1995;6:1-67.
128. Nitta T, Ishii Y, Ogasawara H, et al. Initial experience with the radial incision approach for atrial fibrillation. Ann Thorac Surg. 1999;68:805-10.
129. Defauw JJ, Guiraudon GM, van Hemel NM, et al. Surgical therapy of paroxysmal atrial fibrillation with the "corridor" operation. Ann Thorac Surg. 1992;53:564-70.
130. Mantovan R, Raviele A, Buja G, et al. Left atrial radiofrequency ablation during cardiac surgery in patients with atrial fibrillation. J Cardiovasc Electrophysiol. 2003;14:1289-95.
131. Cox JL, Boineau JP, Schuessler RB, et al. Five-year experience with the maze procedure for atrial fibrillation. Ann Thorac Surg. 1993;56:814-823.
132. Gaynor SL, Schuessler RB, Bailey MS, et al. Surgical treatment of atrial fibrillation: predictors of late recurrence. J Thorac Cardiovasc Surg. 2005;129:104-11.
133. Izumoto H, Kawazoe K, Kitahara H, et al. Operative results after the Cox/maze procedure combined with a mitral valve operation. Ann Thorac Surg. 1998;66:800-4.
134. Pasic M, Musci M, Siniawski H, et al. The Cox maze iii procedure: parallel normalization of sinus node dysfunction, improvement of atrial function, and recovery of the cardiac autonomic nervous system. J Thorac Cardiovasc Surg. 1999;118:287-95.
135. McElderry HT, McGiffin DC, Plumb VJ, et al. Proarrhythmic aspects of atrial fibrillation surgery: mechanisms of postoperative macroreentrant tachycardias. Circulation. 2008;117:155-62.
136. Van Gelder IC, Groenveld HF, Crijns HJ, et al. Lenient versus strict rate control in patients with atrial fibrillation. N Engl J Med. 2010;362:1363-73.
137. Olshansky B, Rosenfeld LE, Warner AL, et al. The Atrial Fibrillation Follow-up Investigation of Rhythm Management (AFFIRM) study: approaches to control rate in atrial fibrillation. J Am Coll Cardiol. 2004;43:1201-8.

138. Curtis AB, Kutalek SP, Prior M, et al. Prevalence and characteristics of escape rhythms after radiofrequency ablation of the atrioventricular junction: results from the registry for AV junction ablation and pacing in atrial fibrillation. Ablate and Pace Trial Investigators. Am Heart J. 2000;139:122-5.
139. Scheinman MM, Huang S. The 1998 NASPE prospective catheter ablation registry. Pacing Clin Electrophysiol. 2000;23:1020-8.
140. Stroke Prevention in Atrial Fibrillation Study. Final results. Circulation. 1991;84:527-39.
141. Hughes M, Lip GY. Risk factors for anticoagulation-related bleeding complications in patients with atrial fibrillation: a systematic review. QJM. 2007;100:599-607.
142. Risk factors for stroke and efficacy of antithrombotic therapy in atrial fibrillation. Analysis of pooled data from five randomized controlled trials. Arch Intern Med. 1994;154:1449-57.
143. Patients with nonvalvular atrial fibrillation at low risk of stroke during treatment with aspirin: stroke Prevention in Atrial Fibrillation III study. The SPAF III Writing Committee for the Stroke Prevention in Atrial Fibrillation Investigators. JAMA. 1998;279:1273-7.
144. Rother J, Crijns H. Prevention of stroke in patients with atrial fibrillation: the role of new antiarrhythmic and antithrombotic drugs. Cerebrovasc Dis. 2010;30:314-22.
145. Johnson WD, Ganjoo AK, Stone CD, et al. The left atrial appendage: our most lethal human attachment! Surgical implications. Eur J Cardiothorac Surg. 2000;17:718-22.
146. Onalan O, Crystal E. Left atrial appendage exclusion for stroke prevention in patients with nonrheumatic atrial fibrillation. Stroke. 2007;38:624-30.
147. Block PC, Burstein S, Casale PN, et al. Percutaneous left atrial appendage occlusion for patients in atrial fibrillation suboptimal for warfarin therapy: 5-year results of the PLAATO (Percutaneous Left Atrial Appendage Transcatheter Occlusion) study. JACC Cardiovasc Interv. 2009;2:594-600.
148. Go AS, Hylek EM, Chang Y, et al. Anticoagulation therapy for stroke prevention in atrial fibrillation: hsow well do randomized trials translate into clinical practice? JAMA. 2003;s290:2685-92.

# Supraventricular Tachycardia

**CHAPTER 6**

*Renee M Sullivan, Wei Wei Li, Brian Olshansky*

## Chapter Outline

- Classification
  - Atria-based AV Nodal Independent SVT
  - AV Nodal Dependent SVT
- Diagnosis
- Electrocardiographic Recordings
- Electrophysiology Studies
- Treatments
  - Acute Care
  - Long-term Management

## INTRODUCTION

Supraventricular tachycardia (SVT) is a heart rhythm disturbance, initiated in the atria or ventricles, with atrial rates exceeding 100 beats per minute (bpm), that requires tissue above the His bundle in order to be perpetuated **(Fig. 1)**. SVTs can be symptomatic or asymptomatic, slow or fast, regular or irregular, sustained or nonsustained, paroxysmal, persistent or permanent, and may be due to various mechanisms involving tissue in the atria, AV node, His Purkinje system and/or the ventricles. SVT is generally not life threatening. Occasionally, SVT impairs hemodynamics, provokes hypotension, precipitates heart failure (either acutely or as a result of long-standing tachycardia), or leads to syncope or causes debilitating symptoms, including palpitations, lightheadedness, dizziness, chest discomfort, dyspnea, or weakness. The treatment depends upon each patient's specific symptom complex, the hemodynamic response to the tachycardia, the relationship of the tachycardia to other comorbidities and the concerns of each patient.

Tremendous advances have occurred in the management of SVT over the past 60 years. No longer do we use deslanoside, lantanoside, atabrine, quinidine, pressors, cholinergics or a host of other therapies[1-9] to attempt to convert episodes of SVT; we have much better therapies. No longer do we have to worry about side effects and long-term treatment of highly

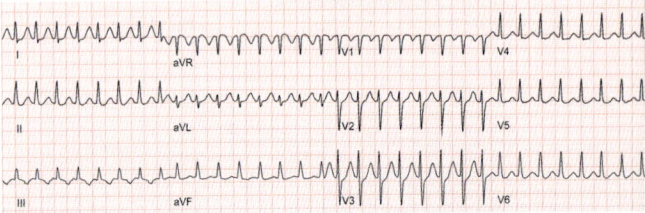

**FIGURE 1:** Typical regular supraventricular tachycardia. No P wave is visible. The most likely diagnosis of this particular supraventricular tachycardia is AV nodal reentry supraventricular tachycardia

symptomatic episodes of SVT as we now have ablation to cure many forms of SVT. While advances continue in the field, most of the attention on the management of SVT has shifted to atrial fibrillation (AFib), leaving few new therapies or modalities to evaluate or manage SVT in the past decade.

This chapter will address a modern approach to the overall evaluation and the management of those patients who have SVT.

## CLASSIFICATION

Supraventricular tachycardias are either AV nodal dependent or AV nodal independent **(Table 1)**. AV nodal dependent SVTs require AV nodal conduction in order to perpetuate. These SVTs generally have a regular ventricular rate. The two common forms of SVT are atrioventricular nodal reentry tachycardia (AVNRT) and atrioventricular reciprocating tachycardia (AVRT). AV nodal independent SVTs require only atrial tissue and do not require AV nodal activation for the tachycardia to occur. They can have a regular ventricular response, as seen in sinoatrial reentry, nonparoxysmal junctional ectopic tachycardia (JET), monomorphic atrial tachycardia (AT), and atrial flutter (AFL) with a fixed or variable AV conduction ratio or an irregular ventricular response as seen with AFib (discussed in detail in another chapter), AFL with variable AV conduction and multifocal atrial tachycardia (MAT). Almost all irregular SVTs are AV nodal independent. AV nodal dependent SVTs can occasionally be irregular, especially at the initiation and

### TABLE 1
**Classification of supraventricular tachycardias**

- AV nodal dependent
  - AV nodal reentry
  - AV reentry
    Orthodromic AV reciprocating tachycardia
    Antidromic AV reciprocating tachycardia
- AV nodal independent
  - Atrial tachycardias
    Sinoatrial reentry
    Focal (triggered, automatic, microreentry)
    Macroreentry (scar mediated, congenital heart disease)
  - Junctional ectopic tachycardia
  - Atria flutter
    Right atrial flutter
      Clockwise
      Counterclockwise
    Left atrial flutter
      Mitral reentry
      Scar mediated
      Pulmonary vein
  - Atrial fibrillation

termination of the tachycardia. AV nodal independent SVTs can be associated with complete AV block such that the ventricular rhythm is a junctional or ventricular escape **(Flowchart 1)**.

## Atria-based AV Nodal Independent SVT

### Sinus Tachycardia

Sinus tachycardia is ubiquitous, occurs with sympathetic activation and may be due to specific triggers such as infection, heart failure, pulmonary embolus, or hyperthyroidism,[10] to name a few. It is not generally considered to be SVT. Sinus tachycardia tends to start with gradual acceleration and usually stops with an even more gradual deceleration. In some instances, it can be difficult to distinguish sinus tachycardia from SVT. The P wave morphology in sinus tachycardia is similar to that in sinus rhythm **(Fig. 2)**, although due to sympathetic stimulation of the sinus node, exit from the sinus node may be more superior and thus the P wave may shift slightly in sinus tachycardia. Rates rarely exceed 200 bpm, except in children or during extreme

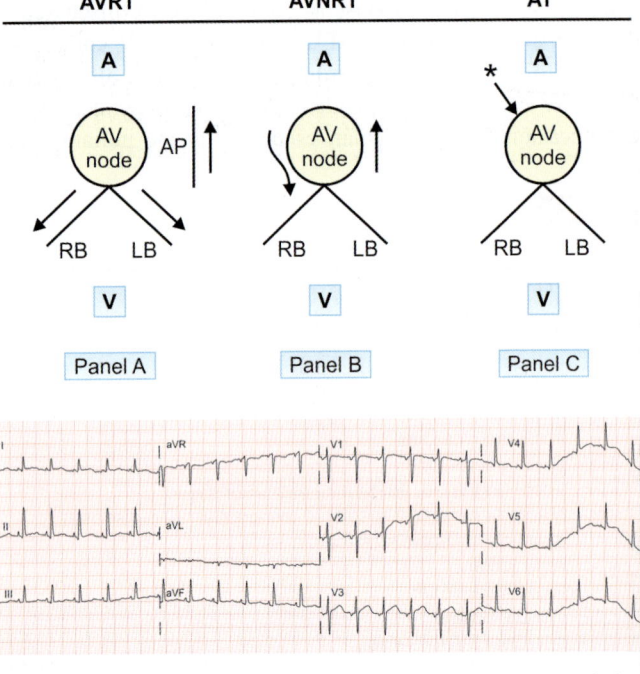

**FLOWCHART 1:** Supraventricular tachycardia—AVRT (Panel A), AVNRT (Panel B) and AT (Panel C)

**FIGURE 2:** In sinus tachycardia, the P wave is similar to that seen during sinus rhythm and the PR interval is normal. In some instances, it can be difficult to distinguish sinus tachycardia from an atrial tachycardia but there is generally more variability in the rate in patients with persistent narrow QRS tachycardia due to sinus tachycardia

physical activity. The P wave normally precedes the QRS complex but this depends on AV nodal conduction. Adenosine may appear to stop the tachycardia, but after slowing, the rate will increase gradually, indicative of sinus tachycardia rather than SVT.

Sinus tachycardia, considered abnormal for the physiological condition, is termed "inappropriate". If extreme sinus tachycardia is dependent upon an upright posture, and unrelated to fluid depletion or other explainable cause, it is termed postural orthostatic tachycardia syndrome (POTS).[11]

In some instances, it can be difficult to distinguish POTS from inappropriate sinus tachycardia or an AT.[12] When SVT persists without change during day or night and is independent of activity, fever or another explainable cause, it is more likely to be AT or AFL with a fixed AV conduction ratio **(Figs 3A and B)**. Vagal maneuvers or adenosine may be required to secure the diagnosis.

## Atrial Flutter

Atrial flutter is a macroreentrant rapid AT typically involving the right atrium.[13] It tends to coexist with AFib (although most AFib originates from the left atrium) and tends to occur in patients with structural heart disease. The atrial rate, without drug therapy, exceeds 200 bpm but can be as high as 350 bpm. In patients treated with antiarrhythmic drugs, such as amiodarone, flecainide or propafenone, and in patients with large atria, the rates of AFL can be slower than 200 bpm. The most common

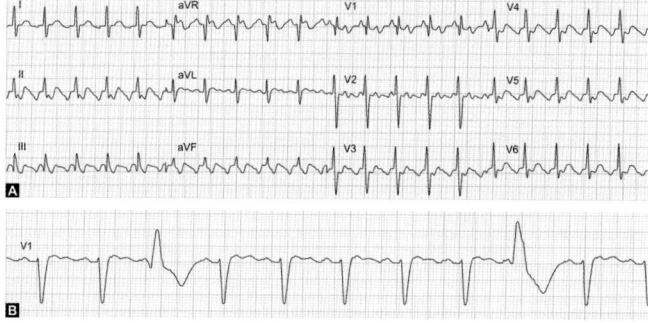

**FIGURES 3A AND B:** (A) In atrial flutter with a 2 to 1 conduction ratio, the ventricular rate is constant at rest and with activity. There appears to be an upright P wave in $V_1$ that may be suspicious for sinus tachycardia but it is clear looking at the inferior leads that this is atrial flutter with 2:1 AV conduction with a flutter wave buried in the ST segment. The fact that the rate does not change is a tipoff that this is not sinus tachycardia as well. (B) This is a constant atrial tachycardia with 2 to 1 conduction and PVCs. The fact that the rate does not change again is an indication that this is not sinus tachycardia

form of AFL, due to counterclockwise electrical activation in the right atrium around the tricuspid ring utilizing an isthmus of tissue, the cavotricuspid isthmus, has a "saw tooth" appearance in the inferior leads with no isoelectric segment between beats **(Fig. 4)**. Approximately 10% of typical AFL is perpetuated by clockwise activation around the tricuspid ring. Atypical forms of right AFL involve upper loop or lower loop reentry mechanisms[14,15] **(Figs 5A and B)**. These flutters do not show the "typical" electrocardiographic appearance or rate and may require an alternative approach during ablation procedures.

Left AFL not only often involves a reentry circuit around the mitral annulus but also can be due to reentry around or in pulmonary veins and/or scar **(Fig. 6)**.[16] Left AFL is often associated with, and may be present after attempts to ablate AFib,[17] requiring further ablation procedures.[18] Some AFL due

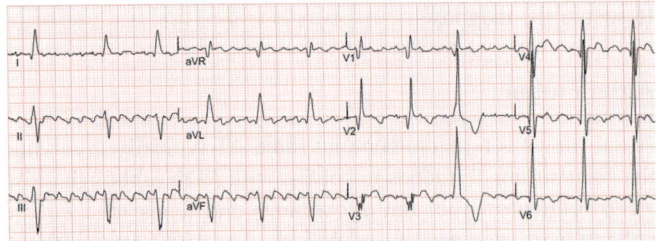

**FIGURE 4:** "Saw tooth" flutter waves are seen in the inferior leads in "typical" counterclockwise, isthmus dependent atrial flutter. Usually, this form of atrial flutter demonstrates upright P waves in lead V1. Here, there is variable AV conduction

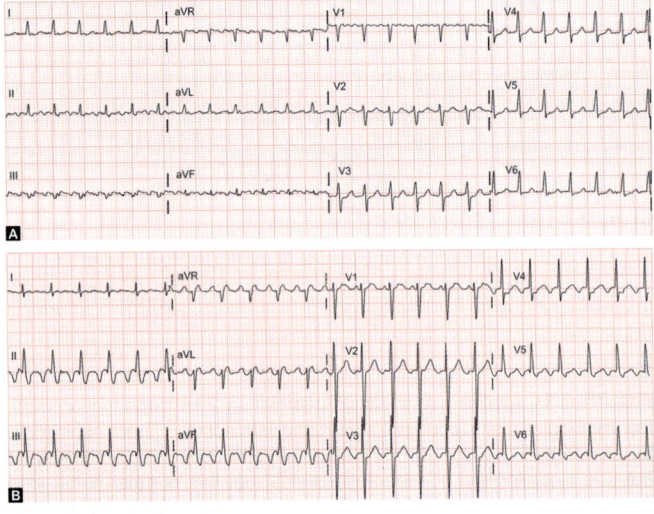

**FIGURES 5A AND B:** Unusual flutter waves in the inferior leads. These flutters may not be isthmus dependent. It can be difficult to distinguish from left atrial flutters and unusual right atrial reentry circuits in some instances

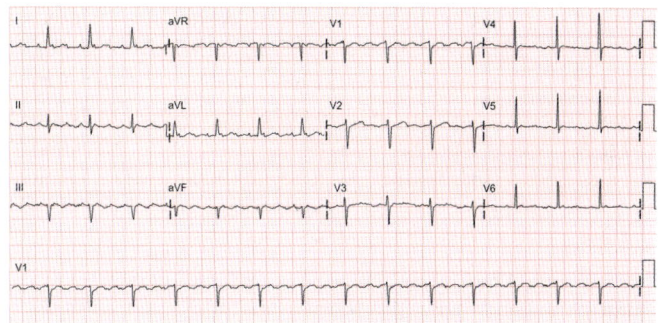

**FIGURE 6:** The left atrial flutter shown here has negative flutter waves in V1 and upright flutter waves in the inferior leads

to scar can be associated with congenital heart disease; these often have very unusual reentrant pathways. The distinction between AFL and AT in complex congenital heart disease (or its repair) is more dependent on the rate than the mechanism or the appearance on the surface electrocardiogram (ECG).

## Atrial Tachycardia

Atrial tachycardia can originate from the left atrium, right atrium, vena cavae or pulmonary veins **[Flowchart 1 (panel C)]**. The tachycardia can be focal or macroreentrant involving large areas of the atria. Focal forms can be microreentrant or due to an automatic or triggered mechanism.

Monomorphic AT represents about 5–10% of all regular SVTs. The P wave precedes the QRS complex but generally has a morphology distinct from the sinus P wave. The PR interval may vary. The atrial rates are generally 120–200 bpm. The conduction can be 1:1 but AV block can be present. An "A-A-V" pattern can be seen during AT.[19] Adenosine may occasionally stop the tachycardia;[20] more commonly, only AV block occurs. Digoxin toxicity may precipitate AT with AV block.

## Focal Atrial Tachycardia

Automatic AT represents less than 2–5% of SVTs. It may have a gradual onset and offset, sometimes similar to sinus tachycardia, in contrast to atrial reentrant tachycardias that start with a premature beat and have a sudden offset. As such, focal ATs may be difficult to distinguish from sinus tachycardia but tend to be faster and occur at rates inappropriate for physiological needs. Furthermore, the P wave morphology is usually distinctly different from that seen in sinus tachycardia.

Triggered ATs have a sudden onset and offset. Some ATs are catecholamine-dependent and begin with exercise. ATs can be associated with acute myocardial infarction (AMI), alcohol

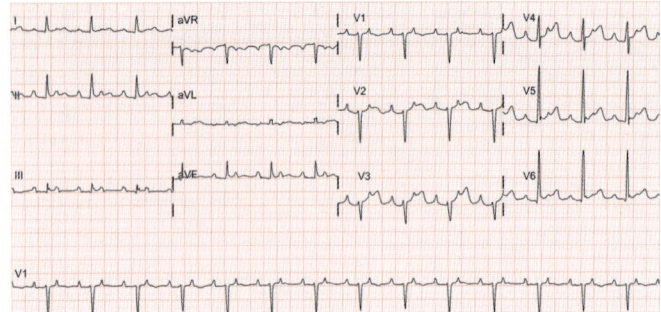

**FIGURE 7:** Atrial tachycardia with 2:1 AV conduction. Discrete P waves are separated by an isoelectric interval, in sharp contrast to atrial flutter in which an isoelectric interval is usual, not present. Occasionally, an atrial flutter is slow, especially if an antiarrhythmic drug is given, but generally, the rate is faster than 250 bpm. This atrial tachycardia is due to attempted ablation of atrial fibrillation in the left atrium and this is a left atrial tachycardia. The P wave morphology can help distinguish the location of atrial tachycardia origination

intoxication, exacerbation of chronic obstructive lung disease, electrolyte abnormalities, and digoxin use. Chronic, persistent, automatic AT, like other forms of persistent SVTs, can cause tachycardia-induced cardiomyopathy.

## Intra-atrial Reentrant Tachycardia

Macroreentry or microreentry AT often utilizes areas of scar at incisions from prior cardiac surgery or corrected congenital heart disease (such as a Fontan procedure) and represents 5–10% of SVTs.[21-24] This type of tachycardia is distinguishable from AFL as there are discrete P waves separated by an isoelectric baseline. Adenosine may terminate atrial reentrant SVTs in 15% of cases.

## Sinoatrial Reentry Tachycardia

Sinoatrial reentry tachycardia (SART) is a unique, uncommon form of regular AT due to a reentrant mechanism involving the sinoatrial node.[25] The P wave morphology is often similar to that in sinus rhythm with the exit point in the right atrium slightly below the sinus node **(Figs 8A and B)** but it can masquerade as other forms of SVT.[26] This tachycardia starts and stops abruptly and tends to be slower and more irregular than other types of SVT. Patients with AVNRT may also have associated SVTs such as sinoatrial reentry.[27]

## Multifocal Atrial Tachycardia

In MAT, atrial activation occurs from multiple locations leading to at least three different morphologies of P waves **(Fig. 9)**. The atrial rate is between 110 bpm and 170 bpm. In some cases, it can be difficult to distinguish from "coarse" AFib **(Fig. 10)**. The vast majority (60–85%) of cases occurs in acutely ill, older individuals and those with severe chronic obstructive lung disease but also can occur in patients with cor pulmonale, pneumonia, sepsis, hypertensive heart disease, and systolic heart failure. Approximately up to 0.40% of hospitalized patients have this arrhythmia.[28] Exacerbating factors include theophylline toxicity, hypokalemia, hypoxia, acidosis and catecholamine infusion. The acute mortality associated with, but not directly due to, MAT is 30–60% but this reflects the underlying disease and not necessarily the arrhythmia itself. Treatment is aimed at the underlying disease and while verapamil has been advocated, it is not particularly effective in all patients.[29]

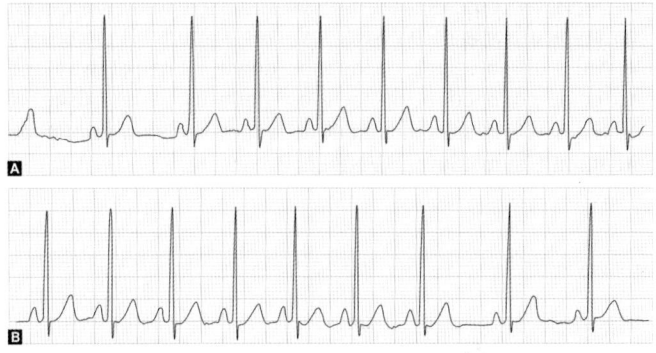

**FIGURES 8A AND B:** This rhythm strip demonstrates an abrupt change (speeding and slowing) in heart rate with an upright P wave in the inferior leads. The P wave morphology does not change and is similar to that in sinus rhythm. This is expected in typical sinoatrial reentrant supraventricular tachycardia

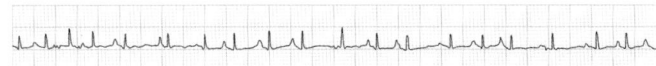

**FIGURE 9:** This rhythm strip shows multifocal atrial tachycardia, with at least three distinct P wave morphologies present. This type of tachycardia is often related to severe pulmonary disease and treatment of the underlying disease is the best way to eliminate the tachycardia. The prognosis is generally poor but not directly related to the atrial tachycardia itself. While there is no specific antiarrhythmic treatment for this tachycardia, amiodarone, or verapamil may be effective

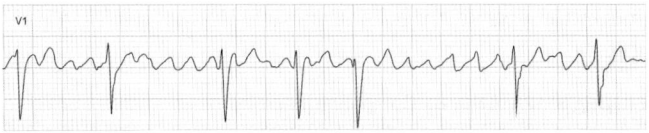

**FIGURE 10:** Coarse atrial fibrillation with atrial activation that continuously changes. This is not flutter or atrial tachycardia

## AV Nodal Dependent SVT

### Atrioventricular Nodal Reentrant Tachycardia

Atrioventricular nodal reentrant tachycardia is due to the presence of two physiological and anatomical ("slow" and "fast") AV nodal pathways.[30] About 65% of all regular SVTs are due to AVNRT. Typically, activation proceeds down the "slow" perinodal pathway and returns via the retrograde "fast" perinodal pathway **[Flowchart 1 (Panel B)]**. The rates of AVNRT are usually between 150 bpm and 200 bpm but it can be as fast as 250 bpm. Slow or fast AVNRT usually begins with a premature atrial depolarization followed by a long PR interval. There can be a pseudo R' in lead V1 **(Fig. 11)** and a pseudo S wave in the inferior leads, a retrograde P wave seen in other leads **(Figs 12A and B)** or not **(Fig. 13)**. AVNRT can be present with a bundle branch block **(Fig. 14)** and this can be tachycardia dependent. The atypical form of AVNRT involves a short PR (long RP) interval with antegrade conduction down a fast pathway and retrograde conduction via the slow pathway. A rare form of AVNRT involves slow antegrade activation and slow retrograde activation ("slow-slow" AVNRT).[31]

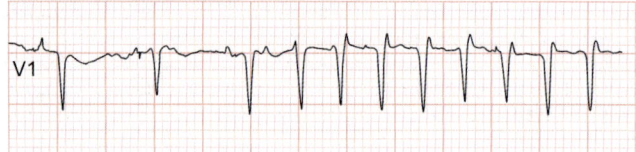

**FIGURE 11:** Atrioventricular nodal reentrant supraventricular tachycardia AVNRT is a narrow QRS complex tachycardia with no obvious P waves present. A pseudo R' can be observed in lead V1 as depicted here. The initiation is with a long PR interval suggesting conduction down a slow AV nodal pathway. This tachycardia can generally be terminated by carotid sinus massage, vagal maneuvers or adenosine

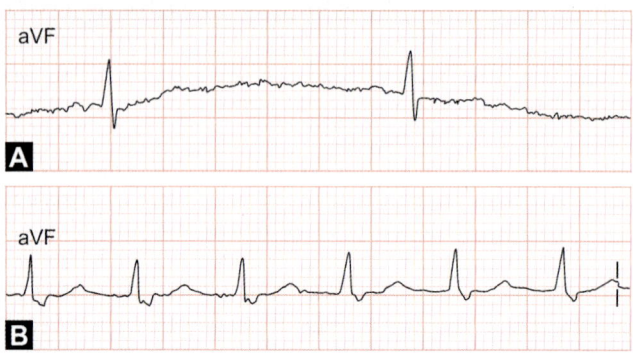

**FIGURES 12A AND B:** Comparing baseline sinus rhythm tracing to that during tachycardia shows a retrograde P wave buried at the end of the QRS complex. In this particular instance, it occurs in lead AVF rather than in V1

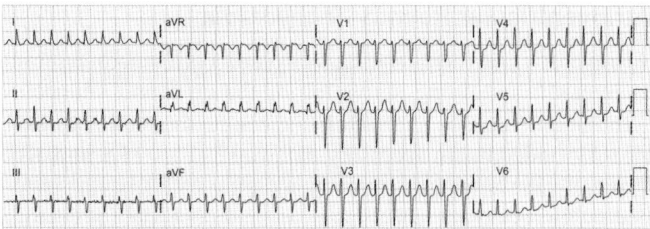

**FIGURE 13:** Typical AVNRT with a faster rate and no obvious P waves associated with the QRS complexes. A regular narrow QRS complex supraventricular tachycardia in which P waves are absent is most likely to be AVNRT

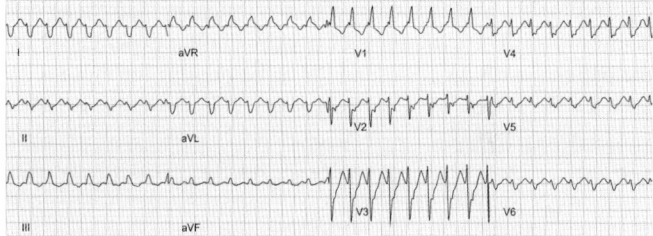

**FIGURE 14:** Example of AVNRT associated with a right bundle branch block. Supraventricular tachycardia can be associated with a wide or narrow QRS complex. In some cases a wide QRS complex supraventricular tachycardia is due to an underlying bundle branch block. Alternatively, there can be tachycardia dependent right bundle branch block aberration. In this case, it is possible to see a retrograde P wave in leads V2–V5

Atrioventricular nodal reentry tachycardia (AVNRT) can begin at any age and occurs more commonly in women than men. It is more likely to occur in the adult population even though dual AV nodal pathways are common in children.[32] Likely, AV nodal pathways change over time. Although dual AV nodal pathways are common, only a small percentage of individuals with dual pathways have AVNRT as specific characteristics are required for the tachycardia to occur: the slow pathway must have a longer refractory period than the fast pathway[33] and there may be specifics about AV nodal pathway conduction and connectedness that play a role.[34,35] High catecholamine states can exacerbate AVNRT. Symptoms, such as palpitations, neck pounding, lightheadedness, weakness, anxiety, shortness of breath, chest discomfort, pulmonary congestion and syncope due to simultaneous atrial and ventricular contraction, may occur during typical forms of AVNRT.[36,37] Although conceivable but rare, AVNRT may occur with conduction block at the lower portion of the AV node or below, demonstrating 1:1 conduction but no evidence for conduction block between the atria and the ventricles. These forms of AVNRT stop abruptly with adenosine, vagal maneuvers, and verapamil.

## Atrioventricular Reentry Tachycardia

Atrioventricular reciprocating tachycardia is a macroreentrant tachycardia involving activation of the atria and ventricles through anterograde and retrograde conducting AV pathways [the AV node and an accessory pathway[38] **(Figs 15A and B)]**. Typically, the antegrade conduction during AVRT is via the AV node with retrograde conduction via an independent accessory pathway. When this occurs, it is known as "orthodromic AV reciprocating tachycardia" **[Flowchart 1 (Panel A)]**, representing approximately 30% of all regular SVTs. It is more common in young males and tends to be faster than AVNRT. During this tachycardia, the P wave is distinctly after the QRS complex and this tachycardia is often termed "long RP tachycardia". The symptoms of neck pounding experienced by the patients with AVNRT tend not to be present for those with

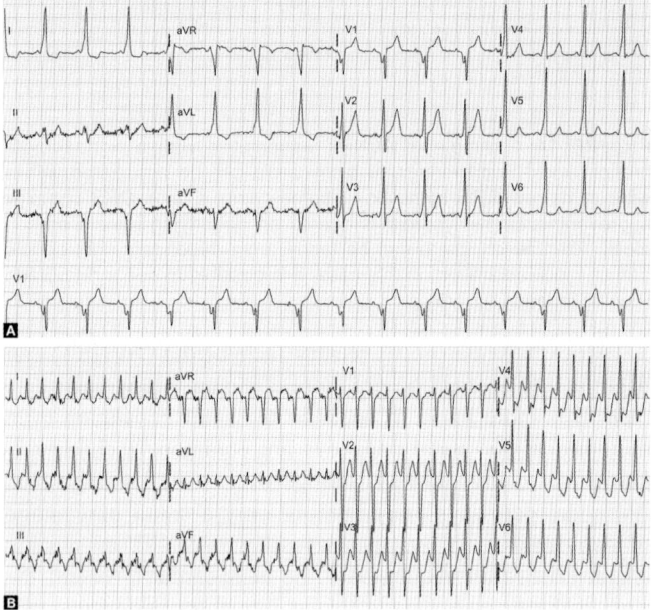

**FIGURES 15A AND B:** (A) This tracing shows evidence for a posteroseptal accessory pathway in sinus rhythm. There is evidence that this is Wolff-Parkinson-White syndrome with a negative delta wave in V1 and a positive delta wave in V2. The delta wave in the inferior leads, i.e. III and AVF, are negative. This delta wave vector is consistent with a posteroseptal accessory pathway (B) This tracing, seen in the same patient who had an EKG in sinus rhythm shown in Figure 15A, shows a rapid narrow QRS complex supraventricular tachycardia with a retrograde inverted P wave present in the ST segment. The P wave is best seen in leads II, III and AVF. This is orthodromic AV reciprocating tachycardia. The supraventricular tachycardia does not demonstrate conduction down the accessory pathway. Since conduction is going down the AV node and up the accessory pathway, there is no evidence for a delta wave during tachycardia. This is typical orthodromic AV reciprocating tachycardia

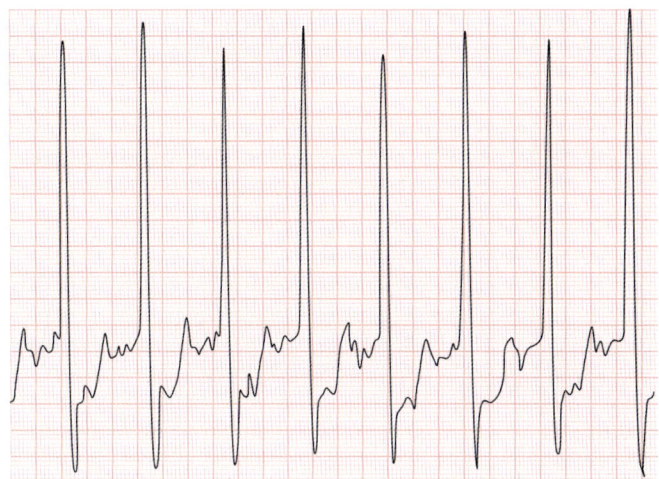

**FIGURE 16:** Example of a recording orthodromic AV reciprocating tachycardia in which there is a retrograde P wave and evidence for QRS alternans

AVRT as atrial and ventricular activation is not simultaneous. Cannon A waves do not tend to occur in AVRT.

As AVRT can be faster than AVNRT, it is more often associated with QRS alternans[39] **(Fig. 16)**. The accessory pathway responsible for AVRT can be "concealed", that is, not present as a "delta wave" on the ECG recording. Like AVNRT, this tachycardia stops abruptly with vagal maneuvers or adenosine due to blockage in the AV node.

## Preexcitation Syndromes

Manifest antegrade conduction through an accessory pathway can "preexcite" the ventricles and cause a fusion complex or complete conduction via the antegrade accessory pathway. The AV connection can occur by way of the left ventricle, the right ventricle, or the septum at virtually any location between the atria and the ventricles. When this is present in sinus rhythm, the pattern on the ECG is known as the "Wolff-Parkinson-White" (WPW) pattern.[40] When this pattern is associated with palpitations, this is known as WPW syndrome.

Orthodromic atrioventricular reentry tachycardia (AVRT) is the most common SVT that occurs with the WPW syndrome and other preexcitation syndromes. In less than 10% of cases of WPW syndrome, antegrade conduction is via an accessory pathway and retrograde conduction is via the AV node ("antidromic tachycardia") **(Fig. 17)**. In rare instances, antegrade conduction proceeds down an accessory pathway and comes up another accessory pathway **(Fig. 18)**. In the case of AVRT, in which conduction proceeds antegrade via an accessory pathway, the QRS complex is bizarre and wide. Irregular and

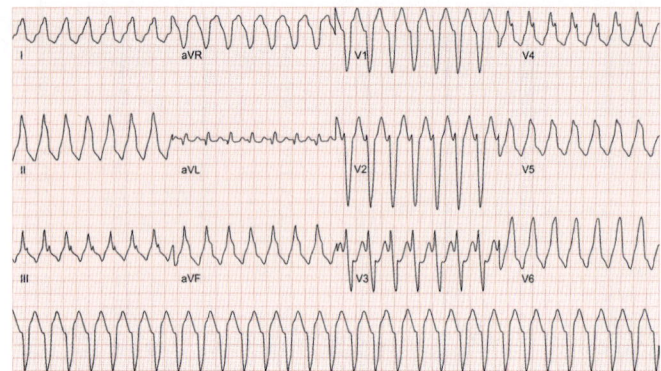

**FIGURE 17:** 12 lead ECG demonstrates antidromic AV reciprocating tachycardia. It is a rare form of supraventricular tachycardia present in patients with preexcitation syndromes. This wide QRS complex tachycardia can be difficult to distinguish from ventricular tachycardia. Preexcitation is likely to be seen in sinus rhythm with a similar QRS complex morphology. An electrophysiology study might be necessary to confirm that this is in fact antidromic tachycardia rather than ventricular tachycardia. During an electrophysiology study, an atrial premature can preexcite the ventricle with a similar QRS morphology when retrograde His bundle activation is refractory during antidromic AVRT. Antegrade conduction proceeds via the accessory pathway rapidly, with a short PR interval (best seen here in lead aVL). There is retrograde conduction via the AV node. This pathway is right-sided based on the QRS morphology (a negative QRS in lead V1). It was located close to the His bundle/AV node region

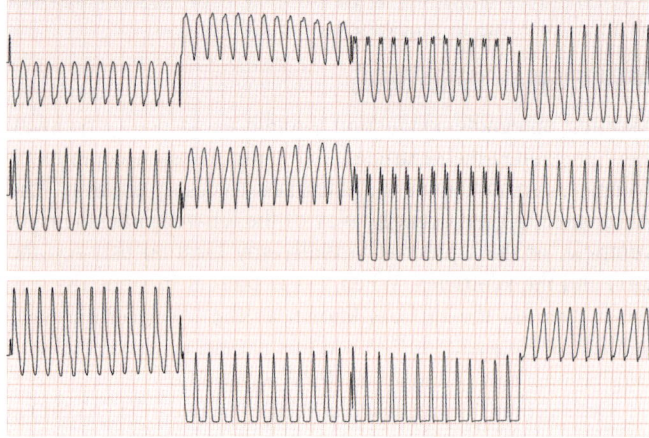

**FIGURE 18:** Very rapid wide QRS complex tachycardia is difficult to distinguish from ventricular tachycardia but is a form of antidromic AV reciprocating tachycardia in which antegrade conduction proceeds down an accessory pathway and goes up another accessory pathway. In this particular case, adenosine changed a slower form of antidromic tachycardia to a faster form after blocking the AV node. In this particular patient, the retrograde accessory pathway was a septal pathway and the antegrade pathway was a left-sided pathway that was anterior. The location of the antegrade conducting accessory pathway is the reason for the morphology of the QRS complex

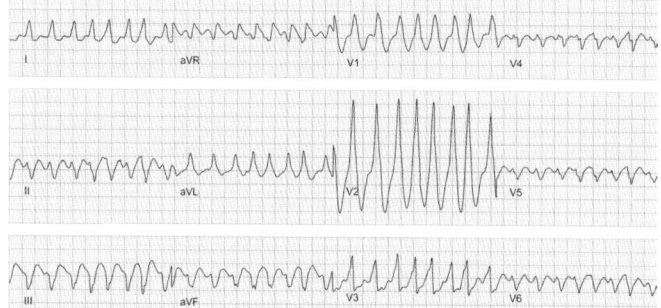

**FIGURE 19:** Atrial fibrillation in a patient with the Wolff-Parkinson-White. This is an irregularly irregular wide QRS complex tachycardia with occasional conducted beats via the AV node. The QRS complexes are due to ventricular activation by way of the left-sided posterior accessory pathway (as determined by the QRS morphology with a positive delta wave in V1). The best long-term treatment is ablation of the accessory pathway. Acutely, cardioversion or drug therapy (procainamide or amiodarone) are the treatments of choice. If the patient is not tolerating this tachycardia hemodynamically, it is important to proceed rapidly to electrical cardioversion. It is important to avoid digoxin, calcium channel blockers and even beta blockers acutely in patients who present with this tachycardia

rapid conduction via an accessory pathway during AFib can be potentially life-threatening and is dependent upon conduction via the AV node and/or the accessory pathway **(Fig. 19)**.

Other "preexcitation syndromes" can be due to His-Purkinje system preexcitation, particularly of the right bundle. An atriofascicular pathway can occur that bypasses the AV node and inserts into or next to the right bundle known as a "Mahaim fiber". Such a pathway tends to have "decremental" properties in which premature beats can be associated with progressively slower conduction through the pathway.[41] These are antegrade only conducting "atriofascicular" pathways **(Fig. 20)**. Mahaim fibers can be "innocent bystanders" whereby AVNRT proceeds down the Mahaim fiber to the ventricles or they can be the antegrade limb of a macroreentrant SVT in which the retrograde limb involves the AV node.[42]

## Permanent Junctional Reciprocating Tachycardia

Permanent junctional reciprocating tachycardia (PJRT) is a persistent form of AVRT in which conduction proceeds down the AV node and up a posteroseptal, slowly conducting accessory pathway **(Fig. 21)**. As this tachycardia is persistent and often rather slow, it may go undetected for years and lead to tachycardia-induced cardiomyopathy.[43,44] In some instances, a macroreentrant SVT involving the AV node is dependent upon slow conducting accessory tissue that is independent of the AV

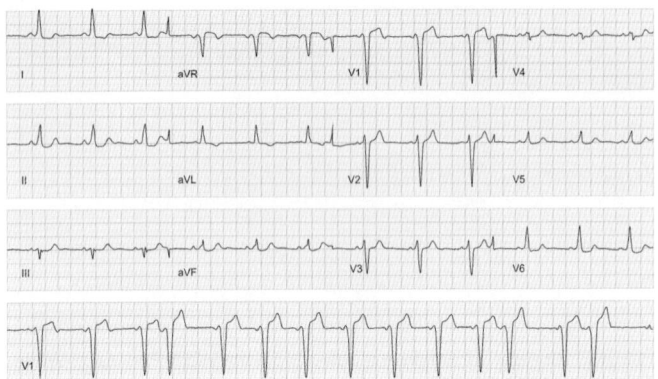

**FIGURE 20:** A Mahaim fiber is noted in sinus rhythm in this tracing from a 19-year-old female who has supraventricular tachycardia. There is evidence for a left bundle branch block without manifest preexcitation but with a short PR interval. There is no obvious delta wave but the QRS morphology becomes slightly wider. There may be slight accentuation of the QRS width with the premature atrial contraction as it may be fused conduction between the atriofascicular ("Mahaim") fiber, connecting the right atrium to the right bundle and normal antegrade AV nodal conduction. A Mahaim fiber may be suspected based on the clinical presentation. It is unlikely for a 19-year-old to have a left bundle branch block with a short PR interval. Patients with this abnormality can have a macroreentrant supraventricular tachycardia involving antegrade conduction down the Mahaim fiber and retrograde conduction up the AV node or they may have AV nodal reentry with "innocent bystander" Mahaim conduction. Mahaim fibers tend to conduct only in the antegrade direction. These are right-sided pathways that insert into the right bundle

node and often (but not always) in a portion of the posterior right atrium.[45]

## Junctional Ectopic Tachycardia

Junctional ectopic tachycardia is an automatic or triggered tachycardia that originates from tissue surrounding the AV node **(Fig. 22)**. The rhythm tends to be persistent and nonparoxysmal and may slow but does not generally terminate with adenosine. This rhythm is more common in children but also tends to occur after cardiac surgery, during AMI, after cardioversion of AFib, with myocarditis, during exercise, and in healthy individuals and those with sinus node dysfunction. This tachycardia may also occur under situations of catecholamine excess or in digoxin toxicity. SVT with AV dissociation is most likely JET. The JET can be associated with a poor prognosis but a recent multicenter report suggests that treatment with antiarrhythmic drugs (particularly amiodarone) or ablation can be associated with good outcomes.[46] Rarely, rapid forms of JET can be seen in specific situations, such as the postoperative period, in which catecholamine stimulation is high **(Fig. 23)**.

# Supraventricular Tachycardia

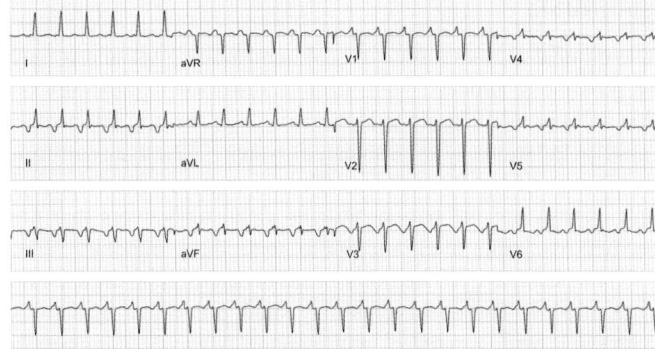

**FIGURE 21:** Supraventricular tachycardia is permanent junctional reciprocating tachycardia (PJRT). The P wave precedes the QRS complex and is inverted in leads II, III and AVF and upright in V1. The differential diagnosis for this tachycardia is atypical AV nodal reentry (antegrade fast pathway and retrograde slow pathway), atrial tachycardia and permanent junctional reciprocating tachycardia. This patient had permanent junctional reciprocating tachycardia with retrograde conduction by way of an accessory pathway that was conducting slowly as determined by electrophysiology testing. This patient developed heart failure and a cardiomyopathy due to persistent tachycardia. With ablation of the slow retrograde accessory pathway, the tachycardia was eliminated, thereby returning the ejection fraction to normal and resolving the heart failure

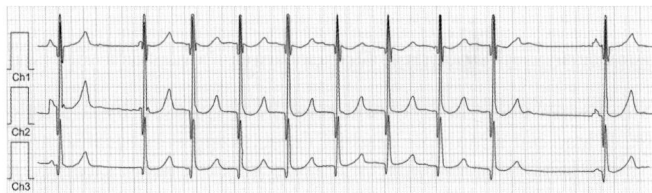

**FIGURE 22:** Tracing shows evidence for a junctional ectopic tachycardia. These tachycardias are most commonly present in younger patients and they originate from tissue surrounding the AV node. This tachycardia can be associated with AV dissociation. The first beat of this tachycardia occurs at the same time as a P wave demonstrating the dissociation of the A and the V during the tachycardia

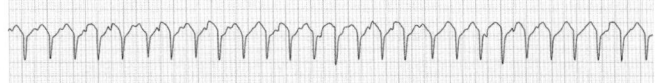

**FIGURE 23:** Rarely, an accelerated junctional tachycardia, faster than junctional ectopic tachycardia, can occur in older people who have high catecholamine states. In this particular case, the patient was postoperative after cardiac surgery. There is a narrow complex tachycardia with retrograde atrial activation but not in a one-to-one fashion. There is retrograde a conduction block

## DIAGNOSIS

The clinical presentation can be diagnostic of SVT and it may be possible to proceed with further evaluation and therapy on this basis alone. Classic symptoms include the abrupt onset of rapid palpitations with associated dyspnea, chest discomfort, dizziness, and lightheadedness. These symptoms often abruptly terminate when the patient utilizes vagal maneuvers that are learned over time in an attempt to abate the symptoms.

The physical examination can aid in determining the specific type of SVT. The neck veins may show prominent pulsations with each beat, consistent with cannon A waves, common in typical AVNRT. Alternatively, patients with AFL may have flutter waves seen as pulsations in the neck at a rate faster than the pulse itself. An irregularly irregular pulse or a pulse deficit would be consistent with AFib. Additionally, important information from the physical examination includes blood pressure recordings as well as evidence for hemodynamic compromise or the presence of congestive heart failure. These findings are indicators that a more aggressive approach is necessary to control the rate and the rhythm.

Bedside maneuvers, such as carotid sinus massage or Valsalva, can terminate tachycardia abruptly. They can also uncover the presence of an AV nodal independent tachycardia, such as AFL **(Fig. 24)**. These maneuvers can also slow down the sinus rate and stop sinoatrial reentry as well. Adenosine can stop some ATs.

### Electrocardiographic Recordings

Supraventricular tachycardia may be diagnosed by a single ECG lead but a multiple lead ECG recording is generally more useful to help distinguish one form of SVT from another. A 12-lead ECG during sinus rhythm can help to determine if there is preexcitation or a bundle branch block. Ambulatory event recorders, including Holter monitors or even an implantable loop recorder, may be necessary to detect intermittent episodes of SVT.

During recorded episodes, clues to the diagnosis of the presence and type of SVT can be discerned from the initiation and termination of the tachycardia **(Fig. 11)**, the relationship between the P waves and the QRS complexes **(Fig. 7)**, the P wave and QRS morphologies, as well as the presence of AV block and the tachycardia regularity and rate. In some cases, the type of SVT may not be clear based on available recordings even with bedside interventions (vagal maneuvers or intravenous adenosine) **(Fig. 24)**. An invasive electrophysiology test may be needed to ascertain the SVT type and mechanism.

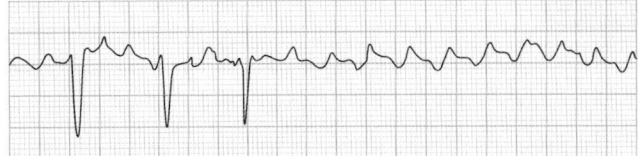

**FIGURE 24:** In this particular tracing, a patient who had atrial flutter was given adenosine to secure the diagnosis. There was transient AV block with a long pause and the presence of flutter waves becoming evident. Prior to the adenosine, there was 2 to 1 conduction during atrial flutter and it was difficult to make the diagnosis for certain. This tracing shows the effect of adenosine uncovering the mechanism of this particular tachycardia, typical atrial flutter. Adenosine will not stop atrial flutter but will stop AV nodal dependent supraventricular tachycardias such as AVNRT and AVRT. It can also stop sinoatrial reentry and some atrial tachycardias. Adenosine can be useful to help distinguish atrial versus AV nodal dependent supraventricular tachycardias

## P Wave Characteristics

In typical (slow antegrade, fast retrograde) AVNRT, the P wave is nearly simultaneous with the QRS complex. Characteristically there is a small upright P wave in lead V1 just at the end of the QRS complex known as a pseudo R' **(Fig. 11)**. At tachycardia initiation, a long PR interval may be seen indicating antegrade conduction via a slow conducting AV nodal pathway. In atypical (fast antegrade, slow retrograde) AVNRT, there is an inverted P wave in the inferior leads often just before the next QRS complex. AVNRT of this type or AVRT should be suspected if inverted P waves are present in leads II, III and aVF.

In monomorphic AT, the P wave is inconsistent with sinus tachycardia. AV block may be present but conduction tends to be at a fixed ratio **(Fig. 7)**. Similarly, for AFL, variable conduction or 2:1 conduction may occur and one of the P waves could be buried in the ST segment (Fig. 3A). In sinus tachycardia, the P wave is before the QRS complex and is upright in the inferior leads **(Fig. 2)**. It may be difficult to distinguish from SART as the P wave morphology can be virtually identical **(Figs 8A and B)**. In JETs, AV dissociation may be present as the tachycardia can be independent of atrial activation. In MAT, at least three morphologies of P waves are present **(Fig. 9)**.

## Wide QRS Tachycardia—Is It SVT?

Wide QRS complex tachycardia (QRS width ≥120) is generally ventricular tachycardia but approximately 10% of wide QRS tachycardias are SVT **(Table 2)**. A wide QRS complex during SVT can be due to a bundle branch block that may be present in sinus rhythm. The baseline ECG can, therefore, be of some help in diagnosing the mechanism as the morphology may be

### TABLE 2

**Differential diagnosis of wide QRS tachycardia**

- SVT with fixed bundle branch block
- SVT with intermittent aberration including concealed perpetuated aberration
- SVT with persistent rate dependant aberration
- SVT with passive conduction down a bypass tract
- SVT due to antidromic AV reciprocating tachycardia
- Ventricular tachycardia

the same as in sinus rhythm but the QRS can change during SVT when there is an underlying bundle branch block[47] and the QRS morphology during ventricular tachycardia can mimic the morphology in sinus rhythm.[48]

There can be rate-dependent ("phase 3") aberration (QRS widening with a bundle branch block pattern) that is due to rate dependent block in one of the bundles due to rate related refractoriness. When this occurs, it is often present when there is an acceleration of the tachycardia or if there is irregular AV conduction such that there is long-short conduction. This is known as "Ashman's phenomenon" which tends to occur during AFib. Tachycardia-dependent aberration can be continuous or intermittent and related to refractoriness in the right or left bundle branch. A phenomenon of "concealed perpetuated aberration" is also possible in which persistent bundle branch block aberration can occur even though there is fluctuation in rate.

In some patients, both wide and narrow QRS SVT coexist. If the rates of the wide and narrow QRS SVT are similar, or if the rate during the wide QRS tachycardia is faster, the cause of the tachycardia remains uncertain. If the wide QRS SVT is due to bundle branch block aberration and it is slower than the narrow QRS SVT, this likely indicates the presence of AVRT with retrograde conduction via the accessory pathway on the same side as the bundle branch block. The slower rate during the wide QRS complex tachycardia is because the bundle branch block causes the contralateral ventricle to activate first and there is conduction delay through the ventricular myocardium before conduction can proceed up the retrograde pathway. A bundle branch block located on the side ipsilateral to the accessory pathway will lengthen the reentry circuit pathway (e.g. a left bundle branch block tachycardia with a left sided retrograde accessory pathway is likely to be slower than SVT with a narrow QRS complex in the same patient) and the VA interval can lengthen with the bundle branch block aberration. Occasionally, tachycardia can begin in a fascicle with a relatively narrow, yet wide, QRS complex, and with AV dissociation. This tachycardia can be confused with SVT.

In rare instances, antegrade conduction via an accessory pathway during tachycardia can present as a wide QRS complex SVT. Antidromic AVRT (conduction down an accessory pathway and up the AV node or another accessory pathway) occurs rarely. Preexcited AFib with intermittent AV conduction down an accessory pathway is also possible.

Adenosine is potentially diagnostic and can be given in a patient with a wide QRS complex tachycardia that is well tolerated.[49] If adenosine or a vagal maneuver stops a wide QRS tachycardia, it is likely SVT even though some idiopathic ventricular tachycardias may stop with adenosine or even a vagal maneuver. Despite careful analysis, and even bedside maneuvers, the diagnosis of the type of SVT may be incorrect in as many as 20% of recorded episodes.

## Electrophysiology Studies

Invasive electrophysiology studies are used either for diagnosis of SVT in patients with classic symptoms or for determination of the mechanism of SVT for those who have recorded episodes or have a wide QRS tachycardia that may be SVT or VT. During the electrophysiology study, 2–5 intravenous catheter sheaths are placed and recording and stimulating catheter electrodes are placed in specific sites of the heart to record electrical activation and to stimulate the heart to initiate tachycardia and understand its mechanism. Catheters can also be used to locate specific tissues that are responsible for the tachycardia. SVTs with regular rate (AVNRT and AVRT), if present, are often readily inducible with delivery of premature atrial or ventricular extrastimuli. In some cases, AT and rare cases of AVNRT or AVRT, the SVTs are not inducible. Increasing the aggressiveness of the atrial and ventricular extrastimuli by pacing at faster rates and adding more extrastimuli may be useful. Sometimes, catecholamine stimulation with isoproterenol, and/or atropine may be necessary to initiate the tachyarrhythmia during extrastimulus testing. In rare instances, a beta-blocker is required to initiate SVT.

After SVT is initiated in the electrophysiology laboratory, the relationships of the atria, ventricles and His bundle during extrastimulus testing and during tachycardia can help to determine the tachycardia mechanism. Transient entrainment may help understand the location and mechanism of the SVT.[50]

During SVT, if specifically timed ventricular extrastimuli activate the atria when the His bundle is refractory, the presence of an accessory pathway is diagnosed and the tachycardia is likely orthodromic AVRT. Similarly, if during SVT specifically timed atrial extrastimuli activate the ventricles when the His bundle is refractory, the tachycardia is like antidromic AVRT. AV relationships can also help to determine if there is an accessory pathway.

In some instances, detailed electroanatomical mapping is necessary to understand the tachycardia mechanism or the origin of the tachycardia (such as focal or reentrant AT). Similarly, simultaneous atrial record map may be helpful to understand the mechanism of the tachycardia. In some instances, it is a transseptal catheterization or even an arterial approach in needed to reach the tissue responsible for tachycardia.

## TREATMENTS

The goal of treatment is to terminate tachycardia acutely, maintain normal sinus rhythm, control ventricular response rate, eliminate symptoms, normalize hemodynamics, and prevent worsening of any underlying cardiovascular conditions due to SVT.

### Acute Care

Acute management depends on the type and severity of symptoms related to the SVT and the type of SVT **(Table 3)**. Acute interventions are designed to slow the ventricular rate (for AV nodal independent SVTs) and/or terminate the tachycardia. Therapies include drugs to cardiovert and prevent recurrence, drugs used to slow the AV conduction and the ventricular rate, and direct current cardioversion. Acute management requires careful electrocardiographic and hemodynamic monitoring. Patients remaining in SVT and having ventricular rates that cannot be controlled require hospital admission. Other indications for admission include frequent recurrences, resistance to initial drug therapy, initiation of new antiarrhythmic drugs, radiofrequency (RF) catheter ablation (elective or urgent) or adverse consequences from SVT (heart failure exacerbation, hypotension, myocardial ischemia) **(Flowchart 2)**.

### AV Nodal Dependent SVT or Regular SVT

AV Nodal dependent SVT or regular SVT for which mechanism is unknown. The first line treatment for an AV nodal dependent tachycardia is a vagal maneuver, such as carotid sinus massage, to create transient AV block and terminate the tachycardia. Patients can learn to perform vagal maneuvers and stop tachycardia on their own without the need for medical intervention.

Carotid sinus massage is an easy and effective methodology to stop AV nodal dependent SVT but should only be used in the absence of a carotid bruit and/or absence of significant carotid disease.[51] In this procedure, the head should be turned away from the side being compressed (usually the right side)

### TABLE 3
**Pharmacologic management for supraventricular tachycardia**

| Drugs | Mechanisms | Dosage | Side effects | Contraindications |
|---|---|---|---|---|
| Adenosine | Purinergic agonist Inhibition sinus node and AV node | 6 mg by rapid IV. If ineffective,12 mg and 18 mg | Nausea, light-headedness, headache, flushing, chest pain, bradycardia, brief asystole | Persantine Cardiac transplant Bronchospasm |
| Verapamil | Slow or block AV nodal conduction and slow sinus rate | 2.5–5 mg over 1–2 min | Negative inotropic effect, hypotension, cardiogenic shock, marked bradycardia | Hypotension Systolic dysfunction Atrial fibrillation with preexcitation |
| Diltiazem | Slow or block AV nodal conduction and slow sinus rate | 0.25 mg/kg IV bolus then 5–15 mg/hour gtt | Negative inotropic effect, hypotension, bradycardia | Hypotension Systolic dysfunction Atrial fibrillation with preexcitation |
| Metoprolol | Block β-sympathetic nervous system at the receptor level Inhibitory effects on sinus node, AV node and myocardial contraction | 2.5–5 mg 3x at 2-min interval | Negative inotropic effect, hypotension | Hypotension Cardiogenic shock Bradycardia Decompensated heart failure Bronchospasm |

*Contd...*

*Contd...*

| Drugs | Mechanisms | Dosage | Side effects | Contraindications |
|---|---|---|---|---|
| Esmolol | Inhibitory effects on sinus node, AV node and myocardial contraction | IV 500 mcg/min loading dose over 1 min before each titration | Negative inotropic effect, hypotension, peripheral ischemia, confusion, bradycardia, bronchospasm | Hypotension Cardiogenic shock Bradycardia Decompensated heart failure Bronchospasm |
| Digoxin | $Na^+/K^+$ ATPase inhibition Parasympathetic activation leading to sinus lowing and AV nodal inhibition | 0.75–1.5 mg in divided doses over 12–24 hours | Nausea, vomiting, diarrhea, fatigue, confusion, colored vision, palpitation, arrhythmia, syncope | WPW syndrome Atrial fibrillation with preexcitation |
| Amiodarone | Class III AAD but with classes I, II and IV activity, block sodium, calcium and potassium channels | Oral: loading 1200–1600 mg daily, maintenance 200–400 mg daily IV: 150 mg over 10 min, then 360 mg over 6 hours, 540 mg over remaining 24 hours, then 0.5 mg/min | Thyroid abnormalities, pulmonary fibrosis, QT prolongation, liver function abnormalities | Severe sinus node dysfunction Hepatic dysfunction Pregnancy |

**FLOWCHART 2:** Management of regular SVT in an acute setting

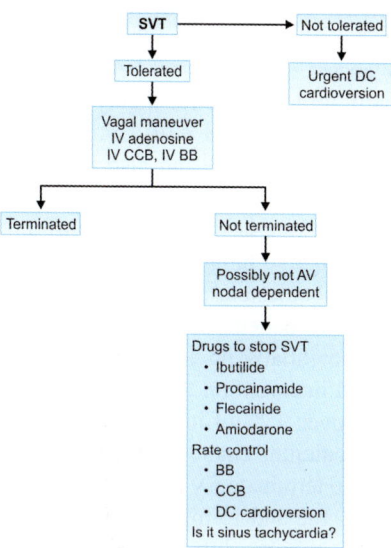

and a firm compression with 2–3 fingers is applied over the bulb of the carotid. A strong arterial impulse must be felt with firm pressure and rubbing. Sometimes, carotid massage can be combined with a Valsalva maneuver and even the Trendelenburg position to facilitate conversion. The success of the carotid sinus massage depends, in part, on the technique. A Valsalva maneuver can similarly increase parasympathetic tone and therefore slow down the conduction in the antegrade slow pathway. Another vagal reflex, the "diving reflex" in which the face is placed in cold water, may be effective.[52] These maneuvers are unlikely to be effective if hypotension is present.

Adenosine **(Flowchart 3)** can differentiate AV nodal independent versus AV nodal dependent SVT and can be used to help to make a diagnosis **(Fig. 24)** but, like AVNRT, SART can respond to autonomic maneuvers and adenosine. Adenosine effectively and rapidly terminates AV nodal dependent SVTs.[53] It is generally effective even if borderline hypotension is present.

**FLOWCHART 3:** Response of SVT to IV adenosine

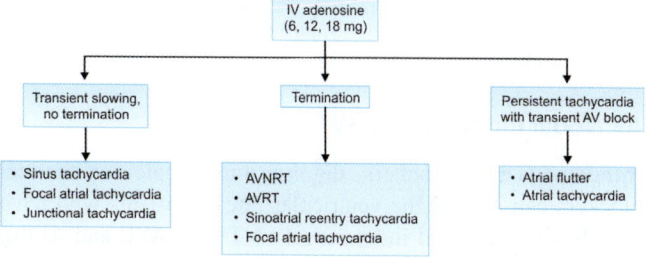

The advantage of adenosine is its rapid onset and short half-life. Adenosine must be given as a rapid intravenous bolus followed by a rapid saline infusion and should be given via a reliable, large bore IV access. The doses are between 6 mg and 12 mg, and occasionally up to 18 mg for highly resistant patients.

Adenosine must be used with caution in patients who are already taking persantine and also in patients who have cardiac transplant because asystole may occur. Furthermore, adenosine can cause long-lasting bronchoconstriction in patients with chronic obstructive lung disease or uncontrolled asthma. Caffeine and phosphodiesterase inhibitors will inhibit the effects of adenosine.[54] Some patients are reticent to have adenosine due to its short but potentially noxious side effects. Nevertheless, it is the preferred intervention to stop AVNRT and AVRT and it is effective in over 95% of individuals.[55,56]

Intravenous calcium channel antagonists, verapamil and diltiazem, can also terminate SVT.[57-59] Intravenous verapamil at doses of 5 mg, 10 mg and 15 mg can be effective; the duration of action is 5–45 minutes. Intravenous diltiazem can be used in doses of 0.15 mg/kg, 0.25 mg/kg and 0.45 mg/kg. Calcium channel blockers have negative inotropic effects and therefore can cause hypotension; use is not recommended when the patient is hypotensive or has ventricular dysfunction. It should be avoided in patients with preexcited AFib (i.e. antegrade activation via an accessory pathway). It should never be used when there is an undiagnosed wide QRS complex tachycardia as the results could be disastrous. Verapamil (and, less commonly, diltiazem) can be used when SVT is terminated with adenosine but recurs or in patients who ingest large amounts of caffeine.

Intravenous beta-adrenergic blockade (metoprolol or esmolol) may be effective in terminating SVT as well.[59] Esmolol has a short half-life of less than 10 minutes. Metoprolol has a longer half-life but is less expensive. Both of these drugs have a negative inotropic effect and may cause hypotension. Beta blockade and intravenous digoxin are third line drugs for termination of AV nodal dependent SVT.

The use of digoxin for the acute management of SVT is now rare. Digoxin requires a loading dose and takes prolonged time to effect. It is less efficacious than other drugs and contraindicated in the WPW syndrome. It may be used in combination with beta-adrenergic blockers or calcium channel blockers to control recurrent episodes of SVT.

## AV Nodal Independent SVT

Intravenous beta-blockers, digoxin and/or calcium channel blockers can control the ventricular rate in patients who have atria-based, AV nodal independent SVT (AT, AFL and AFib). The one exception is SART that responds reliably to adenosine.

The preference of the drug class is related to the underlying conditions, blood pressure and ventricular function. Beta-blockers (in combination with digoxin), for example, are useful in controlling the ventricular response rate in AFL and AFib, especially in the postoperative period. Intravenous diltiazem has less negative inotropic effect than verapamil and can be used to control the ventricular rate when there is borderline low blood pressure. Diltiazem also may be useful when there is concern about bronchospasm. Digoxin may require a large loading dose and a protracted period but is more useful for patients with ventricular dysfunction or hypotension.

For patients with poorly tolerated AV nodal independent SVTs (AFL, AT and AFib), IV amiodarone is used to control the ventricular response rate.[60,61] Amiodarone has little role in the management of SVT otherwise.[62] For children, procainamide appears more effective than amiodarone for SVT.[63] Several drugs can stop AFL and AFib including intravenous procainamide and ibutilide. Acute treatments for AFL and AFib have been discussed in this chapter.

The WPW syndrome, when it is manifest as rapid AFib, should be treated with a drug that blocks the accessory pathway: either procainamide or amiodarone.[64,65] Digoxin, calcium channel blockers and beta-adrenergic blockers are strictly prohibited in these patients during acute management.[65]

Sinus tachycardia and MAT are likely due to underlying conditions. There is no specific treatment for the tachyarrhythmia and the goal is to treat the underlying conditions that are responsible for these problems.[66] When it is uncertain if a wide QRS complex tachycardia is SVT, drugs that block the AV node are not recommended.

## When to Use DC Cardioversion

Cardioversion is the best option for patients with an undiagnosed wide QRS complex tachycardia not tolerated hemodynamically and for any poorly tolerated (hemodynamic instability or evidence of heart failure or myocardial ischemia) SVT in which the rate cannot be controlled and the rhythm cannot be restored to sinus.

In patients who have hemodynamic collapse due to any type of SVT other than sinus tachycardia, synchronized DC cardioversion is recommended. However, it is important to ascertain that the SVT is not sinus tachycardia, as it will not respond to cardioversion.

## Long-term Management

Several issues must be considered before the long-term treatment is contemplated:

- Is it required?
- What are the implications of no treatment?
- How old is the patient? Are there comorbidities?
- Is the rhythm triggered by any acute nonarrhythmic condition such as pneumonia or pulmonary embolus?
- Is there any evidence of worsening congestive heart failure?
- How chronic is the rhythm and what is the rate of the tachycardia?
- Is there hemodynamic compromise?
- How symptomatic is the patient?
- How often does tachycardia occur?
- What are the patient's wishes regarding treatment?

The long-term management of SVT depends on multiple factors, including the symptoms related to SVT, the recurrence rates, the underlying clinical conditions and the presence of structural heart disease. For example, the treatment of a patient who has one episode of a mildly symptomatic SVT terminated by vagal maneuvers or adenosine, treatment will be different from a person with frequent recurrence. Other ensuing factors must be considered such as the necessity for medication to prevent SVT during pregnancy. In this case, more definitive treatment by an ablation would likely be the primary choice.

The choice of drug for long-term management of SVT depends on the mechanism of the SVT and the goal of treatment. For ATs, AFL and AFib a simple strategy is to control the ventricular rate instead of maintaining sinus rhythm **(Fig. 25)**. Beta-adrenergic blockers, calcium channel blockers and digoxin, in combination, can be effective.[66,67] This approach alone may not make sense as symptoms continue. Furthermore, it can be very difficult to control the ventricular response rate for some SVTs, such as in AT and AFL. Therefore, it may be necessary to cardiovert the patient, use an antiarrhythmic drug or even ablate the tachycardia. Antiarrhythmic drugs are often not effective for AFL and ablation may be necessary.[68,69] Occasionally, the ventricular rate cannot be controlled and maintenance of sinus rhythm is not an option; in this case, AV nodal ablation with permanent pacemaker may be required **(Fig. 26)**.

For monomorphic AT, antiarrhythmic drugs can be given to help to maintain the sinus rhythm. The choice of antiarrhythmic drugs is similar to a methodology used for AFib, mainly

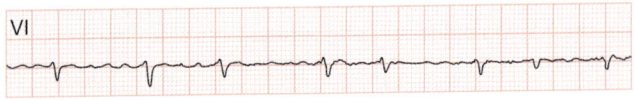

**FIGURE 25:** Atrial fibrillation with a controlled ventricular response rate. Normally, the ventricular response rate to atrial fibrillation is fast as long as there is an intact AV node and the patient is not taking medical therapy to slow or block conduction in the AV node

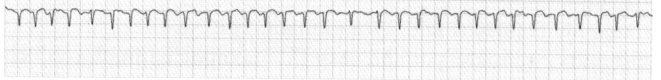

**FIGURE 26:** Irregularly irregular narrow QRS complex tachycardia is atrial fibrillation with a rapid ventricular response rate

based on the underlying structural heart disease. For patients without underlying structural heart disease, sotalol, propafenone and flecainide are possibilities. For patients with underlying structural heart disease, sotalol and amiodarone could be used to maintain sinus rhythm without a proarrhythmic effect.[70]

Like AFib and AFL, ATs can increase the risk of thromboembolic events.[71] Consideration must be given to the use of routine long-term anticoagulation for SVTs at risk for stroke. While guidelines address this issue for patients with AFib, they do not for ATs as data are scarce in this regard. Various antiarrhythmic drugs may suppress AFL. However, the safety and efficacy of antiarrhythmic drugs for AFL have not been well tested in the long term.[71] Furthermore, recent data suggest that ablation techniques especially for isthmus-dependent right AFL are more effective and cost-effective than drug therapy. Similarly, there are no specific guidelines with respect to anticoagulation for AFL. Recent data indicate that AFL has a similar or slightly lower risk of thromboembolic risk when compared with AFib. Therefore, AFL should be considered very much like AFib when contemplating the use of anticoagulation.

## Catheter Ablation

Catheter ablation has emerged as a curative approach for the patients who have SVT.[72-75] The mechanism by which catheter ablation may work is dependent upon the type of tachycardia. Catheter ablation involves the purposeful destruction or isolation of selective tissue responsible for the tachycardia. Even extensive ablation rarely has a significant effect on cardiac function and may only require lesions that are rather small. Delivery of heat to the tissue via RF energy remains the standard approach in the ablation of most arrhythmias. Cryoablation has been used with some success in selected patients who have arrhythmias.[76]

RF ablation can successfully cure AVNRT,[77] AVRT,[74] sinoatrial reentry,[78] WPW,[79] Mahaim tachycardias,[80] focal AT,[81,82] AFL,[83,84] and even AFib.[85] Additionally, ablation may be useful for JETs and occasionally for inappropriate sinus tachycardia.[11,86] Ablation can substantially improve the quality of life in patients with SVT.[87] Candidates for ablation include those patients who want their SVT eliminated and for whom the benefits outweigh the risks. The success rates for RF ablation vary by the rhythm disturbance and its location.

RF ablation is strongly considered for those patients who have frequent, recurrent and symptomatic episodes that are either fast and/or refractory to the drugs.[88] The age distribution of RF ablation is a bell shaped curve with the average being 27 ± 17 years for accessory pathways and 44 ± 18 years for AVNRT.[75] Data suggests that there is a substantial decrease in frequency and severity of arrhythmias and decrease in self-imposed restrictions.[87] Cost-effectiveness data, based on older studies, indicates that the cost per quality adjusted life-year gained is $6,600 to $19,000.[89-91]

Generally, an electrophysiology study is performed as a diagnostic procedure with an ablation.[92] The procedure includes arrhythmia induction, mapping of the pathways or location, ablation and post-ablation attempts at arrhythmia induction. Conscious sedation is used and 3–5 transvenous catheters are placed. A transseptal approach may be needed to ablate a left-sided accessory pathway or for a left AT or AFL. In some instances, retrograde ablation approaches are performed from the aortic route.

Catheter ablation with a 4 mm tip catheter can eliminate the slow pathway in AVNRT, the accessory pathway in AVRT, and select focal ATs safely and effectively. In most cases, the lesions are focal but for AFL and specific ATs associated with complex macroreentry circuits (including patients with congenital heart disease)[93] linear lesions must be delivered to achieve the success. In rare instances, irrigated tipped catheters or an 8 mm tipped catheter is required to deliver extensive or deep lesions.[94] For younger patients and for specific tachycardias that originate from areas directly adjacent to the AV node, such as JET or in some cases, AVNRT, cryoablation may be safer and yet potentially effective.[95] However, RF generally remains the standard for ablation for most SVTs.

The success rate for ablation of AVNRT and many accessory pathway related SVTs exceeds 95%.[96] For typical AVNRT, the success rate is approximately 98% in experienced laboratories. The success rates for accessory pathway ablation may vary depending upon the location and are between 85% and 99% even for PJRT[45] and Mahaim tachycardias.[97] The recurrence rates after ablation, especially for accessory pathways, can be between 3% and 9%. The success rate for typical AFL is greater than 90% and recurrence rates are less than 10%[83,96] but AFib can follow.[98] The success rates for more unusual or AFib ablation-created AFL is less.[99] Ablation of AT has an approximately 70–90% success rate depending on the location and mechanism of the tachycardia.[100] In some instances, the tachycardias are close to the normal conduction system or in unusual locations.[101-105] The efficacy of ablation for ATs is from 80% to 98% with recurrences between 5% and 20% and

### TABLE 4
**Complications of catheter ablation for SVT (depends on SVT type)**

| Complications | Prevalence (%) |
|---|---|
| AV block | 0.67–1 |
| Cardiac tamponade | 0.22–1.1 |
| Pericarditis | 0.31 |
| Pneumothorax | 0.15–0.22 |
| Tricuspid regurgitation | 0.22 |
| Acute myocardial infarction | 0.15 |
| Femoral artery pseudoaneurysm | 0.15 |
| Death | 0.1 |

complications of 1.6% in one series.[106] AFib ablation is not as successful as ablation of AVNRT or AVRT and complication rates are higher.

Complications in ablation include death (0.1%)[107,108] **(Table 4)**. There is approximately 0.4% risk of AV nodal block requiring a pacemaker with AVNRT "slow pathway" ablation. There are also risks of cardiac tamponade, pericarditis, hematoma and deep venous thrombosis. In general, the risks of the procedure are relatively low and all risks are less than 1%. Thus, ablation should be performed in patients who are highly symptomatic and with frequent recurrences, for those who have a high-risk profession, and for those with hemodynamic impairment or cardiomyopathy due to the persistent tachycardia.[92]

## CONCLUSION

Supraventricular tachycardia remains a common and often symptomatic problem for many patients. A wide variety of types and clinical presentations of SVT exist. The diagnosis requires careful observation and interpretation of electrocardiographic recordings. Evaluation involves thoughtful assessment of the relationship between the tachycardia, hemodynamics and symptoms. While rarely life-threatening, treatment is often required. Remarkable advances have been made in the treatment of most forms of SVT.

## REFERENCES

1. Gertler MN, Yohalem SB. The effect of atabrin on auricular fibrillation and supraventricular tachycardia in man. J Mt Sinai Hosp N Y. 1947;13:323-7.
2. Waldman S, Pelner L. The action of neostigmine in supraventricular tachycardias. Ann Intern Med. 1948;29:53-63.
3. Youmans WB, Goodman MJ, Gould J. Neosynephrine in treatment of paroxysmal supraventricular tachycardia. Am Heart J. 1949;37: 359-73.

4. Elek SR, Bernstein JC, Griffith GC. Pressor drugs in the treatment of supraventricular tachycardia. Ann West Med Surg. 1952;6:497-9.
5. Furman RH, Geiger AJ. Use of cholinergic drugs in paroxysmal supraventricular tachycardia; serious untoward reactions and fatality from treatment with methacholine and neostigmine. JAMA. 1952;149:269-72.
6. Levenson RM, Thayer RH. Lanatoside C in the treatment of supraventricular tachycardia. Am Pract Dig Treat. 1952;3:635-7.
7. Levine EB, Blumfield G. Neostigmine bromide orally in prevention of paroxysmal supraventricular tachycardia. Ann West Med Surg. 1952;6:642-7.
8. Chotkowski LA, Powell CP, Rackliffe RL. Methoxamine hydrochloride in the treatment of paroxysmal supraventricular tachycardia; report of three cases. N Engl J Med. 1954;250:674-6.
9. Donegan CK, Townsend CV. Phenylephrine hydrochloride in paroxysmal supraventricular tachycardia. JAMA. 1955;157:716-8.
10. Yusuf S, Camm AJ. The sinus tachycardias. Nat Clin Pract Cardiovasc Med. 2005;2:44-52.
11. Low PA, Opfer-Gehrking TL, Textor SC, et al. Postural tachycardia syndrome (POTS). Neurology. 1995;45:S19-25.
12. Bhatt AG, Monahan KM. Nonreentrant supraventricular tachycardia misdiagnosed as inappropriate sinus tachycardia. Pacing Clin Electrophysiol. 2011;34:e70-3.
13. Olshansky B, Wilber DJ, Hariman RJ. Atrial flutter—update on the mechanism and treatment. Pacing Clin Electrophysiol. 1992;15:2308-35.
14. Garan H. Atypical atrial flutter. Heart Rhythm. 2008;5:618-21.
15. Cosio FG, Martin-Penato A, Pastor A, et al. Atypical flutter: a review. Pacing Clin Electrophysiol. 2003;26:2157-69.
16. Jais P, Hocini M, Weerasoryia R, et al. Atypical left atrial flutters. Card Electrophysiol Rev. 2002;6:371-7.
17. Cummings JE, Schweikert R, Saliba W, et al. Left atrial flutter following pulmonary vein antrum isolation with radiofrequency energy: linear lesions or repeat isolation. J Cardiovasc Electrophysiol. 2005;16:293-7.
18. Chugh A, Oral H, Good E, et al. Catheter ablation of atypical atrial flutter and atrial tachycardia within the coronary sinus after left atrial ablation for atrial fibrillation. J Am Coll Cardiol. 2005;46:83-91.
19. Knight BP, Ebinger M, Oral H, et al. Diagnostic value of tachycardia features and pacing maneuvers during paroxysmal supraventricular tachycardia. J Am Coll Cardiol. 2000;36:574-82.
20. Kall JG, Kopp D, Olshansky B, et al. Adenosine-sensitive atrial tachycardia. Pacing Clin Electrophysiol. 1995;18:300-6.
21. Nakagawa H, Shah N, Matsudaira K, et al. Characterization of reentrant circuit in macroreentrant right atrial tachycardia after surgical repair of congenital heart disease: isolated channels between scars allow "focal" ablation. Circulation. 2001;103:699-709.
22. Triedman JK, Jenkins KJ, Colan SD, et al. Intra-atrial reentrant tachycardia after palliation of congenital heart disease: characterization of multiple macroreentrant circuits using fluoroscopically based three-dimensional endocardial mapping. J Cardiovasc Electrophysiol. 1997;8:259-70.
23. Kalman JM, VanHare GF, Olgin JE, et al. Ablation of 'incisional' reentrant atrial tachycardia complicating surgery for congenital heart

disease. Use of entrainment to define a critical isthmus of conduction. Circulation. 1996;93:502-12.
24. Walsh EP. Arrhythmias in patients with congenital heart disease. Card Electrophysiol Rev. 2002;6:422-30.
25. Gomes JA, Mehta D, Langan MN. Sinus node reentrant tachycardia. Pacing Clin Electrophysiol. 1995;18:1045-57.
26. Gomes JA, Hariman RJ, Kang PS, et al. Sustained symptomatic sinus node reentrant tachycardia: incidence, clinical significance, electrophysiologic observations and the effects of antiarrhythmic agents. J Am Coll Cardiol. 1985;5:45-7.
27. Paulay KL, Ruskin JN, Damato AN. Sinus and atrioventricular nodal reentrant tachycardia in the same patient. Am J Cardiol. 1975;36:810-6.
28. Scher DL, Arsura EL. Multifocal atrial tachycardia: mechanisms, clinical correlates, and treatment. Am Heart J. 1989;118:574-80.
29. Salerno DM, Anderson B, Sharkey PJ, et al. Intravenous verapamil for treatment of multifocal atrial tachycardia with and without calcium pretreatment. Ann Intern Med. 1987;107:623-8.
30. Akhtar M, Jazayeri MR, Sra J, et al. Atrioventricular nodal reentry. Clinical, electrophysiological, and therapeutic considerations. Circulation. 1993;88:282-95.
31. Heidbuchel H, Jackman WM. Characterization of subforms of AV nodal reentrant tachycardia. Europace. 2004;6:316-29.
32. Blaufox AD, Rhodes JF, Fishberger SB. Age related changes in dual AV nodal physiology. Pacing Clin Electrophysiol. 2000;23:477-80.
33. Wu D, Denes P, Dhingra R, et al. Determinants of fast- and slow-pathway conduction in patients with dual atrioventricular nodal pathways. Circ Res. 1975;36:782-90.
34. McGuire MA, Janse MJ, Ross DL. "AV nodal" reentry: Part II: AV nodal, AV junctional, or atrionodal reentry? J Cardiovasc Electrophysiol. 1993;4:573-86.
35. Janse MJ, McGuire MA, Loh P, et al. Electrophysiology of the A-V node in relation to A-V nodal reentry. Jpn Heart J. 1996;37:785-91.
36. Laurent G, Leong-Poi H, Mangat I, et al. Influence of ventriculoatrial timing on hemodynamics and symptoms during supraventricular tachycardia. J Cardiovasc Electrophysiol. 2009;20:176-81.
37. Gonzalez-Torrecilla E, Almendral J, Arenal A, et al. Combined evaluation of bedside clinical variables and the electrocardiogram for the differential diagnosis of paroxysmal atrioventricular reciprocating tachycardias in patients without pre-excitation. J Am Coll Cardiol. 2009;53:2353-8.
38. Obel OA, Camm AJ. Accessory pathway reciprocating tachycardia. Eur Heart J. 1998;19:E13-24, E50-1.
39. Morady F, DiCarlo LA, Baerman JM, et al. Determinants of QRS alternans during narrow QRS tachycardia. J Am Coll Cardiol. 1987;9:489-99.
40. Wolff L, Parkinson J, White PD. Bundle branch block with short P-R interval in healthy young people prone to paroxysmal tachycardia. Am Heart J. 1930;5:685.
41. Sternick EB, Timmermans C, Rodriguez LM, et al. Mahaim fiber: an atriofascicular or a long atrioventricular pathway? Heart Rhythm. 2004;1:724-7.
42. Gallagher JJ, Smith WM, Kasell JH, et al. Role of Mahaim fibers in cardiac arrhythmias in man. Circulation. 1981;64:176-89.

43. Dorostkar PC, Silka MJ, Morady F, et al. Clinical course of persistent junctional reciprocating tachycardia. J Am Coll Cardiol. 1999;33: 366-75.
44. Nerheim P, Birger-Botkin S, Piracha L, et al. Heart failure and sudden death in patients with tachycardia-induced cardiomyopathy and recurrent tachycardia. Circulation. 2004;110:247-52.
45. Meiltz A, Weber R, Halimi F, et al. Permanent form of junctional reciprocating tachycardia in adults: peculiar features and results of radiofrequency catheter ablation. Europace. 2006;8:21-8.
46. Collins KK, Van Hare GF, Kertesz NJ, et al. Pediatric nonpostoperative junctional ectopic tachycardia medical management and interventional therapies. J Am Coll Cardiol. 2009;53:690-7.
47. Datino T, Almendral J, Gonzalez-Torrecilla E, et al. Rate-related changes in QRS morphology in patients with fixed bundle branch block: implications for differential diagnosis of wide QRS complex tachycardia. Eur Heart J. 2008;29:2351-8.
48. Olshansky B. Ventricular tachycardia masquerading as supraventricular tachycardia: a wolf in sheep's clothing. J Electrocardiol. 1988;21:377-84.
49. Marill KA, Wolfram S, Desouza IS, et al. Adenosine for wide-complex tachycardia: efficacy and safety. Crit Care Med. 2009;37: 2512-8.
50. Olshansky B, Okumura K, Hess PG, et al. Demonstration of an area of slow conduction in human atrial flutter. J Am Coll Cardiol. 1990;16:1639-48.
51. Lim SH, Anantharaman V, Teo WS, et al. Comparison of treatment of supraventricular tachycardia by Valsalva maneuver and carotid sinus massage. Ann Emerg Med. 1998;31:30-5.
52. Belz MK, Stambler BS, Wood MA, et al. Effects of enhanced parasympathetic tone on atrioventricular nodal conduction during atrioventricular nodal reentrant tachycardia. Am J Cardiol. 1997;80: 878-82.
53. DiMarco JP, Miles W, Akhtar M, et al. Adenosine for paroxysmal supraventricular tachycardia: dose ranging and comparison with verapamil. Assessment in placebo-controlled, multicenter trials. The Adenosine for PSVT Study Group. Ann Intern Med. 1990;113: 104-10.
54. Cabalag MS, Taylor DM, Knott JC, et al. Recent caffeine ingestion reduces adenosine efficacy in the treatment of paroxysmal supraventricular tachycardia. Acad Emerg Med. 2010;17:44-9.
55. Cairns CB, Niemann JT. Intravenous adenosine in the emergency department management of paroxysmal supraventricular tachycardia. Ann Emerg Med. 1991;20:717-21.
56. Rankin AC, Brooks R, Ruskin JN, et al. Adenosine and the treatment of supraventricular tachycardia. Am J Med. 1992;92:655-64.
57. Sung RJ, Elser B, McAllister RG. Intravenous verapamil for termination of re-entrant supraventricular tachycardias: intracardiac studies correlated with plasma verapamil concentrations. Ann Intern Med. 1980;93:682-9.
58. Dougherty AH, Jackman WM, Naccarelli GV, et al. Acute conversion of paroxysmal supraventricular tachycardia with intravenous diltiazem. IV Diltiazem Study Group. Am J Cardiol. 1992;70:587-92.
59. Das G, Tschida V, Gray R, et al. Efficacy of esmolol in the treatment and transfer of patients with supraventricular tachyarrhythmias to

alternate oral antiarrhythmic agents. J Clin Pharmacol. 1988;28: 746-50.
60. Dilber E, Mutlu M, Dilber B, et al. Intravenous amiodarone used alone or in combination with digoxin for life-threatening supraventricular tachyarrhythmia in neonates and small infants. Pediatr Emerg Care. 2010;26:82-4.
61. Delle Karth G, Geppert A, Neunteufl T, et al. Amiodarone versus diltiazem for rate control in critically ill patients with atrial tachyarrhythmias. Crit Care Med. 2001;29:1149-53.
62. Blomstrom-Lundqvist C, Scheinman MM, Aliot EM, et al. ACC/AHA/ESC guidelines for the management of patients with supraventricular arrhythmias—executive summary: a report of the American College of Cardiology/American Heart Association Task Force on Practice Guidelines and the European Society of Cardiology Committee for Practice Guidelines (Writing Committee to Develop Guidelines for the Management of Patients With Supraventricular Arrhythmias). Circulation. 2003;108:1871-909.
63. Chang PM, Silka MJ, Moromisato DY, et al. Amiodarone versus procainamide for the acute treatment of recurrent supraventricular tachycardia in pediatric patients. Circ Arrhythm Electrophysiol. 2010;3:134-40.
64. Simonian SM, Lotfipour S, Wall C, et al. Challenging the superiority of amiodarone for rate control in Wolff-Parkinson-White and atrial fibrillation. Intern Emerg Med. 2010;5:421-6.
65. Redfearn DP, Krahn AD, Skanes AC, et al. Use of medications in Wolff-Parkinson-White syndrome. Expert Opin Pharmacother. 2005;6:955-63.
66. Kastor JA. Multifocal atrial tachycardia. N Engl J Med. 1990;322): 1713-7.
67. Arsura EL, Solar M, Lefkin AS, et al. Metoprolol in the treatment of multifocal atrial tachycardia. Crit Care Med. 1987;15:591-4.
68. Olshansky B. Dofetilide versus quinidine for atrial flutter: viva la difference!? J Cardiovasc Electrophysiol. 1996;7:828-32.
69. Natale A, Newby KH, Pisano E, et al. Prospective randomized comparison of antiarrhythmic therapy versus first-line radiofrequency ablation in patients with atrial flutter. J Am Coll Cardiol. 2000;35: 1898-904.
70. Chiang CE, Chen SA, Wu TJ, et al. Incidence, significance, and pharmacological responses of catheter-induced mechanical trauma in patients receiving radiofrequency ablation for supraventricular tachycardia. Circulation. 1994;90:1847-54.
71. Fuster V, Ryden LE, Cannom DS, et al. ACC/AHA/ESC 2006 guidelines for the management of patients with atrial fibrillation—executive summary: a report of the American College of Cardiology/ American Heart Association Task Force on Practice Guidelines and the European Society of Cardiology Committee for Practice Guidelines (Writing Committee to Revise the 2001 Guidelines for the Management of Patients With Atrial Fibrillation). J Am Coll Cardiol. 2006;48:854-906.
72. Kay GN, Epstein AE, Dailey SM, et al. Role of radiofrequency ablation in the management of supraventricular arrhythmias: experience in 760 consecutive patients. J Cardiovasc Electrophysiol. 1993;4:371-89.

73. O'Hara GE, Philippon F, Champagne J, et al. Catheter ablation for cardiac arrhythmias: a 14-year experience with 5330 consecutive patients at the Quebec Heart Institute, Laval Hospital. Can J Cardiol. 2007;23:67B-70B.
74. Calkins H, Langberg J, Sousa J, et al. Radiofrequency catheter ablation of accessory atrioventricular connections in 250 patients. Abbreviated therapeutic approach to Wolff-Parkinson-White syndrome. Circulation. 1992;85:1337-46.
75. Calkins H, Yong P, Miller JM, et al. Catheter ablation of accessory pathways, atrioventricular nodal reentrant tachycardia, and the atrioventricular junction: final results of a prospective, multicenter clinical trial. The Atakr Multicenter Investigators Group. Circulation. 1999;99:262-70.
76. Friedman PL, Dubuc M, Green MS, et al. Catheter cryoablation of supraventricular tachycardia: results of the multicenter prospective "frosty" trial. Heart Rhythm. 2004;1:129-38.
77. Jackman WM, Wang XZ, Friday KJ, et al. Catheter ablation of atrioventricular junction using radiofrequency current in 17 patients. Comparison of standard and large-tip catheter electrodes. Circulation. 1991;83:1562-76.
78. Sanders WE, Sorrentino RA, Greenfield RA, et al. Catheter ablation of sinoatrial node reentrant tachycardia. J Am Coll Cardiol. 1994;23:926-34.
79. Jackman WM, Wang XZ, Friday KJ, et al. Catheter ablation of accessory atrioventricular pathways (Wolff-Parkinson-White syndrome) by radiofrequency current. N Engl J Med. 1991;324:1605-11.
80. McClelland JH, Wang X, Beckman KJ, et al. Radiofrequency catheter ablation of right atriofascicular (Mahaim) accessory pathways guided by accessory pathway activation potentials. Circulation. 1994;89:2655-66.
81. Kay GN, Chong F, Epstein AE, et al. Radiofrequency ablation for treatment of primary atrial tachycardias. J Am Coll Cardiol. 1993;21:901-9.
82. Steinbeck G, Hoffmann E. 'True' atrial tachycardia. Eur Heart J. 1998;19:E10-2, E48-9.
83. Feld GK, Fleck RP, Chen PS, et al. Radiofrequency catheter ablation for the treatment of human type 1 atrial flutter. Identification of a critical zone in the reentrant circuit by endocardial mapping techniques. Circulation. 1992;86:1233-40.
84. Cosio FG, Lopez-Gil M, Goicolea A, et al. Radiofrequency ablation of the inferior vena cava-tricuspid valve isthmus in common atrial flutter. Am J Cardiol. 1993;71:705-9.
85. Haissaguerre M, Jais P, Shah DC, et al. Spontaneous initiation of atrial fibrillation by ectopic beats originating in the pulmonary veins. N Engl J Med. 1998;339:659-66.
86. Shen WK. Modification and ablation for inappropriate sinus tachycardia: current status. Card Electrophysiol Rev. 2002;6:349-55.
87. Bubien RS, Knotts-Dolson SM, Plumb VJ, et al. Effect of radiofrequency catheter ablation on health-related quality of life and activities of daily living in patients with recurrent arrhythmias. Circulation. 1996;94:1585-91.
88. Goldberg AS, Bathina MN, Mickelsen S, et al. Long-term outcomes on quality-of-life and health care costs in patients with supraventricular tachycardia (radiofrequency catheter ablation versus medical therapy). Am J Cardiol. 2002;89:1120-3.

89. Bathina MN, Mickelsen S, Brooks C, et al. Radiofrequency catheter ablation versus medical therapy for initial treatment of supraventricular tachycardia and its impact on quality of life and healthcare costs. Am J Cardiol. 1998;82:589-93.
90. Ikeda T, Sugi K, Enjoji Y, et al. Cost effectiveness of radiofrequency catheter ablation versus medical treatment for paroxysmal supraventricular tachycardia in Japan. J Cardiol. 1994;24:461-8.
91. Kertes PJ, Kalman JM, Tonkin AM. Cost effectiveness of radiofrequency catheter ablation in the treatment of symptomatic supraventricular tachyarrhythmias. Aust N Z J Med. 1993;23:433-6.
92. Morady F. Radio-frequency ablation as treatment for cardiac arrhythmias. N Engl J Med. 1999;340:534-44.
93. Yap SC, Harris L, Silversides CK, et al. Outcome of intra-atrial re-entrant tachycardia catheter ablation in adults with congenital heart disease: negative impact of age and complex atrial surgery. J Am Coll Cardiol. 2010;56:1589-96.
94. Feld G, Wharton M, Plumb V, et al. Radiofrequency catheter ablation of type 1 atrial flutter using large-tip 8- or 10-mm electrode catheters and a high-output radiofrequency energy generator: results of a multicenter safety and efficacy study. J Am Coll Cardiol. 2004;43:1466-72.
95. Collins KK, Schaffer MS. Use of cryoablation for treatment of tachyarrhythmias in 2010: survey of current practices of pediatric electrophysiologists. Pacing Clin Electrophysiol. 2011;34:304-8.
96. Spector P, Reynolds MR, Calkins H, et al. Meta-analysis of ablation of atrial flutter and supraventricular tachycardia. Am J Cardiol. 2009;104:671-7.
97. Bohora S, Dora SK, Namboodiri N, et al. Electrophysiology study and radiofrequency catheter ablation of atriofascicular tracts with decremental properties (Mahaim fibre) at the tricuspid annulus. Europace. 2008;10:1428-33.
98. Chinitz JS, Gerstenfeld EP, Marchlinski FE, et al. Atrial fibrillation is common after ablation of isolated atrial flutter during long-term follow-up. Heart Rhythm. 2007;4:1029-33.
99. Satomi K, Bansch D, Tilz R, et al. Left atrial and pulmonary vein macroreentrant tachycardia associated with double conduction gaps: a novel type of man-made tachycardia after circumferential pulmonary vein isolation. Heart Rhythm. 2008;5:43-51.
100. Feld GK. Catheter ablation for the treatment of atrial tachycardia. Prog Cardiovasc Dis. 1995;37:205-24.
101. Rillig A, Meyerfeldt U, Birkemeyer R, et al. Catheter ablation within the sinus of Valsalva—a safe and effective approach for treatment of atrial and ventricular tachycardias. Heart Rhythm. 2008;5:1265-72.
102. Sacher F, Vest J, Raymond JM, et al. Incessant donor-to-recipient atrial tachycardia after bilateral lung transplantation. Heart Rhythm. 2008;5:149-51.
103. Yamada T, Huizar JF, McElderry HT, et al. Atrial tachycardia originating from the noncoronary aortic cusp and musculature connection with the atria: relevance for catheter ablation. Heart Rhythm. 2006;3:1494-6.
104. Iwai S, Badhwar N, Markowitz SM, et al. Electrophysiologic properties of para-hisian atrial tachycardia. Heart Rhythm. 2011. [Epub ahead of print]

105. Ouyang F, Ma J, Ho SY, et al. Focal atrial tachycardia originating from the non-coronary aortic sinus: electrophysiological characteristics and catheter ablation. J Am Coll Cardiol. 2006;48:122-31.
106. Tracy CM. Catheter ablation for patients with atrial tachycardia. Cardiol Clin. 1997;15:607-21.
107. Chen SA, Chiang CE, Tai CT, et al. Complications of diagnostic electrophysiologic studies and radiofrequency catheter ablation in patients with tachyarrhythmias: an eight-year survey of 3,966 consecutive procedures in a tertiary referral center. Am J Cardiol. 1996;77:41-6.
108. Scheinman MM, Huang S. The 1998 NASPE prospective catheter ablation registry. Pacing Clin Electrophysiol. 2000;23:1020-8.

# Clinical Spectrum of Ventricular Tachycardia

**CHAPTER 7**

*Masood Akhtar*

## Chapter Outline

- Monomorphic Ventricular Tachycardia
  - Myocardial VT in Association with Structural Heart Disease
  - Monomorphic VT in Association with Structurally Normal Heart
- Polymorphic Ventricular Tachycardia
  - PVT in Association with Long QT Interval
  - PVT with Normal QT Prolongation
  - PVT in Association with Short QT Syndrome

## INTRODUCTION

In this communication, it is assumed that the clinician has already made the distinction between the various causes of wide QRS tachycardias, of which ventricular tachycardia (VT) is only one, albeit the most common one.[1-4]

As the field of invasive interventional electrophysiology has grown, interest in finding the cellular/molecular basis for arrhythmias has escalated. At this time, however, we clinically deal with myriad complex VTs, often with incomplete understanding and the desire to simplify information for clinical purposes.[5-10] While ultimately VT-VF (ventricular fibrillation) may find a better classification based solely on genetic and cellular knowledge, their definition within the parameters of the current science is still evolving and mostly based on clinical presentation. For all practical purposes, generally only the clinical classifications are used to manage patients. When the word VT is mentioned, a number of natural questions cross one's mind. Is the episode brief or sustained? What are the patient's symptoms? Is there underlying heart disease? In this chapter, we have taken a clinician's approach as practiced today. It is, however, important to realize that, increasingly, new entities are being introduced that may or may not fit into a given classification, and words like miscellaneous, idiopathic and other descriptive terms, will continue to be used.

**Table 1** is an attempt to present a simple and clinically relevant classification. The usual first encounter for an arrhythmologist to a patient with documented VT is a rhythm strip from telemetry, monitor, ambulatory recorder, during device interrogation or, occasionally, a 12-lead ECG **(Fig. 1)** showing a monomorphic (Panel A) or polymorphic (Panel B) VT. This is a striking feature of VT and is seldom missed by a clinician, unless, in a given lead, the polymorphic nature of the VT is not appreciable. There can be serious consequences

## TABLE 1
### Clinical spectrum of ventricular tachycardia

| Monomorphic VT | | | Polymorphic VT | |
|---|---|---|---|---|
| SHD<br>Myocardial fibrosis | No SHD<br>RV outflow | Long QT<br>Congenital<br>(QT–1 to QT–12) | Normal QT<br>Brugada*<br>(Type 1-3) | Short QT<br>Short QT<br>(1–5) |
| BBR<br>(HPS disease) | Idiopathic LV-VT | Acquired:<br>Drugs, electrolyte<br>(Table 2) | Active ischemia | |
| ARVD<br>(RV fatty infiltration)<br>regional and familial<br>forms of ARVD Naxos,<br>venetian | From sinus Valsalva,<br>mitral, pulmonic cusp | | Myocardial hypertrophy | |
| | Bidirectional | | LV Noncompaction | |
| VT Postsurgical scar | Iatrogenic device leads | | Catecholaminergic PVT<br>J-wave syndromes early repolarization<br>syndrome hypothermia<br>Idiopathic VF | |

(Abbreviations: SHD: Structural heart disease; BBR: Bundle branch reentry; ARVD: Arrhythmic RV dysplasia; RV: Right ventricular; LV: Left ventricular; PVT: Polymorphic ventricular tachycardia; VF: Ventricular fibrillation.
- Regional expressions for Brugada-like syndromes:
  — Thailand – Tai Lai (death during sleep)
  — Phillipines – Bangungut (scream followed by sudden death-at night)
  — Japan – Pokkuri – (unexpected sudden death at night)

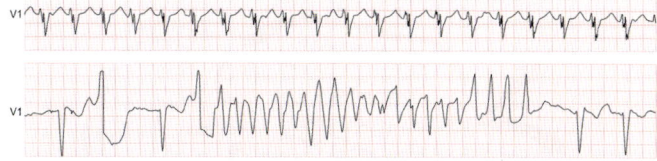

**FIGURE 1:** Rhythm strip (V1) shown. Note the monomorphic appearance of the VT in the top panel. P-wave is not clearly visible but its presence is suggested in some ST-T signals. The bottom tracing shows a prolongation of the QT interval, an episode of torsades de pointes, with rapid polymorphic VT of a constantly changing morphology that appears to be twisting around a central axis, which is the literal meaning of torsades de pointes (twisting of the points)

for not knowing polymorphic versus monomorphic VT (MMVT). For example, administering an additional dose of an antiarrhythmic drug in the presence of a polymorphic variant of VT may aggravate the situation. Hence, it is prudent to emphasize at the outset that the distinction between the monomorphic and the polymorphic nature of the VT is important and can be deciphered by recording two leads perpendicular to each other.

Once that distinction is settled, the usual next line of questioning regards the underlying pathology or structural heart disease (SHD), such as ischemia, myopathy, etc. When there is no SHD detected, attention is then directed to various VT syndromes. These somewhat newer entities are currently hard to classify. In the future, expressions, like channelopathies, repolarization syndromes, J-wave abnormalities or other such terminology, will be used routinely, and it seems this trend has already begun.[11,12]

## MONOMORPHIC VENTRICULAR TACHYCARDIA

### Myocardial VT in Association with Structural Heart Disease

#### Myocardial VT in Association with Fibrosis/Scar[8-10,13-15] (Table 1)

Coronary artery disease (CAD) remains the most common form of VT. Both monomorphic and polymorphic forms exist. However, the monomorphic forms **[Fig. 1 (top panel)]** are better understood and the underlying mechanism is easier to comprehend. Its distinction from other forms of wide QRS tachycardias has been extensively published.[1-4] The classic model used to visualize this circuit of reentry is depicted in **Figure 2**.[9] The fibrotic scar zone, shown as islands, the paths of impulse propagation in various directions, is indicated by arrows. On the

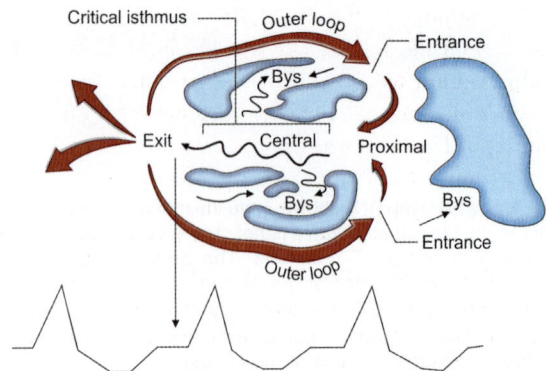

**FIGURE 2:** Classical model used to demonstrate reentry through the surviving muscle bands among the myocardial scar (shown as five islands). The QRS on the surface ECG starts where the impulse exits from the critical isthmus and turns around in a figure-of-eight fashion to reenter from the proximal end. Ablation at sites other than beside the central blue line is unlikely to be successful. This is why it is termed a critical area of slow conduction or as critical isthmus. (*Source:* Modified from Stevenson W, Soejima K. Catheter ablation of ventricular tachycardia. In: Zipes DP, Jalife J (Eds). Cardiac Electrophysiology: From Cell to Bedside, 4th edition. Philadelphia: WB Saunders; 2004. pp. 1087-96)

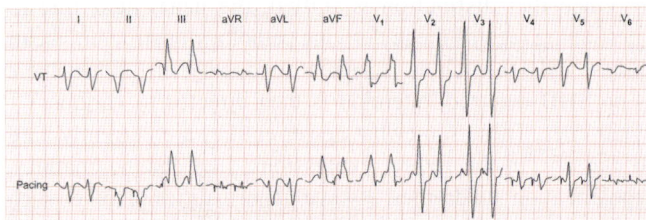

**FIGURE 3:** Example of spontaneous ventricular tachycardia (VT) recorded on a 12-lead ECG. The bottom shows 12-lead pace map. Note that the QRS is identical to spontaneous VT and the stimulus artifact to QRS is relatively short—best appreciated in leads II, $V_{4-6}$

surface ECG, QRS starts when the impulse exits from within the circuit. If one was to electrically stimulate this exit site, the surface ECG QRS would look identical to the spontaneous VT with a short stimulus artifact to QRS interval **(Fig. 3)**. Depending upon the geometry of the scar, the impulse could go in several directions, dictated by the shape of the scar and the state of the myocardium. As an example shown in **Figure 2**, the impulse travels in all the directions in a three-dimensional tissue. During activation of a normal myocardium the impulse will reenter the circuit at the proximal end or some other points. Depending on its length, the location of surviving myocardial cells within the scar and the speed of conduction, the local electrical signal produced may bear a variable relation with the surface QRS. For example, the interval recorded between the electrogram and pacing artifact from the same catheter site in the critical isthmus to the next QRS will measure the same.[9] However, from the adjacent bystander, the interval from the local electrogram to next QRS will be considerably shorter compared to the pacing to the next artifact QRS interval from the same site. These maneuvers are very helpful in finding the right location for catheter ablation.

It should be pointed out that the diastolic interval between the QRS complexes is increased when the reentrant impulse is travelling through the isthmus (surviving muscle band, area of slow conduction), which is critical for the VT to continue. While this electrical activity is not visible on the surface ECG, it can be recorded by placing electrode catheters along the pathway. Penetration of this pathway occurs during sinus rhythm as well, and it can be recorded both by intracardiac recording techniques as well as from the surface by proper magnification and filters (the so-called signal-averaged ECG).[16]

While the actual reentrant circuits may be more complex, the schema shown in **Figure 2** gives one broad concept of how a VT can be mapped. When appropriate location of slow conduction is isolated, it leads to a successful ablation using radiofrequency or another form of energy. The baseline ECG is seldom normal in these patients, and likely to be suggestive of some cardiac pathology.

Therapy for a scar-related VT can be manifold:[17,18]

- In patients with an ejection fraction of less than 35%, implantable cardioverter defibrillator (ICD) is advised. The main reason is the prevention of sudden VT-related death from an existing or new arrhythmia **(Fig. 4)**.[19] In many cases, particularly with slow VT (≥280 ms), antitachycardia pacing will also terminate an organized MMVT, which is more comfortable for the patient

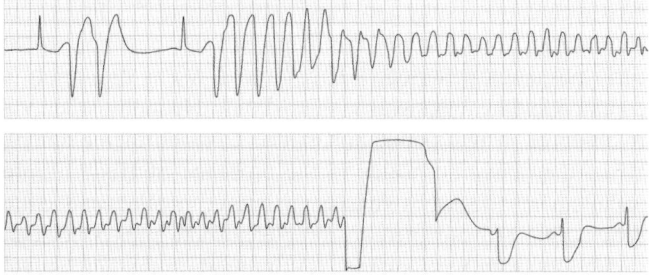

**FIGURE 4:** The ICD shock and defibrillation. The figure displays the onset of a rapid VT and a period of sensing and defibrillation shock with an ICD and restoration of sinus rhythm. The effect of acute injury from ICD can be appreciated in subsequent sinus complexes. This tachycardia was different than previously documented. (*Source:* Tchou PJ, Kadri N, Anderson J, et al. Automatic implantable cardioverter defibrillators and survival of patients with left ventricular dysfunction and malignant ventricular arrhythmias. Ann Intern Med. 1988;109:529-34, with permission)

- With better left ventricular ejection fractions (LVEF), antiarrhythmic agents, particularly Class III drugs, such as sotalol, amiodarone and dronedarone, may be sufficient. In patients with CAD and VT, the addition of beta-blockers is beneficial because the role of ischemia in the initiation and maintenance of VT cannot be excluded with certainty at a given time. In high-risk patients post-myocardial infarction or cardiac surgery, an external defibrillator in the form of a life vest can be recommended
- When VT is incessant, endocardial and/or epicardial mapping and catheter ablation may be necessary
- Although surgical ablation is seldom necessary, it remains an option
- In rare situations, cardiac transplantation with or without ventricular assist devices may be the only option when there are no contraindications.

## Monomorphic VT Due to Bundle Branch Reentry[20-25]

In this form of monomorphic VT, the underlying pathological substrate is the His-Purkinje system, which has markedly prolonged conduction time. More specifically, the right and left bundle branches are used for propagation to the ventricle via one bundle, returning to the His bundle through the contralateral bundle branch. Sometimes the electrical circuit is localized to the two fascicles of the left bundle; in that case, it is termed interfascicular reentry. The schema in **Figure 5** depicts the reentry circuits and **Figures 6A and B** shows bundle branch reentry (BBR)-VT, both a left bundle branch block (LBBB) pattern (A) and a right bundle branch block (RBBB) pattern (B). While the general incidence of BBR as the mechanism of MMVT is 6%, it is much higher in patients with idiopathic dilated cardiomyopathy and aortic valve disease.[24-26] In patients with aortic valve disease, this is particularly common in the early postsurgical period.[24] This form of VT is apparently a common finding in patients with myotonic dystrophy, when VT is observed in that population.[25] The common theme among all of the above scenarios is the presence of His-Purkinje pathology manifested by nonspecific intraventricular conduction defect (IVCD), incomplete-to-complete bundle branch block on surface ECG with prolonged H-V interval on the His bundle electrogram recording.[20] Unlike patients with myocardial disease, where the His bundle potential follows QRS, in BBR it always precedes QRS with equal to or longer H-V than sinus, and H-H cycle length changes precede V-V cycle length changes **(Fig. 6)**.

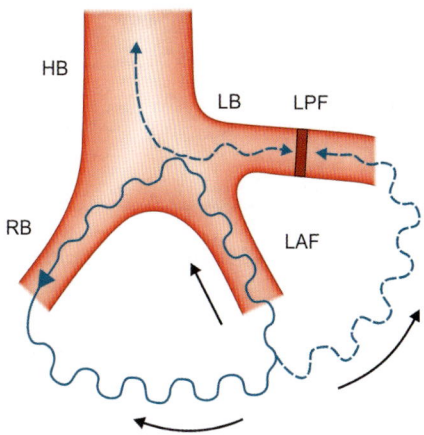

**FIGURE 5:** Schema demonstrates the circuit of bundle branch reentry VT. In this example, the impulse reaches the right ventricle via the right bundle (RB) and returns to the His bundle through the inferior fascicle of the left bundle (LB). While the same impulse approaches the His bundle via the superior fascicle, the two impulses will collide somewhere in that region. Activation of the His bundle will occur as a necessity since the left and right bundle are connected via the His bundle. The arrows depict the direction of impulse propagation. Tracings from top to bottom in each panel are surface ECG leads I, II and $V_1$. The intracardiac tracings are HRA (high right atrial electrogram) and HB (His bundle electrogram). Time lines at the bottom are consecutive (Abbreviations: LAF: Left anterior fascicle; LPF: Left posterior fascicle)

While typically these individuals have low LVEF, BBR can recur in patients without SHD and normal LVEF[27] and in some patients with valvular disease, myocardial dystrophy and preserved left ventricular (LV) function.[24,25] Common to all of the above is the IVCD, and prolonged H-V interval and myocardial damage is not a prerequisite.

The tachycardia morphology is either an LBBB or, less frequently, an RBBB pattern **(Figs 6 and 7)**. Since the reentrant impulse depolarizes the ventricle via the bundle branch, the surface ECG appearance of the QRS has some features common to QRS complex due to aberrant conduction, such as rapid initial inscription of the QRS, unlike the slurred beginning of QRS seen in myocardial VT or preexcited QRS. These tachycardias are often rapid and poorly tolerated by patients with poor LV function, frequently lead to syncope and may degenerate into VF.

Bundle branch ablation is the preferred therapy in patients with good LV function **(Fig. 7)**, while an ICD should be implanted in cases where the LVEF is less than 35%. Although BBR is easily pace-terminable, RBBB ablation may still be necessary in some cases to prevent frequent recurrences.

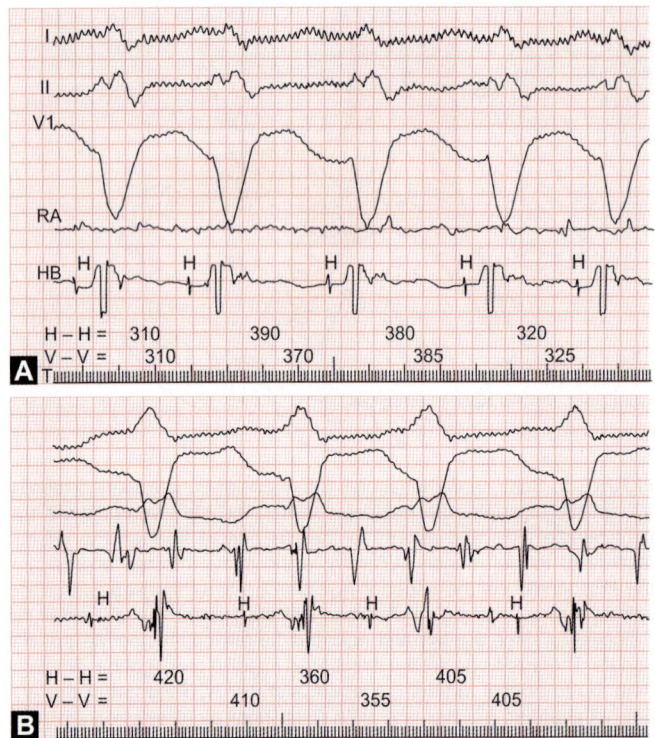

**FIGURES 6A AND B:** Bundle branch reentry with a left bundle branch block pattern (A) and right bundle branch block pattern (B) are shown. The axis is normal in A and leftward in B. The atrial rhythm is atrial fibrillation. Note that His deflection precedes the QRS and the change in the cycle length of H-H (labeled) precedes that of V-V (also labeled). In other words, the His bundle activation drives the ventricle, which is the opposite of what happens in myocardial VT, where the His bundle deflection follows the local V electrogram or is obscured by it. Nonetheless, the V-V cycle drives the H-H cycle in myocardial VT. (*Source:* Blanck Z, Jazayeri M, Akhtar M. Facilitation of sustained bundle branch reentry by atrial fibrillation. J Cardiovasc Electrophysiol. 1996;7:348-52, with permission)

## Monomorphic VT in Association with Arrhythmogenic Right Ventricular Dysplasia[28,29]

In arrhythmogenic right ventricular dysplasia (ARVD), the right ventricular (RV) muscle is replaced by fatty tissue, which in effect creates a model very similar to that shown in **Figure 2**. The disease may be patchy; affecting only RV apex, outflow or other parts, but may be quite extensive, replacing most RV myocardium with fatty infiltrates. At times the LV may also be affected.

For the most part, the main clinical problem these patients have is with nonsustained or sustained VT with a left bundle branch configuration with variable axis. While any axis may

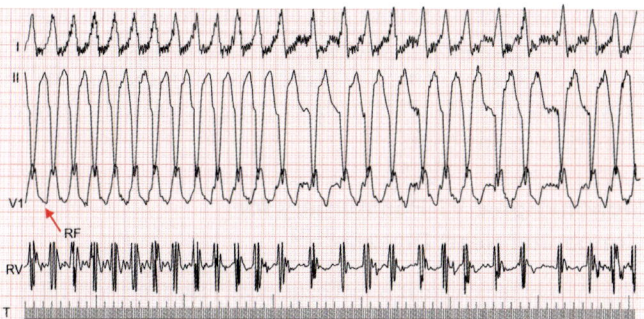

**FIGURE 7:** Ablation of bundle branch reentry (BBR). The first 13 complexes are due to BBR-VT. The underlying atrial rhythm is atrial fibrillation, which becomes obvious when the VT stops due to right bundle ablation. Note, there is no change in QRS configuration but now the ventricular rate is irregular due to AF

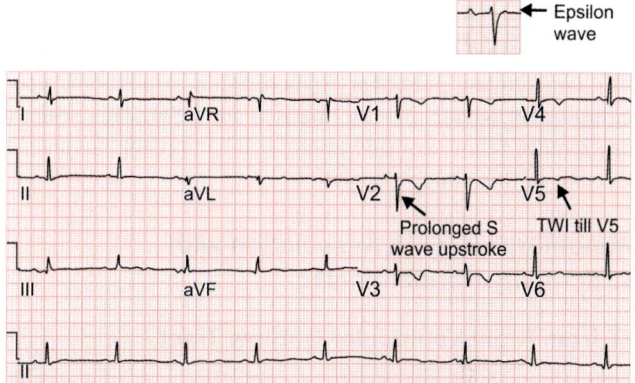

**FIGURE 8:** Twelve-lead ECG in arrhythmic right ventricular dysplasia (ARVD), the T-wave inversion $V_{1-5}$ (usually up to $V_3$) and late wave (epsilon wave in the insert) are characteristic findings in the baseline ECG of these patients. (*Source:* Nasir K, Bomma C, Tandri H, et al. Electrocardiographic features of arrythmogenic right ventricular dysplasia/cardiomyopathy according to disease severity: a need to broaden diagnostic criteria. Circulation. 2004;110:1527-34, with permission)

be noted, left bundle and left atrial (LA) morphology is very suggestive of ARVD; surface ECG also characteristically shows T-wave inversion in $V_1$–$V_3$ **(Fig. 8)** and may extend to $V_4$ or $V_5$. A late small deflection may be seen at the end of the QRS (epsilon wave) (Fig. 8 inset). Signal-averaged ECG is often positive. Diagnostic work should include magnetic resonance imaging (MRI), which is more sensitive than ultrasound, particularly in the early stages when the dysplasia is patchy.

Several genetic abnormalities are associated with this syndrome. In several areas of the world, such as Naxos (Greece) and the Venetian region (Italy), a large prevalence of these VTs is noted among some families.[30] The VTs generally have a monomorphic configuration, but sudden cardiac death

(SCD) may occur. Sotalol and amiodarone have been used to control VT. Catheter ablation is not encouraged due to the risk of perforation. The ICD therapy is recommended for the prevention of SCD. This topic is more extensively covered elsewhere in this book.

## Monomorphic VT Postsurgery for Congenital Heart Disease[31,32]

This type of tachycardia is mechanistically akin to a scar-related reentry. Since most of these incisional scars are in the right ventricle, the morphology is likely to be an LBBB configuration and a variable axis, usually right. Associated congenital heart disease (CHD), adhesions post surgery and other factors, such as development of thorax, etc. may create a somewhat atypical QRS configuration, but endocardial mapping will localize the VT in the neighborhood of the scar. Antiarrhythmic drugs and catheter ablation (while they may be sufficient to control the VT), concomitant pulmonary hypertension or pulmonic valve regurgitation would increase the risk for SCD. This scenario may require serious consideration for ICD implant; not an easy decision in this young population.

## Monomorphic VT in Association with Structurally Normal Heart

### VT from Right Ventricular Outflow Tract

As the name suggests, the most common location of this VT is outflow which produces a characteristic LBBB and right atrial morphology **(Fig. 9)**.[33,34] If the breakthrough occurs on the left side, an RBBB morphology may be seen. The

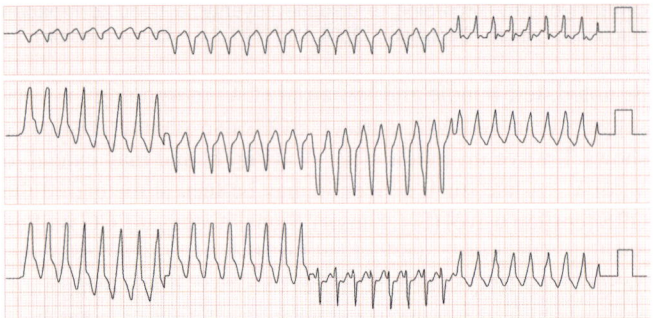

**FIGURE 9:** Right ventricular outflow tract (RVOT) VT. Typically, VT arises in RVOT. The orientation of the 12-lead ECG is leads I, II and III are on the left from top to bottom. The AVR, AVL, AVF are the next three leads, $V_{1-3}$ and $V_{4-6}$ are next. The same designation is used with all of the 12-lead ECGs unless labeled otherwise. Note left bundle branch block pattern and right axis typical of this type of VT

baseline ECG is usually normal. The process could present in the form of isolated premature ventricular complexes, repetitive MMVT, nonsustained or sustained VT. When RV outflow VT is symptomatic, palpitation, lightheadedness and presyncope are common. Syncope may occur, but SCD is rare in these patients. The mechanism is not completely clear, but the tachycardia often is initiated by isoproterenol. Triggered delayed afterdepolarization driven by catecholamines is the prevailing view regarding the arrhythmogenic mechanism. Clinically increased sympathetic drive, such as physical exercise, often triggers the episode and, not infrequently, beta-blockers may be effective in controlling the VT. Thus, antiarrhythmic drugs, such as sotalol and amiodarone, may also be effective. Considering a good long-term outcome, catheter ablation is increasingly used as a preferred form of therapy. Even though the diagnosis is often clear from clinical data, ARVD should be excluded with MRI prior to catheter ablation.

## Idiopathic Left Ventricle VT[35,36]

The electrophysiologic basis of this VT is reentry within the peripheral Purkinje system. The QRS morphology is that of right bundle and LA **(Fig. 10)**, but other ranges of axis are occasionally observed. The baseline ECG and LVEF as a rule are normal. Intravenous verapamil will often terminate the VT but is less effective orally. Class III antiarrhythmic agents, such as amiodarone, are effective, but, at this time, catheter ablation is the first-line treatment considering this VT is easily inducible in the laboratory.

As with RV outflow VT, the ventricle is usually normal. Sudden death is rare but the usual symptoms of arrhythmias, such as palpitations, dizziness and occasional presyncope and syncope, are noted.

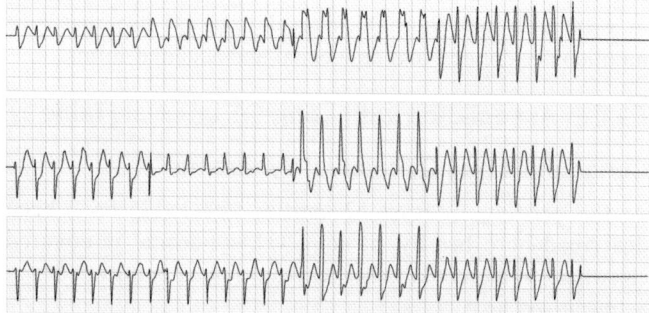

**FIGURE 10:** Idiopathic left ventricular VT. The typical QRS morphology is that of right bundle branch block and left-axis pattern. Note the initial part of the QRS has rapid inscription due to the fascicular origin of this VT. Contrast this with Figure 9 where the initial part of the QRS has a slow inscription

### Aortic Sinus of Valsalva, Pulmonic, Mitral Cusp VT

Monomorphic VT with right axis occasionally arises from structures outside the traditional ventricular myocardium.[37,38] While they mimic the outflow VT, awareness of these loci helps to improve mapping. Catheter ablation is the usual treatment. Although experience with these types of VT is limited, it is likely that the traditional antiarrhythmic drugs may be successful.

### Bidirectional Tachycardia

Bidirectional tachycardia seen with digitalis toxicity is rare, but is seen in the early stages of the exercisein patients with catecholaminergic VT (discussed later). Its classic picture is that of two sets of monomorphic QRS complexes alternating in QRS morphologies. In a sense, bidirectional tachycardia has not a true polymorphic but rather a pleomorphic appearance.

### Iatrogenic VT

Whenever a lead is placed in the ventricle, it is not surprising that a monomorphic VT from mechanical movement may be created that can be mistaken for a spontaneous VT, particularly with ICD leads since these patients also have clinical VT. Whenever a resistant VT is found with morphology that could be generated from the location of the catheter, nothing but catheter withdrawal and repositioning will fix this type of VT, so keeping this in mind is important as part of the differential diagnosis in patients with device implant.

## POLYMORPHIC VENTRICULAR TACHYCARDIA

Polymorphic ventricular tachycardia can be broadly separated into three categories: (i) PVT in association with long QT interval, (ii) PVT in association with normal QT, and (iii) VT in association with short QT interval. Long QT itself has been traditionally divided into congenital and acquired forms.

### PVT in Association with Long QT Interval (Table 1)

### Congenital Long QT Interval Syndrome[39-42]

At this writing, at least 12 entities (QT-1–12) have been described here briefly, as long QT syndrome is covered elsewhere in the book. Other less-understood entities will not be discussed here.

*QT-1*: QT-1 is the most common form **[Fig. 11 (left panel)]**. The recessive variety may be associated with deafness. The PVT is typically triggered with physical activity, emotional stress, diving and swimming. Syncope, presyncope and SCD are the

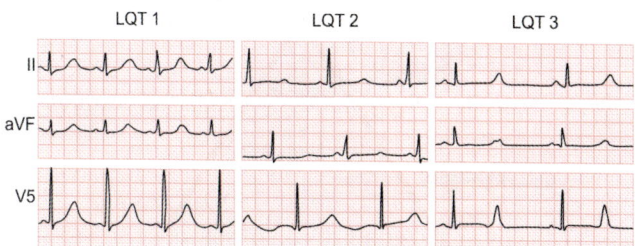

**FIGURE 11:** Congenital long QT (LQT). The three most common genetic varieties are shown. The corresponding chromosomes are labeled. See text for other details. (*Source:* Moss AJ, Zareba W, Benhorin J, et al. ECG T-wave patterns in genetically distinct forms of the hereditary long QT syndrome. Circulation. 1995;92:2929-34, with permission)

most serious clinical manifestations. The surface ECG shows a broad, prolonged T-wave and a long QT interval **(Fig. 11)**. Early onset of symptoms, syncope, excessive QT prolongation is more than or equal to 550 msec in QT-1 and QT-2, and males with QT-3 are associated with a high risk of SCD **(Fig. 11)**. Nonselective beta-blockers, such as propranolol and nadolol, are preferred, but others have been used. A dose of propranolol (5 mg/kg is usual) and avoidance of triggering events, such as adrenergic stress or diving, is highly recommended. Drugs that prolong QT **(Table 2)** and electrolyte imbalance, such as hypokalemia, can be lethal in this population with prolonged QT syndrome.

The gene involved in QT-1 is KCNQ$^1$ and affects K$^+$ current (delayed rectifier current 1Ks), which is reduced, causing the lengthening of QT **[Fig. 11 (left panel)]**. If a proband is found, it is likely that a high percentage of blood relatives may carry the abnormal gene. Genetic screening should be recommended. Beta-blockers are effective in controlling the symptoms. Left stellate ganglionectomy has been successful in controlling symptoms. In patients with malignant manifestation, i.e. SCD or PVT-related syncope, ICD should be seriously considered.

*QT-2:* QT-2 **[Fig. 11 (middle panel)]** is the second most common, and carried by 1Kr (delayed rectifiers), which encodes the gene KCNH2. The QT is prolonged as expected, but the T-wave is somewhat flat and less pronounced than QT-1. While adrenergic stress remains important, auditory stimuli in particular may trigger malignant arrhythmia.

*QT-3:* Compared to QT-1 and QT-2, QT-3 is less common but has a worse outcome. The QT-3 is prolonged **[Fig. 11 (right panel)]** due to increased inward Na$^+$ current. The Na$^+$ channel is encoded by the gene SCN5A and, consequently, the ECG shows a prolonged ST segment, short T-wave and a long QT interval. Prognosis without therapy is poor, particularly in young males **(Fig. 12)**.

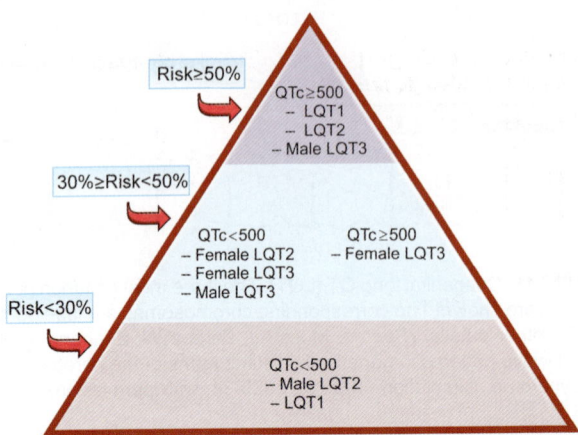

**FIGURE 12:** A pyramid showing risk stratification in congenital long QT (LQT) syndrome. (*Source:* Modified from Schwartz PJ, Priori SG. Long QT syndrome: genotype-phenotype correlations. In: Zipes DP, Jalife J (Eds). Cardiac Electrophysiology: From Cell to Bedside, 4th edition. Philadelphia: WB Saunders; 2004. pp. 651-9)

Bradycardia is one of the main triggers, such that the most fatal events occur at night or during sleep. The role of beta-blockers is still controversial and, without bradycardia support (i.e. pacemaker), somewhat risky. Patients who have exhibited symptoms due to PVT, such as syncope or cardiac arrest, should have ICD implantation. The role of gene-specific drugs and other agents that shorten the QT interval, such as mexiletine, has not been systematically studied sufficiently to be utilized as the sole therapy for prevention of SCD. Risk stratification among patients with congenital long QT is shown in **Figure 12**.

## Acquired Long QT Syndrome

Although the exact underlying etiology of acquired long QT is not understood, it is widely believed and sometimes reported that a genetic basis with low penetrance may account for some of these cases.[43,44] An external trigger, such as an antiarrhythmic drug, electrolyte imbalance, etc. is required for the clinical manifestations of this form, i.e. prolongation of QT and torsades de pointes.

There are an increasing number of pharmaceutical agents that have been documented to be the culprit **(Table 2)**. In addition to the Class I or Class III antiarrhythmic drugs [Fig. 1 (top panel)], many other agents and situations have produced similar adverse affects. Low $Mg^{++}$, $Ca^{++}$ and $K^+$ may trigger torsades de pointes in vulnerable populations. Some drugs, by blocking or binding with certain liver enzymes, may lower metabolism, raising the blood level of the parent compound that in turn may produce prolonged QT and torsades de pointes.

### TABLE 2
**Abbreviated list of drugs reported to cause prolongation of the QT interval or torsades de pointes**

| | |
|---|---|
| Antiarrhythmic: | |
|   Class 1A | Disopyramide, procainamide, quinidine |
|   Class III | Amiodarone, bretylium, sotalol, dofetilide, ibutilide |
| Antimicrobial | Erythromycin, trimethoprim-sulfamethoxazole, clarithromycin |
| Antifungal | Fluconazole, ketoconazole, itraconazole |
| Antimalarial or antiprotozoal | Chloroquine, halofantrine, mefloquine, pentamidine, quinine |
| Antihistamine | Astemizole, terfenadine, diphenhydramine |
| Gastrointestinal prokinetic | Cisapride |
| Psychoactive | Chloral hydrate, haloperidol, lithium, phenothiazines, pimozide, tricyclic antidepressants |
| Antihuman immunodeficiency virus | Efavirenz |
| Miscellaneous | Amantadine, indapamide, probucol, tacrolimus, vasopressin |

*Source:* Modified from El-Sherif N, Turitto G. Torsades de pointes. In: Zipes DP, Jalife J (Eds). Cardiac Electrophysiology: From Cell to Bedside, 4th edition. Philadelphia: WB Saunders; 2004. pp. 687-98

Acute treatment is usually the administration of IV $Mg^{++}$, which often effectively halts torsades, but recognizing and discontinuing the offending agent is usually sufficient. Isoproterenol infusion and overdrive pacing are also used to stop the torsades de pointes. Patients are advised to avoid similar agents and situations, such as over-the-counter medication, where the contents are not clearly labeled. This aspect of long QT is important to realize because the blood relatives of a person with congenital long QT may be prone to the same hazards and a caution regarding this possibility may be wise.

## PVT with Normal QT prolongation

### Brugada Syndrome

Brugada syndrome was initially described by the Brugada brothers as the presence of an injury pattern in leads $V_1$ and $V_2$ with ST elevation followed by T-wave inversion **(Fig. 13)** and a history of syncope and SCD.[45-48] Since then, more has been learned regarding the role of ionic currents, the underlying mechanism and the worldwide prevalence of this potentially malignant syndrome.

The exact preponderance is unknown, but Pokkuri in Japan, Bangungut in the Philippines and Tai Lai in Thailand seem to be the same affliction **(Table 1)**. Mostly seen in young males, death from PVT-VF occurs at night. The ECG abnormalities

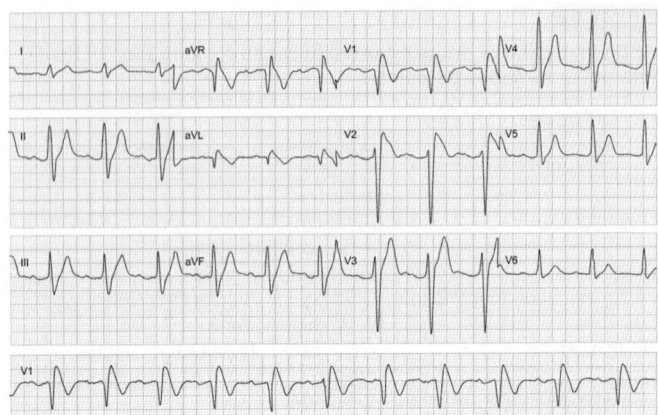

**FIGURE 13:** Typical 12-lead ECG of Brugada syndrome. The most characteristic features are seen in leads $V_1$ and $V_2$. ST elevation and T-wave inversion gives the impression of a right bundle branch block pattern. (*Source:* Modified from Dorian P, Bharati S, Myerburg RJ, et al. Ventricular fibrillation. In: Saksena S, Camm AJ (Eds). Electrophysiological Disorders of the Heart. New York: Churchill-Livingston; 2005. pp. 419-53)

**(Fig. 13)** may not always be present. However, Class I agents, such as ajmaline and procainamide, can unmask the abnormality. In many cases, the genes that encode SCN5A can be detected. Quinidine has been identified as a potentially effective agent to prevent SCD in this population. However, ICD remains the most reliable therapy to prevent PVT-VT related deaths in patients with Brugada syndrome.

## Active Ischemia

In patients with clinically significant CAD, some degree of ischemia may exist at all times, but a critical degree of ischemia with exercise, spontaneously or with spasm, can induce a PVT-VF. This is a highly malignant arrhythmia and usually fatal unless it stops spontaneously or is terminated. Anti-ischemic therapy and revascularization are the preferred treatment modalities. The tracing in **Figure 14** is from a 60-year-old male who underwent coronary artery bypass grafting, was continued on beta-blockers and still had 3–4 episodes of VT per year.[5] He received an ICD, which intervened several times over the years. This occurred primarily due to incomplete revascularization—partly due to some areas of inoperable disease, not an infrequent clinical scenario.

Acute ischemic-related PVT-VF should be addressed promptly since patients have died while waiting for revascularization. Three decades ago, large infarcts resulting in aneurysms were associated with monomorphic VT. With early intervention

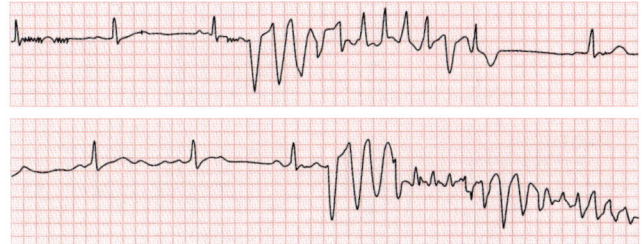

**FIGURE 14:** Polymorphic ventricular tachycardia (top panel) and onset of VF (bottom panel) in a patient with CAD post-coronary artery bypass surgery. Incomplete revascularization may lead to this and, in patients present with VF, ICD may still be necessary. (*Source:* bottom panel Akhtar M. Clinical spectrum of ventricular tachycardia. Circulation. 1990;82:1561-73, with permission)

and the use of several effective anti-ischemic agents, large infarcts and ventricular aneurysms are less frequently encountered. The PVT seems to be more common. Since it degenerates into VF quickly, the true incidence of PVT is difficult to estimate, but it certainly constitutes a significant cause of the SCD from CAD.

## Myocardial Hypertrophy

The VF and consequent SCD remain one of the main causes of cardiovascular death in patients with hypertrophic cardiomyopathy.[49] Apical hypertrophy has a particularly malignant outcome. Syncope, near-syncope and nonsustained VT define a particularly high-risk population. The SCD is seen in both obstructive and nonobstructive forms. High-risk patients should be considered for ICD therapy, both for primary and for secondary prevention of SCD.

## LV Noncompaction[50]

Isolated LV noncompaction (LVNC) is a rare myopathy, primarily of autosomal inheritance. The main structural abnormality is intrauterine failure of LV muscle compaction. The main clinical manifestations are congestive heart failure and ventricular arrhythmias with around 20% incidence of SCD.

## Catecholaminergic PVT[51,52]

Catecholaminergic PVT (CPVT) can be a dominant or recessive inheritance, mostly manifested in childhood or young adults. The heart is structurally normal as is the baseline ECG. Adrenergic drive brings out characteristic bidirectional **[Fig. 15 (4th panel from top)]** or PVT. There is a high incidence of SCD (>30% mortality by age 30). The basis is mutation in the

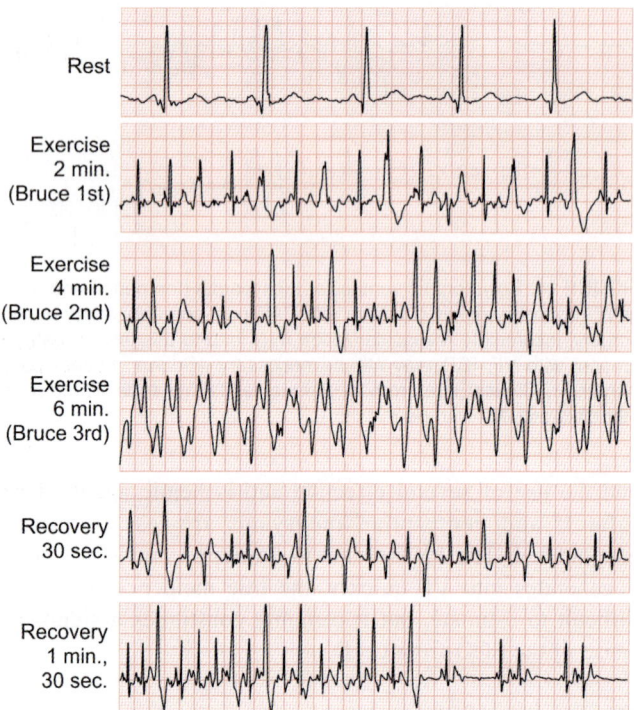

**FIGURE 15:** Progressive increase in ventricular ectopy as the exercises load increases. The 4th panel shows bidirectional tachycardia, and the last two panels show further polymorphism during recovery phase of exercise. (*Source:* Modified from Napolitano C, Priori SG. Catecholaminergic polymorphic ventricular tachycardia and short-coupled Torsades de Pointes. In: Zipes DP, Jalife J (Eds). Cardiac Electrophysiology: From Cell to Bedside, 4th edition. Philadelphia: WB Saunders; 2004. pp. 633-9)

genes, encoding ryanodine receptor 2 (RyR2) or calsequestrin 2 (CASQ2) triggering activity from delayed afterdepolarization due to abnormalities of $Ca^{++}$ handling, which is responsible for arrhythmogenesis. The conversion to PVT is most likely related to transmural dispersion of repolarization between the various myocardial layers. Degeneration to VF is the most likely cause of SCD.

## J-wave Syndromes[12,53,54]

This category includes several entities, where the individuals are prone to arrhythmic death. The primary defect seems to be imbalance of current during Phase 1 of the action potential producing Phase 2 reentry leading to VF. Some examples include Brugada syndrome, early repolarization, short QT syndrome and perhaps many cases of so-called idiopathic VF. Rapid outward current plays a significant role and may be the reason that quinidine, which blocks $K^+$ current, is effective in

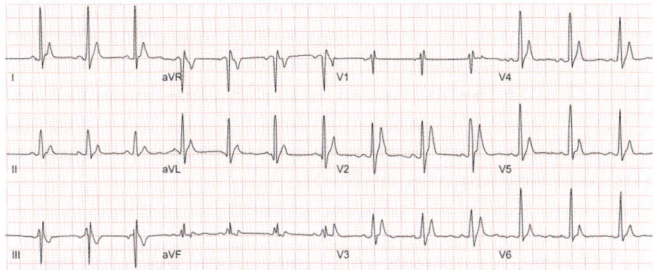

**FIGURE 16:** Example of a short QT, which can lead to fatal arrhythmia. The coexistence of short QT and Brugada has also been described. (*Source:* Bjerregaard P, Gussak I. Short QT syndrome: mechanisms, diagnosis and treatment. Nat Clin Pract Cardiovasc Dis. 2005;2:84-7, with permission)

preventing recurrence. This subject is covered in greater detail elsewhere in this book.

### Idiopathic VF

This fascinating entity is described in greater detail by Belhassan and Viskin, and is associated with identifiable SHD.[55] It is characterized by the occurrence of spontaneous VT with inducible PVT and VF, which respond to oral guideline both in the electrophysiology laboratory (i.e. not inducible after drug) and in the excellent clinical response seen over many years. Nonetheless, as Brugada syndrome and the various J-wave syndrome have been described, it is not clear how many of these cases will continue to be called idiopathic.

### PVT in Association with Short QT Syndrome[56,57]

This topic is somewhat new and with very limited experience and follow-up. **Figure 16** shows an example of short QT, usually described as QT less than 300 msec, but some cases with similar genetic mutation had QT of 320 msec. At least five mutations have been described.

The foregoing outline is a summary of VT as the subject is clinically viewed. For a comprehensive review and ACC/AHA/ESC guidelines on work-up and management of VT, the reader is referred to the most recent literature.[17,18]

## REFERENCES

1. Wellens HJ, Bar FW, Lie KI. The value of the electrocardiogram in the differential diagnosis of a tachycardia with a widened QRS complex. Am J Med. 1978;64:27-33.
2. Akhtar M, Shenasa M, Jazayeri M, et al. Wide QRS complex tachycardia. Reappraisal of a common clinical problem. Ann Intern Med. 1988;109:905-12.

3. Brugada P, Brugada J, Mont L, et al. A new approach to the differential diagnosis of a regular tachycardia with a wide QRS complex. Circulation. 1991;83:1649-59.
4. Miller JM, Das MK, Arora R, et al. Differential diagnosis of wide QRS complex tachycardia. In: Zipes DP, Jalife J (Eds). Cardiac Electrophysiology: From Cell to Bedside, 4th edition. Philadelphia: W.B. Saunders; 2004. pp. 747-57.
5. Akhtar M. Clinical spectrum of ventricular tachycardia. Circulation. 1990;82:1561-73.
6. Dessertenne F. La tachycardie ventriculaire a deux foyers opposes variables. Arch Mal Coeur Vaiss. 1966;59:263-72.
7. Ruan Y, Wang L. Short-coupled variant of torsade de pointes. J Tongji Med Univ. 2001;21:30-1.
8. Josephson ME, Almendral JM, Buxton AE, et al. Mechanisms of ventricular tachycardia. Circulation. 1987;75:III41-7.
9. Stevenson WG, Soejima K. Catheter ablation of ventricular tachycardia. In: Zipes DP, Jalife J (Eds). Cardiac Electrophysiology: From Cell to Bedside, 4th edition. Philadelphia: W.B. Saunders; 2004. pp. 1087-96.
10. Wellens HJ, Schuilenburg RM, Durrer D. Electrical stimulation of the heart in patients with ventricular tachycardia. Circulation. 1972;46:216-26.
11. Priori SG, Rivolta I, Napolitano C. Genetics of long QT, Brugada and other channelopathies. In: Zipes DP, Jalife J (Eds). Cardiac Electrophysiology: From Cell to Bedside, 4th edition. Philadelphia: W.B. Saunders; 2004. p. 462.
12. Tikkanen JT, Anttonen O, Juntilla MJ, et al. Long-term outcome associated with early repolarization on electrocardiography. New Engl J Med. 2009;361:2529-37.
13. de Bakker JM, van Capelle FJ, Janse MJ, et al. Reentry as a cause of ventricular tachycardia in patients with chronic ischemic heart disease: electrophysiologic and anatomic correlation. Circulation. 1968;77:589-606.
14. Buxton AE, Waxman HL, Marchlinski FE, et al. Role of triple extrastimuli during electrophysiologic study of patients with documented sustained ventricular tachyarrhythmias. Circulation. 1984;69:532-40.
15. Luu M, Stevenson WG, Stevenson LW. Diverse mechanisms of unexpected cardiac arrest in advanced heart failure. Circulation. 1989;80:1675-80.
16. Haberl R, Jilge G, Pulter R, et al. Comparison of frequency and time domain analysis of the signal-averaged electrocardiogram in patients with ventricular tachycardia and coronary artery disease: methodologic validation and clinical relevance. J Am Coll Cardiol. 1988;12:150-8.
17. Zipes DP, Camm AJ, Borggrefe M, et al. ACC/AHA/ESC 2006 Guidelines for management of patients with ventricular arrhythmias and the prevention of sudden cardiac death—executive summary: a report of the American College of Cardiology/American Heart Association Task Force and the European Society of Cardiology Committee of Practice Guidelines (Writing Committee to Develop Guidelines for Management of Patients with Ventricular Arrhythmias and the Prevention of Sudden Cardiac Death) Developed in

collaboration with the European Heart Rhythm Association and the Heart Rhythm Society. Eur Heart J. 2006;27:2099-140.
18. Epstein AE, DiMarco JP, Ellenbogen KA, et al. ACC/AHA/HRS 2008 Guidelines for Device-Based Therapy of Cardiac Rhythm Abnormalities: a report of the American College of Cardiology/American Heart Association Task Force on Practice Guidelines (Writing Committee to Revise the ACC/AHA/NASPE 2002 Guideline Update for Implantation of Cardiac Pacemakers and Antiarrhythmia Devices) developed in Collaboration with the American Association for Thoracic Surgery and Society of Thoracic Surgeons. J Am Coll Cardiol. 2008;51:e1-e62.
19. Tchou PJ, Kadri N, Anderson J, et al. Automatic implantable cardioverter defibrillators and survival of patients with left ventricular dysfunction and malignant ventricular arrhythmias. Ann Intern Med. 1988;109:529-34.
20. Akhtar M, Damato AN, Batsford WP, et al. Demonstration of re-entry within the His-Purkinje system in man. Circulation. 1974;50:1150-62.
21. Caceres J, Jazayeri M, McKinnie J, et al. Sustained bundle branch reentry as a mechanism of clinical tachycardia. Circulation. 1989;79:256-70.
22. Tchou P, Jazayeri M, Denker ST, et al. Transcatheter electrical ablation of right bundle branch: a method of treating macroreentrant ventricular tachycardia attributed to bundle branch reentry. Circulation. 1988;78:246-57.
23. Blanck Z, Dhala A, Deshpande S, et al. Bundle branch reentrant ventricular tachycardia: cumulative experience in 48 patients. J Cardiovasc Electrophysiol. 1993;4:253-62.
24. Narasimhan C, Jazayeri MR, Sra J, et al. Ventricular tachycardia in valvular heart disease: facilitation of sustained bundle-branch reentry by valve surgery. Circulation. 1997;96:4307-13.
25. Merino JL, Carmona JR, Fernández-Lozano I, et al. Mechanisms of sustained ventricular tachycardia in myotonic dystrophy: implications for catheter ablation. Circulation. 1998;98:541-6.
26. Cohen TJ, Chien WW, Lurie KG, et al. Radiofrequency catheter ablation for treatment of bundle branch reentrant tachycardia: results and long-term follow-up. J Am Coll Cardiol. 1991:18:1767-73.
27. Blanck Z, Jazayeri M, Dhala A, et al. Bundle branch reentry: a mechanism of ventricular tachycardia in the absence of myocardial or valvular dysfunction. J Am Coll Cardiol. 1993;22:1718-22.
28. Fontaine G, Fontaliran F, Hébert JL, et al. Arrhythmogenic right ventricular dysplasia. Ann Rev Med. 1999;50:17-35.
29. Nasir K, Bomma C, Tandri H, et al. Electrocardiographic features of arrythmogenic right ventricular dysplasia/cardiomyopathy according to disease severity: a need to broaden diagnostic criteria. Circulation. 2004;110:1527-34.
30. Fontaine G, Fornes P, Hebert JL, et al. Ventricular tachycardia in arrhythmogenic right ventricular cardiomyopathies. In: Zipes D, Jalife J (Eds). Cardiac Electrophysiology: From Cell to Bedside, 4th edition. Philadelphia: W.B. Saunders; 2004. pp. 588-600.
31. Gonska BD, Cao K, Raab J, et al. Radiofrequency catheter ablation of right ventricular tachycardia late after repair of congenital heart defects. Circulation. 1996;94:1902-8.

32. Horton RP, Canby RC, Kessler DJ, et al. Ablation of ventricular tachycardia associated with tetralogy of Fallot: demonstration of bidirectional block. J Cardiovasc Electrophysiol. 1997;8:432-5.
33. Lerman BB, Stein KM, Markowitz SM, et al. Recent advances in right ventricular outflow tract tachycardia. Card Electrophysiol Rev. 1999;3:210-4.
34. Rodriguez LM, Smeets JL, Timmermans C, et al. Predictors for successful ablation of right- and left-sided idiopathic ventricular tachycardia. Am J Cardiol. 1997;79:309-14.
35. Ohe T, Shimomura K, Aihara N, et al. Idiopathic sustained left ventricular tachycardia: clinical and electrophysiologic characteristics. Circulation. 1988;77:560-8.
36. Kottkamp H, Chen X, Hindricks G, et al. Radiofrequency catheter ablation of idiopathic left ventricular tachycardia: further evidenced for microreentry as the underlying mechanism. J Cardiovasc Electrophysiol. 1994;5:268-73.
37. Kanagaratnam L, Tomassoni G, Schweikert R, et al. Ventricular tachycardias arising from the aortic sinus of valsalva: an under-recognized variant of left outflow tract ventricular tachycardia. J Am Coll Cardiol. 2001;37:1408-14.
38. Kamakura S, Shimizu W, Matsuo K, et al. Localization of optimal ablation site of idiopathic ventricular tachycardia from right and left ventricular outflow tract by body surface ECG. Circulation. 1998;98:1525-33.
39. Schwartz PJ, Periti M, Malliani A. The long Q-T syndrome. Am Heart J. 1975;89:378-90.
40. Moss AJ, Zareba W, Benhorin J, et al. ECG T-wave patterns in genetically distinct forms of the hereditary long QT syndrome. Circulation. 1995;92:2929-34.
41. Schwartz PF, Priori SG. Long QT. syndrome: genotype-phenotype correlations. In: Zipes DP, Jalife J (Eds). Cardiac Electrophysiology: From Cell to Bedside, 4th edition. Philadelphia: W.B. Saunders; 2004. pp. 651-9.
42. Shimizu W, Antzelevitch C. Sodium channel block with mexiletine is effective in reducing dispersion of repolarization and preventing torsade de pointes in LQT2 and LQT3 models of the long-QT syndrome. Circulation. 1997;96:2038-47.
43. Kay GN, Plumb VJ, Arciniegas JG, et al. Torsades de pointes: the long-short initiating sequence and other clinical features; observations in 32 patients. J Am Coll Cardiol. 1990;2:806-17.
44. El-Sherif N, Turitto G. Torsade de pointes. In: Zipes DP, Jalife J (Eds). Cardiac Electrophysiology: From Cell to Bedside, 4th edition. Philadelphia: WB Saunders; 2004. pp. 687-99.
45. Brugada P, Brugada J. Right bundle branch block, persistent ST segment elevation and sudden cardiac death: a distinct clinical and electrocardiographic syndrome. A multicenter report. J Am Coll Cardiol. 1992;20:1391-6.
46. Brugada P, Brugada R, Brugada J. The Brugada syndrome. In: Saksena S, Camm AJ (Eds). Electrophysiological Disorders of the Heart. New York: Churchill Livingstone; 2005. pp. 697-703.
47. Antzelevitch C, Brugada P, Borggrefe M, et al. Brugada syndrome: report of the second consensus conference: endorsed by the Heart Rhythm Society and the European Heart Rhythm Association. Circulation. 2005;111:659-70.

48. Dorian P, Bharati S, Meyerburg RJ, et al. Ventricular fibrillation. In: Saksena S, Camm AJ (Eds). Electrophysiological Disorders of the Heart. New York: Churchill Livingstone; 2005. pp. 419-53.
49. Maron BJ, McKenna WJ, Danielson GK, et al. American College of Cardiology/European Society of Cardiology clinical expert consensus document on hypertrophic cardiomyopathy: a report of the American College of Cardiology Foundation Task Force on Clinical Expert Consensus Documents and the European Society of Cardiology Committee for Practice Guidelines. J Am Coll Cardiol. 2003;42: 1687-713.
50. Li L, Burke A, Zhang X, et al. Sudden unexpected death due to left ventricular noncompaction of myocardium. Am J Forensic Med Pathol. 2010;31:122-4.
51. Martini B, Buja GF, Canciani B, et al. Bidirectional tachycardia. A sustained form, not related to digitalis intoxication, in an adult without apparent cardiac disease. Jpn Heart J. 1988;29:381-7.
52. Napolitano C, Priori SG. Catecholaminergic polymorphic ventricular tachycardia and short-coupled torsades de pointes. In: Zipes D, Jalife J (Eds). Cardiac Electrophysiology: From Cell to Bedside, 4th edition. Philadelphia: W.B. Saunders; 2004. pp. 633-9.
53. Takagi M, Aihara N, Takahi H, et al. Clinical characteristics of patients with spontaneous or inducible ventricular fibrillation without apparent heart disease presenting with J wave and ST segment elevation in inferior leads. J Cardiovasc Electrophysiol. 2000;11:844-8.
54. Haïssaguerre M, Derval N, Sacher F, et al. Sudden cardiac arrest associated with early repolarization. N Engl J Med. 2008;358:2016-23.
55. Belhassen B, Viskin S. Idiopathic ventricular tachycardia and fibrillation. J Cardiovasc Electrophysiol. 1993;4:356-68.
56. Gussak I, Antzelevitch C. Early repolarization syndrome: clinical characteristics and possible cellular and ionic mechanisms. J Electrocardiol. 2000;33:299-309.
57. Bjerregaard P, Gussak I. Short QT syndrome: mechanisms, diagnosis and treatment. Nat Clin Pract Cardiovasc Med. 2005;2:84-7.

**CHAPTER**

# Bradycardia and Heart Block

# 8

*Arthur C Kendig, James B Martins*

## Chapter Outline

- Conduction System Anatomy and Development
- Bradycardia Syndromes/Diseases
  - Iatrogenic and Noncardiac Causes
  - Familial
  - Vagal Tone
  - Cardiac Transplantation
- Clinical Presentation
- Measurement/Diagnosis
- Sinus Node Disease
  - Sick Sinus Syndrome
- AV Node Disease
  - Pathology
  - First-Degree AV Block
  - Second-Degree AV Block
  - Third-Degree AV Block
  - Paroxysmal AV Block
- Hemiblock
- Bundle Branch Block
  - Left Bundle Branch Block
  - Right Bundle Branch Block
- Treatment

## INTRODUCTION

Bradycardia is generally defined as a heart rate less than 50 beats per minute. However, this simple definition is a gross oversimplification of what is a multifaceted and multifactorial issue. Bradycardia is a dichotomy of sorts, in some cases being a marker of excellent cardiovascular fitness, or conversely, a sign of cardiovascular disease, especially when it is symptomatic.

## CONDUCTION SYSTEM ANATOMY AND DEVELOPMENT

On the most basic level, the normal specialized electrical tissue is comprised of the sinoatrial (SA) node (dominant pacemaker), atrioventricular (AV) node and His-Purkinje system. Embryologically, the conduction system of the human heart begins very early in development, with an ECG recording at 4–6 weeks gestation being similar to what is seen in an adult.

The SA node, first described by Keith and Flack in 1907, is located in the lateral right atrium in the sulcus terminalis. It is first noted at around 20 days into embryonic development, when a primitive SA node is formed in the slow-conducting inflow region of the heart. The inflow region is initially bilateral, with the developing heart's pacemaker on the left side. By 35 weeks, the SA node becomes a more distinct entity in the posterolateral region of the early four-chambered heart's right atrium, with impulses directed with a posterior to anterior vector, toward the AV node. Even after completion of its development, and into adulthood, the SA node is a heterogeneous structure, and essentially a loose collection of cells. Only a portion of these cells act as initiating pacemakers; the rest conduct the signals sent to them. There is also a hierarchy of pacemakers, with overdrive suppression by those with faster rates, of those with lower rates (the AVN and His-Purkinje).

The AV node itself located along the interatrial septum in the Triangle of Koch—which is composed posteriorly of the tendon of Todaro, and anteriorly by the septal leaflet of the tricuspid valve and inferiorly by the coronary sinus os—is also formed around 5 weeks into embryonic development. Like the SA node, it exhibits its conduction properties in its adult role prior to being noted as a distinct identifiable entity.

The His bundle is located anteriorly to the compact AV node along the interatrial septum and near the AV groove. The Purkinje fibers continue from the His into the ventricles and divide into the right, left anterior and left posterior fascicles. Embryologically, the cells making up the His-Purkinje system are derived from already present cardiac myocytes, differentiating into conduction system cells after exposure to endothelin. Interestingly, in animal models, the conduction system is present and functioning prior to coronary vessel formation.[1-5]

## BRADYCARDIA SYNDROMES/DISEASES

### Iatrogenic and Noncardiac Causes

Before considering intrinsic conduction disease, in light of many patients living with multiple medical comorbidities as well as complex medical regimens being commonplace, it is of utmost importance to consider iatrogenic and noncardiac causes of bradycardia and heart block.

First, in terms of medications, the most common agents related to this issue are those already diagnosed for cardiovascular disease including tachycardias. Examples include beta-adrenergic receptor blockers (BARBs, both ophthalmologic and oral),[6] calcium channel blockers, digoxin[7] and clonidine[8,9] **(Table 1)**.

Noncardiac medications as culprits include lidocaine spray or topical such as is used for endoscopic procedures,[10] selective serotonin reuptake inhibitors (SSRIs) such as escitalopram,[11] cholinesterase inhibitors (via enhancement of vagal tone) commonly used for treatment of Alzheimer's disease[12] and succinylcholine.[13]

Propofol is a commonly used sedation agent which, in rare cases, may lead to the development of propofol infusion syndrome. This includes a sudden-onset of bradycardia and possibly asystole, with other effects such as fatty liver, metabolic acidosis or rhabdomyolysis. Risk factors include prolonged (>48 hours) use at high doses, patient age less than 19 years and low carbohydrate reserves.[14]

### TABLE 1

**Common noncardiac and iatrogenic causes of bradycardia and heart block**

| Cardiac medications | Noncardiac medications | Medical disease |
| --- | --- | --- |
| BARBs (including eye drops) | SSRIs | Lyme disease |
| Calcium channel blockers | Opiates | Renal/Hepatic disease |
| Digoxin | Succinylcholine Lithium | Hypothyroidism Carotid hypersensitivity |
| Amiodarone | Cholinesterase inhibitors (e.g. donepezil for Alzheimer's) | Renal/Hepatic disease |
| Sotalol | Propofol | Endovascular cooling after cardiac arrest |
| Sodium channel blockers | | |
| Clonidine | | |

In terms of medical syndromes, potential causes include: hypothyroidism; Lyme disease;[15] endovascular cooling after cardiac arrest;[16] and renal or hepatic disease. The latter two are more related to decreasing medication clearance and metabolism, thus increasing serum drug levels and potentiating a drug's effect. This is especially salient with digoxin and renally cleared BARBs such as atenolol. Again, it is important to note that medication and medical disease-related bradycardia and heart block are potentially reversible, that is why these need to be considered early in the differential diagnosis.

### Familial

Development of the cardiac conduction system does not always occur in a normal fashion. When mutations occur, different syndromes may develop. One example of this (specifically related to sinus bradycardia) is familial sinus bradycardia. In a recent paper, the HCN4 channel (located near the cAMP-binding site) was identified as the mutation causing this syndrome, which acts much like a vagally mediated bradycardia.[17]

### Vagal Tone

Although the cardiac conduction system is primarily derived from myocardial cells and not neural crest cells (which differentiate into neural tissue), there is still a rich two-way interaction with the nervous system communication. In 1867, von Bezold first described cardioinhibitory reflexes initiating from the heart itself. It was found that receptors located primarily in the posterior left ventricle, sensitive to mechanical stretch or certain chemicals, when stimulated, increase parasympathetic and decrease sympathetic tone via the vagus nerve, resulting in—among other reactions—sinus bradycardia.[18]

This mechanism may cause bradycardia commonly in diverse situations, all of which could be called vaso-vagal reactions since they may be triggered by psychiatric stressors like fear or sight of blood as well as volume depletion from many etiologies. The fact that animal models show similar responses with, for example, hemorrhage suggests that this mechanism may be in most or all humans as well. Therefore most physicians will see bradycardia due to this mechanism which is best prevented and treated by increasing vascular volume.

Interestingly, trained athletes, especially elite ones, frequently have asymptomatic bradycardia.[19] Commonly elite athletes may also have orthostatic hypotension. Rarely, vagally mediated bradycardia severe enough to cause presyncope or syncope may also (somewhat paradoxically) be seen in trained athletes, where high vagal tone assists in rapid post-exercise heart rate recovery, but in some cases may go too far in lowering heart rate, causing symptoms.[19] The sinus node as well as the subsidiary pacemakers may be influenced by this vagal tone, at times leading to asystole. Nevertheless, the vagal influence never permanently stops the heart; it only delays its next beat for a matter of seconds. Meantime, in this brief period of time, syncope may occur because of loss of cardiac output as well as peripheral resistance.

## Cardiac Transplantation

In post-cardiac transplant patients, bradycardia—specifically sinus bradycardia—occurs in approximately 18% of patients, and ultimately 4–7% require a permanent pacemaker (usually atrial pacing only). One of the primary risk factors for development of sinus bradycardia is ischemic time.[20] Another related risk factor is surgical technique, with biatrial anastomosis (Shumway-Lower) resulting in less ischemic time, but increased risk of physical damage to the SA node, while the bicaval (Wythenshawe) approach keeps the atria intact, but increases ischemic time.[21–24] Other risk factors for post-transplant bradycardia include SA nodal artery lesions and transplant vasculopathy.[25,26]

Due to vagal and sympathetic denervation resulting from the surgical procedure, these patients have relative bradycardia and some degree of chronotropic incompetence, but are somewhat shielded from bradycardia by the standard definition due to high intrinsic heart rate. The transplanted ventricle increases stroke volume to compensate for the lack of sympathetically increased heart rate with exercise; even though cardiac output may be normal at rest the transplant cannot increase HR enough to make up for the lesser heart rate with exercise. Reinnervation can occur, but the extent to which it happens varies from patient to patient.

## CLINICAL PRESENTATION

Depending on the patient's heart rate and robustness of so-called "back-up" pacemakers, patients may entirely be asymptomatic, symptomatic with exertion only or at rest. Common symptoms include lightheadedness, dizziness, syncope, palpitations, shortness of breath at rest, dyspnea on exertion, angina, and progressive lower extremity edema.

## MEASUREMENT/DIAGNOSIS

The diagnosis of bradycardia, regardless of the mechanism, is primarily made by obtaining an ECG or rhythm strip from surface leads at the time of symptoms. Other times, bradycardia may be found serendipitously by an ECG or telemetry strip performed for other reasons. When a patient's symptoms suggest an arrhythmia of a low rate, occur paroxysmally, and the patient has a normal ECG or rhythm strip at presentation while asymptomatic, monitoring via a Holter (24 or 48 hour) if symptoms occur daily, a 21+ day event recorder if occurring monthly, or if rarely occurring but with significant symptoms at the time of the event, an implantable loop recorder may be appropriate. These methods of measurement are discussed in this textbook in more detail.

Another issue in diagnosing bradycardia is relative bradycardia, also known as chronotropic incompetence. This describes a scenario where at rest, the patient does not demonstrate significant symptoms, but with exertion, is unable to mount an appropriate heart rate to increase cardiac output appropriate for increased needs of exercise. A common definition of this is failure to obtain greater than 85% of age-predicted maximum heart rate with exercise testing. A postulate for an etiology is decreased responsiveness of the heart to increased sympathetic input. Evaluation is primarily via exercise testing. In some cases, direct measurement of cardiac output is performed via right heart catheterization before, during and after temporary pacing at higher rates than baseline; if cardiac output significantly improves with a modest increase in heart rate, then chronotropic incompetence may be assumed.[27]

# SINUS NODE DISEASE

## Sick Sinus Syndrome

Sick sinus syndrome (SSS), in its most rudimentary sense, is defined as sinus node dysfunction which results in an inadequate heart rate physiologically, and may include sinus bradycardia, sinus arrest, or sinus node exit block, and may lead to takeover of cardiac pacemaking and atrial contraction by subsidiary intrinsic cardiac pacemaker sites. The mere present of SSS without symptoms is not an indication for any therapy, but cautious use of agents listed in the **Table 1**.

In a clinicopathological study of six cases of SSS in patients ranging 69–91 years of age, Sugiura et al. found that in patients with SA block and bradycardia-tachycardia syndrome, the SA node and AV node regions showed 70–80% decrease in nodal cells, infiltration of the SA node by connective tissue and left bundle branch (LBB) fibrosis.[28] SSS has also been reported to be caused by cardiac lipomatosis.[29]

During ageing, the sinus node itself changes. A rat model comparing what was equivalent to a young adult and 69 year old showed that in older animals, the heart rate is lower, the pacemaker action potentials were slower, more widely distributed (enlargement of the SA node), and located more toward the inferior vena cava in the RA. Moreover, histologically, the SA nodal cells demonstrated hypertrophy and extracellular matrix remodeling.[30]

The I(f) "funny" current, although first discovered more than 30 years ago, has recently been found, albeit controversially, to have a significant role in spontaneous cardiac pacemaker activity. Basically, these channels determine the slope of phase 4 (depolarization), thereby playing a role in the frequency of cardiac pacing. A mutation in the HCN4 gene, which encodes this channel, demonstrates sinus bradycardia. Thus an abnormality in this channel may be at least partially at fault for some sinus bradycardias.[31]

Recently, genetic factors related to this disease have been elucidated, specifically SCN5A mutation related functional loss of the sodium channel.[32]

In terms of association with other conduction diseases, a recent retrospective study suggests Brugada-type ECG and Brugada syndrome are associated with SSS; 2.87% had Brugada-type ECGs (0.82% with Type I, and 2.05% with Type II). Generally, in the population at large, in recent studies, the prevalence has ranged from 0.07% (7/10,000) in a Danish population[33] and 0.012% (120/10,000)[34] to 0.14%.[35] Moreover, in the above noted study, during a 7-year follow-up period, 50% of those with Type I and none of those with Type II

ECGs experienced VF events. Thus, SSS is associated with an increased prevalence of Brugada-type ECG and Brugada syndrome compared to the general population. This may be associated with the aforementioned SCN5A mutations.

Wu et al. performed EP studies in 38 patients with SSS. The mean sinus pause was 5.6 ± 2.8 sec. Three predominate groups were discovered. Nine patients had SA block, with sinus node function intrinsically noted during pauses. Seven had unidirectional exit block, and two others bidirectional, all of which had evidence of an atrial impulse to conduct into the SA node and inhibit SA node firing. A second group of 22 patients showed slow 1:1 conduction, second degree SA node exit block, with 17/22 patients showing abnormal sinus node recovery times. A third group had no sinus node electrograms measurable.[36]

Frequently patients will have symptoms shown to be due to sinus bradycardia, but careful clinical evaluation will reveal that symptoms began upon institution of BARBs or other drugs listed in the **Table 1**. Usually the symptoms will resolve if the offending drug can be discontinued or replaced by one which does not aggravate SSS.

## AV NODE DISEASE

### Pathology

The most common cause of permanent AV block is idiopathic bundle branch fibrosis. This may include the main AV bundle and LBB, as a result of aging fibrosis of the cardiac skeleton, referred to as Lev's disease, or in fibrosis of the left and right bundles themselves, referred to as Lenègre's disease, occurring in younger people in certain families. Other etiologies include interruption of the AV node related to aortic or mitral valve calcification, myocardial infarction leading to ischemic damage to the AV node, or cardiomyopathy. Other causes, although rare, include: congenital AV block; infiltration by tumor and surgical or other trauma.[37]

### First-degree AV Block

Although a relatively common and seemingly benign finding of a prolonged PR interval (>200 msec) on ECG, it has recently been postulated that this may precede more advanced AV block, and even itself serve as a marker of increased risk of other arrhythmic and mortality concerns. Cheng described long-term outcomes of patients with first-degree AV block.[38] Within the Framingham Heart Study cohort, and found that when compared to control patients with a prolonged PR interval at baseline had twice the risk of developing atrial fibrillation, three times the risk of requiring a pacemaker and almost one-and-a-half times the risk of all-cause mortality. At this point however, despite these findings, management of these patients with only a prolonged PR interval is unclear.

## Second-degree AV Block

Second-degree AV block is commonly divided into two varieties, and was first classified in this manner by Woldemar Mobitz in the early 1920s.[39] Type I (Mobitz I, also "Wenckebach" named for Karel Wenckebach who discovered "Wenckebach periodicity" in 1906) second-degree AV block presents as a normal or near-normal PR interval (120–200 msec) which gradually prolongs with successive beats, until only a P wave is seen, and a "dropped" QRS occurs, which is the absence of AV conduction. After a brief reset, the next cycle begins, again with a normal PR interval, progressively lengthening again until a drop occurs. There may be the same number of beats prior to that which is dropped, or at times, variable numbers. Type I basically demonstrates AV nodal decremental conduction.

Type II (Mobitz II) second-degree AV block, similar to Type I, has dropped ventricular beats due to AV block. However, in this scenario, there is no PR lengthening, only dropped beats, which may occur after one, two or any number of normally conducted beats, and may or may not be a consistent number. An easy way to diagnose this mechanism is to evaluate the PR before and after the regular blocked P wave; if the PR after the block is not shorter by more than 20 msec then Type II block must be considered. False Type II may be diagnosed if irregular P-P intervals, owing to sinus arrhythmia or atrial prematures, or frequent junctional prematures (suggesting concealed His extrasystoles) produce block.[40] Patients with Type II block also tend to have concomitant QRS prolongation.

2:1 AV block, due to 2:1 conduction, may be Type I or II, but which one is not able to be ascertained because there is no PR lengthening—or lack thereof—to be seen because block occurs after only one beat. It may be a progression from either. Anatomically, 80% of the time, 2:1 AV block occurs in the His-Purkinje system, and 20% in the AV node itself. Careful examination of the ECG for more typical 3:2 Type I block will suggest the right mechanism. Interestingly, atropine may increase the degree of AV block.[40–42] The importance of differentiating Type I versus Type II AV block is that even if asymptomatic, the latter requires a permanent pacer, but in the former a permanent pacemaker is not needed unless symptoms are present.

### Third-degree AV Block

Third-degree AV block is defined as a complete electrical dissociation of the atria and ventricles. The ventricular rhythm in this case, referred to as an escape rhythm, may either be a junctional, which is a normal, narrow-complex rhythm emanating more proximally to the AV node and generally faster (40–60 bpm), or ventricular where the rhythm is abnormal, wide complex and slower (<30 bpm). Rarely no escape rhythm is present whatsoever.

Treatment entails eliminating or neutralizing reversible causes, such as AV nodal blockers (e.g. BARBs). If a patient has complete heart block and is unstable with symptoms, such as chest pain, shortness of breath, lightheadedness and/or hypotension, then atropine should be given, especially if a junctional escape is present. If not effective, temporary pacing (first transcutaneous, then transvenous) is recommended. If the patient has what appears to be a permanent complete heart block, then permanent pacing is recommended. Of note, some sources state that an escape rhythm of greater than 40 beats per minute is adequate, and is not an indication for permanent pacing. However, as noted in the recent ACC/AHA device therapy guidelines, this arbitrary number is not based on strong data.[43]

### Paroxysmal AV Block

Paroxysmal AV block has been defined by Lee et al. as "a sudden," "paroxysmal pause-dependent phase 4 AV block occurring in diseased conduction system". It is essentially the change from 1:1 AV conduction, suddenly to complete heart block. There is no official definition for this type of AV block and the prevalence may be underestimated because of this and difficulty in recording this arrhythmia. The most common risk factor is apparently right bundle branch (RBB) block. Since paroxysmal AV block originates in the distal portion of the AV node, it most often is seen in older patients, with the commonest presentation being syncope. As noted above, the mechanism has been postulated to be block due to phase 4 depolarization of the distal AV node, more specifically in the His-Purkinje system itself. Factors differentiating this from vagally mediated complete heart block include rapid or sudden onset, no change in P-R interval, and infranodal versus nodal level of block. Treatment includes pacemaker implantation and removal of culprit AV nodal blocking agents if present.[44]

## HEMIBLOCK

Although not necessarily a bradycardia per se, or AV block, such as those described above, the hemiblocks (first described by

Rosenbaum, et al. in 1968 are nevertheless important findings which may be intricately related to AV block.[45]

First of all, building on the concept of the right and left bundles in the His-Purkinje system, the LBB is generally found to divide into two discrete—yet somewhat interconnected—branches, the anterior and the posterior fascicles. Thus, the ventricular system, including the right bundle, is essentially trifascicular.

The left anterior fascicle, being anteriorly located as the name implies, is the more vulnerable of the two LBBs. For one, there is only a single coronary artery distribution (anterior descending) for blood supply. In addition, this fascicle is smaller and thinner than its posterior counterpart. A left anterior hemiblock (LAH) (also known as a fascicular block) is defined as a leftward axis of less than −45 degrees, an "rS" pattern in II, III, aVF and with a narrow QRS complex. The most common causes include hypertension, cardiomyopathy, VSD closure (spontaneous or iatrogenic), and Lev and Lenegre disease. In the case of the latter two causes, the RBB is commonly affected concurrently, and by the nature of these progressive diseases, patients often eventually develop complete heart block. Generally, without significant coronary artery disease or infarction, the finding of isolated LAH does not appear to portend an increased risk of morbidity or mortality, although LAH is associated with risk of disease in the Framingham study.

The left posterior fascicle is supplied by both the anterior and the posterior descending coronary arteries. This dual blood supply source is beneficial especially in the setting of myocardial infarction, where if one of the supplying coronary arteries is the site of occlusion or stenosis, the fascicle is unlikely to sustain significant ischemic damage, and thus is less prone to infarction and hemiblock. Left posterior hemiblock is defined as a rightward axis of greater than +100 degrees, narrow QRS and an rS in I and aVL. A pure left posterior fascicular block is rare, but tends to be found coexisting with a RBB block. This combination, if occurring in the setting of a myocardial infarction, significantly increases mortality to over 80% within a few weeks after infarction. The risk of progression to complete heart block with this bifascicular block is 42%. Because of the dual-coronary artery supply, however, the cause of the block is more likely due to Lenegre disease, or even Chagas disease in endemic regions.[46]

That said, the concept of trifascicular block is uncommon as opposed to bifascicular block (LAH and RBB block) and concomitant first degree AVB which is usually in the AVN.[40] Patients with LAH and RBB block frequently have no symptoms or VT inducible.[47] Differentiation from trifascicular block is usually easy (lack of symptoms) but exercise testing or EPS may be necessary if a certain diagnosis is necessary.

## BUNDLE BRANCH BLOCK

The concept of the existence of the LBB and RBB coming off of the His was first published by Eppinger and Rothberger in 1909.[48]

### left bundle branch Block

Left bundle branch (LBB) block is defined grossly as a QRS duration greater than or equal to 120 msec with left axis deviation. As early as 1940, Rasmussen and Moe determined via clinical, ECG, radiography, and necropsy in 100 patients with LBB block, that the most common cause, comprising approximately 72% of cases was left ventricular hypertrophy and/or enlargement (with associated increased weight of the heart), primarily due to hypertension or aortic valve disease.[49] Early data from the Framingham Study had similar findings, including hypertension, "cardiac enlargement" and/or coronary artery disease. It was further noted that the mean age of onset of new LBB block was 62 years of age. The 10-year prognosis was poor, with one-half of these patients expiring due to cardiovascular disease.[50] Other etiologies for LBB block include myocardial infarction, electrolyte abnormalities, and fibrosis (Lev's and Lenegre's disease).[51]

A salient issue with LBB block is in myocardial infarction. It may be the presenting ECG finding for MI. If an old finding, it may hinder interpretation of the ECG in the setting of suspected acute coronary syndrome. The specific method of diagnosing an MI in this setting is discussed in this book.[52]

Moreover, and discussed earlier, LBB block is important in selecting patients with interventricular dyssynchrony who would be candidates for cardiac resynchronization therapy, in that most studies suggest that interventricular dyssynchrony due to LBB block responds better to CRT than those with RBB block.[53,54]

### Right bundle branch Block

Right bundle branch (RBB) block is grossly defined as a QRS duration greater than or equal to 120 msec with right axis deviation, and commonly an RSR pattern in lead $V_1$.

From the same follow-up data from the Framingham Study noted above, patients with new onset RBB block over an 18-year-follow-up period were identified. These patients were more likely to have hypertension prior to the development of RBB block; these patients were 2.5–4 times more likely to have coronary artery disease or congestive heart failure than their age-matched controls. A QRS of greater than or equal to 130 ms with an axis −45° to −90° was associated with increased risk of cardiovascular issues. Common causes of or association with RBB block include pulmonary embolism (acute or chronic), pulmonary hypertension, left sided heart failure causing RV volume or pressure overload, severe MR, pulmonic valve stenosis.

In patients with a new RBB or LBB block associated with high grade AV block during an acute myocardial infarction, permanent pacing is indicated as is infarction with bilateral BB block.

## TREATMENT

Treatment of bradycardia and heart block is primarily initiated if there are symptoms such as syncope, lightheadedness, chest pain, shortness of breath and/or evidence of hemodynamic compromise or low cardiac output. Obviously, if offending agents—be it cardiac drugs such as AV nodal blocking agents, digoxin or noncardiac such as opiates—are in use and thought to be the perpetrators, these should be discontinued, if possible. However, if cessation of drug therapy for a reasonable duration of time does not result in improvement, or drug therapy is needed for an indication such as a tachyarrhythmia, then permanent pacing should be considered.

If cessation of an agent does not result in improvement, and a more rapid means of treating bradycardia is needed, the next step would be pharmacologic therapy such as atropine. There are special circumstances, such as BARB toxicity or digoxin toxicity where an antidote of sorts is available, such as glucagon or Digibind, respectively. For more acutely decompensated patients, beta-agonists, such as isoproterenol, may be required until pacing may be initiated.

Beyond pharmacologic measures are the electromechanical therapies. External pacing is considered, but is only of benefit in a very short period of time, due to the unpredictable transcutaneous capture and is poorly tolerated by the patient unless sedated, which itself may perpetuate bradycardia. Temporary pacing is indicated if the above noted therapies are unhelpful and the patient requires more time for conservative measures (holding medications, pharmacologic therapy) to work, or as a bridge to permanent pacing which is commonly not as rapidly available. Permanent pacing is discussed in this book in detail.[43]

## REFERENCES

1. Keith A, Flack M. The form and nature of the muscular connections between the primary divisions of the vertebrate heart. J Anat Physiol. 1907;41:172-89.

2. Baruscotti M, Robinson RB. Electrophysiology and pacemaker function of the developing sinoatrial node. Am J Physiol Heart Circ Physiol. 2007;293:H2613-23.
3. Moorman AF, de Jong F, Denyn MM, et al. Development of the cardiac conduction system. Circ Res. 1998;82:629-44.
4. Boullin J, Morgan JM. The development of cardiac rhythm. Heart. 2005;91:874-5.
5. Rossi L. Anatomopathology of the normal and abnormal AV conduction system. Pacing Clin Electrophysiol. 1984;7:1101-7.
6. Calvo-Romero JM, Lima-Rodriguez EM. Bradycardia associated with ophthalmic beta-blockers. J Postgrad Med. 2003;49:186.
7. Mills TA, Kawji MM, Cataldo VD, et al. Profound sinus bradycardia due to diltiazem, verapamil, and/or beta-adrenergic blocking drugs. J La State Med Soc. 2004;156:327-31.
8. Byrd BF 3rd, Collins HW, Primm RK. Risk factors for severe bradycardia during oral clonidine therapy for hypertension. Arch Intern Med. 1988;148:729-33.
9. Golusinski LL Jr, Blount BW. Clonidine-induced bradycardia. J Fam Pract. 1995;41:399-401.
10. Amornyotin S, Srikureja W, Chalayonnavin W, et al. Endoscopy. 2009;41:581-6.
11. van Gorp F, Whyte IM, Isbister GK. Clinical and ECG effects of escitalopram overdose. Ann Emerg Med. 2009;54:404-8.
12. Bordier P, Garrigue S, Barold SS, et al. Significance of syncope in patients with Alzheimer's disease treated with cholinesterase inhibitors. Europace. 2003;5:429-431.
13. Birkenhäger TK, Pluijms EM, Groenland TH, et al. Severe bradycardia after anesthesia before electroconvulsive therapy. J ECT. 2010;26:53-4.
14. Fudickar A, Bein B. Propofol infusion syndrome: update of clinical manifestations and pathophysiology. Minerva Anestesiol. 2009;75:339-44.
15. McAlister HF, Klementowicz PT, Andrews C, et al. Lyme carditis: an important cause of reversible heart block. Ann Intern Med. 1989;110:339-45.
16. Holzer M, Müllner M, Sterz F, et al. Efficacy and safety of endovascular cooling after cardiac arrest: cohort study and Bayesian approach. Stroke. 2006;37:1792-7.
17. Milanesi R, Baruscotti M, Gnecchi-Ruscone T, et al. Familial sinus bradycardia associated with a mutation in the cardiac pacemaker channel. N Engl J Med. 2006;354:151-7.
18. Mark AL. The Bezold-Jarisch reflex revisited: clinical implications of inhibitory reflexes originating in the heart. J Am Coll Cardiol. 1983;1:90-102.
19. Link MS, Estes NAM. Athletes and arrhythmias. J Cardiovasc Electrophysiol. 2010;21:1-6.
20. Jacquet L, Ziady G, Stein K, et al. J Am Coll Cardiol. 1990;16:832-7.
21. Miyamoto Y, Curtiss El, Kormos RL, et al. Bradyarrhythmias after heart transplantation. Incidence, time, course, and outcome. Circulation. 1990;82:IV313-7.
22. Bernardi L, Valenti C, Wdowczyck-Szuluc J, et al. Influence of type of surgery on the occurrence of parasympathetic reinnervation after cardiac transplantation. Circulation. 1998;97:1368-74.

23. Deleuze PH, Benvenuti C, Mazzucotelli JP, et al. Orthotopic cardiac transplantation with direct caval anastomosis: is it the optimal procedure? J Thorac Cardiovasc Surg. 1995;109:731-7.
24. el Gamel A, Yonan NA, Grant S, et al. Orthotopic cardiac transplantation: a comparison of standard and bicaval Wythenshawe techniques. J Thorac Cardiovasc Surg. 1995;109:721-30.
25. Rothman SA, Jeevanandam V, Combs WG, et al. Eliminating bradyarrrhythmias after orthotopic heart transplantation. Circulation. 1996;94:II278-82.
26. DiBiase A, Tse TM, Schnittger I, et al. Frequency and mechanism of bradycardia in cardiac transplant recipients and need for pacemakers. Am J Cardiol. 1991;67:1385-9.
27. Kawasaki T, Kaimoto S, Sakatani T, et al. Chronotropic incompetence and autonomic dysfunction in patients without structural heart disease. Europace. 2010;12:561-6.
28. Sugiura M, Ohkawa S, Hiraoka K, et al. A clinicopathological study on the sick sinus syndrome. Jpn Heart J. 1976;17:731-41.
29. Kadmon E, Paz R, Kusniec J, et al. Sick sinus syndrome in a patient with extensive cardiac lipomatosis (sinus node dysfunction in lipomatosis). Pacing Clin Electrophysiol 2010;33:513-5.
30. Yanni J, Tellez JO, Sutyagin PV, et al. Structural remodeling of the sinoatrial node in obese old rats. J Mol Cell Cardiol 2010;48:653-62.
31. Verkerk AO, van Ginneken AC, Wilders R. Pacemaker activity of the human sinoatrial node: role of the hyperpolarization-activated current, I(f). Int J Cardiol. 2009;132:318-36.
32. Butters TD, Aslanidi OV, Inada S, et al. Mechanistic links between $Na^+$ channel (SCN5A) mutations and impaired cardiac pacemaking in sick sinus syndrome. Circ Res. 2010;107:126-37.
33. Pedcini R, Cedergreen P, Theilade S, et al. The prevalence and relevance of the Brugada-type electrocardiogram in the Danish general population: data from the Copenhagen city heart study. Europace. 2010;12:982-6.
34. Patel SS, Anees S, Ferrick KJ. Prevalence of a Brugada pattern electrocardiogram in an urban population in the United States. Pacing Clin Electrophysiol. 2009;32:704-8.
35. Donohoe D, Tehrani F, Jamehdor R, et al. Am Heart Hosp J. 2008;6:48-50.
36. Wu DL, Yeh SJ, Lin FC, et al. Sinus automaticity and sinoatrial conduction in severe symptomatic sick sinus syndrome. J Am Coll Cardiol. 1992;19:355-64.
37. Davies MJ. Pathology of chronic A-V block. Acta Cardiol. 1976;21:19-30.
38. Cheng S, Keyes MJ, Larson MG, et al. Long-term outcomes in individuals with prolonged pr interval or first-degree atrioventricular block. JAMA. 2009;301:2571-7.
39. Silverman ME, Upshaw CB Jr, Lange HW. Woldemar Mobitz and his 1924 classification of second-degree atrioventricular block. Circulation. 2004;110:1162-7.
40. Zipes DP. Second degree atrioventricular block. Circulation. 1979;60:465-72.
41. Barold SS, Hayes DL. Second-degree atrioventricular block: a reappraisal. Mayo Clin Proc. 2001;76:44-57.
42. Barold SS. 2:1 Atrioventricular block: order from chaos. Am J Emerg Med. 2001;19:214-7.

43. Epstein AE, DiMarco JP, Ellenbogen KA, et al. ACC/AHA/HRS 2008 Guidelines for device based therapy of cardiac rhythm abnormalities. Circulation. 2008;117:e350-e408.
44. Lee S, Wellens HJ, Josephson ME. Paroxysmal atrioventricular block. Heart Rhythm. 2009;6:1229-34.
45. Rosenbaum MB, Elizari MV, Lázzari JO. Los Hemibloqueos. Buenos Aires, Argentina: Paidós;1968.
46. Elizari MV, Acunzo RS, Ferreiro M. Hemiblocks revisited. Circulation. 2007;115:1154-63.
47. McAnulty JH, Rahimtoola SH, Murphy E, et al. Natural history of "high-risk" bundle-branch block: final report of a prospective study. NEJM. 1982;307:137-43.
48. Eppinger H, Rothberger CJ. Zur analyse des elektrokardiogramms. Wien Klin Wochenschr. 1909;22;1091-8.
49. Rasmussen H, Moe T. Pathogenesis of left bundle branch block. Br Heart J. 1948;10:141-7.
50. Schneider JF, Thomas Jr HE, Kreger BE, et al. Newly acquired left bundle-branch block: the Framingham study. Annals Int Med. 1979;90:303-10.
51. Haft JI, Herman MV, Gorlin R. Left bundle branch block: etiologic, hemodynamic, and ventriculographic considerations. Circulation. 1971;43:279-87.
52. Sgarbossa EB, Pinski SL, Barbagelata A, et al. Electrocardiographic diagnosis of evolving acute myocardial infarction in the presence of left bundle-branch block. GUSTO-1 (Global utilization of streptokinase and tissue plasminogen activator for occluded coronary arteries) investigators. N Engl J Med. 1996;22:334:481-7.
53. Riccckard J, Kumbhani DJ, Gorodeski EZ, et al. Cardiac resynchronization therapy in non-left bundle branch block morphologies. Pacing Clin Electrophysiol. 2010;33:590-5.
54. Wokhlu A, Rea RF, Asirvatham SJ, et al. Upgrade and de novo cardiac resynchronization therapy: impact of paced or intrinsic QRS morphology on outcomes and survival. Heart Rhythm. 2009;6:1439-47.

# Arrhythmogenic Right Ventricular Dysplasia/Cardiomyopathy

**CHAPTER 9**

*Richard NW Hauer, Frank I Marcus, Moniek GJP Cox*

## Chapter Outline

- Molecular and Genetic Background
  - Desmosome Structure and Function
  - Desmosomal Dysfunction and ARVD/C Pathophysiology
  - Autosomal Recessive Disease
  - Autosomal Dominant Disease
  - Other Non-desmosomal Genes
- Epidemiology
- Clinical Presentation
- Clinical Diagnosis
  - Global and/or Regional Dysfunction and Structural Alterations
  - Endomyocardial Biopsy
  - ECG Criteria
  - Depolarization Abnormalities
  - Repolarization Abnormalities
  - Arrhythmias
  - Family History
- Non-classical ARVD/C Subtypes
  - Naxos Disease
  - Carvajal Syndrome
  - Left Dominant ARVD/C (LDAC)
- Differential Diagnosis
- Molecular Genetic Analysis
- Prognosis and Therapy

## INTRODUCTION

Arrhythmogenic right ventricular dysplasia/cardiomyopathy (ARVD/C) is a disease characterized histopathologically by progressive fibrofatty replacement of the myocardium, primarily of the right ventricle (RV).[1-3] Affected individuals typically present between the second and the fourth decade of life with monomorphic ventricular tachycardia (VT) originating from the RV. ARVD/C can be the cause of sudden death in all stages of the disease, but particularly in adolescence.[4] From autopsy studies, it is known that fibrofatty tissue can replace major parts of normal myocardium in teenagers **(Fig. 1)**. Sudden death may occur in the early concealed phase of the disease.

The first series of ARVD/C patients was published in 1982. It was described as a disease in which "the right ventricular musculature is partially or totally absent and is replaced by fatty and fibrous tissue".[1] This disease was initially thought to be a defect in RV development, which is why it was first called "dysplasia". In the past 25 years, increased insight in the development of the disease as well as the discovery of pathogenic mutations involved led to our current understanding that ARVD/C is a genetically determined "cardiomyopathy".[3,5] However, since non-familial sporadic cases occur even after extensive family screening, nongenetic causes cannot be excluded. It is now clear that ARVD/C is a desmosomal disease resulting from defective cell adhesion proteins. Desmosomes maintain mechanical coupling of cardiomyocytes. The first disease-causing gene, encoding the desmosomal protein plakoglobin (JUP), was identified in patients with Naxos

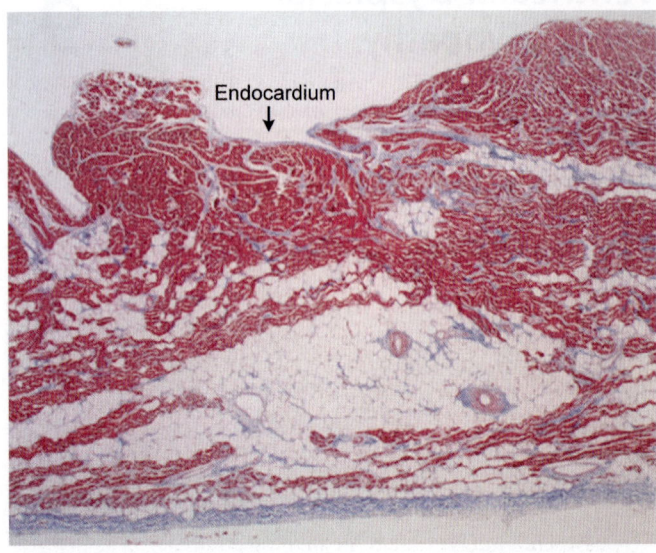

**FIGURE 1:** Histology of right ventricular wall (x400) of a 13-year-old girl who died suddenly during exercise. AZAN stain with cardiac myocytes (red), collagen (blue) and adipocytes (white). Shown is the typical pattern of ARVD/C with strands of fibrosis reaching all the way to the endocardium (particularly just to the right of the arrow). Bundles of cardiac myocytes are embedded in between the fibrotic strands, particularly in the subendocardial layers. These interconnecting bundles of myocytes give rise to activation delay and re-entrant circuits, the typical electrophysiologic substrate for ventricular arrhythmias in ARVD/C. The large homogeneous subepicardial area of adipose tissue is not arrhythmogenic, although it may be observed in ARVD/C. However, it is a typical feature of the cor adiposum, a non-arrhythmogenic condition

disease, an autosomal recessive variant of ARVD/C, reported from the Greek island of Naxos.[6] This discovery stimulated research in the direction of other desmosomal genes. Until 2004, only three genes were identified as responsible for the autosomal dominantly inherited ARVD/C.[7-15] Since *RyR2* mutations are typically associated with catecholaminergic polymorphic VT, it is less certain that these *RyR2* mutations are a cause of ARVD/C. The desmoplakin gene (*DSP*) was the first desmosomal protein gene associated with the autosomal dominant form of ARVD/C.[15] It was followed by discovery of mutations in plakophilin-2 (*PKP2*), desmoglein-2 (*DSG2*) and desmocollin-2 (*DSC2*), all components of the cardiac desmosome.[16-18] Impaired desmosomal function results in myocardial cell-to-cell uncoupling, followed by cell death and fibrofatty replacement, and thus disruption of the myocardial architecture leading to activation delay and arrhythmias. In a few rare cases, autosomal dominant ARVD/C has been linked to other genes unrelated to the cell adhesion complex, i.e. the genes encoding the cardiac ryanodine receptor (*RyR2*), the transforming growth factor-β3 gene (*TGFβ3*), and transmembrane protein 43 (*TMEM43*).[13,14,19] Since *RyR2* mutations are typically associated with catecholaminergic polymorphic VT, it is less certain that these *RyR2* mutations are a cause of ARVD/C.

With mutations found in about half of the patients, mainly in desmosomal genes and *PKP2* in particular, ARVD/C is currently considered a genetically determined desmosomal disease.

This chapter provides an overview of ARVD/C, starting from the genetic defects that are responsible for the pathophysiologic mechanisms to clinical diagnosis, treatment and prognosis.

## MOLECULAR AND GENETIC BACKGROUND

### Desmosome Structure and Function

The cellular adhesion junctions in the intercalated disk are vital for the structural and functional integrity of cardiac myocytes. Intercalated disks are located between cardiomyocytes at their longitudinal ends and contain three different kinds of intercellular connections: (1) Desmosomes; (2) Adherens junctions and (3) Gap junctions.

Desmosomes are important for cell-to-cell adhesion and are predominantly found in tissues that experience mechanical stress—the heart and the epidermis. They couple cytoskeletal elements to the plasma membrane. Desmosomes also protect the other components of the intercalated disk from mechanical stress and are involved in structural organization of the intercalated disk. Desmosomes consist of multiple proteins which belong to the following three different families:
1. Transmembranous cadherins (desmogleins and desmocollins)
2. Linker armadillo repeat proteins (plakoglobin and plakophilin)
3. Plakins (desmoplakin and plectin).

**Figure 2** schematically represents the organization of the various proteins in the cardiac desmosome.

Within desmosomes, cadherins are connected to armadillo proteins which interact with plakins. The plakins anchor the desmosomes to intermediate filaments, mainly desmin. They form a three-dimensional scaffold providing mechanical support.

Adherens junctions act as bridges that link the actin filaments within sarcomeres of neighboring cells. These junctions are involved in force transmission and, together with desmosomes, these mechanical junctions act as "spot welds" to create membrane domains that are protected from shear stress caused by contraction of the neighboring cells. Furthermore, they

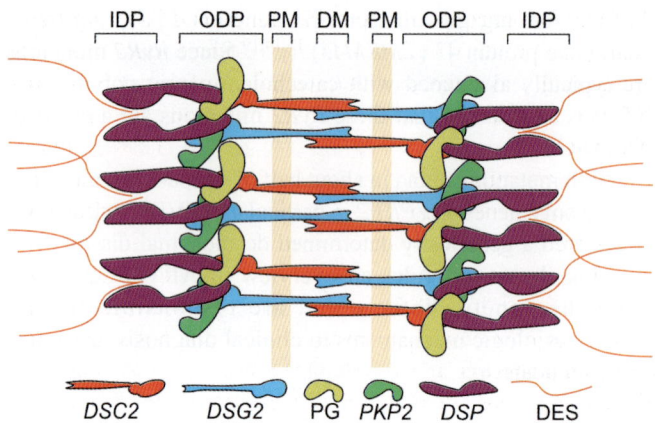

**FIGURE 2:** Schematic representation of the molecular organization of cardiac desmosomes. The plasma membrane (PM) spanning proteins desmocollin-2 (DSC2) and desmoglein-2 (DSG2) interact in the extracellular space at the dense midline (DM). At the cytoplasmic side, they interact with plakoglobin (PG) and plakophilin-2 (PKP2) at the outer dense plaque (ODP). The PKP2 and PG also interact with desmoplakin (DSP). At the inner dense plaque (IDP), the C-terminus of DSP anchors the intermediate filament desmin (DES) (*Source:* Modified from: Van Tintelen et al. Curr Opin Cardiol. 2007;22:185-92)

facilitate assembly and maintenance of gap junctions, securing intercellular electrical coupling.

Cardiomyocytes are individually bordered by a lipid bilayer which gives a high degree of electrical insulation. The electrical current that forms the impulse for mechanic contraction travels from one cell to the other via gap junctions. Gap junctions provide electrical coupling by enabling ion transfer between cells. The number, size and distribution of gap junctions all influence impulse propagation in cardiac muscle. Consequently, alterations in function of gap junctions can lead to intercellular propagation disturbances and arrhythmogenesis.[20]

The intercalated disk is an intercellular structure, where desmosomes and adherens junctions not only provide mechanical strength but also protect the interspersed gap junctions, enabling electrical coupling between cells.

## Desmosomal Dysfunction and ARVD/C Pathophysiology

It is not well-known how mutations of desmosomal protein genes are related to the ARVD/C phenotype. Several mechanisms have been proposed.

First, alterations in desmosomal proteins are thought to lead to mechanical uncoupling of myocytes at the intercalated disks, particularly under mechanical stress (e.g. exercise, sports activities, etc.). Mechanical uncoupling will be followed by:

(1) Electrical uncoupling due to dysfunction of gap junctions and (2) Cell death with fibrofatty replacement. Both electrical uncoupling and interconnecting bundles of surviving myocardium embedded in the fibrofatty tissue lead to lengthening of conduction pathways and load mismatch. This results in marked activation delay and conduction block, which are pivotal mechanisms for re-entry and thereby VT. Invasive electrophysiologic studies have confirmed that VT in patients with ARVD/C is due to re-entrant circuits in areas of abnormal myocardium.[21] In addition, environmental factors, such as exercise or inflammation from viral infection, could aggravate impaired adhesion and accelerate disease progression. The RV may be more vulnerable to histopathologic alteration than the left ventricle (LV) due to its thinner walls and its normal dilatory response to exercise.

Secondly, recent studies have shown that impairment of cell-to-cell adhesion due to changes in desmosomal components may affect the amount and distribution of other intercalated disk proteins, including connexin 43, the major protein forming gap junctions in the ventricular myocardium.[22-24] This was shown for *DSP* and *JUP* by Western blotting and confocal immunofluorescence techniques, but alterations in other desmosomal components, such as *PKP2, DSG2* and *DSC2*, are thought to have similar effects. Changes in number and function of gap junctions will diminish intercellular electrical coupling contributing to intraventricular activation delay.

The third hypothesis involves the canonical Wnt/β-catenin signaling pathway. Plakoglobin can localize both to the plasma membrane and the nucleus. It was demonstrated that disruption of desmoplakin frees plakoglobin from the plasma membrane allowing it to translocate to the nucleus and suppress canonical Wnt/β-catenin signaling. Wnt signaling can inhibit adipogenesis by preventing mesodermal precursors from differentiating into adipocytes.[25] Suppression of Wnt signaling by plakoglobin nuclear localization could, therefore, promote the differentiation to adipose tissue in the cardiac myocardium in patients with ARVD/C.[26]

Finally, since ion channels, like the $Na^+$ channel, are also located in the intercalated disk, they might be disrupted and contribute to arrhythmogenicity.

The pathophysiological mechanisms proposed above are not mutually exclusive and could occur simultaneously.

Two patterns of inheritance have been described in ARVD/C. The most common or classical form of ARVD/C is inherited as an autosomal dominant trait. Naxos disease and Carvajal syndrome are rare, inherited as autosomal recessive. **Table 1** summarizes the different genes involved in ARVD/C with the corresponding phenotypes.

### TABLE 1

**Mutated genes and concurrent types of autosomal dominant ARVD/C**

|  | Gene | Type of disease |
|---|---|---|
| Desmosomal | PKP2 | ARVD/C |
|  | DSG2 | ARVD/C |
|  | DSC2 | ARVD/C |
|  | JUP | Naxos disease* |
|  | DSP | Carvajal syndrome* |
|  |  | ARVD/C |
|  |  | LDAC |
| Non-desmosomal | RyR2 | CPVT ARVD/C |
|  | TGF-β | ARVD/C |
|  | TMEM43 | ARVD/C |

*Autosomal recessive inheritance;
(*Abbreviations:* CPVT: Catecholaminergic polymorphic VT; LDAC: Left dominant arrhythmogenic cardiomyopathy. See text for other abbreviations. (*Source:* Modified from: Van Tintelen et al. Curr Opin Cardiol. 2007;22:185-92)

## Autosomal Recessive Disease

In Naxos disease, affected individuals were found to be homozygous for a 2-base pair deletion in the *JUP* gene.[6] All patients who are homozygous for this mutation have diffuse palmoplantar keratosis and woolly hair in infancy. Children usually have no cardiac symptoms, but may have electrocardiographic abnormalities and nonsustained ventricular arrhythmias.[27] In one report, an Arab family was found to have an autosomal recessive mutation in the desmoplakin gene that caused ARVD/C with a classical ARVD/C cardiac phenotype, that was also associated with woolly hair, and a pemphigus-like skin disorder.[28] A different autosomal recessive disease, Carvajal syndrome, is associated with a desmoplakin gene mutation. It manifests by woolly hair, epidermolytic palmoplantar keratoderma and cardiomyopathy.[29] The cardiomyopathy of Carvajal syndrome was thought to have a predilection for the LV, but subsequent evaluation of a deceased child revealed typical ARVD/C changes in both ventricles.[24]

## Autosomal Dominant Disease

Mutations in the gene encoding the intracellular desmosomal component desmoplakin can cause "classic ARVD/C" with a clinical presentation of VT, sudden death as well as LV involvement as the disease progresses.[15,30,31] Desmoplakin gene mutations have also been associated with predominantly left-sided ARVD/C and, as noted above, with autosomal recessive disease.

Various authors identified mutations in the *PKP2* gene as the most frequently observed genetic abnormality. **Figure 3** shows the pedigree of a family with a *PKP2* mutation. Incomplete penetrance and clinical variability are well documented. In four studies from different countries, analyzing 56–100 ARVD/C patients each, the following observations were made.[16,32-34] *PKP2* mutations were found in 11–43% of unrelated index patients who fulfilled diagnostic task force criteria for ARVD/C. In a Dutch ARVD/C cohort, 78 of 149 (52%) probands had a pathogenic *PKP2* mutation. This high yield of *PKP2* mutations is partly due to occurrence of founder mutations in the Netherlands. Haplotype analysis previously performed suggested founder mutations were responsible for 4 of the 14 different mutations identified.[34] Among index patients with a positive family history of ARVD/C, 70% had a *PKP2* mutation.[34]

Pilichou et al. screened patients with ARVD/C for mutations in the transmembranous desmosomal component *DSG2*.[17] Among 80 unrelated probands, 26 were found to have *DSP* or *PKP2* mutations. Direct sequencing of *DSG2* in the other 54 patients revealed nine distinct mutations in eight individuals. These individuals demonstrated typical clinical characteristics of ARVD/C. An analogous study of 86 ARVD/C probands identified eight novel *DSG2* mutations in nine probands. Clinical evaluation of family members with *DSG2* mutations revealed a penetrance of 58% using task force criteria from 1994 and 75% using proposed modified criteria.[35] Morphological abnormalities

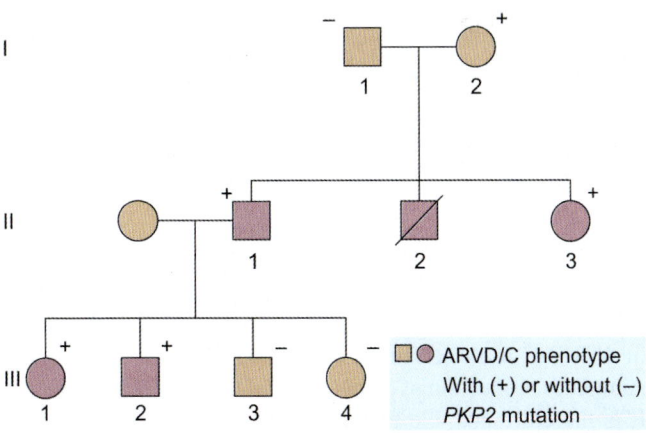

**FIGURE 3:** Pedigree of family with ARVD/C and *PKP2* mutation. This figure shows incomplete penetrance and variability of clinical expression. Both the 72-year-old grandmother (I:2) and 20-year-old grandson (III:2) are free of any signs of disease, despite carrying the mutation. The proband (II:1) was resuscitated at age 35, his brother (II:2) died suddenly at age 18. Both the proband's sister (II:3) and daughter (III:1) were diagnosed with the disease due to a positive family history, arrhythmias and RV structural abnormalities. The sister (II:3) of the proband has structural and ECG abnormalities, but no symptomatic arrhythmias

of the RV were present in 66% of gene carriers, LV involvement in 25% and classical right precordial T-wave inversion in only 26%. The authors noted that disease expression of *DSG2* mutations was of variable severity, but that overall penetrance was high and LV involvement prominent.[36]

In *DSC2*, another important transmembranous desmosomal cadherin, two heterozygous mutations (a deletion and an insertion) were identified in 4 of 77 probands with ARVD/C.[18] Finally, a dominant mutation in the plakoglobin (*JUP*) gene has been identified.[37] The identification of so many desmosomal cell adhesion gene abnormalities supports the hypothesis that ARVD/C is predominantly a disease of cell-to-cell coupling.

## Other Non-Desmosomal Genes

Mutations in the cardiac ryanodine receptor *RyR2*, which is responsible for calcium release from the sarcoplasmic reticulum, have been described in only one Italian ARVD/C family.[13] Affected patients have exercise-induced polymorphic VT.[38] Mutations in *RyR2* have primarily been associated with familial catecholaminergic polymorphic VT without ARVD/C.[19,39] Although the general opinion is that *RyR2* mutations lead to catecholaminergic polymorphic VT without structural abnormalities, the mutations in ARVD/C have been advocated to act differently from those in familial polymorphic VT without ARVD/C.[40-42]

The *TGFβ3* regulates the production of extracellular matrix components and modulates expression of genes encoding desmosomal proteins. The gene has been mapped to chromosome 14. Sequencing studies failed to identify any disease-causing mutations in the exonic regions of *TGFβ3*. This led to screening of the promoter and untranslated regions, where a mutation of the *TGFβ3* gene was found in all clinically affected members of a large family with ARVD/C.[14] The mutation is predicted to produce an amino acid substitution in a short peptide with an inhibitory role in *TGFβ3* regulation. The implication of these observations is that regulatory mutations resulting in overexpression of *TGFβ3* may contribute to the development of ARVD/C in these families. The *TGFβ* family of cytokines stimulates production of components of the extracellular matrix. It is therefore possible that enhanced *TGFβ* activity can lead to myocardial fibrosis. However, genetic analysis of two other families with ARVD/C failed to identify mutations in any of the regions of the *TGFβ3* gene.

A missense mutation in the *TMEM43* gene was found in 15 unrelated ARVD/C families from a genetically isolated population in New Foundland and caused a fully penetrant, sex-influenced, high risk form of ARVD/C.[19] The *TMEM43* gene contains the response element for peroxisome proliferator-activated receptor gamma (PPAR-γ), an adipogenic transcription factor. The *TMEM43* gene mutation is thought to cause dysregulation of an adipogenic pathway regulated by PPAR-γ, which may explain the fibrofatty replacement of myocardium in ARVD/C patients.

## EPIDEMIOLOGY

Estimations of the prevalence of ARVD/C in the general population vary from 1:2000 to 1:5000.[43] The real prevalence of ARVD/C, however, is unknown and is presumably higher due to many non-diagnosed and misdiagnosed cases.

The disease appears to be especially common in adolescents and young adults in northern Italy, accounting for approximately 11% of cases of sudden cardiac death overall and 22% in athletes.[44,45] In as many as 20% of sudden deaths occurring in people under 35 years of age, features of ARVD/C were detected at postmortem evaluation.[45] In nearly half of them, no prior symptoms had been reported.

ARVD/C has incomplete penetrance and extremely variable clinical expression. For instance, family screening has identified pathogenic mutation carriers, who had remained free of any sign of disease up to or over 70 years of age **(Fig. 3)**.

From the genetic aspect, both men and women should be equally affected. However, men are more frequently diagnosed with ARVD/C than women. In a recent large multicenter study, 57% of affected individuals were male.[46] As many women as men show at least some signs of disease, but women more often do not fulfill criteria to meet the diagnosis. Factors explaining this difference in severity of disease expression have not yet been elucidated. It is speculated that (sports) activity or hormonal factors may play a role. Familial disease has been demonstrated in greater than 50% of ARVD/C cases.

## CLINICAL PRESENTATION

ARVD/C patients typically present between the second and the fourth decade of life with monomorphic VT originating from RV. However, in a minority of patients, sudden death, frequently at a young age, or RV failure are the first signs. Based on clinicopathologic and patient follow-up studies, four different disease phases have been described for the classical form of ARVD/C, i.e. primarily affecting the RV **(Table 2)**:
1. Early ARVD/C is often described as "concealed" owing to the frequent absence of clinical findings, although minor ventricular arrhythmias and subtle structural changes may be found. Although patients tend to be asymptomatic, they may nonetheless be at risk of sudden death, mainly during vigourous exercise.

**TABLE 2**
**Different phases of disease severity**

| Phase | Characteristics |
|---|---|
| 1. Concealed | Asymptomatic patients with possibly only minor ventricular arrhythmia and subtle structural changes |
| 2. Overt | Symptoms due to LBBB VT or multiple premature complexes, with more obvious structural RV abnormalities |
| 3. RV failure | With relatively preserved LV function |
| 4. Biventricular | Significant overt LV involvement |

2. The overt phase follows, in which patients suffer from palpitations, syncope and ventricular arrhythmias of left bundle branch block (LBBB) morphology, ranging from isolated ventricular premature complexes to sustained VT and ventricular fibrillation (VF).
3. The third phase is characterized by RV failure due to progressive loss of myocardium with severe dilatation and systolic dysfunction, in the presence of preserved LV function.
4. Biventricular failure occurs due to LV involvement at a later stage. This phase may mimic dilated cardiomyopathy (DCM) and may require cardiac transplantation.

In the initially described classical form of ARVD/C, the RV is primarily affected with possibly (in a later stage) some LV involvement. Two additional patterns of disease have been identified by clinicogenetic characterization of families. These are the left dominant phenotype, with early and predominant LV manifestations, and the biventricular phenotype with equal involvement of both ventricles.

Recent immunohistochemical analysis of human myocardial samples demonstrated that both ventricles are affected by the disease.[47] A marked reduction in immunoreactive signal levels for plakoglobin was observed both in RV and LV, independent of genotype. Thus, at a molecular level ARVD/C is a global biventricular disease. However, histologically and functionally overt manifestations of the disease usually start in the RV. The reason for this is still unclear. The most commonly advocated hypothesis is that the thin walled RV is less able to withstand pressure (over)load in the presence of impaired function of mechanical junctions.

## CLINICAL DIAGNOSIS

Diagnosis of ARVD/C can be very challenging. Although VF and sudden death may be the first manifestations of

ARVD/C, symptomatic patients typically present with sustained VT with LBBB morphology, thus originating from the RV. The occurrence of VT episodes is usually induced by adrenergic stimuli mainly during exercise, especially competitive sports. The ARVD/C is a disease that shows progression over time.

In ARVD/C demonstration of transmural fibrofatty replacement primarily of right ventricular myocardium can be determined at surgery or autopsy **(Fig. 1)**. Predilection sites for these structural abnormalities are the so-called triangle of dysplasia formed by the RV outflow tract, the apex and the subtricuspid region.[1] In clinical practice, diagnosis based on cardiac pathology is not practical. Endomyocardial biopsies have major limitations. Tissue sampling from the affected often thin RV free wall, directed by imaging techniques or voltage mapping, is associated with a slight risk of perforation. Sampling from the interventricular septum is relatively safe. However, the septum is histopathogically rarely affected in ARVD/C. In addition, histology may be classified as normal due to the focal nature of the lesions. Finally, since subendocardial layers are usually not affected in an early stage of the disease, histological diagnosis may be hampered by the non-transmural nature of endomyocardial biopsies.[48]

Clinical diagnosis has been facilitated by a set of clinically applicable criteria for ARVD/C diagnosis defined by a Task Force based on consensus in 1994, and modified in 2010.[49,50] The current Task Force criteria are the essential standard for classification of individuals suspected of ARVD/C. In addition, its universal acceptance contributes importantly to unambiguous interpretation of clinical studies and facilitates comparison of results. The Task Force criteria included six different categories. They are derived into: (1) Global and regional dysfunction and structural alterations; (2) Tissue characterization; (3) Depolarization abnormalities; (4) Repolarization abnormalities; (5) Arrhythmias and (6) Family history, including pathogenic mutations. Within these groups, diagnostic criteria are categorized as major or minor according to their specificity for the disease. In order to fulfill ARVD/C diagnosis, it is required to have either two major or one major plus two minor or four minor criteria. From each different group, only one criterion can be counted for diagnosis, even when multiple criteria in one group are present. **Table 3** gives an overview of the Task Force criteria which is defined in 2010.

Specific evaluations are recommended in all patients suspected of ARVD/C. Detailed history and family history, physical examination, 12-lead electrocardiogram (ECG), signal averaged ECG (SAECG), 24-hours Holter monitoring,

## TABLE 3
**New task force criteria**

I. Global and/or regional dysfunction and structural alterations

  *Major (2D echo)*
  Regional RV akinesia, dyskinesia or aneurysm and one of:
  - Parasternal long axis view
    | | |
    |---|---|
    | RVOT (PLAX) | $\geq$32 mm |
    | Corrected for body size (PLAX/BSA) | $\geq$19 mm/m$^2$ |
  - Parasternal short axis view
    | | |
    |---|---|
    | RVOT (PSAX) | $\geq$36 mm |
    | Corrected for body size (PSAX/BSA) | $\geq$21 mm/m$^2$ |
  - Fractional area change (FAC) — $\leq$33%

  *Major (MRI)*
  Regional RV akinesia or dyskinesia or dyssynchronous RV contraction
  And one of:
  - RV end diastolic volume (RVEDV/BSA) — $\geq$110 mL/m$^2$ male; $\geq$100 mL/m$^2$ female
  - RV ejection fraction (RVEF) — $\leq$40%

  *Major (RV cineangiography)*
  Regional RV akinesia, dyskinesia or aneurysm

  *Minor (2D echo)*
  Regional RV akinesia or dyskinesia
  And one of:
  - Parasternal long axis view
    | | |
    |---|---|
    | RVOT (PLAX) | $\geq$29–$\leq$31 mm |
    | Corrected for body size (PLAX/BSA) | $\geq$16–$\leq$18 mm/m$^{2i}$ |
  - Parasternal short axis view
    | | |
    |---|---|
    | RVOT (PSAX) | $\geq$32–$\leq$35 mm |
    | Corrected for body size (PSAX/BSA) | $\geq$18–$\leq$20 mm/m$^2$ |
  - Fractional area change (FAC) — $\leq$40%

  *Minor (MRI)*
  Regional RV akinesia or dyskinesia or dyssynchronous RV contraction
  And one of:
  - RV end diastolic volume/BSA — $\geq$100 mL/m$^2$ male; $\geq$90 mL/m$^2$ female
  - RV ejection fraction (RVEF) — $\leq$45%

II. Tissue characterization of wall

  *Major*
  Residual myocytes <60% by morphometric analysis, (or <50%, if estimated), with fibrous replacement of the RV free wall myocardium in at least 1 sample, with or without fatty tissue replacement

  *Minor*
  Residual myocytes 60–75% by morphometric analysis, (or 50–65%, if estimated), with fibrous replacement of the RV free wall myocardium in at least 1 sample, with or without fatty tissue replacement

*Contd...*

*Contd...*

| |
|---|
| **III. Repolarization abnormalities** |
| *Major* |
|     Negative T waves in at least leads V1–3 |
| *Minor* |
|     Negative T waves only in leads V1 and V2 or in V4–6 |
|     In case of complete right bundle branch block: negative T waves in leads V1–4 |
| **IV. Depolarization/conduction abnormalities** |
| *Major* |
|     Epsilon wave in one of leads V1–3 |
| *Minor* |
|     Late potentials by signal averaged ECG in at least one of three parameters in the absence of a QRS duration of $\geq$110 msec on the standard ECG |
|     • Filtered QRS duration (fQRS)     $\geq$114 msec |
|     • Duration of terminal QRS < 40 μV (LAS)     $\geq$38 msecs |
|     • RMS voltage of terminal 40 msecs     $\leq$20 μV |
|     • Terminal activation duration $\geq$55 msec |
| **V. Arrhythmias** |
| *Major* |
|     • (Non-)sustained VT of left bundle branch block morphology with superior axis |
| *Minor* |
|     • (Non-)sustained VT of left bundle branch block morphology with inferior axis or unknown axis |
|     • >500 ventricular extrasystoles/24 hours by Holter |
| **VI. Family history** |
| *Major* |
|     • ARVD/C confirmed in a first-degree relative who meets current task force criteria |
|     • ARVD/C confirmed pathologically at autopsy or surgery in a first-degree relative |
|     • Identification of a pathogenic mutation associated with ARVD/C |
| *Minor* |
|     • History of ARVC/D in a first-degree relative in whom it is not possible or practical to determine if the family member meets current task force criteria |
|     • Premature sudden death (<35 years) due to suspected ARVD/C in a first-degree relative |
| (*Source:* Modified from: Marcus FI et al. Circulation. 2010;121:1533-41). |

exercise testing and 2D echocardiography with quantitative wall motion analysis. When appropriate, more detailed analysis of the RV can be done by cardiac magnetic resonance imaging (MRI). Invasive tests are also useful for diagnostic purposes: RV and LV cineangiography, electrophysiological testing, and endomyocardial biopsies.

## Global and/or Regional Dysfunction and Structural Alterations

Evaluation of RV size and function can be done by various imaging modalities, including echocardiography, cardiac MRI, computed tomography and/or cineangiography. According to the Task Force criteria, major criteria are defined as presence of an akinetic or dyskinetic areas in the RV **(Fig. 4)** combined with severe dilatation of the RV or RV ejection fraction 40% or lower.[50] With RV cineangiography the finding of only regional akinesia, dyskinesia or aneurysm is considered sufficient for qualification as a major criterion. RV cineangiography has historically been considered the most sensitive method to visualize RV structural abnormalities, with a high specificity of 90%.[51] Compared to cineangiography, the non-invasive technique of echocardiography is widely used and serves as the first-line imaging technique in evaluating patients suspected of ARVD/C and in family screening. Especially with improvement of echocardiographic modalities, such as 3-dimensional echocardiography, strain and tissue Doppler, the sensitivity and specificity of echocardiography have improved in recent years. Cardiac MRI has the unique ability to characterize tissue composition, by differentiating fat from fibrous tissue by using delayed enhancement. However, this technique is expensive, not widely available and requires great expertise to prevent misdiagnosis of ARVD/C.[52] Also; this technique cannot be applied in patients with an implantable cardioverter-defibrillator (ICD). Incorrect interpretation of cardiac MRI is the most common cause of overdiagnosis and physicians should be reluctant to diagnose ARVD/C when structural abnormalities

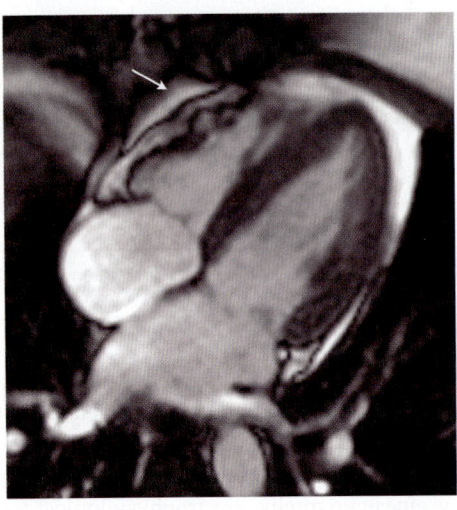

**FIGURE 4:** An MRI image of ARVD/C patient at the end of systole. Dyskinetic areas are visible in the RV free wall (arrow)

are present only on MRI.[53] Furthermore, it is important to note that the presence of fat in the epimyocardial and midmyocardial layers (without fibrosis) can be a normal finding and should not be considered diagnostic of ARVD/C **(Fig. 1)**.

## Endomyocardial Biopsy

For reasons previously noted, undirected endomyocardial biopsies are infrequently diagnostic. However, it had been included as a major criterion by the Task Force, since the finding of fibrofatty replacement was considered to strongly support any findings derived from other clinical investigations. The rather vague terminology of any "fibrofatty replacement of myocardium" has been quantified. Diagnostic values according to the new Task Force criteria are considered major if histomorphometric analysis of endomyocardial biopsies shows that the number of residual myocytes is below 60% or below 50% by estimation, with fibrous replacement of the RV free wall in at least one sample, with or without fatty tissue replacement.[54] If the number of residual myocytes is higher but still below 75% (morphometric) or below 65% (estimated), only a minor criterion is fulfilled.

## ECG Criteria

The 12-lead ECG is most important for diagnosis of ARVD/C. Consistent with early electrical uncoupling, ECG changes and arrhythmias may develop before histologic evidence of myocyte loss or clinical evidence of ARVD/C. ECG criteria on depolarization and repolarization have to be obtained during sinus rhythm and while off antiarrhythmic drugs. These drugs may cause misinterpretation of ECG criteria due to their contribution on activation delay and repolarization abnormalities.

## Depolarization Abnormalities

The RV activation delay is a hallmark of ARVD/C. This delay is reflected by the presence of an epsilon wave, prolonged terminal activation duration (TAD) in the terminal part and after the QRS complex, and also by recording of late potentials on SAECG.

Epsilon waves are defined as low amplitude potentials after and clearly separated from the QRS complex, in at least one of precordial leads, V1–V3 **(Fig. 5)**.[55] This highly specific major criterion is observed in only a small minority of patients.[56,57] TAD has been defined as the longest value measured from the nadir of the S wave to the end of all depolarization deflections in V1–V3, thereby including not only the S wave upstroke but also both late and fractionated signals and epsilon waves **(Figs 5 and 6)**.[58] Thus, total activation delay presumably from the

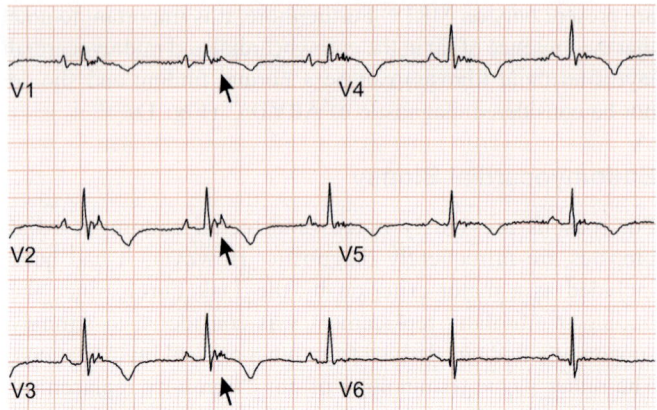

**FIGURE 5:** Epsilon waves indicated by arrows (also prolonged terminal activation duration; 120 ms) and negative T waves in V1–5

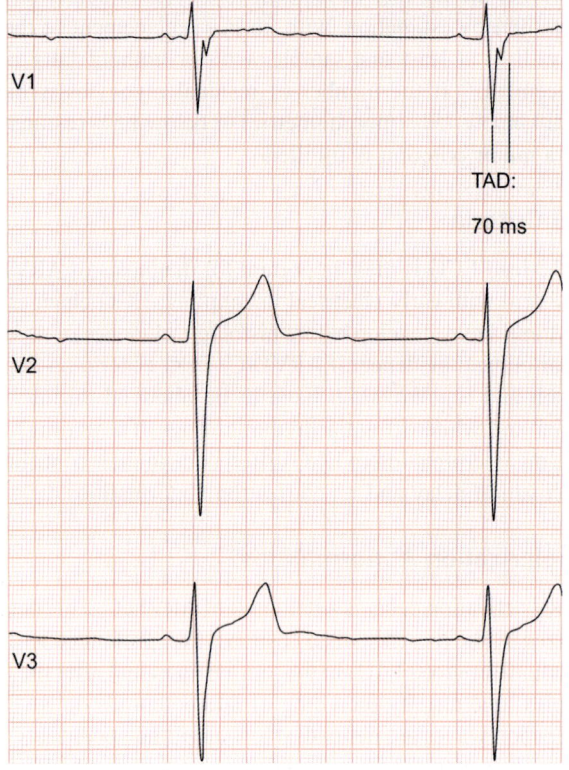

**FIGURE 6:** Prolonged terminal activation duration (70 ms from nadir of S wave to end of depolarization). Paper speed 25 mm/s

RV is conveyed by this new parameter. The TAD is considered prolonged if greater than or equal to 55 ms, and only applicable in the absence of complete right bundle branch block (RBBB). Prolonged TAD, introduced as minor criterion, appears to be equally sensitive as the presence of late potentials and much more sensitive than epsilon waves. Prolonged TAD was recorded in 30 of 42 ARVD/C patients and in only 1 of 27 patients with idiopathic VT.[58] Both epsilon waves and prolonged TAD are measured only in V1–V3, which face the RV outflow tract. Activation delay in other areas of RV is not reflected by these criteria.

The detection of late potentials on SAECG is the surface counterpart of delayed activation or late potentials detected during endocardial mapping in electrophysiologic studies. They are frequently found in patients with documented VT. However, these late potentials can also be observed after myocardial infarction and with other structural heart diseases. Due to this lack of specificity, SAECG abnormalities were considered a minor criterion. For all the depolarization criteria, it is apparent that they will correlate with disease severity. For instance, a positive correlation has been found between late potentials and the extent of RV fibrosis, reduced RV systolic function and significant morphological abnormalities on imaging.[59-61]

## Repolarization Abnormalities

In the new Task Force criteria negative T waves in leads V1, V2 and V3 form a major ECG criterion in the absence of complete RBBB, and only if the patient is older than 14 years of age **(Fig. 5)**. Studies have reported variable prevalences of right precordial T wave inversion, ranging from 19 to 94%.[49,55-57,62] The lower rates are often due to evaluation of family members, while higher rates are seen in series consisting of unrelated index patients. In the recent study by Cox et al., this criterion was identified in 67% of exclusively ARVD/C index patients and in none of the patients with idiopathic VT.[58] T wave inversion can be a normal feature of the ECG in children and in early adolescence. Therefore, this finding is not considered abnormal in persons at the age of 14 years and younger.

In the new Task Force criteria, two minor repolarization criteria were included:
1. Inverted T waves only in leads V1–V2 or in V4–V6 in individuals older than 14 years of age and in the absence of complete RBBB.
2. Inverted T waves in leads V1–V4 in individuals older than 14 years of age in the presence of RBBB. This was included since T wave inversion in RBBB seldom extends to V4 in otherwise healthy individuals.

## Arrhythmias

In ARVD/C, ventricular arrhythmias range from premature ventricular complexes to sustained VT and VF, leading to cardiac arrest.[58,63] Due to their typical origin in the RV, QRS complexes of ventricular arrhythmias show a LBBB morphology. Moreover, the QRS axis indicates the VT origin, i.e. superior axis from the RV inferior wall or apex (major criterion) and inferior axis (minor criterion) from the RV outflow tract (RVOT) **(Figs 7 and 8)**. The VT of LBBB configuration with an unknown axis counts as minor criterion. Patients with extensively affected RV often show multiple VT morphologies.[58,64]

VF is the mechanism of instantaneous sudden death especially occurring in young people and athletes with ARVD/C, who were often previously asymptomatic. In this subset of patients, VF may occur from deterioration of monomorphic VT, or in a phase of acute disease progression, due to myocyte death and reactive inflammation.[3]

Finally, in the new Task Force criteria, the number of premature ventricular complexes on 24-hour Holter recordings is reduced to 500 or more for a minor criterion.

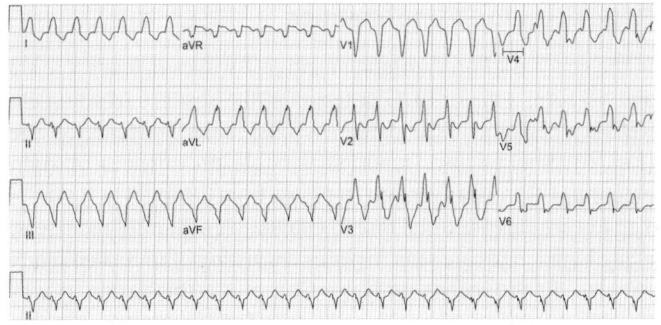

**FIGURE 7:** An ECG (25 mm/s) from ARVD/C patient with *PKP2* mutation. This VT has an LBBB morphology and superior axis (with positive QRS complex in aVL), thus originating inferiorly from the RV

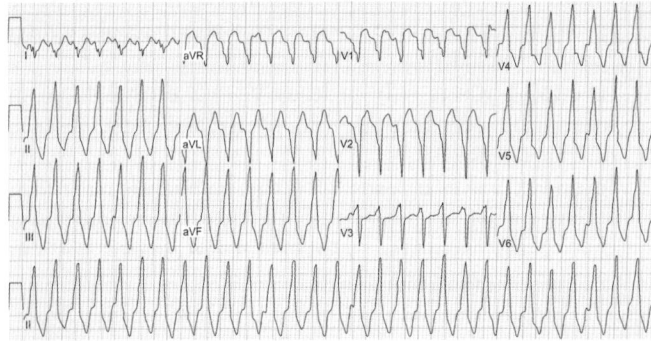

**FIGURE 8:** An ECG (25 mm/s) from ARVD/C patient without identified mutation. This VT has also LBBB morphology, but with inferior axis, originating from RV outflow tract. Note the typical negative QRS complex in aVL

## Family History

Before the discovery of pathogenic mutations underlying the disease, it was recognized that ARVD/C often occurs in family members.[1] Having a family member with proven ARVD/C is considered an increased risk for other family members to be affected. Therefore, having a first-degree relative who meets the current Task Force criteria, or having ARVD/C confirmed pathologically at autopsy or during surgery, or identification of a pathogenic mutation in the family, is included as major diagnostic criteria. If a first-degree relative is diagnosed with ARVD/C but does not fulfill the diagnostic criteria, only a minor criterion is counted. Sudden death of a family member under the age of 35 years, presumably but not proven to be due to ARVD/C related arrhythmias, is a minor criterion. Pathologic confirmation of transmural fibrofatty replacement of the RV at autopsy or after surgical resection is considered a major criterion for the diagnosis.[50]

## NONCLASSICAL ARVD/C SUBTYPES

### Naxos Disease

All patients who homozygously carry the recessive *JUP* mutation for Naxos disease have diffuse palmoplantar keratosis and woolly hair in infancy. Children usually have no cardiac symptoms, but may have ECG abnormalities and nonsustained ventricular arrhythmias.[6,27] The cardiac disease is 100% penetrant by adolescence, being manifested by symptomatic arrhythmias, ECG abnormalities, right ventricular structural alterations and LV involvement. In one series of 26 patients followed for 10 years, 62% had structural progression of right ventricular abnormalities and 27% developed heart failure due to LV involvement.[27] Almost half of the patients developed symptomatic arrhythmias and the annual cardiac and SCD mortality were 3% and 2.3% respectively, which are slightly higher than seen in autosomal dominant forms of ARVD/C. A minority of heterozygotes has minor ECG and ECG changes, but clinically significant disease is not present.

### Carvajal Syndrome

Carvajal syndrome is associated with a *DSP* gene mutation, and is also a recessive disease manifested by woolly hair, epidermolytic palmoplantar keratoderma and cardiomyopathy.[29] All diagnosed patients have been from Ecuador. The cardiomyopathy of Carvajal syndrome was first thought to be mainly left ventricular, with dilated left ventricular cardiomyopathy. A number of the patients with Carvajal syndrome had heart failure in their teenage years, resulting in early morbidity. Further research revealed that the disease is characterized mainly by ventricular hypertrophy, ventricular dilatation and discrete focal ventricular aneurysms. In the RV, focal wall thinning and aneurysmal dilatation were identified in the triangle of dysplasia.

## Left Dominant ARVD/C (LDAC)

As previously mentioned, in classic ARVD/C the histologic process predominantly involves the RV and extends to the LV in more advanced stages.[52,62,65-67] In contrast, patients with left-dominant arrhythmogenic cardiomyopathy (LDAC, also known as left-sided ARVD/C or arrhythmogenic left ventricular cardiomyopathy) have fibrofatty changes that predominantly involve the LV.[68] Clinically, this disease entity is characterized by (infero)lateral T-wave inversion, arrhythmias of LV origin and/or proven LDAC.

Patients may present with arrhythmias or chest pain at ages ranging from adolescence to over 80 years. By cardiac MRI about one-third of patients show a LV ejection fraction less than 50%. Furthermore, MRI with late gadolinium enhancement (LGE) of the LV demonstrated late enhancement in a subepicardial/midwall distribution. Similar to ARVD/C, some patients with LDAC have desmosomal gene mutations (see below).

## DIFFERENTIAL DIAGNOSIS

Although the diagnosis in an overt case of ARVD/C is not difficult, early and occasionally late stages of the disease may show similarities with a few other diseases. In particular, differentiation from idiopathic VT originating from the RVOT can be challenging. However, idiopathic RVOT VT is a benign non-familial condition, in which the ECG shows no depolarization or repolarization abnormalities and no RV structural changes can be detected. Furthermore, VT episodes have a single morphology (LBBB morphology with inferior axis) and are usually not reproducibly inducible by premature extrastimuli at programmed stimulation during electrophysiologic studies.[69,70] Idiopathic RVOT VT may be inducible by regular burst pacing and isoproterenol infusion. It is important to differentiate idiopathic RVOT VT from ARVD/C for several reasons. The first is the known genetic etiology in ARVD/C. A genetic abnormality is not present in patients with idiopathic VT originating from the RVOT. Therefore, it has implications with regards to screening of family members. The prognosis of RVOT tachycardia is uniformly excellent with sudden death occurring rarely. Finally, in contrast to ARVD/C, catheter ablation is usually curative in idiopathic RVOT tachycardia.

Another disease mimicking ARVD/C is cardiac sarcoidosis. Sarcoidosis is a disease of unknown etiology, characterized by the presence of noncaseating granulomas in affected tissues, mainly lungs, but heart, skin, eyes, reticuloendothelial system, kidneys and central nervous system can also be affected. The prevalence of this condition varies in different geographical regions, and the disease may also be familial and occurs in specific racial subgroups.[71] Clinical symptoms of cardiac involvement are present in about 5% of all patients with sarcoidosis. The clinical manifestations of cardiac sarcoidosis depend upon the location and extent of granulomatous inflammation and include conduction abnormalities, ventricular arrhythmias, valvular dysfunction and congestive heart failure. Myocardial sarcoid granulomas or areas of myocardial scarring are typically present in the LV and septum of patients with this condition, and the RV can be predominantly affected. The VT associated with right ventricular abnormalities can, therefore, result in diagnostic confusion, especially if there is no systemic evidence of sarcoidosis. Patients can present with clinical features similar to those of ARVD/C including arrhythmias and sudden cardiac death.[72] Cardiac sarcoidosis can be diagnosed definitively by endomyocardial biopsy if granulomas are visualized.[73] To strengthen differentiation from ARVD/C, gadolinium-enhanced MRI may be beneficial by detecting located abnormalities in the septum, which is typical for sarcoidosis but seldom seen in ARVD/C. Active foci of sarcoidosis can be visualized by positron emission tomography (PET) scan. Therapy with corticosteroids is recommended for patients diagnosed with cardiac sarcoidosis. Treatment aims to control inflammation and fibrosis in order to maintain cardiac structure and function.

Myocarditis has to be excluded in patients suspected of ARVD/C. Myocarditis may arise from viral or other pathogens as well as toxic or immunologic insult. In general, endomyocardial biopsy is required to distinguish ARVD/C from myocarditis.

ARVD/C may mimic DCM, especially in the more advanced stages of disease. Patients with DCM usually present with heart failure or thromboembolic disease, including stroke. Since it is uncommon to have sustained VT or sudden death as the initial presenting symptom of DCM, patients with these symptoms should be first suspected of having ARVD/C.

## MOLECULAR GENETIC ANALYSIS

It is important to realize that the clinical diagnosis of ARVD/C is based exclusively on fulfillment of the diagnostic Task Force criteria. Mutations underlying the disease show incomplete penetrance and variable clinical expression. Some genetically affected patients may have no signs or symptoms whatsoever, whereas no mutations can be identified in a large minority of clinically diagnosed patients. Therefore, genetic analysis may not be of any critical diagnostic value for the index patient who meets Task Force criteria, but can be used to identify if family members are predisposed to disease development.

**The strategy for genetic testing in ARVD/C is as follows:**

Individuals with clinical diagnosis of ARVD/C are the first to be tested. The detection of a pathogenic mutation does not make a clinical diagnosis of ARVD/C. In contrast, if no mutation can be identified in a patient diagnosed with ARVD/C, the clinical diagnosis of ARVD/C is still applicable. If a pathogenic mutation is identified in the proband, parents, siblings and children of this patient can be tested for the mutation via the cascade method. When an (asymptomatic) relative is found to carry a pathogenic mutation, periodic cardiologic screening is required.

**Table 1** shows the different genes related to ARVD/C. Currently, DNA analysis for *PKP2, DSG2, DSC2, DSP* and *JUP* is recommended in ARVD/C patients with an appropriate indication for this analysis.

## PROGNOSIS AND THERAPY

The prognosis of classical ARVD/C is considerably better than that of patients with sustained VT from left ventricular structural heart disease. However, ARVD/C is a progressive disease and may lead to RV and also LV failure or sudden cardiac death. The death rate for patients with ARVD/C has been estimated at 2.5% per year.[74] Retrospective analysis of clinical and pathologic studies identified several risk factors for sudden death, such as previously aborted sudden death, syncope, young age, malignant family history, severe RV dysfunction and LV involvement.[75,76]

Electrophysiologic induction of VT with LBBB morphology and superior axis is a major diagnostic criterion.[58,64] However, electrophysiologic studies have not proven to be useful in risk stratifying patients with ARVD/C. This was illustrated in a multicenter study of 132 patients with ARVD/C in whom electrophysiologic study was performed prior to ICD implantation.[77] The positive and negative predictive values of VT inducibility for subsequent appropriate device therapy were 49% and 54% respectively.

In addition to symptomatic treatment, prevention of sudden death is the most important therapeutic goal in ARVD/C. Most data on effective treatment strategies refer to retrospective analyzes in single centers with only limited numbers of patients, and results are difficult to compare due to different patient selection and treatment strategies. There is limited data on long-term outcomes and no controlled randomized trials have been performed. International registries have been established, but have not yet reported results on treatment.

Evidence suggests that asymptomatic patients and healthy mutation carriers do not require prophylactic treatment. They should undergo regular cardiac evaluations including 12-lead ECG, 24-hours Holter monitoring, echocardiography and exercise testing for early identification of unfavorable signs. In patients diagnosed with or have signs or symptoms of ARVD/C as well as mutation carriers, specific lifestyle advice is advisable. Sports participation has been shown to increase the risk of sudden death fivefold in ARVD/C patients.[78] Furthermore, excessive mechanical stress, such as during competitive sports activity and training, may aggravate the underlying myocardial abnormalities and accelerate disease progression. Therefore, patients with ARVD/C should be advised against practicing competitive and endurance sports, such as running marathons.

Therapeutic options in patients with ARVD/C include antiarrhythmic drugs, catheter ablation and ICD.

Patients with VT have a favorable outcome when they are treated medically and therefore pharmacologic treatment is the first choice. This concerns not only patients who have presented with sustained VT but also patients and family members with nonsustained VT or greater than 500 ventricular extrasystoles on 24-hours Holter monitoring. Since ventricular arrhythmias and cardiac arrest occur frequently during or after physical exercise or may be triggered by catecholamines, antiadrenergic β-blockers are recommended. Sotalol is the drug of first choice. Alternatively, other β-receptor blocking agents, amiodarone and flecainide have all been reported as useful.[79] Efficacy of drug treatment has to be evaluated by serial Holter monitoring and/or exercise testing. This strategy has proven to have better long-term outcome when compared to standard empirical treatment.[79]

Catheter ablation is an alternative in patients who are refractory to drug treatment and have frequent VT episodes with a predominantly single morphology. Marchlinski et al. performed VT ablation in 19 ARVD/C patients by the use of focal and/or linear lesions; in 17 no VT recurred during the subsequent $7 \pm 22$ months.[80] In a series of 50 consecutive patients studied during 16 years, Fontaine et al. reported a 40% success rate by radiofrequency ablation after multiple ablation sessions, that increased to 81% when fulguration was used additionally.[81] However, these reports are from single centers with highly experienced electrophysiologists, and may not be reproducible in general practice. Catheter ablation is generally considered to be palliative and not curative. Long-term success rates are poor. Due to disease progression, new VTs with different morphologies will usually occur.[82]

Although antiarrhythmic drugs and catheter ablation may reduce VT burden, there is no proof from prospective trials that these therapies will also prevent sudden death. The ICD implantation is indicated in patients who are intolerant of antiarrhythmic drug therapy and who are at serious risk for sudden death. Implantation of an ICD has to be considered in ARVD/C patients with aborted cardiac arrest, intolerable fast VT and those with risk factors as mentioned above.

## SUMMARY

Arrhythmogenic right ventricular dysplasia cardiomyopathy is most often a genetically determined disease characterized by fibrofatty replacement of myocardial tissue. Primarily affecting the RV, but extension to the LV occurs, especially in more advanced stages of the disease. At the molecular level, both ventricles are affected, presumably in all stages of the disease. Its prevalence has been estimated to vary from 1:2000 to 1:5000. Patients typically present between the second and the fourth decade of life with exercise induced tachycardia episodes originating from the RV. It is also a major cause of sudden death in the young and athletes.

The causative genes encode proteins of mechanical cell junctions (e.g. plakoglobin, plakophilin-2, desmoglein-2, desmocollin-2, desmoplakin) and account for intercalated disk remodeling. The classical form of ARVD/C is inherited in an autosomal dominant trait, but has variable expression. The rare recessively inherited variants are often associated with palmoplantar keratoderma and woolly hair. The diagnosis is made according to a set of Task Force criteria, based on family history, depolarization and repolarization abnormalities, ventricular arrhythmias with an LBBB morphology, functional and structural alterations of the RV, and fibrofatty replacement in endomyocardial biopsy. Two dimensional echocardiography, cineangiography and magnetic resonance are the imaging tools to visualize structural-functional abnormalities. The main differential diagnoses are idiopathic right ventricular outflow tract tachycardia, myocarditis and sarcoidosis. Palliative therapy consists of antiarrhythmic drugs, catheter ablation and implantable cardioverter defibrillator. Young age, family history of juvenile sudden death, overt left ventricular involvement, VT, syncope and previous cardiac arrest are the major risk factors for adverse prognosis.

## REFERENCES

1. Marcus FI, Fontaine GH, Guiraudon G, et al. Right ventricular dysplasia: a report of 24 adult cases. Circulation. 1982;65:384-98.
2. Corrado D, Basso C, Thiene G, et al. Spectrum of clinicopathologic manifestations of arrhythmogenic right ventricular cardiomyopathy/dysplasia: a multicenter study. J Am Coll Cardiol. 1997;30:1512-20.
3. Basso C, Thiene G, Corrado D, et al. Arrhythmogenic right ventricular cardiomyopathy: dysplasia, dystrophy, or myocarditis? Circulation. 1996;94:983-91.

4. Thiene G, Nava A, Corrado D, et al. Right ventricular cardiomyopathy and sudden death in young people. N Engl J Med. 1988;318:129-33.
5. Richardson P, McKenna W, Bristow M, et al. Report of the 1995 World Health Organization/International Society and Federation of Cardiology Task Force on the Definition and Classification of cardiomyopathies. Circulation. 1996;93:841-2.
6. McKoy G, Protonotarios N, Crosby A, et al. Identification of a deletion in plakoglobin in arrhythmogenic right ventricular cardiomyopathy with palmoplantar keratoderma and woolly hair (Naxos disease). Lancet. 2000;355:2119-24.
7. Rampazzo A, Nava A, Miorin M, et al. ARVD4: a new locus for arrhythmogenic right ventricular cardiomyopathy, maps to chromosome 2 long arm. Genomics. 1997;45:259-63.
8. Ahmad F, Li D, Karibe A, et al. Localization of a gene responsible for arrhythmogenic right ventricular dysplasia to chromosome 3p23. Circulation. 1998;98:2791-5.
9. Li D, Ahmad F, Gardner MJ, et al. The locus of a novel gene responsible for arrhythmogenic right-ventricular dysplasia characterized by early onset and high penetrance maps to chromosome 10p12–p14. Am J Hum Genet. 2000;66:148-56.
10. Melberg A, Oldfors A, Blomstrom-Lundqvist C, et al. Autosomal dominant myofibrillar myopathy with arrhythmogenic right ventricular cardiomyopathy linked to chromosome 10q. Ann Neurol. 1999;46:684-92.
11. Rampazzo A, Nava A, Danieli GA, et al. The gene for arrhythmogenic right ventricular cardiomyopathy maps to chromosome 14q23–q24. Hum Mol Genet. 1994;3:959-62.
12. Severini GM, Krajinovic M, Pinamonti B, et al. A new locus for arrhythmogenic right ventricular dysplasia on the long arm of chromosome 14. Genomics. 1996;31:193-200.
13. Tiso N, Stephan DA, Nava A, et al. Identification of mutations in the cardiac ryanodine receptor gene in families affected with arrhythmogenic right ventricular cardiomyopathy type 2 (ARVD2). Hum Mol Genet. 2001;10:189-94.
14. Beffagna G, Occhi G, Nava A, et al. Regulatory mutations in transforming growth factor-beta3 gene cause arrhythmogenic right ventricular cardiomyopathy type 1. Cardiovasc Res. 2005;65:366-73.
15. Rampazzo A, Nava A, Malacrida S, et al. Mutation in human desmoplakin domain binding to plakoglobin causes a dominant form of arrhythmogenic right ventricular cardiomyopathy. Am J Hum Genet. 2002;71:1200-6.
16. Gerull B, Heuser A, Wichter T, et al. Mutations in the desmosomal protein plakophilin-2 are common in arrhythmogenic right ventricular cardiomyopathy. Nat Genet. 2004;36:1162-4.
17. Pilichou K, Nava A, Basso C, et al. Mutations in desmoglein-2 gene are associated to arrhythmogenic right ventricular cardiomyopathy. Circulation. 2006;113:1171-9.
18. Syrris P, Ward D, Evans A, et al. Arrhythmogenic right ventricular dysplasia/cardiomyopathy associated with mutations in the desmosomal gene desmocollin-2. Am J Hum Genet. 2006;79:978-84.
19. Merner ND, Hodgkinson KA, Haywood AF, et al. Arrhythmogenic right ventricular cardiomyopathy type 5 is a fully penetrant, lethal arrhythmic disorder caused by a missense mutation in the TMEM43 gene. Am J Hum Genet. 2008;82:809-21.
20. Bernstein SA, Morley GE. Gap junctions and propagation of the cardiac action potential. Adv Cardiol. 2006;42:71-85.

21. Ellison KE, Friedman PL, Ganz LI, et al. Entrainment mapping and radiofrequency catheter ablation of ventricular tachycardia in right ventricular dysplasia. J Am Coll Cardiol. 1998;32:724-8.
22. Saffitz JE. Dependence of electrical coupling on mechanical coupling in cardiac myocytes: insights gained from cardiomyopathies caused by defects in cell-cell connections. Ann N Y Acad Sci. 2005;1047:336-44.
23. Kaplan SR, Gard JJ, Protonotarios N, et al. Remodeling of myocyte gap junctions in arrhythmogenic right ventricular cardiomyopathy due to a deletion in plakoglobin (Naxos disease). Heart Rhythm. 2004;1:3-11.
24. Kaplan SR, Gard JJ, Carvajal-Huerta L, et al. Structural and molecular pathology of the heart in Carvajal syndrome. Cardiovasc Pathol. 2004;13:26-32.
25. Ross SE, Hemati N, Longof KA, et al. Inhibition of adipogenesis by Wnt signaling. Science. 2000;289:950-3.
26. Garcia-Gras E, Lombardi R, Giocondo MJ, et al. Suppression of canonical Wnt/beta-catenin signaling by nuclear plakoglobin recapitulates phenotype of arrhythmogenic right ventricular cardiomyopathy. J Clin Invest. 2006;116:2012-21.
27. Protonotarios N, Tsatsopoulou A, Anastasakis A, et al. Genotype-phenotype assessment in autosomal recessive arrhythmogenic right ventricular cardiomyopathy (Naxos disease) caused by a deletion in plakoglobin. J Am Coll Cardiol. 2001;38:1477-84.
28. Alcalai R, Metzger S, Rosenheck S, et al. A recessive mutation in desmoplakin causes arrhythmogenic right ventricular dysplasia, skin disorder, and woolly hair. J Am Coll Cardiol. 2003;42:319-27.
29. Norgett EE, Hatsell SJ, Carvajal-Huerta L, et al. Recessive mutation in desmoplakin disrupts desmoplakin-intermediate filament interactions and causes dilated cardiomyopathy, woolly hair and keratoderma. Hum Mol Genet. 2000;9:2761-6.
30. Bauce B, Basso C, Rampazzo A, et al. Clinical profile of four families with arrhythmogenic right ventricular cardiomyopathy caused by dominant desmoplakin mutations. Eur Heart J. 2005;26:1666-75.
31. Sen-Chowdhry S, Syrris P, McKenna WJ. Desmoplakin disease in arrhythmogenic right ventricular cardiomyopathy: early genotype-phenotype studies. Eur Heart J. 2005;26:1582-4.
32. Syrris P, Ward D, Asimaki A, et al. Clinical expression of plakophilin-2 mutations in familial arrhythmogenic right ventricular cardiomyopathy. Circulation. 2006;113:356-64.
33. Dalal D, Molin LH, Piccini J, et al. Clinical features of arrhythmogenic right ventricular dysplasia/cardiomyopathy associated with mutations in plakophilin-2. Circulation. 2006;113:1641-9.
34. Van Tintelen JP, Entius MM, Bhuiyan ZA, et al. Plakophilin-2 mutations are the major determinant of familial arrhythmogenic right ventricular dysplasia/cardiomyopathy. Circulation. 2006;113:1650-8.
35. Hamid MS, Norman M, Quraishi A, et al. Prospective evaluation of relatives for familial arrhythmogenic right ventricular cardiomyopathy/dysplasia reveals a need to broaden diagnostic criteria. J Am Coll Cardiol. 2002;40:1445-50.
36. Syrris P, Ward D, Asimaki A, et al. Desmoglein-2 mutations in arrhythmogenic right ventricular cardiomyopathy: a genotype-phenotype characterization of familial disease. Eur Heart J. 2007;28:581-8.

37. Asimaki A, Syrris P, Wichter T, et al. A novel dominant mutation in plakoglobin causes arrhythmogenic right ventricular cardiomyopathy. Am J Hum Genet. 2007;81:964-73.
38. Rampazzo A, Beffagna G, Nava A, et al. Arrhythmogenic right ventricular cardiomyopathy type 1 (ARVD1): confirmation of locus assignment and mutation screening of four candidate genes. Eur J Hum Genet. 2003;11:69-76.
39. Wehrens XH, Lehnart SE, Huang F, et al. FKBP12.6 deficiency and defective calcium release channel (ryanodine receptor) function linked to exercise-induced sudden cardiac death. Cell. 2003;113:829-40.
40. Bauce B, Nava A, Rampazzo A, et al. Familial effort polymorphic ventricular arrhythmias in arrhythmogenic right ventricular cardiomyopathy map to chromosome 1q42-43. Am J Cardiol. 2000;85:573-9.
41. Priori SG, Napolitano C, Memmi M, et al. Clinical and molecular characterization of patients with catecholaminergic polymorphic ventricular tachycardia. Circulation. 2002;106:69-74.
42. Tiso N, Salamon M, Bagattin A, et al. The binding of the RyR2 calcium channel to its gating protein FKBP12.6 is oppositely affected by ARVD2 and VTSIP mutations. Biochem Biophys Res Commun. 2002;299:594-8.
43. Gemayel C, Pelliccia A, Thompson PD. Arrhythmogenic right ventricular cardiomyopathy. J Am Coll Cardiol. 2001;38:1773-81.
44. Corrado D, Pelliccia A, Bjørnstad HH, et al. Cardiovascular pre-participation screening of young competitive athletes for prevention of sudden death: proposal for a common European protocol. Consensus Statement of the Study Group of Sport Cardiology of the Working Group of Cardiac Rehabilitation and Exercise Physiology and the Working Group of Myocardial and Pericardial Diseases of the European Society of Cardiology. Eur Heart J. 2005;26:516-24.
45. Basso C, Corrado D, Thiene G. Cardiovascular causes of sudden death in young individuals including athletes. Cardiol Rev. 1999;7:127-35.
46. Marcus FI, Zareba W, Calkins HG, et al. Arrhythmogenic right ventricular dysplasia/cardiomyopathy, clinical presentation and diagnostic evaluation: results from the North American multidisciplinary study. Heart Rhythm. 2009;6:984-92.
47. Asimaki A, Tandri H, Huang H, et al. A new diagnostic test for arrhythmogenic right ventricular cardiomyopathy. N Engl J Med. 2009;360:1075-84.
48. Corrado D, Basso C, Thiene G. Arrhythmogenic right ventricular cardiomyopathy: diagnosis, prognosis, and treatment. Heart. 2000;83:588-95.
49. McKenna WJ, Thiene G, Nava A, et al. Diagnosis of arrhythmogenic right ventricular dysplasia/cardiomyopathy. Task Force of the Working Group Myocardial and Pericardial Disease of the European Society of Cardiology and of the Scientific Council on Cardiomyopathies of the International Society and Federation of Cardiology. Br Heart J. 1994;71:215-8.
50. Marcus FI, McKenna WJ, Sherrill D, et al. Diagnosis of arrhythmogenic right ventricular cardiomyopathy/dysplasia: proposed modification of the task force criteria. Circulation. 2010;121:1533-41, Eur Heart J. 2010;31:801-14.
51. White JB, Razmi R, Nath H, et al. Relative utility of magnetic resonance imaging and right ventricular angiography to diagnose

arrhythmogenic right ventricular cardiomyopathy. J Interv Card Electrophysiol. 2004;10:19-26.
52. Bluemke DA, Krupinski EA, Ovitt T, et al. MR Imaging of arrhythmogenic right ventricular cardiomyopathy: morphologic findings and interobserver reliability. Cardiology. 2003;99:153-62.
53. Tandri H, Calkins H, Nasir K, et al. Magnetic resonance imaging findings in patients meeting task force criteria for arrhythmogenic right ventricular dysplasia. J Cardiovasc Electrophysiol. 2003;14:476-82.
54. Basso C, Ronco F, Marcus F, et al. Quantitative assessment of endomyocardial biopsy in arrhythmogenic right ventricular cardiomyopathy/dysplasia: an in vitro validation of diagnostic criteria. Eur Heart J. 2008;29:2760-71.
55. Fontaine G, Umemura J, Di Donna P, et al. Duration of QRS complexes in arrhythmogenic right ventricular dysplasia. A new non-invasive diagnostic marker. Ann Cardiol Angeiol (Paris). 1993;42:399-405.
56. Peters S, Trümmel M. Diagnosis of arrhythmogenic right ventricular dysplasia-cardiomyopathy: value of standard ECG revisited. Ann Noninvasive Electrocardiol. 2003;8:238-45.
57. Pinamonti B, Sinagra G, Salvi A, et al. Left ventricular involvement in right ventricular dysplasia. Am Heart J. 1992;123:711-24.
58. Cox MG, Nelen MR, Wilde AA, et al. Activation delay and VT parameters in arrhythmogenic right ventricular dysplasia/cardiomyopathy: toward improvement of diagnostic ECG criteria. J Cardiovasc Electrophysiol. 2008;19:775-81.
59. Nasir K, Rutberg J, Tandri H, et al. Utility of SAECG in arrhythmogenic right ventricle dysplasia. Ann Noninvasive Electrocardiol. 2003;8:112-20.
60. Oselladore L, Nava A, Buja G, et al. Signal-averaged electrocardiography in familial form of arrhythmogenic right ventricular cardiomyopathy. Am J Cardiol. 1995;75:1038-41.
61. Turrini P, Angelini A, Thiene G, et al. Late potentials and ventricular arrhythmias in arrhythmogenic right ventricular cardiomyopathy. Am J Cardiol. 1999;83:1214-9.
62. Nava A, Bauce B, Basso C, et al. Clinical profile and long-term follow-up of 37 families with arrhythmogenic right ventricular cardiomyopathy. J Am Coll Cardiol. 2000;36:2226-33.
63. Zareba W, Piotrowicz K, Turrini P. Electrocardiographic manifestations. In: Marcus FI, Nava A, Thiene G (Eds). Arrhythmogenic Right Ventricular Dysplasia/Cardiomyopathy, Recent Advances. Milano: Springer Verlag;2007. pp. 121-8.
64. Cox MG, Van der Smagt JJ, Wilde AA, et al. New ECG criteria in arrhythmogenic right ventricular dysplasia/cardiomyopathy. Circ Arrhythm Electrophysiol. 2009;2:524-30.
65. Tandri H, Saranathan M, Rodriguez ER, et al. Noninvasive detection of myocardial fibrosis in arrhythmogenic right ventricular cardiomyopathy using delayed-enhancement magnetic resonance imaging. J Am Coll Cardiol. 2005;45:98-103.
66. Sen-Chowdhry S, Prasad SK, Syrris P, et al. Cardiovascular magnetic resonance in arrhythmogenic right ventricular cardiomyopathy revisited: comparison with task force criteria and genotype. J Am Coll Cardiol. 2006;48:2132-40.
67. Corrado D, Basso C, Thiene G, et al. Spectrum of clinicopathologic manifestations of arrhythmogenic right ventricular cardiomyopathy/dysplasia: a multicenter study. J Am Coll Cardiol. 1997;30:1512-20.

68. Sen-Chowdhry S, Syrris P, Prasad SK, et al. Left-dominant arrhythmogenic cardiomyopathy: an under-recognized clinical entity. J Am Coll Cardiol. 2008;52:2175-87.
69. Lerman BB, Stein KM, Markowitz SM. Idiopathic right ventricular outflow tract tachycardia: a clinical approach. PACE. 1996;19:2120-37.
70. Markowitz SM, Litvak BL, Ramirez de Arellano EA, et al. Adenosine-sensitive ventricular tachycardia, right ventricular abnormalities delineated by magnetic resonance imaging. Circulation. 1997;96:1192-200.
71. Thomas KW, Hunninghake GW. Sarcoidosis. JAMA. 2003;289:3300-3.
72. Chapelon C, Piette JC, Uzzan B, et al. The advantages of histological samples in sarcoidosis. Retrospective multicenter analysis of 618 biopsies performed on 416 patients. Rev Med Interne. 1987;8:181-5.
73. Ladyjanskaia GA, Basso C, Hobbelink MG, et al. Sarcoid myocarditis with ventricular tachycardia mimicking ARVD/C. J Cardiovasc Electrophysiol. 2010;21:94-8.
74. Fontaine G, Fontaliran F, Hebert J, et al. Arrhythmogenic right ventricular dysplasia. Annu Rev Med. 1999;50:17-35.
75. Hulot JS, Jouven X, Empana JP, et al. Natural history and risk stratification of arrhythmogenic right ventricular dysplasia/cardiomyopathy. Circulation. 2004;110:1879-84.
76. Peters S. Long-term follow-up and risk assessment of arrhythmogenic right ventricular dysplasia/cardiomyopathy: personal experience from different primary and tertiary centres. J Cardiovasc Med. 2007;8:521-6.
77. Corrado D, Leoni L, Link MS, et al. Implantable cardioverter-defibrillator therapy for prevention of sudden death in patients with arrhythmogenic right ventricular cardiomyopathy/dysplasia. Circulation. 2003;108:3084-91.
78. Corrado D, Basso C, Rizzoli G, et al. Does sports activity enhance the risk of sudden death in adolescents and young adults? J Am Coll Cardiol. 2003;42:1959-63.
79. Wichter T, Paul TM, Eckardt L, et al. Arrhythmogenic right ventricular cardiomyopathy. Antiarrhythmic drugs, catheter ablation, or ICD? Herz. 2005;30:91-101.
80. Marchlinski FE, Zado E, Dixit S, et al. Electroanatomic substrate and outcome of catheter ablative therapy for ventricular tachycardia in setting of right ventricular cardiomyopathy. Circulation. 2004;110:2293-8.
81. Fontaine G, Tonet J, Gallais Y, et al. Ventricular tachycardia catheter ablation in arrhythmogenic right ventricular dysplasia: a 16-year experience. Curr Cardiol Rep. 2000;2:498-506.
82. Dalal D, Jain R, Tandri H, et al. Long-term efficacy of catheter ablation of ventricular tachycardia in patients with arrhythmogenic right ventricular dysplasia/cardiomyopathy. J Am Coll Cardiol. 2007;50:432-40.

**CHAPTER**

# Long QT, Short QT and Brugada Syndromes

# 10

*Seyed Hashemi, Peter J Mohler*

## Chapter Outline

- LQT Syndrome
  - Clinical Manifestations
  - Pathogenesis
  - Molecular Genetics
  - Genotype-Phenotype Correlation Studies and Risk Stratification Strategies
  - Diagnosis
  - Genetic Testing
  - Therapy
  - ICD Therapy
  - Left Cardiac Sympathetic Denervation
  - Genotype-Specific Therapy
- SQT Syndrome
  - Clinical Manifestations
  - Molecular Genetics
  - Pathogenesis
  - Diagnosis
  - Therapy
- Brugada Syndrome
  - Clinical Manifestations
  - Genetics
  - Pathogenesis
  - Diagnosis
  - Prognosis, Risk Stratification and Therapy

## INTRODUCTION

Over the past two decades, ample information has been accumulated on cellular mechanisms and genetics of arrhythmias in structurally normal heart. The basic pathogenic mechanism for these arrhythmias may involve hereditary disturbances in ionic currents at the cellular level while the heart remains grossly normal. The high rate of sudden death (especially in the young) due to congenital arrhythmias, coupled with the potential availability of preventive measures, mandate the need for higher awareness of the medical community of these potentially lethal arrhythmia syndromes. In this chapter, we will review the current state of understanding of inherited arrhythmias including long QT (LQT) syndrome, short QT (SQT) syndrome and Brugada syndrome. This review focuses on inherited arrhythmias and will not cover acquired LQT syndrome.

## LQT SYNDROME

Jervell and Lange-Nielsen, in 1957, firstly described the congenital LQT syndrome in a Norwegian family with four members suffering from prolonged QT, syncope and congenital deafness.[1] Three of the four affected patients died suddenly at the age of 4, 5 and 9 years.[1] Jervell and Lange-Nielsen syndrome, is inherited in an autosomal recessive pattern. Several years later, Romano et al. and Ward et al. independently described a similar syndrome but without deafness and with an autosomal dominant pattern of inheritance.[2,3] The underlying genes for LQT syndrome, however, were not discovered until more recently; in 1995 and 1996, the first three genes associated with the most common forms of the LQT syndromes (types 1, 2 and 3) were identified.[4–6] Since then, the scientific and medical community has witnessed discovery of hundreds of variants in nearly a dozen genes associated with a wide variety of LQT or related arrhythmia syndromes.

## Clinical Manifestations

The congenital LQT syndrome is a common identifiable cause of sudden death in the presence of structurally normal heart.[7] The natural history of LQT syndrome is highly variable.[8–12] The majority of patients may be entirely asymptomatic with the only abnormality being QT prolongation in the ECG.[8–12] Some gene variant carriers of LQT syndromes may not even display the prolonged QT interval (silent carriers).[13,14] Symptomatic patients typically, present in the first two decades of life including the neonatal period, with recurrent attacks of syncope precipitated by *torsade de pointes* type of ventricular arrhythmias.[8,11] This form of tachycardia is characterized by cyclical changes in the amplitude and, polarity of QRS complexes such that their peak appears to be twisting around an imaginary isoelectric baseline. *Torsade de pointes* may resolve spontaneously, however, it has a great potential to degenerate into ventricular fibrillation and is an important cause of sudden death.[9]

## Pathogenesis

As the QT interval represents a combination of action potential (AP) depolarization and repolarization, variations in QT interval may arise from the dysfunction of ion channel, responsible for the timely execution of the cardiac AP. A decrease in the outward repolarizing currents (mainly potassium currents) or an increase in the inward depolarizing currents (mainly sodium and calcium) may increase action potential duration (APD) and QT prolongation. The increases in APD result in lengthening of effective refractory period (ERP) that in turn predisposes to the occurrence of early after depolarizations (EADs), due to enhancement of the sodium-calcium exchanger (NCX) current and reactivation of the L-type calcium channels.[15–18] These EADs are known to support ventricular arrhythmias.[16–18]

## Molecular Genetics

Over the last fifteen years, gain- or loss-of-function variants in nearly a dozen genes have been associated with development of LQTS. LQT1 is the most common form of the LQT syndrome and results from loss-of-function variants in *KCNQ1*, which encodes the alpha subunit of $I_{Ks}$, the cardiac slowly activating delayed-rectifier potassium channel current.[6] The mechanism(s) by which, each variant causes decreased $I_{Ks}$ current varies among the gene variant carriers. Variant sub-units may co-assemble with the wild-type protein and render them defective causing more than 50% loss-of-function (i.e. dominant-negative effect).[19] Alternatively, the variants may result in haploinsufficiency with ~ 50% reduction in protein expression and the resultant current.[19] In addition to the biophysical function (dominant-negative vs haploinsufficiency), the location of variants appears to significantly influence the severity of phenotype. For example, Moss et al. demonstrated significantly higher cardiac event rates in patients with transmembrane variants in *KCNQ1* gene[19] **(Fig. 1)**.

LQT2 results from loss-of-function variants in *KCNH2* (also known as *HERG*), which encodes the alpha-subunit of $I_{Kr}$, the rapidly activating delayed-rectifier potassium current in the heart.[5] The loss-of-function in the genes responsible for $I_{Ks}$ and $I_{Kr}$ reduces the outward potassium current and prolongs APD, leading to QT prolongation in LQT1 and LQT2, respectively[5,6] **(Fig. 2)**.

LQT3 arises from variants in *SCN5A* that encodes the alpha-subunit of $Na_V1.5$, the primary cardiac voltage-gated

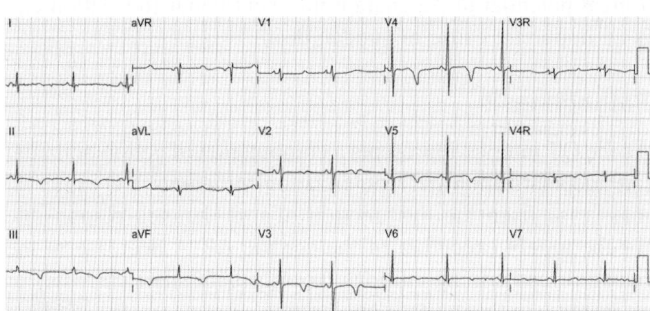

**FIGURE 1:** LQT1 ECG belongs to a 7-year-old boy with history of cardiac arrest during swimming. Note the prolonged QT with inverted, broad-based and T-wave pattern

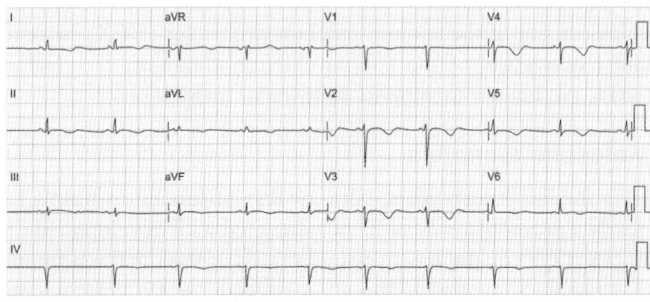

**FIGURE 2:** LQT2 ECG belongs to a 19-year-old female with history syncope and polymorphic ventricular tachycardia. ECG shows QT prolongation with low-amplitude inverted T-waves

sodium-channel.[4] These variants disrupt fast inactivation of $Na_V1.5$ leading to excess late inward sodium current that in turn results in prolonged repolarization and APD.[4] The three most common LQTS, i.e. LQT 1–3, vary significantly in their natural history and clinical presentation, which will be discussed later in this chapter.

Unlike LQT1–3, LQT4 is not caused by an ion channel gene variant. LQT4 arises from variants in *ANK2*, which encodes ankyrin-B in cardiomyocytes.[20] The human *ANK2* gene was the first LQT syndrome gene that was discovered to encode a membrane associated protein (ankyrin-B) rather than an ion channel or channel subunit.[20] Ankyrin-B is an adaptor protein that interacts with several membrane-associated ion channels and transporters in ventricular myocytes including $Na^+/K^+$ ATPase, $Na^+/Ca^{2+}$ exchanger-1 (NCX1) and IP3 receptors.[20] Dysfunction of Na/K ATPase and NCX1 are associated with a significant increase in $[Ca^{2+}]_i$ transient amplitude, SR calcium load and catecholamine-induced after depolarizations.[20] Abnormal intracellular calcium homeostasis is thought to be the central mechanisms underlying ventricular arrhythmias.[20] Symptomatic patients with specific *ANK2* variants may display significant QT prolongation (mean QTc: $490 \pm 30$ ms), ventricular tachycardia, syncope and sudden death.[21] However, many variant carriers do not display prolonged QTc, but display other ventricular phenotypes with risk of syncope and death. Additionally, *ANK2* variant carriers may manifest with sinus node dysfunction and/or atrial fibrillation in addition to ventricular arrhythmias and sudden death, hence, the name ankyrin-B syndrome.[20,21] Notably, ventricular phenotypes are often triggered by catecholamines, and thus, ankyrin-B syndrome may ultimately be more appropriately described as a class of catecholaminergic polymorphic ventricular tachycardia (CPVT).

LQT5 and LQT6 arise from loss-of-function variants in *KCNE1* and *KCNE2*, that encode the beta subunit of $I_{Ks}$ and $I_{Kr}$, respectively (same currents in which the alpha subunit variants cause LQT1 and LQT2).[22–24] Akin to LQT1 and LQT2, these variants reduce outward potassium current leading to subsequent QT prolongation.[22–24]

LQT7 arises from loss-of-function variants in *KCNJ2* that encodes inward rectifying potassium channels (Kir2.1), responsible for $I_{K1}$.[25] $I_{K1}$ represents the major ion conductance in the later stages of repolarization and during diastole, and reduced $I_{K1}$ is associated with QT prolongation. Linkage studies on patients with LQT7 variants demonstrate a wide range of extra-cardiac findings associated with this form of LQTS.[25,26] These patients suffer from an autosomal dominant multisystem disease, also known as Andersen-Tawil syndrome, characterized by a combination of potassium-sensitive periodic paralysis, cardiac arrhythmia and distinctive facial or skeletal dysmorphic features such as low set ears and micrognathia.[25,26]

LQT8 is related to variants in *CACNA1c* that encodes the alpha-1C subunit of the voltage-gated calcium channel (CaV1.2) responsible for L-type calcium current ($I_{Ca,L}$) in myocytes.[27] These variants are associated with loss of voltage-dependent CaV1.2 inactivation, leading to $Ca^{2+}$ overload and delayed repolarization due to prolonged inward, $Ca^{2+}$ current during the plateau phase of the AP.[27] Similar to LQT7 syndrome, patients with LQT8 variants display a variety of extra-cardiac signs and symptoms (also termed Timothy syndrome) including syndactyly, abnormal teeth, immune deficiency, intermittent hypoglycemia, cognitive abnormalities, autism and baldness at birth[27] consistent with the critical role of $I_{Ca,L}$ in other tissues. Cardiac manifestations include patent foramen ovale (PFO) and septal defects, in addition to ventricular arrhythmias.[28] The condition is severe, with most affected patients dying in early childhood.[27,28]

LQT9 is associated with variants in *CaV3*, that encodes caveolin-3.[29] Caveolins are the principal proteins required for the assembly of caveolae, 50–100 nm membrane invaginations involved in the localization of membrane proteins including $Na_v1.5$ (LQT3 associated channel).[29,30] These variants interfere with the regulatory pathways between caveolin-3 and $Na_v1.5$, disrupting inactivation of $Na_v1.5$, resulting in a gain-of-function effect on late $I_{Na}$; the same pathological mechanism that underlies LQT3.[29]

LQT10 is linked to variants in *SCN4B*, which encodes $Na_v1.5$ one of four auxiliary subunits of $Na_v1.5$.[31] $Na_v\beta$ dysfunction is associated with a significant increase in late sodium current that affects the terminal repolarization phase of the AP, and prolongs the QT interval by a similar mechanism as LQT3—associated variants in the alpha subunit of $Na_v1.5$.[31]

LQT11 is associated with variants in *AKAP9*, that encodes A-kinase anchoring protein (AKAP), also known as yotiao, involved in the subcellular targeting of protein kinase A (PKA).[32] Yotiao is a PKA targeting protein for multiple cardiac ion channel complexes including the ryanodine receptor, the L-type calcium channel, and the slowly activating delayed rectifier $I_{Ks}$ potassium channel (*KCNQ1*).[32,33] Variants in the *AKAP9* are associated with disruption of the interaction between *KCNQ1* and yotiao, reducing the cAMP-induced phosphorylation of the channel, that in turn eliminates the functional response of the $I_{Ks}$ channel to cAMP, prolongs the APD and QT interval.[32,33]

LQT12 is associated with variants in *SNTA1*, which encodes for α1-syntrophin, a scaffolding protein with multiple molecular interactions including $Na_v1.5$, plasma membrane $Ca^{2+}$—ATPase

(PMCA4b) and neuronal nitric oxide synthase (nNOS).[34] The variants in *SNTA1* are associated with increased direct nitrosylation of Na$_v$1.5 and increased late I$_{Na}$.[34] Akin to the mechanism in LQT3 syndrome, the increase in late sodium current causes prolonged QT interval.

## Genotype–Phenotype Correlation Studies and Risk Stratification Strategies

The pattern of inheritance of LQTS varies depending on the type of the syndrome. Most LQTS are inherited as autosomal dominant Romano-Ward syndrome. LQT syndrome types 1 and 5 (representing variants in alpha and beta subunit of I$_{Ks}$) are inherited as either autosomal recessive Jervell and Lange-Nielsen or autosomal dominant Romano-Ward syndrome.[35] Additionally, a host of factors may influence disease severity. Recently, the genotype-phenotype correlation studies on the most common forms of LQTS (type 1–3) have allowed for more in-depth understanding of natural history of each variant. For example, Priori et al. prospectively studied a large data base of unselected, consecutively, genotyped patients with LQTS (n = 647) and developed a risk stratification scheme based on gender, genotype and QTc interval after a mean observation period of 28 years.[13] The authors showed that different genotypes may manifest differently in males versus females. For example, the incidence of a first cardiac arrest or sudden death was greater among LQT2 females than LQT2 males and LQT3 males than LQT3 females.[13]

The duration of QT interval may be influenced by the genetic locus, and may also predict the likelihood of future cardiac events (defined as syncope, cardiac arrest or sudden death). In the Priori study, mean QTc was 466 ± 44 msec in LQT1, 490 ± 49 msec in LQT2 and 496 ± 49 msec in LQT3.[13] Event free survival was higher in LQT1 than LQT2 and LQT3.[13] Within each LQTS category, QTc of patients with cardiac events was significantly, longer than asymptomatic patients.[13] Amongst LQT1 patients, mean QTc was 488 ± 47 msec in those with cardiac events versus 459 ± 40 msec in asymptomatic subjects.[13] These data suggest that LQTS may have a normal or near normal QTc and sustain a cardiac event (albeit at a very low rate) and vice versa. However, irrespective of the genotype, the risk of becoming symptomatic was associated with QTc duration; a QTc of 500 msec or more was the most significant predictor of potential cardiac events.[13]

Notably, the percentage of silent variant carriers (those with gene variants but normal QT interval) was higher in the LQT1 (36%) than LQT2 (19%) or LQT3 (10%).[13] Higher percentage of silent carriers in LQT1 may at least partly explain the lower rate of cardiac events in patients with LQT1 compared to LQT2 and LQT3.[14,36-38] The fact that silent variant carriers may have normal QT interval, yet to be at increased risk of cardiac events indicates that LQTS cannot be excluded solely based on ECG findings. Furthermore, the silent carrier state may confer susceptibility of drug-induced QT prolongation and Torsade de pointes arrhythmias.[36,38,39]

Triggers of cardiac events in LQT syndrome have been shown to be largely gene specific. Schwartz et al. studied specific triggers of cardiac events in 670 LQTS patients (types 1, 2 and 3) with known genotype.[40] In LQT1, nearly 80% of cardiac events occurred during physical or emotional stress, whereas LQT3 patients experience 40% of their events at rest or during sleep and only 13% during exercise.[40] In LQT2 patients, the events occurred during emotional stress in 43% of patients. For lethal cardiac events (cardiac arrest and sudden death), the difference among the groups were more dramatic. In LQT1, 68% of lethal events occurred during exercise, whereas this rarely occurred for LQT2 and occurred in only 4% of cases for LQT3 patients.[40] In contrast, 49% and 64% of lethal events occurred during rest/sleep without arousal for LQT2 and LQT3 patients, respectively, whereas this occurred in only 9% of cases for LQT1 patients.[40] Auditory stimuli particularly clustered among LQT2 patients, whereas swimming as a trigger was more frequent in LQT1 patients.[40] A stunning percentage of patients who experienced their cardiac events during swimming were LQT1.[40]

| Criteria | Point |
|---|---|
| *ECG criteria* | |
|   QTc | |
|     >480 | 3 |
|     460–479 | 2 |
|     450–459 | 1 |
|   Torsade de pointes | 2 |
|   T-wave alternans | 1 |
|   Notched T-wave in 3 leads | 1 |
|   Low heart rate for age | 0.5 |
| *Clinical history* | |
|   Syncope with stress | 2 |
|   Syncope without stress | 1 |
|   Congenital deafness | 1 |
| *Family history* | |
|   Definite LQT syndrome in family | 1 |
|   Unexplained SCD <30 y/o in immediate family | 0.5 |
| *Scoring* | |
| ≤1 point = low probability for LQTS | |
| 2–3 points = intermediate probability for LQTS | |
| ≥3.5 points = high probability for LQTS | |

The T-wave repolarization pattern varies according to genotype. Patients with LQT1 variant positive genotype display a distinct, inverted, broad-based, prolonged T-wave pattern that is different from the low-amplitude and sometimes, notched T-wave observed in LQT2 patients.[41] Both of these repolarization patterns are different from late-appearing T-wave seen in LQT3 patients.[41] Patients with LQT4 genotype display a characteristic notched, biphasic T-wave morphology in ECG.[21]

## Diagnosis

The typical case of LQTS, characterized by syncope or cardiac arrest associated with QT prolongation on ECG is fairly straightforward to diagnose. However, borderline cases may be more complex and pose a diagnostic challenge to the practicing clinician. Schwartz and his colleagues devised a diagnostic criteria based on a scoring system first in 1985 and then, updated in 1993.[37,42] Based on this scoring system, a score of one or less indicates low probability for LQTS; 2–3 denotes intermediate probability and higher than 3.5 indicates high probability for LQTS. If a patient receives a score of 2–3, serial ECG and 24-h Holter monitoring may be obtained as the QT interval may vary from time to time.[38] Short-term variability of QT interval has recently been demonstrated to correlate with high risk LQT syndrome.[43]

## Genetic Testing

The diagnostic criteria based on ECG and clinical history were primarily devised before the human genome project era and therefore, may not always account for many new advances in molecular genetics. As mentioned earlier, individuals may harbor disease-associated variants and yet have normal ECG parameters and QT interval (silent carriers). In select cases, genetic testing and molecular diagnostic methods may complement the ECG and clinical criteria; allowing for screening of proband family members to detect silent variant carriers that may predispose individuals to potential events.[36,39,44,45] For example, HERG inhibition is commonly the mechanism associated with drug-induced QT prolongation, and variants in other ion channel/ion channel modulator genes may also predispose individuals to QT prolongation and ventricular arrhythmias.[36,45,46] Therefore, identifying gene variants that promote arrhythmia susceptibility (either congenital or acquired) may provide important information to a physician in their clinical practice (i.e. avoiding QT prolonging drugs in patients harboring specific channel variants). It is important to note that current genetic testing for arrhythmias may harbor its own drawbacks. For example, false negative results may occur when the patient has a variant in a gene not covered in the testing panel (the relevant gene or gene variant may not have even been discovered!). Moreover, the significance of a positive test result may often be difficult to ascertain. As reviewed by others, it will be critical to continue to define genotype-phenotype relationships to provide additional new data that can be carefully considered when utilizing patient genotype to predict and/or manage clinical phenotypes.

## Therapy

As the risk of cardiac events in LQTS is genotype, age and gender dependent, therapy should be carefully tailored to the individual patients according to their risk factors. According to a recently published study from the International LQTS Registry, beta blocker therapy, significantly, reduces the risk of cardiac events in LQT1 and LQT2 patients.[47] This is not surprising as the most common triggers of cardiac events in LQT1 and LQT2 patients are exercise and emotional stress, respectively.[40] Furthermore, LQT1 patients harbor $I_{Ks}$ dysfunction, which has been shown to activate in higher heart rates and is necessary for QT interval shortening with tachycardia.[6] In contrast, beta blockers may offer limited efficacy among LQT3 patients; as they display further QT prolongation at slower heart rates.[48] Moreover, according to the International LQT Registry data, beta blocker therapy reduces the risk to similar extent in LQT1 and LQT2 patients (67% and 71% risk reduction, respectively).[47] Different beta blockers displayed differential effects in each category of LQTS. Atenolol, but not nadolol, reduced the risk significantly in LQT1 patients, whereas nadolol, but not atenolol was associated with a significant risk reduction in LQT2 patients.[47] Higher risk patients, such as LQT1 males and LQT2 females gained more benefit from beta blocker therapy compared to lower risk subsets. Despite the significant risk reduction with beta blocker therapy, high risk patients experienced considerable residual event rates during beta blocker therapy.[47] History of syncope during beta blocker therapy was associated with higher event rates.[47] LQT2 genotype was associated with significantly higher residual event rates while taking beta blockers compared to LQT1.[47,49]

## Implantable Cardioverter Defibrillator (ICD) Therapy

Insofar, as high risk patients with LQT syndrome continue to have a residual event rate while receiving beta blocker therapy, there may be a need for additional protection against potentially fatal arrhythmias. Current guidelines recommend ICD therapy as a class IIa indication for primary prevention of cardiac events in LQTS patients who experience syncope or ventricular tachycardias during beta blocker therapy.[50] These guidelines provide a class IIb recommendation for ICD therapy in patients with risk factors for SCD, irrespective of medical therapy.[50]

## Left Cardiac Sympathetic Denervation

Left cardiac sympathetic denervation (LCSD) was introduced in 1971, as the first therapy for LQT syndrome.[51] The contemporary LCSD techniques use extrapleural approach and obviate the need for thoracotomy.[52] A recent study of 147 very high-risk LQTS patients, who underwent LCSD over a span of 35 years (average follow-up period of 8 years) demonstrated that LCSD reduced the number of cardiac events by 91% per patient per year.[52] According to the result from this study, LCSD may be considered in patients with recurrent syncope despite beta-blockade, and in patients, who experience arrhythmia storms with ICD therapy.[52]

## Genotype-specific Therapy

As cardiac events may be clustered around exercise or emotional stress in LQT1 patients, these individuals may be advised to avoid competitive sports and/or stressful situations. For example, swimming has previously been particularly discouraged in LQT1 patients. Beta blockers remain the mainstay of therapy in LQT1 syndrome.

In patients with LQT2, maintaining adequate serum potassium level is essential, as $I_{Kr}$ activity may vary with serum potassium levels.[53] Therefore, use of potassium supplements in combination with potassium sparing diuretics may be recommended in LQT2 patients.[53] Since arousal from sleep, especially with a sudden noise may be a triggering a risk factor in LQT2 patients, the use of alarm clock or telephone in the patient's bedroom should also be carefully considered.[40]

Sodium channel blockers have been proposed for gene-specific treatments in LQT3, which is associated with variants in the sodium channel gene (*SCN5A*).[48] Early clinical studies demonstrated efficacy of mexiletine or flecainide in shortening of repolarization period and QT interval.[48] Indeed, ACC/AHA 2006 guidelines for management of patients with ventricular arrhythmias and the prevention of sudden cardiac death recommended sodium channel blockers for treatment of LQT3 patients as a class IIb indication.[54] However, more recently, Ruan et al. in an elegant study, provided *in vitro* cellular evidence that different *SCN5A* variants may display heterogeneous biophysical properties; and the use of sodium channel blockers may be deleterious in selected group of LQT3 patients.[55] The study was prompted by the death of a young child affected by an *SCN5A* variant whose QT interval not only shorten, but also prolonged in response to mexiletine treatment.

## SQT SYNDROME

SQT syndrome is a rare channelopathy associated with increased risk of atrial and ventricular arrhythmias. The association of SQT interval with sudden cardiac death was first described near two decades ago by Algra and his colleagues.[56] They reported a two-fold risk of sudden death in patients with a QTc less than 400 milliseconds, as compared with patients with a QTc between 400 and 440 milliseconds.[56] In the year 2000, Gussak et al. reported the first familial cases of idiopathic SQT syndrome associated with paroxysmal atrial fibrillation. A few years later, Gaita et al. described additional cases of SQT syndrome associated with sudden cardiac death.[57] To date, the number of identified patients with SQT syndrome is low.[58,59] However, with increasing awareness of medical community of the relationship of SQT with AF and sudden cardiac death, the prevalence is expected to rise.

### Clinical Manifestations

The clinical manifestations of SQT syndrome include propensity to AF, syncope and sudden death.[57,60,61] In most reported cases, the QTc was less than 320 ms and often less than 340 ms.[62,63] Therefore, it is prudent to suspect SQT syndrome in patients with a QT interval of less than 340 ms and personal and/or family history of lone AF, ventricular fibrillation, syncope or sudden cardiac death. To date, there is no gender predilection for SQT syndrome.[63] Age at onset of symptoms vary widely with reported cases from one year old (sudden infant death syndrome) to age 80 year old.[63] One study reported the mean age at diagnosis of 30 years.[63] Cardiac arrest has been reported to occur both at rest and under stress.[63,64]

### Molecular Genetics

To date, three genes with an association with SQT syndrome have been identified. All three genes encode potassium channel proteins. SQT1 is associated with variants in *KCNH2* (also LQT2 gene), that result in increases in $I_{Kr}$.[60] SQT2 is associated with variants in *KCNQ1* (also LQT1 gene) that result in increased $I_{Ks}$.[65] SQT3 is associated with variants in *KCNJ2* (also LQT7 gene) that encodes the inwardly rectifying potassium channel protein, Kir2.1.[66] Gain-of-function variants in *KCNJ2* may result in increased outward $I_{K1}$ current and SQT syndrome type 3.[66]

### Pathogenesis

Gain-of-function variants in specific cardiac potassium channels may cause acceleration of repolarization and abbreviation of APD leading to shortening of ERP.[60,61,65,66] Shortened refractory period is a well established substrate for re-entrant tachycardias; hence, predisposition to atrial fibrillation and ventricular tachycardias in patients with SQT syndrome.[67] A second proposed mechanism for predisposition to re-entrant arrhythmias in SQT syndrome is the increases in transmural dispersion of repolarization. The ECG of affected individuals has distinctive features including tall, peaked, symmetrical T-waves with prolonged $T_{peak}$-$T_{end}$.[68] Prolonged $T_{peak}$-$T_{end}$ has been proposed to be indicative of augmented transmural dispersion of repolarization.[68] Exaggerated transmural heterogeneity during repolarization forms the substrate for the development of re-entrant arrhythmias.[68] Extramiana and colleagues demonstrated that QT-interval abbreviation in the absence of transmural dispersion of repolarization was not sufficient to induce ventricular arrhythmias.[68] Therefore, the combination of short refractory periods and increased dispersion of refractoriness may result in patients with SQT syndrome vulnerable to arrhythmias.

## Diagnosis

The precise cut-off point for QT interval in SQT syndrome is still somewhat debated. Currently, based on several reports, the upper limit of QT interval suggestive of SQT syndrome is considered 320–340 ms.[62,63] However, the mere presence of SQT interval does not necessarily appear to be sufficient to make the diagnosis. Anttonen et al. screened a population of over 1000 healthy volunteers for SQT interval and followed them up for a mean of 29 years.[69] The prevalence of QTc interval less than 320 ms (very short) and less than 340 ms (short) was 0.10% and 0.4%, respectively.[69] All cause or cardiovascular mortality did not differ between subjects with a very short or SQT interval and those with normal QT intervals (360–450 ms).[69] There were no sudden cardiac deaths, aborted sudden cardiac deaths, or documented ventricular tachyarrhythmias among subjects with SQT interval.[69]

In addition to shortened QT interval, patients with SQT syndrome may display a peculiar ECG morphology.[62,70,71] Affected patients often demonstrate absent ST segment with the T-wave attached to the S-wave.[71,72] A second finding, that is seen in at least about half of the patients, is a tall, peaked, narrow-based T-waves in the right precordial leads.[69,70,72] Another distinctive ECG feature of patients with SQT syndrome is the relatively prolonged $T_{peak}$-$T_{end}$ interval which may indicate enhanced transmural dispersion of repolarization.[68] Electrophysiological studies have been reported in a limited number of patients with SQT syndrome. Both atrial and ventricular ERP were reported to be shortened.[61,63,73] Furthermore, ventricular tachycardias were inducible in nearly all patients.[61,63,73]

As part of the diagnostic evaluation of SQT syndrome, acquired causes of QT interval shortening are often excluded. Electrolyte/acid-base abnormalities, such as hyperkalemia, hypercalcemia and acidosis, are well known to shorten QT interval. Other causes include hyperthermia and QT shortening medications such as digoxin and mexiletine. Finally, QT measurements are commonly made at heart rates <80 beat/min, as the QT interval in SQT syndrome patients may fail to adapt to increase heart rates.

## Therapy

The paucity of SQT syndrome cases may limit the opportunity to systematically study treatment of this recently recognized arrhythmia syndrome. Nonetheless, drugs that block outward potassium current and prolong repolarization seem attractive and have been tested in a limited number of cases. The class Ia anti-arrhythmic agents, quinidine and disopyramide have been demonstrated to prolong QT interval and ventricular ERP and reduce inducibility of ventricular arrhythmias.[63,74–76]

The high incidence of fatal cardiac events associated with SQT suggests the use of ICD therapy, early on, in the management of the symptomatic patients.[62] In asymptomatic patients, however, the indications for ICD may be less clear. Patients with SQT interval and implanted ICD may be at increased risk for inappropriate therapy due to oversensing as a result of the detection of short-coupled and prominent T-waves.[77] Reprogramming of the ICD with adaptation of sensing levels and decay delays without sacrificing correct arrhythmia detection may be helpful in these patients.[77]

## BRUGADA SYNDROME

In 1992, Brugada and Brugada described a hereditary arrhythmia syndrome characterized by ST segment elevation in the right precordial leads, right bundle branch block and increased vulnerability to ventricular tachycardias and sudden death in the absence of any structural heart disease.[78] Although the Brugada brothers are the first to formally describe and characterize the syndrome, the history of the syndrome dates back to several decades prior. A similar syndrome manifested as sudden death during sleep frequently after a heavy meal, most often affecting young men, has long been noted in the south Asian culture. The terms sudden unexplained nocturnal deaths (SUND) or sudden unexplained death in sleep (SUDS) are used to explain this folk illness with various local names including *Bangungot* (in Philippines), *Pokkuri* (in Japan) or *Lai Tai* (in Thailand). Although Brugada syndrome seems to be endemic in south-east Asian countries, cohorts of the syndrome have been reported across the world.[79] Currently, Brugada syndrome is considered as a major cause of sudden cardiac death in the young. Timely identification of symptomatic Brugada syndrome patients is important, as implantable cardioverter defibrillators (ICD) may be life-saving in these individuals.

## Clinical Manifestations

Brugada syndrome is characterized by the occurrence of polymorphic ventricular tachycardias in patients with the ECG patterns of a peculiar ST-segment elevation in right precordial leads and right bundle branch block (RBBB).[78] An increased propensity to atrial fibrillation and supraventricular arrhythmias has also been reported.[80] Patients with Brugada syndrome have structurally normal hearts; and are typically, otherwise healthy and active.[80] Notwithstanding, recent research suggests that with the use of high resolution magnetic resonance imaging, subclinical structural abnormalities in right ventricle may be identified.[81] Many patients with the syndrome may have the characteristic ECG findings; however, remain asymptomatic until the first arrhythmic episode that may lead to syncope or sudden death. On the other hand, the symptomatic patients with positive ECG findings may transiently display normal ECG which makes the diagnosis more challenging.

## Genetics

Brugada syndrome is a familial arrhythmia syndrome with autosomal dominant pattern of inheritance, incomplete and gender-dependent penetrance. The mean age of clinical manifestations is 40 years with a wide range from infancy to the eighth decade of life.[82,83] Men are affected much more commonly than women with a male to female ratio of 3/1.[82,83] The true prevalence of the disease is unknown. A great deal of work has been published during the last two decades, since the Brugada brothers' authored the initial report.

In 1998, Chen et al. identified the first loss-of-function gene variant related to the Brugada syndrome on *SCN5A,* that encodes cardiac voltage gated sodium channels.[84] Since then, over 100 associated variants have been reported in the literature with 15–30% of them located on *SCN5A* gene.[85,86] Another 11–12% have been attributed to *CACNA1C* and *CACNB2*.[85] Variants in other genes (*GPD1L, SCN1B, KCNE3* and *SCN3B*) likely contribute to the Brugada phenotype, although to a lesser extent.[85] Notably, all the genes discovered to date explain only one-third of Brugada syndrome cases, indicating that there is, still an important amount of work to be done to unravel the genetic basis of this lethal disease.

## Pathogenesis

In 2001, Antzelevitch et al. proposed the dispersion of repolarization model, which hypothesizes a pathophysiological mechanism of re-entrant arrhythmias in Brugada syndrome.[87] This model is based on the demonstration that the density and kinetics of currents underlying phase-1 of AP (Ito), exhibit transmural dispersal.[88] The Ito current density is more profound in epicardium compared to endocardium.[88] In Brugada syndrome, the impaired sodium influx in epicardial cells is subject to exaggerated Ito defect leading to accentuated AP morphology variability between epicardial cells and endocardial cells. The arrhythmic substrate is, therefore, the result of increased transmural heterogeneity of the currents involved in the phase-I depolarization of the ventricle, enabling local re-excitation via re-entry.[87,89]

## Diagnosis

Electrocardiographic signs of Brugada syndrome are classified into three types as follows:[80]
- Type I: Coved ST-segment elevation greater than 2 mm followed by negative T-wave in greater than 1 mm right precordial lead (V1–V3)
- Type 2: Saddleback ST-segment elevation with a high takeoff ST-segment elevation of greater than 2 mm, a trough displaying greater than 1 mm ST-elevation followed by a positive or biphasic T-wave
- Type 3: Saddleback or coved appearance of ST-segment elevation less than 1 mm, present in greater than 1 mm right precordial lead (V1–V3).

Type 2 ST-segment elevation is less specific and more common in general healthy population.[80] Type 1 (coved type) ST-segment elevation is more specific and more predictive of future arrhythmic events, and is considered the diagnostic ECG abnormality for Brugada syndrome.[80] The coved type ST-elevation is less sensitive owing to its dynamic nature. In up to 50% of patients with coved ST-segment elevation, the ECG may normalize or the ST-segment elevation may convert from the coved type to the saddle type periodically.[80] However, the coved-type ECG pattern, can be unmasked by administration of sodium channel blockers, ajmaline, flecainide or procainamide in the electrophysiology laboratory.[90] Additionally, vagotonic agents and fever are known to bring about the ECG signs when concealed.[91,92]

Brugada syndrome is diagnosed on the basis of a spontaneous or drug-induced type 1 (coved-type), ST-segment elevation in the right precordial leads plus one of the following conditions:[80]

- Documented VF or polymorphic VT
- Unexplained syncope
- Nocturnal agonal respiration
- Inducibility of VT/VF with programmed electrical stimulation
- A family history of SCD at a young age (<45 years) or a coved-type ECG pattern.

The differential diagnosis of syncope and the ECG abnormalities is broad and the following conditions may be considered and ruled out: atypical right bundle branch block, left ventricular hypertrophy, early repolarization, acute pericarditis, acute myocardial ischemia or infarction, pulmonary embolism, printzmetal angina, dissecting aortic aneurysm, central or peripheral nervous system abnormalities, duchenne muscular dystrophy, thiamine deficiency, hyperkalemia, hypercalcemia, arrhythmogenic right ventricular cardiomyopathy, pectus excavatum, hypothermia, or mechanical compression of the right outflow tract (RVOT) as seen with mediastinal tumors or hemopericardium.[80]

## Prognosis, Risk Stratification and Therapy

Patients displaying the Brugada syndrome, ECG pattern were initially thought to carry a high risk of cardiac events. The second consensus conference report on Brugada syndrome recommended electrophysiology studies (EPS) as a valuable tool in risk stratifying asymptomatic patients with spontaneous type 1 ECG pattern or with drug induced type 1 ECG pattern plus positive family history of SCD.[80] Subsequent studies, however, have questioned the role of EPS in risk stratification of asymptomatic patients.[93] The role inducibility of ventricular arrhythmias by EPS remains debatable. Recently, the investigators of the FINGER Brugada syndrome registry addressed the long-term prognosis of Brugada syndrome and the role of EPS in risk stratifying asymptomatic patients.[93] In the largest cohort of symptomatic and asymptomatic patients with Brugada syndrome to date, following a 32-month follow-up period of the cohort, they demonstrated the following results:[93]

- The risk of arrhythmic events is low in asymptomatic patients (0.5% event rate per year)
- The presence of symptoms and a spontaneous type 1 ECG are the only independent predictors of arrhythmic events
- Genders, family history of SCD, inducibility of ventricular tachyarrhythmias during EPS and presence of a variant in the *SCN5A* gene, have no predictive value.

In view of these results, the risk stratification strategy proposed in the second consensus report may be revised to reflect the decreased value of EPS as a predictor of future cardiac events. Recommendations to implant ICD at the present time may be limited to symptomatic patients with type 1 ECG pattern. To date, no pharmacologic intervention has been approved for the treatment of Brugada syndrome. However, active research is underway to define potential pharmacologic options to treat this potentially lethal arrhythmia syndrome.[94–96]

## ACKNOWLEDGMENTS

The authors thank Drs Ian Law and Nicholas Von Bergen of the University of Iowa Carver College of Medicine for LQT1 and LQT2 ECGs.

## REFERENCES

1. Jervell A, Lange-Nielsen F. Congenital deaf-mutism, functional heart disease with prolongation of the Q-T interval and sudden death. Am Heart J. 1957;54:59-68.
2. Romano C, Gemme G, Pongiglione R. Rare cardiac arrhythmias of the pediatric age. II. Syncopal attacks due to paroxysmal ventricular fibrillation (presentation of 1st case in Italian pediatric literature). Clin Pediatr (Bologna). 1963;45:656-83.
3. Ward OC. A new familial cardiac syndrome in children. J Ir Med Assoc. 1964;54:103-6.
4. Wang Q, et al. SCN5A mutations associated with an inherited cardiac arrhythmia, long QT syndrome. Cell. 1995;80:805-11.
5. Curran ME, et al. A molecular basis for cardiac arrhythmia: HERG mutations cause long QT syndrome. Cell. 1995;80:795-803.
6. Wang Q, et al. Positional cloning of a novel potassium channel gene: KVLQT1 mutations cause cardiac arrhythmias. Nat Genet. 1996;12:17-23.
7. Tester DJ, Ackerman MJ. Postmortem long QT syndrome genetic testing for sudden unexplained death in the young. J Am Coll Cardiol. 2007;49:240-6.
8. Schwartz PJ, Periti M, Malliani A. The long Q-T syndrome. Am Heart J. 1975;89:378-90.
9. Moss AJ, Schwartz PJ. Sudden death and the idiopathic long Q-T syndrome. Am J Med. 1979;66:6-7.
10. Moss AJ, Schwartz PJ. Delayed repolarization (QT or QTU prolongation) and malignant ventricular arrhythmias. Mod Concepts Cardiovasc Dis. 1982;51:85-90.
11. Moss AJ, et al. The long QT syndrome: a prospective international study. Circulation. 1985;71:17-21.
12. Moss AJ, et al. The long QT syndrome. Prospective longitudinal study of 328 families. Circulation. 1991;84:1136-44.
13. Priori SG, et al. Risk stratification in the long-QT syndrome. N Engl J Med. 2003;348:1866-74.
14. Mohler PJ, et al. A cardiac arrhythmia syndrome caused by loss of ankyrin-B function. Proc Natl Acad Sci USA. 2004;101:9137-42.
15. Viswanathan PC, Rudy Y. Pause induced early afterdepolarizations in the long QT syndrome: a simulation study. Cardiovasc Res. 1999;42:530-42.

16. Szabo B, et al. Role of $Na^+$:$Ca^{2+}$ exchange current in Cs(+)-induced early afterdepolarizations in Purkinje fibers. J Cardiovasc Electrophysiol. 1994;5:933-44.
17. Keating MT, Sanguinetti MC. Molecular and cellular mechanisms of cardiac arrhythmias. Cell. 2001;104:569-80.
18. Marban E, Robinson SW, Wier WG. Mechanisms of arrhythmogenic delayed and early afterdepolarizations in ferret ventricular muscle. J Clin Invest. 1986;78:1185-92.
19. Moss AJ, et al. Clinical aspects of type-1 long-QT syndrome by location, coding type, and biophysical function of mutations involving the KCNQ1 gene. Circulation. 2007;115:2481-9.
20. Mohler PJ, et al. Ankyrin-B mutation causes type 4 long-QT cardiac arrhythmia and sudden cardiac death. Nature. 2003;421:634-9.
21. Schott JJ, et al. Mapping of a gene for long QT syndrome to chromosome 4q25-27. Am J Hum Genet. 1995;57:1114-22.
22. Schulze-Bahr E, et al. KCNE1 mutations cause jervell and Lange-Nielsen syndrome. Nat Genet. 1997;17:267-8.
23. Splawski I, et al. Mutations in the hminK gene cause long QT syndrome and suppress IKs function. Nat Genet. 1997;17:338-40.
24. Abbott GW, et al. MiRP1 forms IKr potassium channels with HERG and is associated with cardiac arrhythmia. Cell. 1999;97:175-87.
25. Tristani-Firouzi M, et al. Functional and clinical characterization of KCNJ2 mutations associated with LQT7 (Andersen syndrome). J Clin Invest. 2002;110:381-8.
26. Lucet V, Lupoglazoff JM, Fontaine B. Andersen syndrome, ventricular arrhythmias and channelopathy (a case report). Arch Pediatr. 2002;9:1256-9.
27. Splawski I, et al. Ca(V)1.2 calcium channel dysfunction causes a multisystem disorder including arrhythmia and autism. Cell. 2004;119:19-31.
28. Splawski I, et al. Severe arrhythmia disorder caused by cardiac L-type calcium channel mutations. Proc Natl Acad Sci USA. 2005;102:8089-96; discussion 8086-8.
29. Vatta M, et al. Mutant caveolin-3 induces persistent late sodium current and is associated with long-QT syndrome. Circulation. 2006;114:2104-12.
30. Palygin OA, Pettus JM, Shibata EF. Regulation of caveolar cardiac sodium current by a single Gsalpha histidine residue. Am J Physiol Heart Circ Physiol. 2008;294:H1693-9.
31. Medeiros-Domingo A, et al. SCN4B-encoded sodium channel beta4 subunit in congenital long-QT syndrome. Circulation. 2007;116:134-42.
32. Chen L, et al. Mutation of an A-kinase-anchoring protein causes long-QT syndrome. Proc Natl Acad Sci USA. 2007;104:20990-5.
33. Summers KM, et al. Mutations at KCNQ1 and an unknown locus cause long QT syndrome in a large Australian family: implications for genetic testing. Am J Med Genet A. 2010;152A(3):613-21.
34. Ueda K, et al. Syntrophin mutation associated with long QT syndrome through activation of the nNOS-SCN5A macromolecular complex. Proc Natl Acad Sci USA. 2008;105:9355-60.
35. Schwartz PJ, et al. The Jervell and Lange-Nielsen syndrome: natural history, molecular basis, and clinical outcome. Circulation. 2006;113:783-90.
36. Mohler PJ, et al. Defining the cellular phenotype of "ankyrin-B syndrome" variants: human ANK2 variants associated with clinical

phenotypes display a spectrum of activities in cardiomyocytes. Circulation. 2007;115:432-41.
37. Schwartz PJ. Idiopathic long QT syndrome: progress and questions. Am Heart J. 1985;109:399-411.
38. Schwartz PJ. The congenital long QT syndromes from genotype to phenotype: clinical implications. J Intern Med. 2006;259:39-47.
39. Donger C, et al. KVLQT1 C-terminal missense mutation causes a forme fruste long-QT syndrome. Circulation. 1997;96:2778-81.
40. Schwartz PJ, et al. Genotype-phenotype correlation in the long-QT syndrome: gene-specific triggers for life-threatening arrhythmias. Circulation. 2001;103:89-95.
41. Moss AJ, et al. ECG T-wave patterns in genetically distinct forms of the hereditary long QT syndrome. Circulation. 1995;92:2929-34.
42. Schwartz PJ, et al. Diagnostic criteria for the long QT syndrome. An update. Circulation. 1993;88:782-4.
43. Hinterseer M, et al. Relation of increased short-term variability of QT interval to congenital long-QT syndrome. Am J Cardiol. 2009;103:1244-8.
44. Napolitano C, et al. Evidence for a cardiac ion channel mutation underlying drug-induced QT prolongation and life-threatening arrhythmias. J Cardiovasc Electrophysiol. 2000;11:691-6.
45. Yang P, et al. Allelic variants in long-QT disease genes in patients with drug-associated torsades de pointes. Circulation. 2002;105:1943-8.
46. Sesti F, et al. A common polymorphism associated with antibiotic-induced cardiac arrhythmia. Proc Natl Acad Sci USA. 2000;97:10613-8.
47. Goldenberg I, et al. Beta-blocker efficacy in high-risk patients with the congenital long-QT syndrome types 1 and 2: implications for patient management. J Cardiovasc Electrophysiol, 2010.
48. Schwartz PJ, et al. Long QT syndrome patients with mutations of the SCN5A and HERG genes have differential responses to $Na^+$ channel blockade and to increases in heart rate. Implications for gene-specific therapy. Circulation. 1995;92:3381-6.
49. Priori SG, et al. Association of long QT syndrome loci and cardiac events among patients treated with beta-blockers. JAMA. 2004;292:1341-4.
50. Epstein AE, et al. ACC/AHA/HRS 2008 Guidelines for device-based therapy of cardiac rhythm abnormalities: a report of the American College of Cardiology/American Heart Association Task Force on Practice Guidelines (writing committee to revise the ACC/AHA/NASPE 2002 guideline update for implantation of cardiac pacemakers and antiarrhythmia devices): developed in collaboration with the American Association for Thoracic Surgery and Society of Thoracic Surgeons. Circulation. 2008;117:e350-408.
51. Moss AJ, McDonald J. Unilateral cervicothoracic sympathetic ganglionectomy for the treatment of long QT interval syndrome. N Engl J Med. 1971;285:903-4.
52. Schwartz PJ, et al. Left cardiac sympathetic denervation in the management of high-risk patients affected by the long-QT syndrome. Circulation. 2004;109:1826-33.
53. Tan HL, et al. Long-term (subacute) potassium treatment in congenital HERG-related long QT syndrome (LQTS2). J Cardiovasc Electrophysiol. 1999;10:229-33.

54. Zipes DP, et al. Guidelines for management of patients with ventricular arrhythmias and the prevention of sudden cardiac death. Executive summary. Rev Esp Cardiol. 2006;59:1328.
55. Ruan Y, et al. Trafficking defects and gating abnormalities of a novel SCN5A mutation question gene-specific therapy in long QT syndrome type 3. Circ Res. 2010;106:1374-83.
56. Algra A, et al. QT interval variables from 24 hour electrocardiography and the two year risk of sudden death. Br Heart J. 1993;70:43-8.
57. Gussak I, et al. Idiopathic short QT interval: a new clinical syndrome? Cardiology. 2000;94:99-102.
58. Patel U, Pavri BB. Short QT syndrome: a review. Cardiol Rev. 2009;17:300-3.
59. Crotti L, et al. Congenital short QT syndrome. Indian Pacing Electrophysiol J. 2010;10:86-95.
60. Brugada R, et al. Sudden death associated with short-QT syndrome linked to mutations in HERG. Circulation. 2004;109:30-5.
61. Hong K, et al. Short QT syndrome and atrial fibrillation caused by mutation in KCNH2. J Cardiovasc Electrophysiol. 2005;16:394-6.
62. Schimpf R, et al. Short QT syndrome. Cardiovasc Res. 2005;67:357-66.
63. Giustetto C, et al. Short QT syndrome: clinical findings and diagnostic-therapeutic implications. Eur Heart J. 2006;27:2440-7.
64. Wolpert C, et al. Clinical characteristics and treatment of short QT syndrome. Expert Rev Cardiovasc Ther. 2005;3:611-7.
65. Bellocq C, et al. Mutation in the KCNQ1 gene leading to the short QT-interval syndrome. Circulation. 2004;109:2394-7.
66. Priori SG, et al. A novel form of short QT syndrome (SQT3) is caused by a mutation in the KCNJ2 gene. Circ Res. 2005;96:800-7.
67. Weiss JN, et al. The dynamics of cardiac fibrillation. Circulation. 2005;112:1232-40.
68. Extramiana F, Antzelevitch C. Amplified transmural dispersion of repolarization as the basis for arrhythmogenesis in a canine ventricular-wedge model of short-QT syndrome. Circulation. 2004;110:3661-6.
69. Anttonen O, et al. Prevalence and prognostic significance of short QT interval in a middle-aged Finnish population. Circulation. 2007;116:714-20.
70. Anttonen O, et al. Differences in twelve-lead electrocardiogram between symptomatic and asymptomatic subjects with short QT interval. Heart Rhythm. 2009;6:267-71.
71. Gussak I, et al. ECG phenomenon of idiopathic and paradoxical short QT intervals. Card Electrophysiol Rev. 2002;6:49-53.
72. Bjerregaard P, Gussak I. Short QT syndrome: mechanisms, diagnosis and treatment. Nat Clin Pract Cardiovasc Med. 2005;2:84-7.
73. Gaita F, et al. Short QT syndrome: a familial cause of sudden death. Circulation. 2003;108:965-70.
74. Gaita F, et al. Short QT syndrome: pharmacological treatment. J Am Coll Cardiol. 2004;43:1494-9.
75. Wolpert C, et al. Further insights into the effect of quinidine in short QT syndrome caused by a mutation in HERG. J Cardiovasc Electrophysiol. 2005;16:54-8.
76. Schimpf R, et al. In vivo effects of mutant HERG K$^+$ channel inhibition by disopyramide in patients with a short QT-1 syndrome: a pilot study. J Cardiovasc Electrophysiol. 2007;18:1157-60.

77. Schimpf R, et al. Congenital short QT syndrome and implantable cardioverter defibrillator treatment: inherent risk for inappropriate shock delivery. J Cardiovasc Electrophysiol. 2003;14:1273-7.
78. Brugada P, Brugada J. Right bundle branch block, persistent ST segment elevation and sudden cardiac death: a distinct clinical and electrocardiographic syndrome. A multicenter report. J Am Coll Cardiol. 1992;20:1391-6.
79. Antzelevitch C, et al. Brugada syndrome: a decade of progress. Circ Res. 2002;91:1114-8.
80. Antzelevitch C, et al. Brugada syndrome: report of the second consensus conference: endorsed by the Heart Rhythm Society and the European Heart Rhythm Association. Circulation. 2005;111:659-70.
81. Catalano O, et al. Magnetic resonance investigations in Brugada syndrome reveal unexpectedly high rate of structural abnormalities. Eur Heart J. 2009;30:2241-8.
82. Brugada J, et al. Long-term follow-up of individuals with the electrocardiographic pattern of right bundle-branch block and ST-segment elevation in precordial leads V1 to V3. Circulation. 2002;105:73-8.
83. Brugada J, Brugada R, Brugada P. Determinants of sudden cardiac death in individuals with the electrocardiographic pattern of Brugada syndrome and no previous cardiac arrest. Circulation. 2003;108:3092-6.
84. Chen Q, et al. Genetic basis and molecular mechanism for idiopathic ventricular fibrillation. Nature. 1998;392:293-6.
85. Campuzano O, Brugada R, Iglesias A. Genetics of Brugada syndrome. Curr Opin Cardiol. 2008;23:176-83.
86. Mohler PJ, et al. Nav1.5 E1053K mutation causing Brugada syndrome blocks binding to ankyrin-G and expression of Nav1.5 on the surface of cardiomyocytes. Proc Natl Acad Sci USA. 2004;101:17533-8.
87. Antzelevitch C. Molecular biology and cellular mechanisms of Brugada and long QT syndromes in infants and young children. J Electrocardiol. 2001;34:177-81.
88. Antzelevitch C. Transmural dispersion of repolarization and the T wave. Cardiovasc Res. 2001;50:426-31.
89. Antzelevitch C, Fish J. Electrical heterogeneity within the ventricular wall. Basic Res Cardiol. 2001;96:517-27.
90. Hong K, et al. Value of electrocardiographic parameters and ajmaline test in the diagnosis of Brugada syndrome caused by SCN5A mutations. Circulation. 2004;110:3023-7.
91. Antzelevitch C, Brugada R. Fever and Brugada syndrome. Pacing Clin Electrophysiol. 2002;25:1537-9.
92. Brugada P, Brugada J, Brugada R. Arrhythmia induction by anti-arrhythmic drugs. Pacing Clin Electrophysiol. 2000;23:291-2.
93. Probst V, et al. Long-term prognosis of patients diagnosed with Brugada syndrome: results from the FINGER Brugada Syndrome Registry. Circulation. 2010;121:635-43.
94. Hermida JS, et al. Hydroquinidine therapy in Brugada syndrome. J Am Coll Cardiol. 2004;43:1853-60.
95. Belhassen B, Glick A, Viskin S. Efficacy of quinidine in high-risk patients with Brugada syndrome. Circulation. 2004;110:1731-7.
96. Viskin S, et al. Empiric quinidine therapy for asymptomatic Brugada syndrome: time for a prospective registry. Heart Rhythm. 2009;6:401-4.

# Surgical and Catheter Ablation of Cardiac Arrhythmias

**CHAPTER 11**

Yanfei Yang, David Singh, Nitish Badhwar, Melvin Scheinman

## Chapter Outline

- Supraventricular Tachycardia
  - Introduction
  - History of Clinical Electrophysiologic Studies
  - Cardiac-Surgical Ablation
  - Catheter Ablation
- Atrioventricular Nodal Re-entrant Tachycardia
  - Electrophysiology of AVNRT
  - Surgical Ablation of AVNRT
  - Catheter Ablation of AVNRT
- Wolff-Parkinson-White Syndrome and Atrioventricular Re-entrant Tachycardia
  - Historical Evolution of Ventricular Pre-excitation and AVNRT
  - Cardiac-Surgical Contribution
  - Development of Catheter Ablation
  - Clinical Implications of WPW Syndrome and AVRT
  - Classification and Localization of Accessory Pathways
  - Efficacy and Challenges of Catheter Ablation for Accessory Pathways
  - Complications of Catheter Ablation
- Focal Atrial Tachycardia
  - Mechanisms and Classification of AT
  - Differentiation of the Mechanisms of AT
  - Indications of Catheter Ablation for Focal AT
  - Techniques of Catheter Ablation for Focal AT
  - Efficacy of Catheter Ablation of AT
- Atrial Flutter
  - Clinical Implications of AFL and Indication for Catheter Ablation
  - History of Nonpharmacologic Treatment in Patients with AFL
  - Ablation of CTI Dependent AFLs
  - End-point of CTI Ablation
  - Ablation of Non-CTI Dependent AFLs
  - Right Atrial Flutter Circuits
  - Left Atrial Flutter Circuits
- Ablation of Ventricular Tachycardia in Patients with Structural Cardiac Disease
  - Anatomic Substrate
  - Patient Selection
  - Prior to Ablation
  - 12-Lead Localization
  - Approach to Ablation
  - Activation Mapping (Focal Tachycardias)
  - Re-entrant Tachycardia
  - Entrainment Mapping
  - Electroanatomic Three-dimensional Mapping
  - Voltage Mapping
  - Pace Mapping
  - Substrate-based Ablation
  - Safety
  - Epicardial VT
- Idiopathic Ventricular Tachycardia
  - Outflow Tract-Ventricular Tachycardia
  - RVOT VT
  - VT Arising from the Pulmonary Artery
  - LVOT VT
  - Cusp VT
  - Epicardial VT
  - Management
  - Catheter Ablation
  - Idiopathic Left Ventricular Tachycardia (ILVT) or Fascicular VT
  - ECG Recognition
  - Management
  - Catheter Ablation
  - Mitral Annular VT
  - ECG Recognition
  - Catheter Ablation
  - Tricuspid Annular VT

## SUPRAVENTRICULAR TACHYCARDIA

### Introduction

Supraventricular tachycardias (SVTs) arise from the atrium or atrioventricular (AV) junction and include atrial tachycardia (AT), AV nodal re-entrant tachycardia (AVNRT), AV re-entrant tachycardia (AVRT), atrial flutter (AFL) and atrial fibrillation

(AF). Re-entry is the mechanism for the majority of SVTs, while triggered activity and abnormal automaticity are the mechanisms for the others.[1] Paroxysmal SVT (PSVT) denotes a clinical syndrome characterized by SVT associated with sudden onset and termination. The most common causes of PSVT are AVNRT (56%), AVRT (27%) and AT (17%).[2]

Pharmacological management of SVT was used as a first-line approach in the past. However, as knowledge of tachycardia mechanisms and technology advanced, nonpharmacological therapy allows for safe and curative treatment. Current guidelines consider ablation as first-line therapy for most forms of SVT.[3]

## History of Clinical Electrophysiologic Studies

The modern era of invasive electrophysiologic studies begin with the work of Drs Durrer and Wellens[4,5] who were the first to use programmed electrical stimulation in the heart to define the mechanism(s) of arrhythmias and Dr Scherlag and his colleagues[6] were the first to systematically record the His bundle activity in humans. Drs Durrer and Wellens showed that reciprocating tachycardia could be induced by premature atrial or ventricular stimulation and could be either orthodromic or antidromic; they also defined the relationship of the accessory pathway refractory period to the ventricular response during AF. These workers provided the framework for the use of intracardiac electrophysiological studies to define re-entrant circuit in patients with SVT.[7,8]

## Cardiac-Surgical Ablation

Prior to the era of catheter ablation, patients with SVT that were refractory to medical therapy underwent direct surgical ablation of the AV junction.[9,10] This approach, however, is not appropriate for the management of the patient with AF with rapid conduction over a bypass tract. In 1960s, Durrer and Roos[11] were the first to perform intraoperative mapping and cooling to locate an accessory pathway. Later, using intraoperative mapping, Burchell et al.[12] showed that the accessory pathway conduction could be abolished by injection of procainamide (1967). Sealy and the Duke team were the first to successfully ablate a right free-wall pathway (1968).[13] Dr Iwa of Japan also concurrently demonstrated the effectiveness of cardiac electrosurgery for these patients.[14]

## Catheter Ablation

The technique of catheter ablation of the AV junction was introduced by Scheinman et al. in 1981.[15] The initial attempts used high energy DC countershocks to destroy cardiac tissue, but expansion of its use to other arrhythmias was limited due to risk of causing diffuse damage from barotrauma. In 1984, Morady and Scheinman introduced a catheter technique for disruption of posteroseptal accessory pathways.[16] This technique was associated with 65% efficacy.[17] Later, successful ablation of nonseptal pathways was reported by Warin et al.[18] The introduction of radiofrequency (RF) energy in the late 1980s[19,20] completely altered catheter ablation procedures. The salient advances in addition to RF energy included much better catheter design, together with better understanding in the mechanism of SVTs.[20-22] A variety of both registry and prospective studies have documented the safety and efficacy of ablative procedures for these patients.[23,24]

## ATRIOVENTRICULAR NODAL RE-ENTRANT TACHYCARDIA

Atrioventricular nodal re-entrant tachycardia (AVNRT) is the most common regular, narrow-complex tachycardia. In order to better diagnose this tachycardia and guide the ablation procedure, it is important to understand the anatomy of AVN and the pathophysiology of AVNRT.

### Electrophysiology of AVNRT

The seminal findings by Moe and Mendez[25,26] of reciprocal beats in animal models were rapidly applied to humans and introduced just as the field of clinical invasive electrophysiology began to emerge. Early invasive electrophysiologic studies[27,28] attributed AV nodal re-entry as cause of paroxysmal SVT. The work of Dr Ken Rosen and his colleagues[28] demonstrated evidence for dual AV nodal physiology manifest by an abrupt increase in AV nodal conduction time in response to critically timed atrial premature depolarizations. These data served as an excellent supportive compliment to the original observations of Moe and Mendez.[25,26] By the end of the 1970s, the concept of dual AV nodal conduction in humans had been well established.

The working model used to explain the electrophysiological behavior of the AVNRT circuit involves two pathways: one is the so-called "fast pathway" which conducts more rapidly and has a relatively longer refractory period; while the other is the "slow pathway" which conducts slower than the fast pathway but has a relatively shorter refractory period **(Fig. 1)**. The fast pathway constitutes the normal, physiological AV conduction axis.

Traditionally AVNRT has been categorized into typical and atypical forms. Such categorization is based on the retrograde limb of the re-entrant circuit **(Fig. 1)**. Typical AVNRT has antegrade conduction through slow pathway and the retrograde

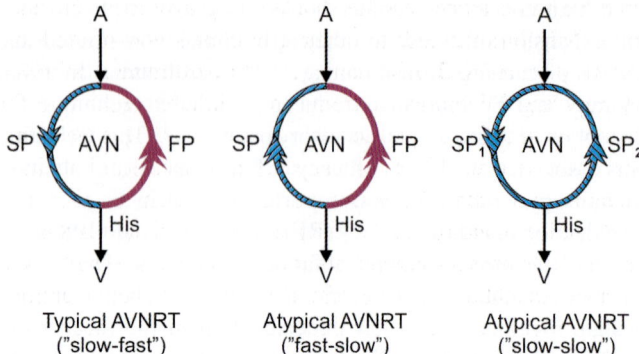

**FIGURE 1:** A schema of different AVNRT circuits. The broken line indicates the slow pathway (SP) and the solid line represent the fast pathway (FP). (*Abbreviations:* A: Atrium; V: Ventricle; AVN: Atrioventricular node; His: His bundle)

limb is the fast pathway (so-called "slow-fast"); whereas atypical AVNRT shows retrograde conduction via slow pathway, which is less common and includes "fast-slow" and "slow-slow" variants.

In addition, there are several case reports that documented the need to ablate AVNRT from the left annulus or left posteroseptal area.[29,30] One source of LA input is via the left-sided posterior nodal extension.

## Surgical Ablation of AVNRT

Ross et al.[31] first introduced nonpharmacologic therapy of AVNRT that involved surgical dissection in Koch's triangle, and their results were confirmed by a number of surgical groups.[32-34] In most patients the retrograde fast pathway (either during tachycardia or ventricular pacing) showed earliest atrial activation over the apex of Koch's triangle while in the minority earliest atrial activation occurred near the CS. This observation nicely compliment the current designation of AVNRT subforms.[35]

## Catheter Ablation of AVNRT

In 1989, two groups[36,37] almost simultaneously reported success using high energy discharge in the region of slow pathway. The subsequent use of RF energy completely revolutionized catheter cure of AVNRT. The initial attempts targeted the fast pathway by applying RF energy superior and posterior to the His bundle region (so-called anterior approach) until the prolongation of AV nodal conduction occurred. Initial studies[36-38] showed a success rate of 80–90%, but the risk of AV block was up to 21%. Due to the high-risk of developing AV block, fast pathway ablation is no longer used as the primary approach. Jackman et al.[39] first introduced the technique of ablation of the slow pathway for AVNRT. Ablation of the slow pathway is achieved by applying RF energy at the posterior-inferior septum in the region of the $CS_{OS}$. This technique can be guided by either discrete potentials[39,40] or via an anatomic approach,[41] both have equal success rate. The safest and most effective approach is to combine anatomic and eletrogram approaches together, in which RF lesions are applied at the posteroseptal sites with slow pathway potentials **(Fig. 2)**. The RF energy is usually applied until junctional ectopics appear and diminish, but at times successful slow pathway ablation may result without eliciting the junctional ectopic complexes. The end point for slow pathway ablation involves the proof either that the slow pathway has been eliminated of which there is no more evidence of dual AV nodal physiology (i.e. no AH "jump" with atrial programmed stimulus) or that no more than one AV nodal echo is present.[39]

Among experienced centers the current acute success rate for this procedure is 99% with a recurrence rate of 1.3%, and a 0.4% incidence of AV block requiring a pacemaker.[42] Although the risk of AV block from selective slow pathway ablation in patients with normal baseline PR interval is very low, some reports have suggested that the risk may be higher in patients with pre-existing PR prolongation and/or older age (>70 years old).[43] In those patients at higher risk, delayed onset of symptomatic AV block can develop and vigilant follow-up may be needed.[43,44] An approach of retrograde fast pathway ablation has been used in patients with baseline PR prolongation and is associated with no delayed development of AV block.[45]

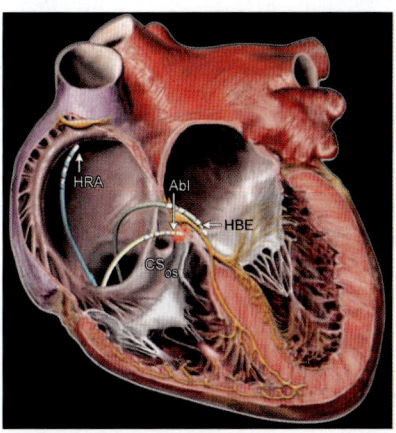

**FIGURE 2:** Typical slow pathway ablation site. This diagram shows catheter positions for slow pathway ablation in patients with typical AVNRT. The ablation catheter is positioned at the posterior septum just above the $CS_{OS}$. (*Abbreviations:* HRA: High right atrium; HBE: His bundle electrogram; Abl: Ablation catheter; $CS_{OS}$: The ostium of coronary sinus)

Technologic advances continue to improve the safety of ablation procedures. Besides significantly reducing radiation time to both patient and operator, the development of sophisticated, real-time, 3D mapping systems has allowed for precise localization of the His bundle, reducing the risk of AV block. In addition, cryoablation may be used for slow pathway ablation.[46] The advantage of this technology includes catheter sticking to adjacent endocardium during application of energy, avoiding inadvertent catheter displacement and damage to the node or His bundle. In addition, any AV conduction delay during test ablation is reversible.

## WOLFF-PARKINSON-WHITE SYNDROME AND ATRIOVENTRICULAR RE-ENTRANT TACHYCARDIA

### Historical Evolution of Ventricular Pre-excitation and AVNRT

The first complete description of WPW syndrome was by Drs Wolff, Parkinson and White in 1930s.[47] They reported 11 patients without structural heart disease who had a short P-R interval, "bundle branch block (BBB)" ECG pattern and episodes of PSVT. At the time, the wide QRS patterns seen in ventricular pre-excitation were thought to be related to a short P-R interval and BBB. Discrete extranodal AV connections accounting for ventricular pre-excitation were initially proposed by Kent[48] and later confirmed by Wood,[49] Öhnell[50] and others.

### Cardiac-Surgical Contribution

Sealy et al.[13] were the first to successfully ablate a right free-wall pathway. Their subsequent results conclusively showed that a vast majority of patients with the WPW syndrome could be cured by either direct surgical or cryoablation of these accessory pathways. Simultaneously, Iwa et al. also demonstrated the efficacy of cardiac electrosurgery in these patients.[14] He should be credited for being among the first to use an endocardial approach for accessory pathway ablation. The endocardial approach was independently used by the Duke team of Sealy and Cox. Only later was the "closed" epicardial approach reintroduced by Guiraudon.

### Development of Catheter Ablation

The technique of catheter ablation was first introduced by Scheinman and his colleagues in the early 1980s,[15-17] but ablation using DC shocks was limited due to its high-risk of causing diffuse damage from barotrauma. The introduction of

RF energy in the late 1980s[19,20] along with better catheter design and the demonstration of accessory pathway (AP) potential for facilitating localization of AP have dramatically improved the safety and efficacy of catheter ablation. The remarkable work of Jackman,[20] Kuck[21] and Calkins[22] ushered in the modern era of ablative therapy for patients with accessory pathways in all locations. A variety of both registry and prospective studies have documented the safety and efficacy of ablative procedures for these patients.[23,24] Nowadays, catheter ablation is the procedure of choice for patients with symptomatic WPW syndrome. In most experienced centers, the success rate is 95–97% with a recurrence rate of approximately 6%.

## Clinical Implications of WPW Syndrome and AVRT

Patients with WPW syndrome may experience very rapid conduction over the AP during AF. In some patients, ventricular fibrillation (VF) may be the first manifestation of this syndrome.[51] In a symptomatic patient with WPW syndrome, the lifetime incidence of sudden cardiac death (SCD) has been estimated to be approximately 3–4%.[52]

## Classification and Localization of Accessory Pathways

The accessory pathways (APs) are classified into three different types: (1) manifest APs which show a typical WPW pattern on surface ECG; (2) concealed APs are those that lack antegrade conduction but only show retrograde conduction over the APs and (3) a third group known as latent WPW syndrome shows pre-excitation when pacing close to the atrial insertion of the AP.

Precise mapping of APs is critical to the success of ablation procedure. The delta waves and QRS morphologies of the 12-lead ECG in patients with WPW syndrome can help predict the AP location and guide ablation. A successful ablation site can be identified an AP potential **(Fig. 3)**, early onset of local

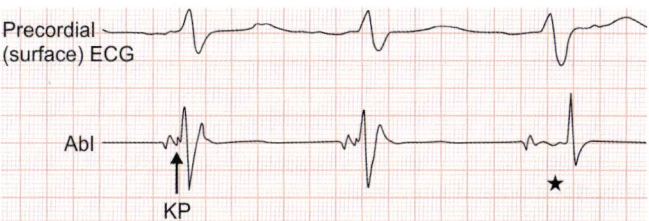

**FIGURE 3:** Electrogram in sinus rhythm during application of radiofrequency energy. Kent potential (AP potential) on ablation catheter (Abl) disappears (*) and there is abrupt local A-V interval prolongation and a subtle change in the surface QRS, indicating loss of pre-excitation. (*Abbreviations:* Abl: Ablation catheter; KP: Kent potential)

ventricular activation compared to the onset of delta waves on surface ECG during antegrade pre-excitation and fused local atrial and ventricular electrogram.

## Efficacy and Challenges of Catheter Ablation for Accessory Pathways

The majority of the APs are located at the left free wall, 20–30% are located in the posteroseptum, 10–20% along the right free wall and 5–10% at the anteroseptum. The left free-wall APs can be mapped and ablated along the mitral annulus (MA) via either a transseptal or a retrograde transaortic approach. Overall, catheter ablation of left free-wall APs are associated with a high success rate (95%); while ablation of the right free-wall APs is associated with a lower success rate (90%) and a recurrence rate of 14%.[53] The relatively low success rate of right-sided AP ablation is due to the more poorly formed tricuspid annulus (TA) resulting in problems with catheter stability and lack of an accessible right-sided CS-like structure that parallels the TA to facilitate AP localization. Ablation of right-sided APs may be improved by using long deflectable sheaths and a small multipolar mapping catheter placed in the right coronary artery to assist AP mapping.

Ablation of septal APs can be challenging due to the anatomic relationship to the normal conduction system. Therefore, catheter ablation in these areas has the potential risk of producing AV block. The electrogram recorded from the ablation catheter should be carefully assessed and monitored before and during RF delivery. Using 3D electroanatomic mapping (EAM) system to localize the His bundle and track the ablation catheter may prevent or reduce the risk of AV block. Lately cryomapping and cryoablation have improved the safety in difficult cases.[54] Most posteroseptal APs can be ablated from the right side, although up to 20% of the cases require a left-side approach.[55] About 5–17% of the posteroseptal and left posterior APs are located epicardially and require ablation within the CS or middle cardiac vein.[56] Coronary sinus diverticulum may harbor the posteroseptal APs, and CS angiography can confirm such an anomaly. In some patients RF ablation at the neck of the diverticulum may be required to eliminate the APs.[57,58] Applying RF ablation within the CS should be initiated with low energy in order to prevent the risk of perforation and tamponade.

A small percentage of APs are epicardial, suggested by the finding of small or no AP potential during endocardial mapping but with a large AP potential recorded within the CS.[59] Left-side epicardial AP can be successfully ablated within the CS. However, ablation of some epicardial APs may require a percutaneous epicardial approach.[60]

## Complications of Catheter Ablation

Overall, catheter ablation of APs is associated with a complication rate of 1–4%, including life-threatening complications (such as perforation, tamponade and embolism) (0.6–0.7%), and procedure-related death (approximately 0.2%).[22,56,61] Complete AV block occurs in about 1% of the patients and is mostly associated with the ablation procedures for septal APs.

## FOCAL ATRIAL TACHYCARDIA

Atrial tachycardia (AT) is a group of SVT that is confined to the atrium without involvement of AV node. It is a relatively uncommon arrhythmia, comprising less than 10% of symptomatic SVTs encountered in the adult electrophysiological laboratory.[62] However, AT is more common in children (up to 14–23%).[63]

### Mechanisms and Classifications of AT

The AT can be classified into two types: (1) focal AT and (2) macro-re-entry. The mechanism of focal AT can be due to abnormal automaticity or triggered activity. In adults, macro-re-entry is the most common mechanism for AT,[62] while automatic or triggered mechanisms are more common in children.[63]

### Differentiation of the Mechanisms of AT

Distinguishing the mechanisms of focal AT may be difficult. In general, a focal AT due to abnormal automaticity tends to have spontaneous initiations or initiation with isoproterenol. It can be suppressed but not terminated by atrial overdrive pacing, and lacks response to adenosine, verapamil or vagal maneuvers.[64,65] The AT with triggered activity can be initiated or terminated by rapid atrial overdrive pacing, and it is sensitive to large-dose of adenosine or vagal maneuvers.[65]

Differentiating focal from macro-re-entrant AT is important to the ablation procedure. Ablation of focal AT is accomplished by targeting the discharging focus (usually it is a single source, except for multifocal AT); whereas ablation of macro-re-entrant AT requires delineation of a critical isthmus that allows for tachycardia perpetuation. Detailed atrial activation mapping, including electrogram and EAM mapping, can distinguish focal from macro-re-entrant AT.

### Indications of Catheter Ablation for Focal AT

Pharmacologic therapy in patients with focal AT is often ineffective. The proarrhythmia effects of these drugs also limit the long-term efficacy of pharmacologic therapy. Therefore, catheter ablation of focal AT may be considered as a first-line option, especially in those patients who have incessant tachycardia or baseline ventricular dysfunction.[3]

## Techniques of Catheter Ablation for Focal AT

Most focal ATs arise from the right atrium (67%), especially from the crista terminalis and TA.[66] Left atrial focal ATs mostly involve the pulmonary veins (PVs) and MA, and less often from the CS, atrial appendages and atrial septum. The surface P wave morphology facilitates the mapping and ablation of AT **(Fig. 4)**. Left-sided ATs require a transseptal approach.

Successful ablation of AT relies on detailed atrial activation mapping during the tachycardia, and use of multipolar catheters and/or 3D EAM systems **(Fig. 5)**.[67,68] A successful ablation site can be identified by early local endocardial activation (usually

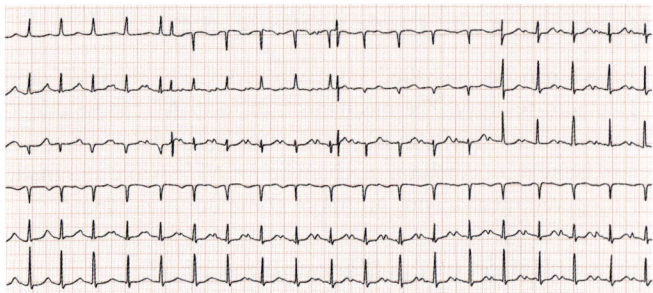

**FIGURE 4:** Surface ECG in a patient with focal AT arising from the high crista terminalis. Note the P waves in the inferior leads (II, III and aVF) are positive, and negative in V1

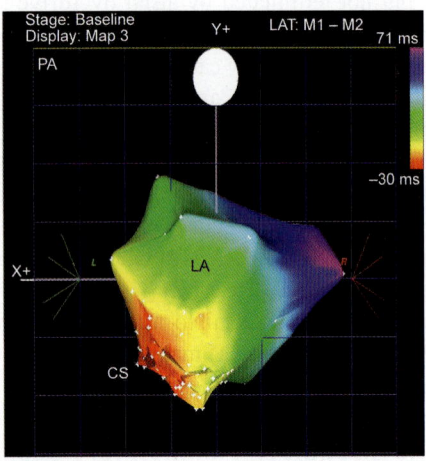

**FIGURE 5:** A 3D activation map (by CARTO system) of the left atrium (LA) during tachycardia in a patient with a focal AT originating from the CS musculature. The posteroanterior projection (PA) view showed the earliest activation (red area) at the posterior lateral wall

preceding the surface P wave by ≥ 30 ms) and/or low-amplitude, fractionated electrograms **(Fig. 6)**. The RF energy is typically delivered during tachycardia. Acceleration of the tachycardia during ablation is usually a reliable predictor for successful ablation of automatic AT,[69] and noninducibility is the end-point of ablation procedure for focal AT.

Caution should be taken during ablation of focal ATs originating from the areas where important anatomic structures situated such as sinus node and AV node. Lately, cryoablation has been used for ATs originating from the region of His bundle to reduce the potential risks of AV block.[70]

## Efficacy of Catheter Ablation of AT

The success rate of ablation for focal AT is about 93% with a recurrence rate of 7%.[71] Left-sided ATs have a lower success rate than the right-sided ATs. Patients with multifocal AT have

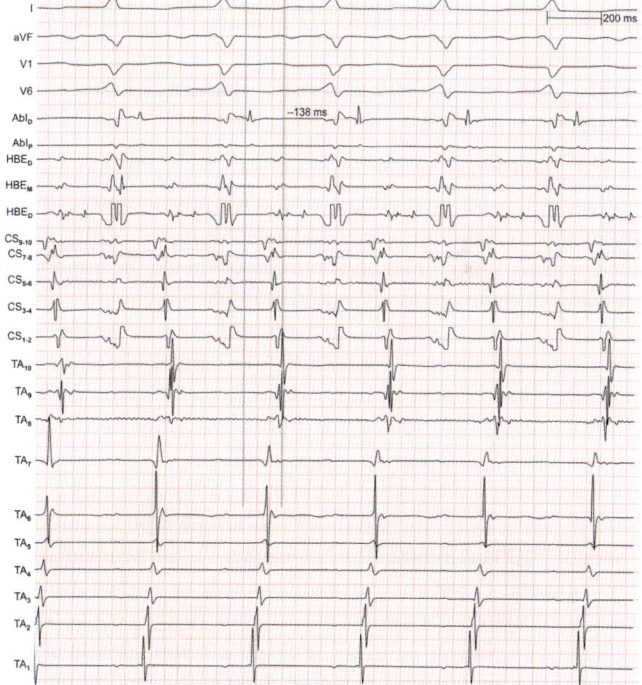

**FIGURE 6:** Simultaneous recordings from surface leads and catheters placed at ablation site (Abl), His bundle region (HBE), the CS and a 20-pole catheter around the TA with its distal pair of electrodes (TA1) at low lateral TA and proximal at the high septum during tachycardia in a patient with a focal AT originating from inferior TA. Note the earliest atrial activation, which was recorded by the distal ablation catheter, was 138 ms earlier than the onset of surface P waves. The RF delivered at this site abolished the tachycardia without inducibility

a higher recurrence rate than those with single tachycardia foci. Also, elder patients and patients with structural heart disease tend to have a higher recurrence rate after initial "successful" ablation.

## ATRIAL FLUTTER

### Clinical Implications of AFL and Indication for Catheter Ablation

Atrial flutter is a rapid macro-re-entrant circuit that is confined to either atrium, and bounded by either functional or anatomic barriers. Due to its rapid and regular atrial rate, AFL often produces more rapid ventricular responses. Hence, chronic AFL can result in tachycardia-mediated cardiomyopathy and heart failure. It also predisposes to intracardiac thrombus formation and the risk for stroke. Although antiarrhythmic agents can suppress paroxysmal AFL, the long-term efficacy is poor.[72] Therefore, with technological advances in catheter ablation and better understanding of locating re-entrant circuits, catheter ablation should be considered as first-line treatment for AFL.

### History of Nonpharmacologic Treatment in Patients with AFL

In the late 1970s, the seminal observations by Waldo and his colleagues, who studied patients with postoperative flutter by means of fixed atrial electrodes, confirmed re-entry as the mechanism of AFL in humans and demonstrated the importance of using entrainment for detection of re-entrant circuits.[73] Klein and Guiraudon mapped two patients with AFL in the operating room found evidence of a large RA re-entrant circuit and the narrowest part of the circuit lay between the TA and the IVC.[74] They successfully treated the flutter by using cryoablation around the CS and surrounding atrium.

Following the report of Klein et al., there appeared several studies using high-energy shocks in an attempt to cure AFL (Saoudi,[75] Chauvin and Brechenmacher[76]). Subsequently both Drs Feld and Cosio almost simultaneously described using RF energy to disrupt cavotricuspid isthmus (CTI) conduction in order to cure patients with AFL. Feld et al. contributed an elegant study using endocardial mapping techniques and entrainment pacing to prove that the area posterior or inferior to the CS was a critical part of the flutter circuit and application of RF energy to this site terminated AFL.[77] Cosio et al. used similar techniques but placed the ablative lesion at the area between the TA and IVC.[78] The latter technique forms the basis for current ablation of CTI dependent flutter.

## Ablation of CTI Dependent AFLs

In the majority of patients with RA flutter, the CTI is a critical part of the re-entrant circuit. The CTI dependent AFL circuits include those with counterclockwise (CCW) and clockwise (CW) re-entrant circuits around the TA;[79] double-wave re-entry (DWR) which has two wavefronts traveling around the TA simultaneously;[80] lower-loop re-entry (LLR) around the inferior vena cava (IVC)[81–83] and intraisthmus re-entry (IIR).[84,85]

Detailed electrogram mappings as well as entrainment techniques are required to diagnose the flutter circuits. Electrograms recorded from the multielectrode catheter placed around the TA demonstrate the RA activation sequence such as CCW or CW pattern **(Fig. 7A)**. Entrainment pacing at different atrial sites can help identify the re-entrant circuit and its critical isthmus **(Fig. 7B)**. In addition, using 3D EAM mapping systems can facilitate illustrating the re-entrant circuit and guide catheter ablation over the CTI. A complete linear lesion from TA to IVC during AFL results in interrupting the CTI-dependent flutter circuit and terminating the tachycardia.

## End-Point of CTI Ablation

Initially it was felt that a good end point for successful CTI ablation was tachycardia termination during RF application. However, many patients suffered recurrences, and eventually it was recognized that it was important to achieve true bidirectional block in the isthmus. Many studies have shown that recurrence rates of AFL are much improved when bidirectional block is achieved.[86] Currently there are many techniques for assessing bidirectional isthmus block.[87-89]

## Ablation of Non-CTI Dependent AFLs

As shown in **Flow chart 1**, non-CTI dependent AFL circuit can be classified into two categories: (1) RA and (2) LA flutter circuits. Ablation of non-CTI dependent AFLs can sometimes be challenging, but using 3D EAM system can facilitate the procedure.

## Right Atrial Flutter Circuits

In the RA, non-CTI dependent AFL includes scar-related macro-re-entrant tachycardia and upper loop re-entry (ULR). It has been shown that macro-re-entrant AT can occur in patients with or without atriotomy or congenital heart disease.[82,90,91] In these patients, the 3D electroanatomic voltage maps from the RA often show "scar(s)" or low-voltage area(s) ($< 0.2$ mV) which act(s) as the central obstacle or channels for the re-entrant circuit.

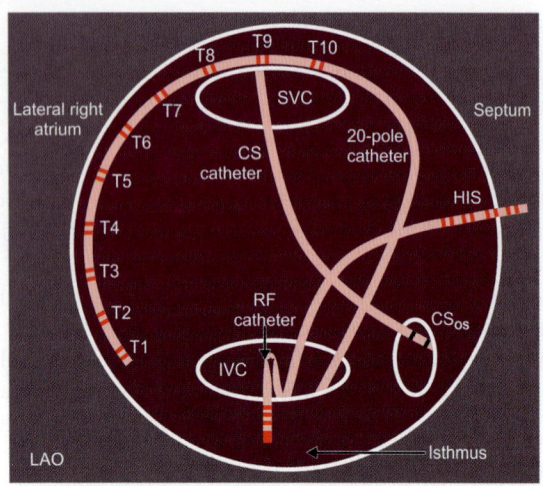
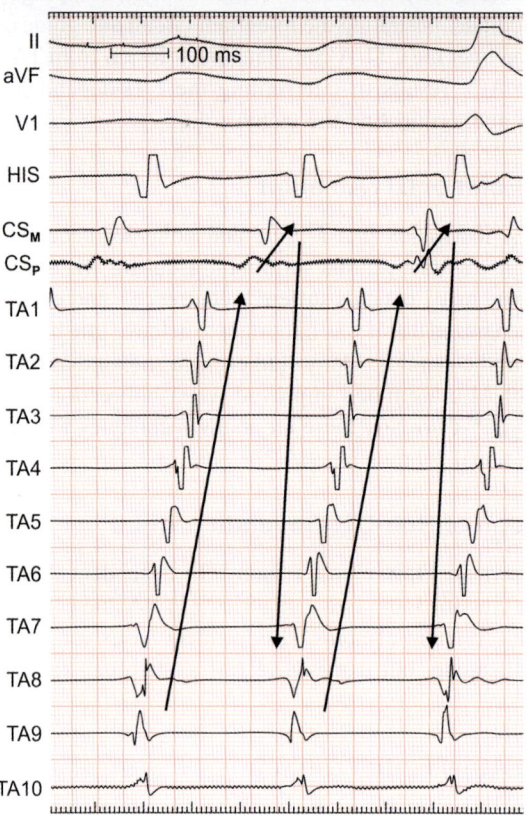

**FIGURE 7A:** Left panel shows the schema of catheter positions in the left anterior oblique projection (LAO) view during ablation for CTI dependent AFL. A duo-decapolar catheter is positioned along the TA, as well as a quadrupolar catheter at His bundle region and a decapolar catheter inside of the CS. Right panel shows the simultaneous recordings from surface ECG and these catheters. The intracardiac electrogram demonstrates a counterclockwise activation sequence (as shown by the arrows) around the TA

The morphology of surface ECG varies depending on where the scar(s) and low-voltage area(s) are and how the wavefronts exit the circuits. The critical isthmus of the re-entrant circuit can be identified by entrainment pacing, and the electrogram recorded at such a site often shows low-amplitude, fractionated, long duration mid-diastolic potentials. Catheter ablation of scar-related macro-re-entrant tachycardia involves deliver RF energy within the critical channel/isthmus or linear lesion connecting from the scar to an anatomic barrier, such as IVC or super vena cava (SVC).

The ULR is a form of AFL only involving the upper portion of RA with transverse conduction over the CT and wavefront collision occurring at the lower part of RA or within the CTI.[82,92] It was initially felt to involve a re-entrant circuit using the channel between the superior vena cava (SVC), fossa ovalis (FO) and CT.[82] A study by Tai et al. using noncontact mapping

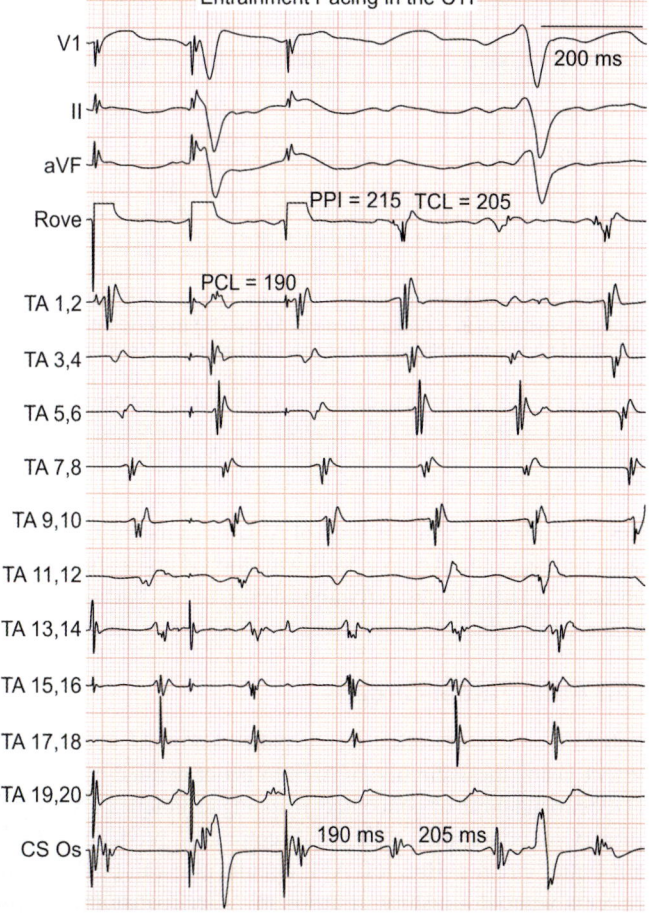

**FIGURE 7B:** Contd...

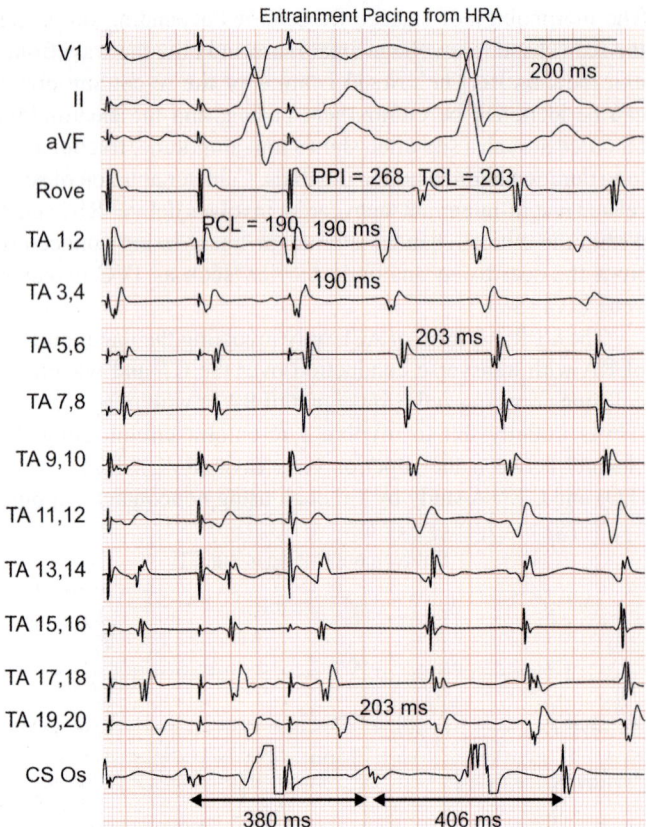

**FIGURE 7B:** Entrainment pacing from the mapping catheter (Rove) during tachycardia in a patient with clockwise CTI dependent AFL. The left panel shows the difference between PPI and TCL (< 30 ms) when pacing within the CTI, and the atrial activation sequence was same compared to that of the tachycardia, which indicated that the CTI is the critical part of the flutter circuit. The right panel showed the "PPI-TCL" was greater than 30 ms when pacing from the high right atrium (HRA), which suggested that this area is out of the circuit

technique showed that this form of AFL was a macro-re-entrant tachycardia in the RA free wall with the CT as its functional obstacle.[92] They successfully abolished ULR by linear ablation of the gap in the CT.

## Left Atrial Flutter Circuits

Left AFL circuits are often seen in patients post-AF ablation. In recent years, these circuits have been better defined by the use of electroanatomic or noncontact mapping techniques.[93] Cardiac surgery involving the LA or atrial septum can produce various left flutter circuits. But, left AFL circuits also can be found in patients without a history of atriotomy. Electroanatomic maps in these patients often show low voltage or scar areas in the

LA, which act as a central obstacle in the circuit. There are several subgroups of left AFLs **(Flowchart 1)**.

Mitral annular AFL involves re-entry around the MA either in a CCW or CW direction **(Fig. 8)**. The surface ECG of MA flutter can mimic CTI-dependent CCW or CW flutter, but with low-amplitude flutter waves in most of the 12 leads.[94] This arrhythmia is more common in patients with structural heart disease. However, it has been described in patients without obvious structural heart disease.[93,94] Electroanatomic voltage

**FLOWCHART 1:** Nomenclature of atrial flutter (AFL)

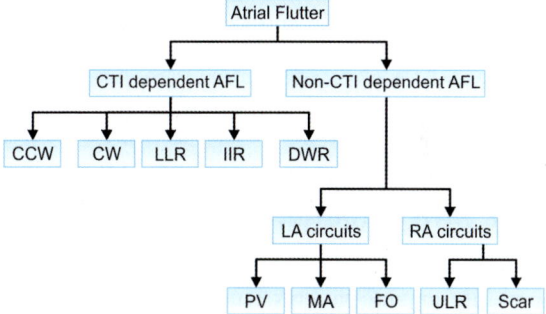

(*Abbreviations:* CTI: Cavotricuspid isthmus; CCW: Counterclockwise AFL around the tricuspid annulus (TA); CW: Clockwise AFL around the TA; LLR: Lower loop re-entry around inferior vena cava; IIR: Intraisthmus re-entry; DWR: Double-wave Re-entry around the TA; LA: Left atrium; RA: Right atrium; PV: Re-entrant circuit around the pulmonary vein (s) with or without scar(s) in the LA; MA: re-entrant circuit around mitral annulus; FO: Re-entrant circuit around the fossa ovalis; ULR: Upper loop re-entry in the RA)

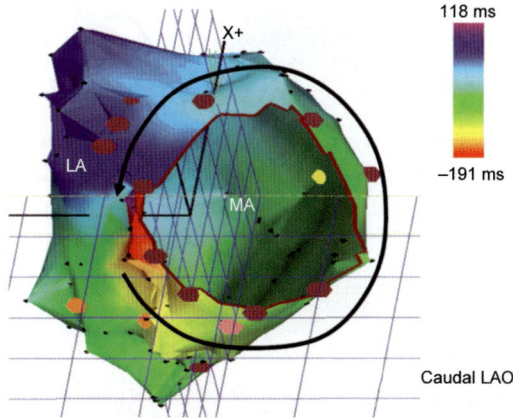

**FIGURE 8:** A CARTO activation map of the left atrium in a caudal LAO view in a patient with CCW AFL around the mitral annulus (MA). The map shows "early meets late activation" at the spetal MA and the mapped cycle length spanned the TCL. Ablation was completed with a line from the left inferior pulmonary vein (PV) to the MA

map from the LA often shows scar(s) or low-voltage area(s) at the posterior wall as a posterior boundary of this circuit. A linear RF lesion is usually applied at the mitral isthmus, i.e. from the ostium of left lower PV to the lateral MA.[94] Bidirectional mitral isthmus block should be assessed after completing the ablation line.

Various left AFL circuits involve the PVs, especially in those patients who underwent AF ablation or those with mitral valve disease. Re-entry can circle around one or more PVs and/or posterior scar or low-voltage area(s).[93,94] In order to cure these complex circuits, 3D EAM is required to reveal the circuit and guide ablation **(Fig. 9)**. Since these circuits are related to low voltage or scar area(s), the surface ECG usually shows low amplitude or flat flutter waves.

In summary, modern mapping techniques allow for identification and successful ablation of complex AFL circuits.

## ABLATION OF VENTRICULAR TACHYCARDIA IN PATIENTS WITH STRUCTURAL CARDIAC DISEASE

Ventricular tachycardia (VT) is an important source of morbidity and mortality among patients with ischemic heart disease. Patients with VT and a history of myocardial infarction are at high-risk of recurrent VT, VF and SCD. Internal cardiac defibrillators (ICDs) have become the mainstay of therapy in

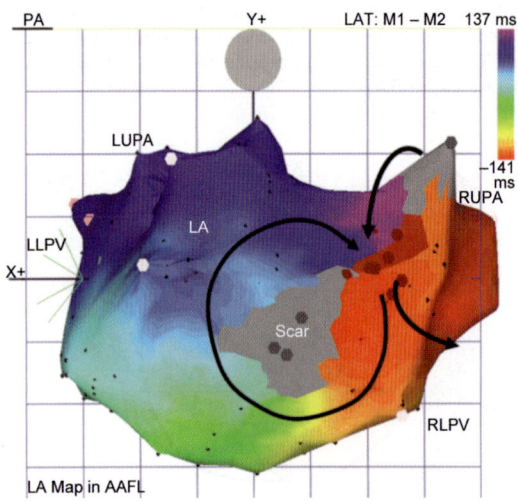

**FIGURE 9:** A CARTO activation map of the LA in a patient with LA AFL. The map shows a scar over the posterior LA wall. The tachycardia wave front traveled in a "Figure-of-8" pattern around the scar and the right upper pulmonary vein (RUPV) respectively and through the common channel between the scar and the right upper pulmonary vein (RUPV). Successful ablation was achieved with an RF line from the RUPV to the scar. (*Abbreviations:* LUPV: Left upper pulmonary vein; LLPV: Left lower pulmonary vein; RLPV: Right lower pulmonary vein)

this patient population and are effective at terminating episodes of VT and VF. Among patients at high-risk for VT and SCD, ICD therapy has been shown to reduce SCD and all-cause mortality.[95-98] Although ICDs are highly effective, they do not prevent VT or VF, and ICD shocks have been associated with decreased quality of life, increased anxiety and depression and increased mortality.[99-105] While antiarrhythmic therapy is frequently used to prevent ICD shocks, its efficacy is limited and frequently associated with untoward side effects.[106,107]

Catheter ablation for scar-based VT has emerged as an important treatment option, particularly among individuals who have received recurrent ICD shocks. Several studies have demonstrated that this approach can reduce the incidence of ICD shocks and/or VT burden.[108-110] In the case of incessant VT or VT storm (three or more episodes within a 24 hour period), catheter ablation can be a lifesaving measure. However, catheter ablation in patients with ischemic heart disease can be technically challenging. Patients with ischemic heart disease and VT are by definition, a vulnerable population, and are often unable to tolerate long procedure-times and VT rates frequently induced during ablation. This section will provide an overview of catheter ablation for patients with scar-related VT. It will review the mechanisms of scar-related VT, indications for ablation and describe the various mapping and ablation techniques commonly employed.

## Anatomic Substrate

The vast majority of VT in patients with ischemic heart disease is due to re-entry involving a healed scar. Unidirectional block is a necessary condition for re-entry. Areas of conduction block can be anatomically fixed (present during tachycardia and sinus rhythm) or can be functional (present only during tachycardia).[111] The sites of VT origin are frequently located adjacent to and within scar locations where surviving bundles of muscle fiber can be found. These muscle bundles are isolated from neighboring bundles by strands of fibrous tissue. Endocardial recordings form these sites demonstrate fractionated (low-amplitude and disorganized) potentials which serve as regions of slow conduction and provide the substrate for re-entrant VT (**Fig. 10**).[112] Although scar based re-entry is the most common arrhythmia associated with ischemic heart disease, other clinical VTs, such as focal tachycardia, bundle branch re-entry and fascicular re-entry, are also observed on occasion.

## Patient Selection

In general, ablation for scar-related VT is reserved for patients with recurrent monomorphic VT and/or frequent ICD shocks.

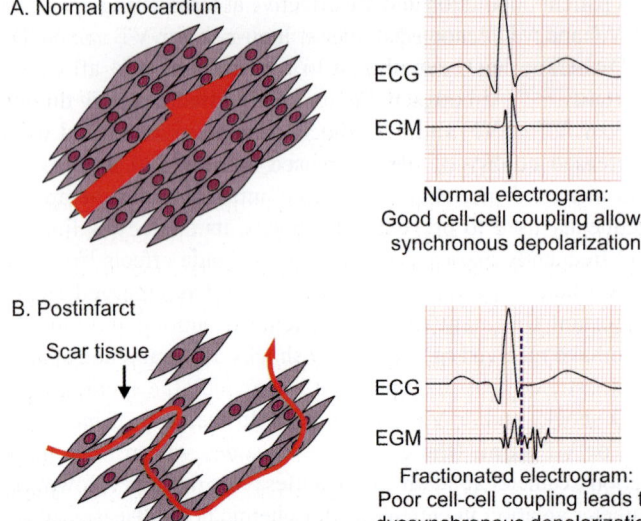

**FIGURE 10:** Slow conduction through scarred myocardium provides the substrate for re-entry. This is often accompanied by the presence of fractionated electrograms as seen on the right (*Source:* John Miller)

There is, however, a growing interest in performing early or even prophylactic ablation to prevent VT episodes and ICD discharges.[109,110] In general, ablation is not considered for patients with recurrent polymorphic VT. However, it is important to recognize the occasional patient with idiopathic polymorphous VT or VF due to short coupled premature ventricular complexes (PVCs). If the triggering PVC can be identified and ablated in these patients, it may result in cure or significant attenuation of the VT/VF burden.[113]

The most recent American College of Cardiology/American Heart Association Task Force/European Society of Cardiology guidelines for management of patients with ventricular arrhythmias and the prevention of sudden death recommends catheter ablation for VT as adjunctive therapy for patients with an ICD who have had multiple shocks due to sustained VT, not amenable to ICD reprogramming or drug therapy (Class I, level of evidence C).[114] In addition, ablation may be considered as an alternative to long-term drug therapy. A more recent consensus document expands on these guidelines and recommends catheter ablation for patients with structural heart disease in each of the following conditions:

- Symptomatic sustained monomorphic VT (SMVT), including VT terminated by an ICD, that recurs despite antiarrhythmic drug therapy or when antiarrhythmic drugs are not tolerated or not desired.
- Incessant SMVT or VT storm that is not due to a transient reversible cause.

- Patients with frequent PVCs, nonsustained VT (NSVT) or VT that is presumed to cause ventricular dysfunction.
- Bundle branch re-entrant or interfascicular VTs.
- Recurrent sustained polymorphic VT and VF that is refractory to antiarrhythmic therapy when there is a suspected trigger that can be targeted for ablation.[115]

It is generally agreed that the role of catheter ablation for scar-related VT is to reduce a patient's arrhythmic burden. As such, even successful ablations do not obviate the need for an ICD. There have been several studies that have prospectively evaluated the role of catheter ablation for VT.

The multicenter thermocool ventricular tachycardia ablation trial examined the role of catheter ablation for VT in patients with reduced ejection fraction (EF) and recurrent monomorphic VT.[108] Around 231 patients were enrolled and underwent ablation. The median number of VT morphologies per patient was three. All inducible VTs with rates near to or less than the clinical VT were targeted. Ablation abolished all inducible VTs in 49% of patients. At six months, 53% of patients achieved the primary endpoint of freedom from recurrent incessant or sustained VT. In 142 patients with ICDs VT episodes were reduced from a median of 11.5 to 0 ($p < 0.0011$). The 1-year mortality rate was 18%, with 72.5% of deaths attributed to ventricular arrhythmias or heart failure. The procedure mortality rate was 3%, with no strokes. Although this was a nonrandomized trial, it demonstrated moderate success for VT ablation in carefully selected patients.

Two prospective trials have evaluated the role of VT ablation for the prevention of SMVT. The SMASH-VT trial enrolled 128 subjects with an ICD placed either for secondary prevention (for VF or hemodynamically unstable VT or syncope with inducible VT) or for primary prevention with subsequent delivery of an appropriate ICD therapy.[110] Patients were randomly assigned to defibrillator implantation alone or defibrillator implantation with adjunctive catheter ablation (64 patients in each group). Ablation was performed using a substrate-based approach (see below). The primary end point was survival free from any appropriate ICD therapy. During the mean 23 month follow-up period, appropriate ICD therapy occurred more frequently in the ICD alone group and than in the ablation group (33 vs 12%, P = 0.007). There was a trend toward decreased mortality in the ablation group (9 vs 17%, P = 0.29). Ablation-related complications occurred in three patients (pericardial tamponade, HF requiring prolonged hospitalization and deep venous thrombosis) and there were no procedural deaths.

The VTACH trial, enrolled 110 subjects with stable VT, prior MI and LVEF less than or equal to 50% who were randomly assigned to either catheter ablation plus an ICD or ICD alone.[109] The primary endpoint was the time to first recurrence of VT or VF. After a mean follow-up duration of 22.5 months, time to VF or recurrent VT was longer in the ablation group than in the control group (median 18.6 vs 5.9 months p = 0·045). According to the Kaplan-Meier analysis, 59% of patients in the ablation group and 40% of patients in the control group were free from any VT or VF episode after 12 months. No significant difference in quality of life was found. Ablation-related complications occurred in two patients (transient ST-segment elevation in one patient and a transient cerebral ischemic event in the other) and there were no procedural deaths.

## Prior to Ablation

Since patients with scar-based VT frequently have coronary artery disease (CAD), it is important to understand a patient's coronary anatomy and ischemic burden prior to ablation. Patients with unrevascularized CAD are unlikely to tolerate prolonged periods of VT. It is therefore prudent to obtain a noninvasive or invasive assessment of a patient's coronary anatomy prior to the procedure. Ideally, a patient should be revascularized before catheter ablation. In some instances, revascularization itself may reduce a patient's arrhythmic burden. For some patients, inotropic support, balloon pump or other forms of hemodynamic support (e.g. left ventricular assist-device or extracorporeal membrane oxygenation) may be required to perform the case.[116-118]

Prior to ablation, all patients should undergo a transthoracic echocardiogram to assess for left ventricular thrombus. The presence of thrombus is a contraindication to endocardial VT ablation due to the risk of thrombus dislodgement. In patients without ICDs, preprocedural magnetic resonance imaging (MRI) can be used to guide ablation by identifying areas of scar. Finally patients should be assessed for peripheral arterial disease (PAD) prior to ablation. Frequently, access to the LV is achieved retrogradely across the aortic valve. For patients with extensive PAD a transseptal approach may be more desirable.

## 12-Lead Localization

Planning for scar-based monomorphic VT ablation first requires analysis of the 12-lead ECG during tachycardia (providing it is available). For re-entrant VT, the QRS morphology represents the exit point of the circuit. In general, VTs arising in the septum (or fascicular system) are more narrow than VTs originating on the free wall. Positive concordance (dominant R wave in all precordial leads) is associated with VTs that exit from the posterior base of the heart. Negative concordance (QS in all precordial leads) is associated with VTs that exit from the anterior left ventricular apex.[119] A left bundle branch block (LBBB) pattern (R < S in lead V1) is observed with VTs from the right ventricle or ventricular septum. Right bundle branch block (RBBB) morphologies (R > S in lead V1) are almost invariably associated with VTs that arise in the left ventricle. In patients with prior infarction, the frontal axis of the VT is also influenced by the location of the VT exit site. The VTs that exit on the inferior wall produce a superiorly directed axis. An inferiorly directed axis usually reflects a VT from the anterior wall. A negative deflection in lead I and AVL indicates a lateral exit.[111] The VTs of epicardial origin typically have a longer QRS duration and slurred QRS upstrokes in the precordial leads.[120]

## Approach to Ablation

The term "mapping" is a broad term that refers to number of electrophysiological techniques that are used to gain insight about the nature and location of an arrhythmia. There are a variety of mapping techniques that are used in scar-based VT ablations. Many of these techniques require the presence of sustained VT. In many instances patients are unable to tolerate sustained VT, and alternative approaches must be taken (see substrate-based ablation below). In order to induce VT, an operator usually performs a pacing stimulation protocol from the right ventricular apex or right ventricular outflow tract (RVOT). Frequently, the resulting VT is not the patient's "clinical VT" (e.g. it may have a different cycle length or morphology). Moreover, many patients will have multiple inducible VTs. Ideally all VTs that are reproducibly inducible and hemodynamically tolerated should be mapped and ablated.

## Activation Mapping (Focal Tachycardias)

One common form of mapping used in the electrophysiology lab is "activation mapping". This form of mapping is performed while a patient is in VT. To accomplish this, a mapping catheter is maneuvered to different sites in the chamber of interest, focusing on areas suggested by the 12-lead ECG or over scars detected by Echo or MRI. The recording electrode at the catheter tip reflects local myocardial activation. By comparing the timing of the local electrogram to a standard reference (typically the onset of the QRS for VT), a great deal can be learned about the arrhythmia. Activation mapping is of particular use in focal tachycardias where the local electrogram (EGM) typically precedes the onset of the QRS by 20–30 ms at site of the focus. By locating areas with very early EGM-QRS timings the arrhythmia focus can be localized and ablated.

## Re-entrant Tachycardia

In contrast to focal VTs, re-entrant VTs demonstrate continuous electrical activity as the wavefront propagates around a circuit. Strictly speaking, there is no "earliest" point in the circuit.[121] Operators frequently use this form of mapping to identify a critical isthmus (or zone of slow conduction) where the VT is likely to be terminated **(Fig. 11)**. Typically, these sites are characterized by a diastolic electrogram that is low amplitude and fractionated. However, the specificity of isolated diastolic potentials is limited as they can be observed in regions other than a critical isthmus.[122]

## Entrainment Mapping

During activation mapping it is not uncommon to encounter areas with fractionated diastolic potentials that may or may not be participating in the VT circuit. In order to determine whether such a site is participating in the VT circuit, a technique known as entrainment mapping can be used. Entrainment refers to the continuous resetting of a tachycardia circuit by pacing at a cycle length slightly faster than the tachycardia cycle length (TCL). To demonstrate entrainment, the tachycardia must be accelerated to the pacing cycle length and tachycardia resumption upon cessation of pacing. When pacing from a site remote from the circuit, the pacing impulse travels toward the circuit and penetrates it in two directions. In the antidromic direction, the impulse collides with the previous circulating orthodromic wavefront. In the orthodromic direction, the

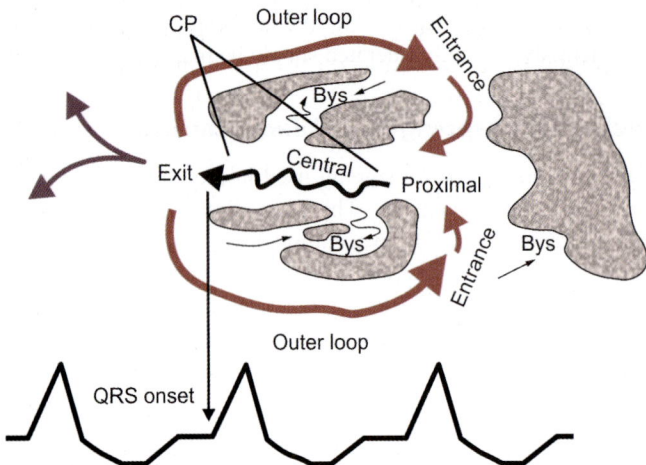

**FIGURE 11:** Hypothetical VT circuit consisting of a central isthmus with entrance and exit zones. The outer loops course around the scarred areas (border zones) and bystander loops (Bys) are found within the scar. (*Source:* Modified from reference 123)

impulse travels around circuit, resetting the tachycardia. The result of entrainment from a site remote to the VT circuit is QRS fusion where the resulting QRS morphology represents a fusion between the tachycardia circuit and a purely paced beat.

If entrainment is performed from a site within the circuit, the resulting QRS morphology should replicate the VT morphology exactly. This is known as concealed entrainment. If concealed entrainment is observed, several other features can be examined to confirm that the pacing catheter is located within the tachycardia circuit. The stimulus-QRS interval can be compared to the electrogram-QRS interval during VT. These intervals should be similar if pacing is taking place along the circuit (generally within 10 ms).

The post-pacing interval (PPI) refers to the duration between the last pacing stimulus artifact and the return of next local EGM at that pacing site. This represents the time that it takes for the last orthodromically stimulated wavefront to revolve around the circuit. If pacing is performed from a site within the circuit, the PPI should equal the TCL. A PPI-TCL less than or equal to 30 ms suggests that pacing site is within the VT circuit. Pacing at a site remote from the re-entry circuit can also entrain tachycardia, but in this instance the PPI is equal to the conduction time from the pacing site to the circuit through the circuit and back to the pacing site and therefore exceeds the TCL.[123]

The VT model proposed by Stevenson and others have greatly enhanced out understanding about the entrainment of these circuits.[124] The hypothetical VT shown in **Figures 11 and 12** depicts a "figure of eight" circuit with a common central isthmus.[123] The common isthmus has an entrance and exit as well as central regions. The QRS complex is inscribed after the wavefront leaves the exit site and begins propagating around the border of the scar around two outer loops. The wavefronts then enter the infarct region through entrances to reach the entrance of the isthmus. Regions that are within scar do not participate in the circuit are labeled as bystanders.

Analysis of the PPI can reveal a great deal about the location of the pacing catheter with respect to this hypothetical circuit. As above, when pacing from within the VT circuit (i.e. from the exit, or critical isthmus) the QRS morphology will be identical to the VT and the PPI-TCL should be less than or equal to 30 ms **(Figs 13A to C)**. Pacing from a bystander site will, however, produce a PPI-TCL that is greater than or equal to 30 ms as it reflects the time for the stimulus to leave the bystander region, propagate around the circuit and return to the site of pacing. In this instance the QRS morphology will resemble the VT so long as the bystander is insulated by scar from

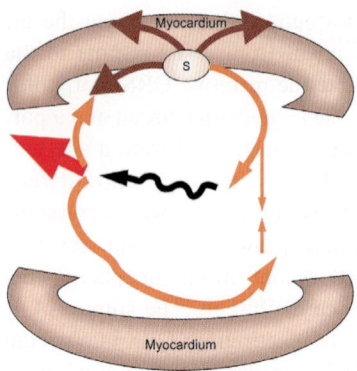

**FIGURE 12:** Stimulation is performed at an outer loop site (S) in the re-entry circuit that shown in Figure 12. The stimulated orthodromic wavefront propagates through the circuit, resetting the re-entry circuit. After the last stimulus, the pacing site is next depolarized by the orthodromic wavefront that has made one revolution through the circuit. The PPI therefore approximates the TCL. Pacing at this site also directly stimulates surround myocardial tissue. The QRS morphology therefore differs from that of the VT. (*Source:* Modified from reference 123)

surrounding myocardium. Entrainment from the true isthmus will result in stimulus to QRS = mid-diastolic EGM to QRS. Pacing from the innerloop bystander shows that the stimulus to QRS will be greater than EGM to QRS onset. Entrainment from an outerloop site will generally produce a PPI-TCL less than or equal to 30 ms (as it is within the circuit), however, the resulting QRS morphology will represent fusion between the circuit's propagating wavefront and direct depolarization of surrounding myocardium **(Figs 13A to C)**.

Entrainment can also be used to differentiate a diastolic potential that reflects activation of the VT circuit and a "far field potential" which is due to depolarization of tissue remote from the circuit. During entrainment, a potential that is participating the VT circuit will be captured by the entrainment maneuver and will be obscured by the pacing artifact. In contrast, the inability to "entrain" the potential whereby it appears dissociated from the pacing suggests that the potential is a far field electrogram and not part of the VT circuit.

It can be challenging to localize an ideal target for ablation even under the best of circumstances. The size and location of an isthmus can vary widely depending on a patient's substrate. On most occasions, ablation of a critical portion of a VT circuit requires multiple lesions depending on the volume of the isthmus. While any one of the above mentioned findings (i.e. mid-diastolic potential, concealed entrainment, Stim-QRS ~ EGM-QRS, PPI-TCL < 30 ms) does not itself predict an ideal site for ablation, the presence of multiple findings is likely to increase the rate of success.[125]

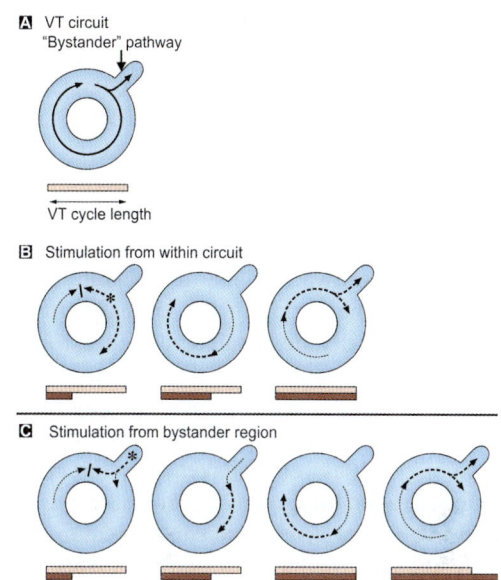

**FIGURES 13A TO C:** Stimulation during VT is shown: (A) A VT circuit is depicted with an arrow showing the direction of impulse propagation. The impulse also enters a bystander or dead-end pathway to the side. The length of one complete VT cycle is shown as a white bar below; (B) A stimulus (*) is delivered from within the circuit; propagation occurs in the same direction around the circuit path as during VT as well as in the opposite direction, where it collides with the advancing "head" of the prior VT beat (dimmed arrow). A black bar beneath tracks the progress of the stimulated wavefront during the figure. At right, the wavefront has continued propagating around the circuit until it reaches the site of stimulation. The time required to return to the site of stimulation (the postpacing interval) just equals the VT cycle length (black bar = white bar); (C) A stimulus is given from a bystander pathway. The impulse must travel a short distance to arrive at the circuit, after which it propagates as in (B). However, the PPI (black bar) exceeds the VT cycle length by the time required to exit from and return to the point of stimulation. (*Source:* Modified from reference 138)

## Electroanatomic Three-dimensional Mapping

The advent of three-dimensional EAM systems, such as CARTO (Biosense Webster, Baldwin Park, CA) or EnSite/NavX (St. Jude Medical, St. Paul, MN), has greatly enhanced the ability to perform complex ablations. Whereas conventional mapping relies on interpretation of intracardiac electrograms and two-dimensional fluoroscopic imaging, EAM provides additional real-time three-dimensional data about a patient's arrhythmogenic substrate. Although there are differences in their underlying technologies, in general, these systems allow operators to determine the spatial orientation of catheters in three-dimensions, delineate anatomic areas of interest, define cardiac chamber geometry and locate areas of scar. In addition, they can be used to create detailed activation maps that help to accurately locate the region and nature of the arrhythmia (e.g. focal versus macro-re-entrant). Given the complexity of the procedure, EAM systems are particularly useful for ablation in patients with scar-based VT.

Both CARTO and EnSite can be used to perform activation mapping during VT. To perform this, a designated catheter is moved to various locations in the chamber of interest. At each point, the mapping software is able to integrate the position of the catheter and the timing of the local electrogram with respect to an arbitrarily designated reference (or fiducial) point. The temporal relationship of the local activation to the reference point is color coded with red generally representing early activation sites, purple representing late activation sites and other colors, such as yellow, green and blue, representing intermediate activation times. This isochronal map is displayed in real-time on a review screen and can be used to distinguish focal from re-entrant arrhythmias. In addition to this, it localizes the source or circuit in the cardiac chamber. Focal arrhythmias generally have a small site of early activation with centrifugal spread away from the source. In contrast, macro-re-entrant rhythms display transitions from early (red) to intermediate (yellow-green-blue) to late (purple) as the wavefront propagates around the circuit with characteristic early-meets-late patterns. By tracing this propagation map around the chamber, the location of the circuit can be visualized.

## Voltage Mapping

Another powerful application of EAM is its ability to delineate areas of scar. Areas of scar typically exhibit low voltage electrograms. A cardiac chamber can be readily mapped by placing a catheter at various locations and recording the signal amplitude. As with endocardial activation, a color-coded voltage scale can be used to display areas of low voltage amplitude, to distinguish between areas of scar, dense scar and relatively normal tissue **(Figs 14A and B)**. Voltage mapping is critical for re-entrant VT as the majority of these circuits are intimately related to scar border zones and other anatomic obstacles. Understanding the location and density of scar can be used to perform VT ablations in patients during sinus rhythm (see substrate-based ablation below).

## Pace Mapping

Pace mapping is a useful technique for individuals with hemodynamically unstable VT in whom entrainment or activation mapping is not possible. The 12-lead ECG of the

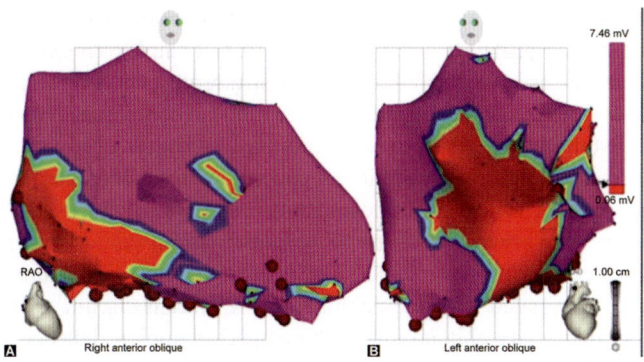

**FIGURES 14A AND B:** Voltage map of the left ventricle (LV), RAO and LAO view, in an individual with large, healed inferolateral myocardial infarction. Local bipolar electrogram voltage of less than or equal to 0.5 mV has been arbitrarily selected as the threshold for delineating low electrogram amplitude, as seen on the voltage scale on the right. Areas of red coloration represent sites with the lowest electrogram amplitude, with the area of next lowest voltage demarcated by yellow, followed by green, etc. The area colored in magenta indicates normal electrogram voltage. (*Source:* Modified from reference 139)

paced site is compared to the 12 lead of the VT. When the paced and comparison QRS morphologies are identical, this is referred to as a 12/12 match. When performing pace mapping, it is essential to confirm that the lead placement for the pace map is identical to the 12 lead used for comparison. Often times, the precordial leads are placed in modified locations during EP studies. If the comparison ECG has been performed outside of the EP lab, alternate lead placement may render an inaccurate pace map.

Pace mapping can be useful for focal VTs where a 12/12 match is presumed to represent the site of origin.[126] However, its utility is somewhat limited in patients with macro-re-entrant VT. In patients with scar-related VT, a perfect or near perfect pace map usually indicates that the catheter is located near the VT exit site. However, ablation at a VT exit site may merely result in shifting the exit to a new location and fail to eliminate the circuit. Ablation at isthmus sites are more likely to be successful, however this can be difficult to perform with pace mapping alone. Regions of functional block that may define activation pathways during VT may not be present while pacing during sinus rhythm. When pace mapping in a defined isthmus, the stimulated wavefront can propagate in both antidromic and orthodromic directions.[127] The resulting QRS may be a fusion between these two wavefronts and differ from the morphology of the VT circuit under investigation.

Pace mapping can be used to identify areas of slow conduction which are typically found within infarct zones, and may represent positioning in a critical isthmus. When pacing normal myocardium, there is typically little or no delay between the pacing stimulus and the surface ECG complex (S-QRS). Since the isthmus of VT circuit usually represents a zone of slow conduction, pacing from this region can sometimes result in S-QRS delay. Brunckhorst and his colleagues performed pace mapping at 890 sits in 12 patients with postmyocardial infarction VT. Their data demonstrated that areas with S-QRS delay were always localized to an infarct region (as identified by electrogram voltage) and 13 of 14 areas of conduction delay were associated with the isthmus of a re-entrant circuit.[128] In a similar study, this group combined pace mapping data with S-QRS data to localize successful sites of ablation in patients with scar-related VT.[127]

Pace mapping within scar at sites with an isolated diastolic potential during sinus rhythm has also been shown to be an indicator of a critical isthmus **(Fig. 15)**. In one study, application of this strategy resulted in freedom from recurrent VT in 16 of 19 patients (84%) during a mean follow-up period of 10 months.[129] Finally, pace mapping can be useful to define regions within scar that are electrically unexcitable. Soejima and his colleagues

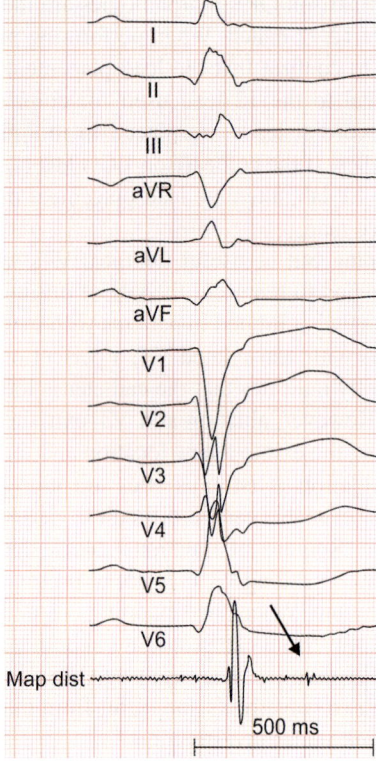

**FIGURE 15:** Example of an isolated diastolic potential recorded during sinus rhythm within left ventricular scar. (*Source:* Modified from reference 129)

demonstrated that in some patients, re-entry circuit isthmuses can be identified by delineation of electrically unexcitable scar that defines their borders.[130] The RF ablation in these regions abolished inducible VT in 10 of 14 patients and abolished or markedly reduced spontaneous VT during follow-up.

## Substrate-based Ablation

The term "substrate-based ablation" has been used to describe the combination of techniques used to perform VT ablations in patients with hemodynamically unstable VT who cannot tolerate entrainment or activation mapping. The specific protocols for substrate ablation vary from center to center. However, in general, the following steps are taken **(Flowchart 2 and Fig. 16)**:

- A voltage map of the chamber of interest is created such that areas of scar can be defined.
- Areas with diastolic potentials and fractionated electrograms are tagged and noted on the electroanatomic map.
- Pace mapping is performed in multiple areas with particular attention to QRS morphology and S-QRS duration and electrically unexcitable scar.
- Probable VT exit sites as well as isthmuses are identified based on the above information.
- Ablations in regions of potential isthmus sites are undertaken.

Although substrate-based ablation is often performed in patients with poorly tolerated VT, it can also be used in conjunction with other mapping techniques such as entrainment mapping. Substrate mapping may help to operators to identify areas likely to be of particular interest for re-entrant circuits. Following this an operator may induce VT for short periods of time and further localize potential ablation sites.

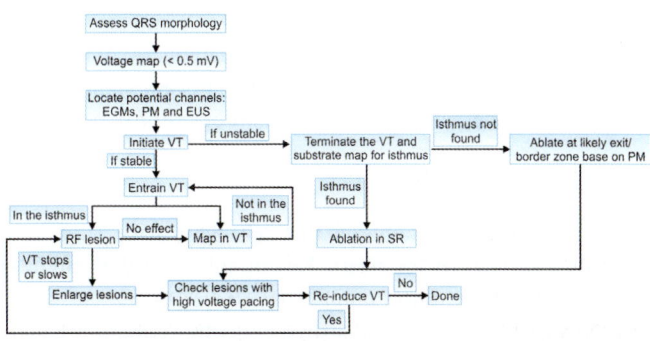

**FLOWCHART 2:** Steps of VT ablation

(*Abbreviations:* EGM: Electrogram; EUS: Electrically unexcitable Scar; PM: Pace-map). (*Source:* Modified from reference 111)

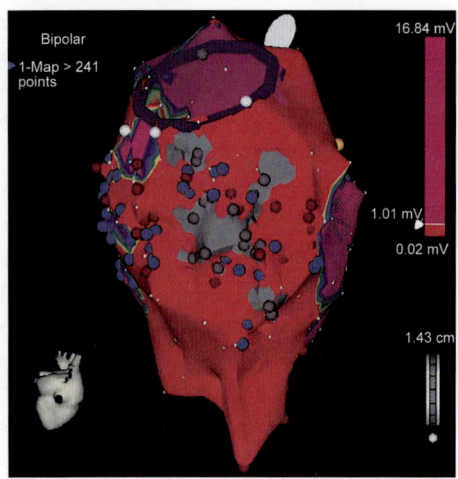

**FIGURE 16:** Voltage map of the posterior aspect of the left ventricle. Radiofrequency energy was delivered at points marked in red. Sites with an isolated potential during sinus rhythm are marked in blue and sites where there was noncapture are marked in grey[129]

## Safety

The VT ablation is complex procedure performed in patients who frequently have advanced illness. These features coupled with long procedure times and extensive ablation in the left-sided circulation creates risk for a myriad of serious complications, including stroke, valvular injury, major bleeding, tamponade, hemodynamic collapse and death. In two prospective trials in which VT ablation was performed for recurrent VT the periprocedural mortality was 2.7–3%.[108,131] The risk of a major complication, such as the ones mentioned above, was 8–10%. The risk of minor complications was 6–7.3%.

The 1998 North American Society of Pacing and Electrophysiology (NASPE) Prospective Catheter Ablation Registry, significant procedural complications were observed in 3.8% of patients undergoing a VT ablation.[132] However, this registry included idiopathic VT patients and the higher complication rates should be expected in sicker patient populations. As with most complex procedures, complication rates may vary significantly depending on the patient population and operator experience.

## Epicardial VT

It is increasingly recognized that VT may arise from the epicardial surface of the heart. Re-entrant epicardial circuits are particularly common among patients with nonischemic cardiomyopathies as well as arrhythmogenic right ventricular dysplasia (ARVD).[133] However, epicardial re-entry has also been documented in 15–23% of patients with postmyocardial infarction VT.[133,134] The presence of an epicardial circuit may be the cause of failure for many endocardially based VT ablations. Various ECG criteria have been proposed to recognize the presence of an epicardial VT circuit.[120,135,136] In general, epicardial circuits manifest a QRS onset that is often slurred, with pseudodelta wave appearance. A pseudodelta wave of more than 34 ms is quite specific for epicardial VT (95%), but less sensitive than an intrinsicoid deflection time of more than 85 ms (defined as the interval measured from the earliest ventricular activation to the peak of the R wave in V2).[120]

Access to the epicardial space can be accomplished by the approach described by Sosa and his colleagues.[134] Using a needle originally designed to enter to epidural space, fluoroscopy is used to approach the epicardium. Contrast injection confirms proper placement of the needle and an introducer sheath is subsequently advanced over a guidewire into the pericardial space **(Figs 17A to C)**. Once access to the pericardial space is achieved, mapping of VT can be performed using methods similar to endocardial ablation. Access to the pericardial space can be challenging in patients with prior cardiac surgery or pericarditis due to the presence of dense adhesions. In these instances, hybrid surgical approaches, such as the use of a

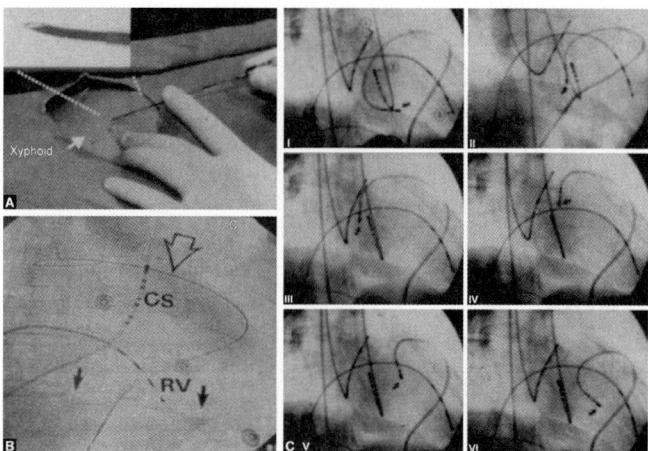

**FIGURES 17A TO C:** Technique used to insert the mapping and ablation catheter in the pericardial space. (A) A touchy needle is advanced into the pericardial space with the aid of fluoroscopy and contrast injection; (B) A soft guidewire is introduced into the pericardial space (large arrow), where contrast is also present (panel 2). An introducer is then advanced, the guidewire is removed and an ablation catheter is introduced into the pericardial sac to perform epicardial mapping and/or ablation; (C) Demonstrates a right anterior oblique view at 60° obtained by fluoroscopy during epicardial mapping procedure. The epicardial catheter (arrow) is manipulated and placed in different locations of the epicardial space (I to VI), where epicardial electrograms are obtained. (*Source:* Modified from reference 132)

subxiphoid window or limited anterior thoracotomy, can be employed to access the epicardium.[137]

The risk of injury to the coronary arteries is a significant concern with epicardial VT ablation and coronary angiography is usually performed to ensure that an ablation site is safe. In addition, the left phrenic nerve courses along the lateral aspect of the left ventricle and pacing should be performed at all ablation sites to confirm the absence of phrenic nerve capture. Sacher and his colleagues recently evaluated the safety of epicardial ablation through retrospective analysis of 156 epicardial VT procedures at three tertiary care centers.[133] Major periprocedural complications occurred in 9% of patients approximately half of which were due to epicardial bleeding. One patient developed asymptomatic right coronary artery stenosis due to cryoablation. There were no periprocedural deaths. It is important to note that these procedures were performed at experienced centers in carefully selected patients. The safety and efficacy of this procedure will likely continue to evolve over the next several years.

## IDIOPATHIC VENTRICULAR TACHYCARDIA

Most causes of VT are due to underlying structural heart disease (mainly ischemia or cardiomyopathy). Idiopathic VT (VT in patients without structural heart disease) accounts for 10–20% of VT cases evaluated by electrophysiology centers.[140,141] It is important to recognize these patients as they frequently respond to nonpharmacologic ablative techniques. Idiopathic VT can be broadly classified as polymorphic VT and monomorphic VT **(Table 1)**. In this chapter we will describe the clinical, electrocardiographic (ECG) and electrophysiologic findings for patients with monomorphic idiopathic VT.

### TABLE 1
**Idiopathic ventricular tachycardia**

| *Monomorphic ventricular tachycardia* | *Polymorphic ventricular tachycardia* |
|---|---|
| • Outflow tract VT-RVOT-VT, LVOT-VT, aortic cusp VT, Epicardial VT<br>• Fascicular VT-LAF-VT, LPF-VT, Septal VT<br>• Adrenergic monomorphic VT<br>• Annular VT-mitral annular, tricuspid annular | • Long QT syndrome<br>• Brugada syndrome<br>• Short coupled torsades<br>• Short QT syndrome<br>• Catecholaminergic polymorphic VT<br>• Idiopathic VF |

(Abbreviations: LAF: Left anterior fascicular; LPF: Left posterior fascicular; LVOT: Left ventricular outflow Tract; RVOT: Right ventricular outflow tract; VF: Ventricular fibrillation; VT: Ventricular tachycardia)

## Outflow Tract-Ventricular Tachycardia (OT-VT)

This form of idiopathic VT arises from the outflow tract of the right or the left ventricle. Based on its origin it can be classified as VT that arises from the right ventricular outflow tract (RVOT VT), the left ventricular outflow tract (LVOT VT) and the aortic cusps (Cusp VT) **(Table 2)**.

The RVOT VT is more common in females and is usually seen in third to fifth decade of life while LVOT VT is equally distributed between males and females.[142] Symptoms include palpitations, dizziness, atypical chest pain and syncope. There are three predominant clinical forms of this syndrome: (i) non-sustained repetitive monomorphic VT alternating with periods of sinus rhythm; (ii) paroxysmal exercise-induced sustained VT[143] and (iii) frequent premature ventricular contractions (PVCs) often in a bigeminal fashion. There are also reports of tachycardia-induced cardiomyopathy in patients with a high PVC burden.[144]

## RVOT VT

This form of VT is associated with a characteristic ECG morphology of LBBB with inferior axis suggesting origin from the right ventricle outflow tract **(Fig. 18)**. Jadonath et al.[145] evaluated the utility of 12-lead ECG in localizing the site of

### TABLE 2
**Electrocardiographic findings in outflow tract VT**

| RVOT VT | LVOT/aortic cusp VT |
|---|---|
| • QS in avR<br>• Monophasic R wave in II, III, avF, $V_6$<br>• Septal sites have negative qrs in avL<br>• Free-wall sites have wider notched qrs in inferior leads and later precordial transition ($V_4$)<br>• Lead I shows Q wave in anterior sites and dominant R wave in posterior sites<br>• Phase analysis as measured from the earliest QRS onset to:<br>  a. Earliest QRS onset is $V_2$<br>  b. Initial peak/nadir in III $\geq$ 120 ms<br>  c. Initial peak/nadir in $V_2 \geq$ 78 ms | • Early precordial transition ($V_2$ $V_3$)<br>• Taller and broader R wave or RBBB in $V_1$ $V_2$<br>• Septal LVOT has dominant Q in $V_1$, qrs II/qrs III > 1<br>• Aortomitral LVOT has qR in $V_1$, qrs II/qrs III < 1<br>• Cusp VT – notch in $V_5$, lack of S in $V_5$-$V_6$, taller R in inferior leads<br>• Lead I – rS in left cusp VT, notched R in noncoronary cusp VT<br>• Phase analysis (Cusp VT) as measured from the earliest QRS onset to:<br>  d. Earliest QRS onset not in $V_2$<br>  e. Initial peak / nadir in III $\leq$ 120 ms<br>  f. Initial peak/nadir in $V_2 \leq$ 78 ms |

(Abbreviations: LVOT: Left ventricular outflow tract; RVOT: Right ventricular outflow tract)

origin of RVOT VT. A QS pattern in lead avR and monophasic R waves in leads II, III, avF and V6 were noted in each patient at all pacing sites. The anterior sites showed a dominant Q wave or a qR complex in lead I and QS complex in avL. Pacing at the posterior sites produced a dominant R wave in lead I, QS or R wave in avL and an early precordial transition (R/S ≥ 1 by V3). Coggins et al. showed that septal RVOT VT was associated with negative QRS complex in avL while RVOT VT arising from the lateral wall produced a positive QRS complex in avL.[146] Dixit et al.[147] used pace-mapping techniques to differentiate between RVOT VT arising from the free wall versus the septal wall. They showed that free-wall RVOT VT was associated with wider and notched QRS complexes in the inferior leads and the precordial R wave transition was late (R/S ≥ 1 by V4). Recently OT-VTs arising from near the His-bundle region have been described.[148,149] The characteristic ECG abnormalities for VT arising from this site include a R/RSR' pattern in avL, taller R wave in I, small R waves in inferior leads, taller R waves in V5, V6 and QS pattern in V1 **(Fig. 19)**. Representative example of RVOT VT is shown in **Figure 18**.

## VT Arising from the Pulmonary Artery

Idiopathic VT may arise just proximal to the outflow tract in the para-Hisian region. This ECG will mimic that of RVOT except for the characteristic rSr1 in avL **(Fig. 19)**.

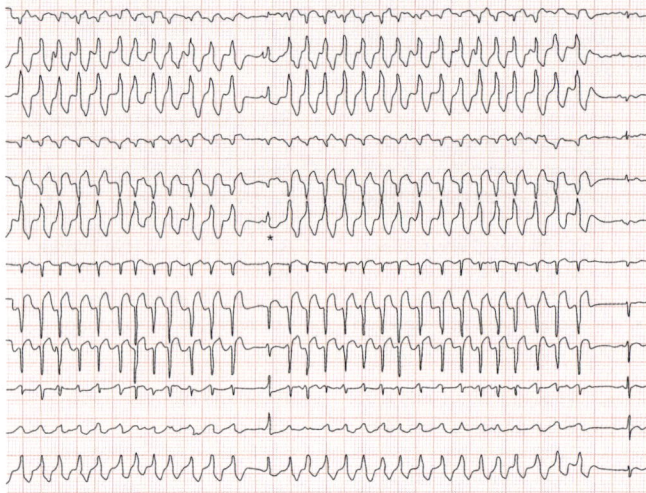

**FIGURE 18:** Twelve-lead ECG of ventricular tachycardia arising from the right ventricular outflow tract. There is left bundle branch block morphology with late transition (V4) in the precordial leads and an inferior axis in the limb leads. Negative QRS complex in lead avL suggests a septal origin while a q wave in lead I points to anterior focus. Fusion beat is noted in the middle of the tracing (*)

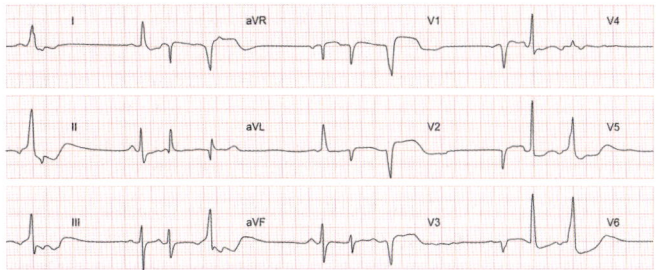

**FIGURE 19:** Twelve-lead ECG during sinus rhythm showing PVC arising from the para-Hisian region. There is tall R wave in lead I, QS in V1, R in lead II > R in lead III, taller R waves in V5 and V6 and a characteristic rSR in lead avL. Local intracardiac signal at this site preceded the QRS onset by 25 ms

Idiopathic VT arising from the pulmonary artery (PA) has also been described.[150,151] The origin of VT from the PA is thought to be from remnants of embryonic muscle sleeves that have been noted in amphibian and mammalian outflow tract.[152] This is supported by the presence of a sharp potential at these sites that preceded the onset of ventricular activation during VT.[150] Sekiguchi et al.[151] noted the following ECG characteristics during VT that favor PA VT as compared to RVOT VT: (i) Larger R-wave amplitude in inferior leads; (ii) The ratio of the Q-wave amplitude in avL/avR was larger in the PA group; (iii) Significantly larger R/S amplitude in lead V2 in patients with PA VT than in those with RVOT VT.

## LVOT VT

The VT arising from the LVOT shares similar characteristics to the RVOT VT due to a common embryonic origin. This form of VT can be differentiated from the RVOT VT by differences in QRS morphology **(Fig. 20)**. The LVOT VT is suggested by LBBB morphology with inferior axis with small R waves in V1 and early precordial transition (R/S $\geq$ 1 by V2 or V3) or RBBB morphology with inferior axis[153-156] and presence of S wave in V6.[157,158] The LVOT VT arising from the septal para-Hisian region has an ECG pattern of QS or Qr in V1 with early precordial transition and ratio of QRS in leads II/III greater than 1 while LVOT VT arising from the aortomitral continuity has a characteristic qR pattern in V1 with a ratio of QRS in leads II/III less than or equal to 1.[159]

## Cusp VT

Case reports of idiopathic VT with LBBB morphology and inferior axis that failed ablation in the RVOT but were successfully ablated in the left coronary cusp[160,161] were

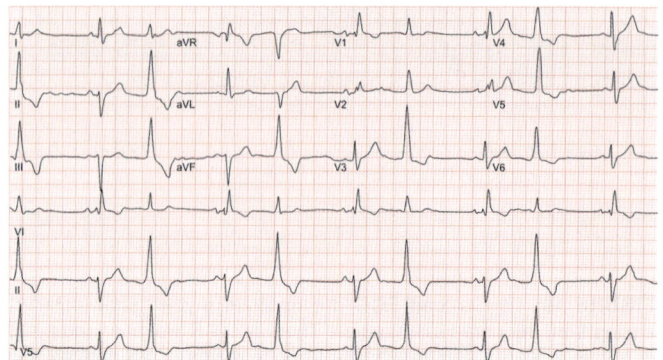

**FIGURE 20:** Twelve-lead ECG during sinus rhythm showing PVC arising from the left ventricular outflow tract that was successfully ablated in the aortomitral continuity region. There is qR wave in lead V1 that differentiates this from a septal LVOT focus that has a QS or Qr pattern in lead V1

described in 1999. Kanagaratnam et al.[162] reported the ECG characteristics of 12 patients with outflow tract VT that required ablation in the aortic sinus of Valsalva. All patients had LBBB inferior axis morphology with taller monophasic R waves in inferior leads and an early precordial R wave transition by V2 or V3. The VT arising from the left cusp had an rS pattern in lead I, and VT arising from the noncoronary cusp had a notched R wave in lead I. Ouyang et al.[163] evaluated the ECG differences between 8 patients with RVOT VT and 7 patients with VT arising from the aortic sinus cusp (5 from left sinus and 2 from right sinus). They found that a broader R wave duration and a taller R/S wave amplitude in V1 and V2 favored VT arising from the aortic cusp. Yang[164] used phase differences in the 12-lead ECG to differentiate between RVOT VT (32 patients) and aortic cusp VT (15 patients). They showed that RVOT VT was associated with earliest ventricular activation in V2 and a shorter time from onset of ECG to peak/nadir in lead III and V2.

## Epicardial VT

Outflow tract VT occasionally arises from the epicardial surface of the heart that requires ablation in the great cardiac vein or via pericardial approach.[148,165-167] Ito et al.[148] showed that the Q wave ratio of avL to avR greater than 1.4 or an S wave amplitude in V1 greater than 1.2 mV was useful in differentiating between epicardial VT from aortic cusp VT. Daniels et al.[167] showed that 9% of the patients with idiopathic VT referred to their institution had an epicardial focus. They found that a delayed precordial maximum deflection index greater than or equal to 55 (calculated by measuring the time from the QRS onset to the earliest maximum deflection (nadir or peak) in precordial leads and dividing it by the total QRS duration) differentiates this form of idiopathic epicardial VT from other forms of outflow tract VT **(Figs 21A and B)**. Recently idiopathic VT arising from the crux of the heart has been described that presents as rapid VT usually triggered by exercise and can lead to hemodynamic compromise.[168]

The 12-lead ECG morphology of this VT is similar to that of maximally pre-excited posteroseptal accessory pathways with QRS transition from V1 to V2 and QS complexes in inferior leads **(Fig. 22)**. This can be successfully ablated in the middle cardiac vein or via an epicardial approach.

## Management

The majority of the patients with outflow tract VT have a benign course with a very low-risk of sudden death.[169-171] It can be associated with tachycardia-induced cardiomyopathy that improves after successful treatment.[144,172] It is important

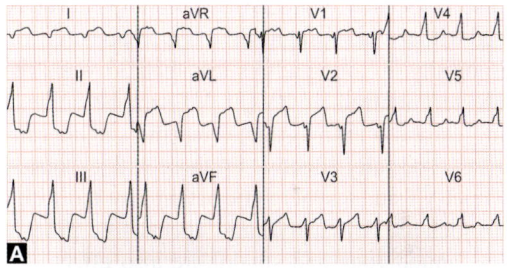

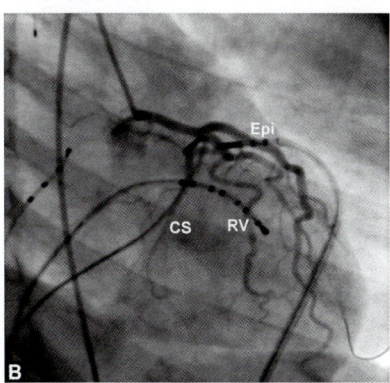

**FIGURES 21A AND B:** (A) Twelve-lead ECG of VT that was successfully ablated on the epicardial surface of the heart. The precordial maximum deflection index is greater than 55 (calculated by measuring the time from the QRS onset to the earliest maximum deflection (nadir or peak) in precordial leads and dividing it by the total QRS duration). (B) Fluoroscopic view of the catheter position with ablation catheter (Epi) on the epicardial surface of the heart, right ventricular (RV) catheter and coronary sinus catheter (CS). The site of origin of the VT is very close to the left anterior descending (LAD) coronary artery

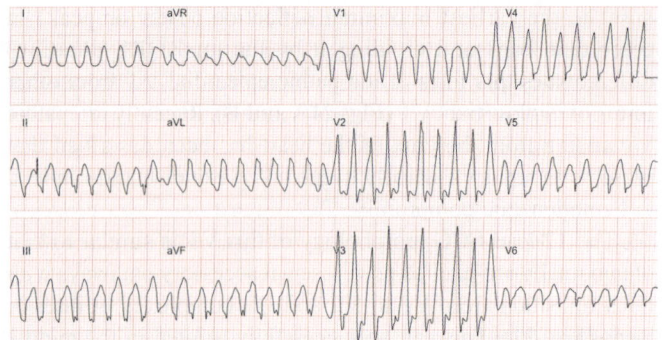

**FIGURE 22:** Twelve-lead ECG of ventricular tachycardia arising from the crux of the heart. There is left bundle branch block morphology with early transition (V2) in the precordial leads and qs complex in the inferior leads. This matches the 12-lead ECG morphology of maximally pre-excited posteroseptal accessory pathway. This VT was successfully ablated via an epicardial approach

to differentiate this form of VT from VT associated with arrhythmogenic right ventricular cardiomyopathy (ARVC) that is also associated with LBBB morphology. The ARVC is associated with a worse prognosis and is responsible for sudden death especially in young adults less than 35 years old.[173,174]

Acute termination of RVOT VT can be achieved with adenosine, carotid sinus massage, verapamil and lidocaine. Beta-blockers are especially effective for those with exercise-induced outflow tract VT and a synergistic action is noted with calcium channel blockers. Antiarrhythmic agents like procainamide, flecainide, amiodarone and sotalol are also effective in these patients. There was a trend toward greater efficacy with sotalol in a study of 23 patients with RVOT VT.[175] Nicorandil, a potassium channel opener, has been reported to terminate and suppress adenosine-sensitive VT.[176]

## Catheter Ablation

Catheter ablation using RF energy to cure patients with outflow tract VT is associated with a high success rate due to the focal origin of this form of VT. The 12-lead ECG is a useful initial guide to localize the site of origin of the tachycardia. Intracardiac mapping to select the optimal site for ablation include activation mapping (earliest local intracardiac electrogram that precedes the onset of surface QRS during VT) and pace mapping (pacing the ventricle from a selected site in sinus rhythm to match the 12-lead morphology of the spontaneous or induced VT). The use of three dimensional (3D) electroanatomical mapping systems reduces fluoroscopic exposure and improves the efficacy of catheter ablation.[177,178] The success rate of catheter ablation for outflow tract VT reported from various series is greater than 90%[132,179] with a recurrence rate of 5% (mainly in the first year). Serious complications include induction of RBBB (2%), cardiac perforation and tamponade (1%); there has been one death reported secondary to complications from RVOT perforation.[146] Ablation for LVOT VT has been associated with occlusion of a coronary artery.[180] Failure of endocardial RF ablation can be due to epicardial location of the VT focus **(Fig. 21)**. Epicardial ablation can be achieved by a subxiphoid technique of pericardial puncture as described by Sosa[181] or by ablating within the great cardiac vein. Coronary angiograms are performed prior to epicardial ablation and ablation in the aortic sinus to avoid ablation close to the coronary arteries that can lead to arterial damage and thrombus formation. Intracardiac echocardiography has also been used to provide real time visualization of the relationship between the ablation site and the coronary arteries during ablation in the aortic sinus.[182]

## Idiopathic Left Ventricular Tachycardia (ILVT) or Fascicular VT

This form of VT arises from the fascicles in the left ventricle. Based on the QRS morphology and the site of origin it can be classified as left posterior fascicular VT (LPF VT), left anterior fascicular VT (LAF VT) and left upper septal VT (Septal VT).[183]

This form of idiopathic VT was first described by Zipes et al.[184] in 1979 with the following characteristic triad: (i) induction with atrial pacing; (ii) the RBBB morphology with left axis deviation and (iii) occurrence in patients without structural heart disease. In 1981, Belhassen et al.[185] showed that this form of VT could be terminated by verapamil, the fourth identifying feature. In 1988, Ohe et al.[186] described another form of this VT with RBBB and right axis deviation that required ablation in the region of the left anterior fascicle. Shimoike[187] and Nogami[183] described a form of idiopathic VT with narrow QRS that required ablation in the upper LV septum.

The ILVT is typically seen in patients between the ages of 15–40 years with an earlier presentation in females.[142,188-193] Most of the affected patients are males (60–70%). The symptoms include palpitations, fatigue, dyspnea, dizziness and presyncope. Syncope and sudden death are very rare.[188] Most of the episodes occur at rest; however, this form of VT can be triggered by exercise and emotional stress.[193,194] Incessant tachycardia leading to tachycardia-induced cardiomyopathy has also been described.[191]

## ECG Recognition

The baseline 12-lead ECG is normal in most patients or it may show transient T wave inversions related to T wave memory shortly after a tachycardia episode terminates. The 12-lead ECG of left posterior fascicular VT (LPF VT) shows RBBB with left axis deviation suggesting an exit site from the inferoposterior ventricular septum **(Fig. 23)**. Nogami et al.[195] reported 6 patients with left anterior fascicular VT (LAF VT) that showed a RBBB with right axis deviation with the earliest ventricular activation in the anterolateral wall of the LV in all patients. Three patients had a distal type of LAF VT with QS or rS morphology in leads I, V5 and V6 and the other 3 had proximal type of LAF VT with RS or Rs morphology in the same leads. **Figure 24** shows PVCs that were successfully ablated at the distal LAF. The QRS duration in fascicular VT varies 140–150 msec and the duration from the beginning of the QRS onset to the nadir of the S-wave (RS interval) in the precordial leads is 60–80 msec unlike VT associated with structural heart disease that is usually associated with longer duration of QRS and RS intervals. This makes it difficult to differentiate fascicular VT from SVT with aberrancy using the criteria based on QRS morphology and RS interval.[196-198]

## Management

The long-term prognosis of patients with fascicular VT without structural heart disease is very good. Patients with mild symptoms without medical therapy did not show progression of their arrhythmias in a study of 37 patients with verapamil sensitive VT during an average follow-up of 5.8 years,[190] however those with incessant tachycardia can develop a

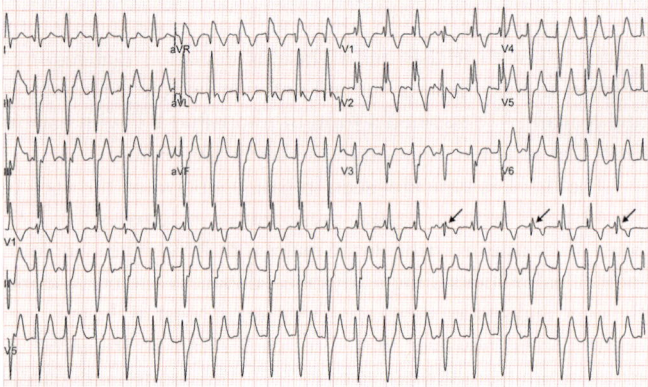

**FIGURE 23:** Twelve-lead ECG of VT arising from the left posterior fascicle (ILVT) with an RBBB superior axis morphology. The duration of the QRS complex and the RS interval are narrower than that of noted in VT associated with structural heart disease. However, the presence of AV dissociation and fusion beats (arrows) is diagnostic of VT. (*Source:* Modified from Badhwar et al. Idiopathic ventricular tachycardia: diagnosis and management. Curr Probl Cardiol. 2007;32:7-43)

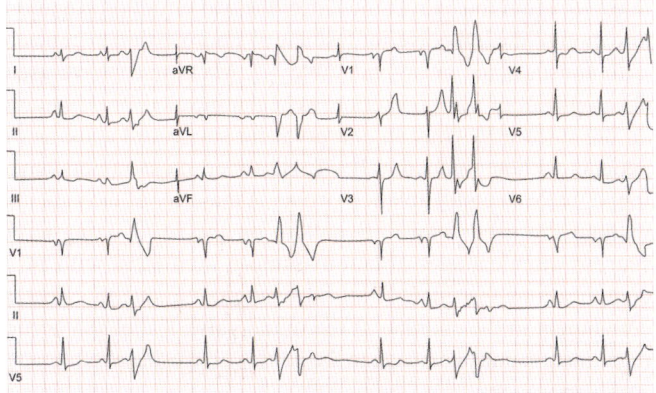

**FIGURE 24:** Twelve-lead ECG showing PVCs and couplets with RBBB inferior axis morphology suggestive of origin from the left anterior fascicle. The QRS complex in leads I, V5 and V6 show an rS morphology that is consistent with an exit site in the distal part of the left anterior fascicle where this VT was successfully ablated (*Source:* Modified from Badhwar et al. Idiopathic ventricular tachycardia: diagnosis and management. Curr Probl Cardiol. 2007;32:7-43)

cardiomyopathy.[191] Acute termination of VT can be achieved with intravenous verapamil (adenosine and Valsalva maneuvers are ineffective). Patients with moderate symptoms can be treated with oral verapamil (120–480 mg/day).

## Catheter Ablation

The RF catheter ablation is an appropriate management strategy for patients with severe symptoms or those intolerant or resistant to antiarrhythmic therapy. Nakagawa[194] and Wen[199] showed that successful ablation could be performed by targeting the earliest high-frequency Purkinje potential (and not the earliest ventricular activation) during VT and this could be recorded far from the exit site that shows the perfect pace map. Tsuchiya et al.[200] targeted a site recording both a late diastolic potential and a presystolic potential and showed tachycardia termination with catheter pressure at these sites. Nogami et al.[201] used an octapolar catheter to record double potentials during VT from the mid-septal LV and successfully terminated VT during catheter ablation at these sites. Ouyang et al.[202] used a 3D electroanatomical mapping system to record sites with retrograde Purkinje (retro PP) potential during sinus rhythm (sharp low amplitude potentials that followed a Purkinje potential and local ventricular electrogram) in patients with ILVT. They showed that the site recording the earliest retro PP during sinus rhythm correlated with early diastolic potential during VT. They suggest that use of the earliest retro PP as a target for ablation when VT cannot be induced in the electrophysiology lab. Chen et al.[203] used a noncontact mapping system to create a successful linear ablation line perpendicular to the wavefront propagation direction of the left posterior fascicle in sinus rhythm in patients with nonsustained or noninducible VT. Ma et al.[204] have used development of a left posterior fascicular block (LPFB) pattern on the surface ECG as an end point for successful ablation in patients with noninducible ILVT. However, most authors have found that successful ablation occurs in majority of the patients without the need for LPFB pattern on the ECG. Long-term success after catheter ablation is more than 92% with rare complications that include mitral regurgitation due to catheter entrapment in the chordae of the mitral valve leaflet and aortic regurgitation due to damage to the aortic valve using a retrograde aortic approach.[146,193,194,199, 205-207]

## Mitral Annular VT

There have been case reports of adenosine sensitive monomorphic VT that was successfully ablated at the anterobasal LV.[208-210] Tada et al.[211] were the first group to describe the prevalence and ECG characteristics of mitral annular VT (MAVT). Their definition was based on the ratio of atrial to ventricular electrograms less than 1 and the amplitude of the atrial and ventricular electrograms greater than 0.08 and 0.5 mV respectively at the successful ablation site.

Tada et al. reported that MAVT was noted in 5% of all the cases of idiopathic VT while Kumagai et al. showed that MAVT accounts for 49% of idiopathic repetitive monomorphic VT arising from the left ventricle (other sites included coronary cusps and inferoseptal region). Patients presented with palpitations and were noted to have repetitive monomorphic VT or frequent monomorphic PVCs. Tachycardia was noted spontaneously or initiated with isoproterenol. Termination was noted with adenosine (10–40 mg) and intravenous verapamil in some patients. The VT entrainment was not observed in any of the sustained episodes.

## ECG Recognition

Tada et al.[211] showed that the surface ECG in all patients with MAVT had an RBBB pattern (transition in V1 or V2); S-wave in V6 and monophasic R or Rs in leads V2–V6. They further classified MAVT into three categories depending on the site of origin as being anterolateral (AL) MAVT (58%), posterior (Pos) MAVT (11%) and posteroseptal (PS) MAVT (31%). In AL-MAVT, the polarity of the QRS complex in leads I and avL was negative and positive in the inferior leads; Pos-MAVT and PS-MAVT showed a negative polarity in the inferior leads and positive in leads I and avL. The AL-MAVT and Pos-MAVT showed a longer QRS duration and "notching" in the late phase of the R wave/Q wave in the inferior leads suggesting an origin from the free wall. This feature was not observed in PS-MAVT. Pos-MAVT showed a dominant R in V1 while PS-MAVT had a negative QRS component in V1 (qR, qr, rs, rS or QS). The Q-wave amplitude ratio of lead III to lead II was greater in PS-MAVT than in Pos-MAVT. **Figure 25** shows the representative ECG from PVCs arising from the lateral MA. Kumagai et al.[212] illustrated the delta-wave like beginning on the QRS complex during VT and showed a similarity between the MAVT and the maximally pre-excited left-sided accessory pathways in terms of QRS morphology.

## Catheter Ablation

Electrophysiology mapping was performed using activation mapping and pace mapping to localize the site of origin of the VT. All successful sites had an adequate atrial and ventricular electrogram satisfying the criteria for mitral annular origin and a potential was noted before the local ventricular electrogram in most of the patients. Pace mapping was useful in patients with nonsustained tachycardia. Acute success was obtained in all the patients in both the series; however, there was a recurrence rate of 8% in one series.[212]

## Tricuspid Annular VT

Recently VT arising from the TA has been described. This form of VT was noted in 8% of the patients presenting with idiopathic VT.[213] This was preferentially seen to originate from the septal region (74%) than the free wall (26%). Most of the septal VT

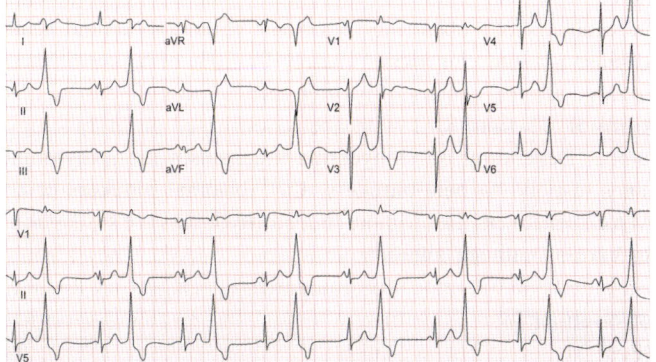

**FIGURE 25:** Twelve-lead ECG during sinus rhythm showing PVCs arising from the lateral mitral annulus. There is precordial concordance with negative QRS complexes in leads I, avL and inferior axis which is similar to the ECG morphology of a maximally pre-excited accessory pathway located on the lateral mitral annulus

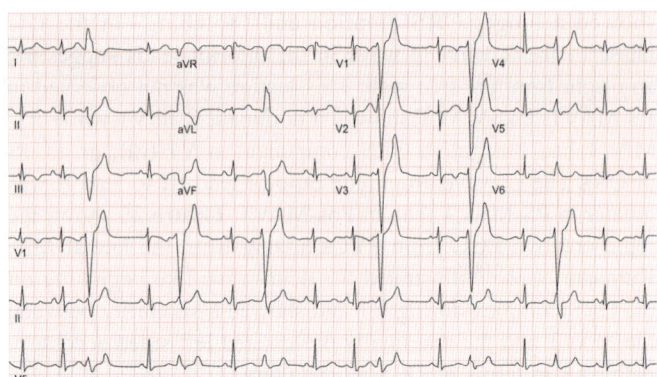

**FIGURE 26:** Twelve-lead ECG during sinus rhythm showing PVCs arising from the posterolateral tricuspid annulus. There is LBBB in V1 with late transition, left axis and notching in the inferior leads

was seen to arise from the anteroseptal region (72%). The septal VT had an early transition in precordial leads (V3), narrower QRS complexes, Qs in lead V1 with absence of "notching" in the inferior leads while the free-wall VT was associated with late precordial transition (> V3), wider QRS complexes, absence of Q wave in lead V1 and "notching" in the inferior leads (the timing of the second peak of the "notched" QRS complex in the inferior leads corresponded precisely with the left ventricular free-wall activation). **Figure 26** shows ECG characteristics of PVCs originating from posterolateral TA. The success rate for catheter ablation of the free-wall VT was 90% as compared to 57% in the septal group due to the presence of junctional rhythm and the likelihood of impairing AV nodal conduction with catheter ablation.

## SUMMARY

In summary, VT occurring in patients without structural heart disease accounts for approximately 10–20% of VTs evaluated at large referral centers. It is often difficult to differentiate this form of VT from SVT with aberration based on morphology alone. Depending on tachycardia mechanism the arrhythmia may respond to beta-blockers, $Ca^{++}$ channel blockers or to vagal maneuvers. In addition, these arrhythmias are susceptible to cure by catheter ablation.

## REFERENCES

1. Waldo AL, Wit AL. Mechanisms of cardiac arrhythmias. Lancet. 1993;341:1189-93.
2. Porter MJ, Morton JB, Denman R, et al. Influence of age and gender on the mechanism of supraventricular tachycardia. Heart Rhythm. 2004;1:393-6.

3. Blomstrom-Lundqvist C, Scheinman MM, Aliot EM, et al. ACC/AHA/ESC Guidelines for the Management of Patients with Supraventricular Arrhythmias–Executive Summary. A Report of the American College of Cardiology/American Heart Association Task Force on Practice Guidelines and the European Society of Cardiology Committee for Practice Guidelines (Writing Committee to Develop Guidelines for the Management of Patients with Supraventricular Arrhythmias) developed in collaboration with NASPE-Heart Rhythm Society. J Am Coll Cardiol. 2003;42:1493-531.
4. Durrer D, Schoo L, Schuilenburg RM, et al. The role of premature beats in the initiation and the termination of supraventricular tachycardia in the Wolff-Parkinson-White syndrome. Circulation. 1967;36:644-62.
5. Wellens HJ, Schuilenburg RM, Durrer D. Electrical stimulation of the heart in patients with Wolff-Parkinson-White syndrome, type A. Circulation. 1971;43:99-114.
6. Scherlag BJ, Lau SH, Helfant RH, et al. Catheter technique for recording His bundle activity in man. Circulation. 1969;39:13-8.
7. Gallagher JJ, Pritchett ELC, Sealy WC, et al. The preexcitation syndrome. Circulation. 1976;54:571-91.
8. Jackman WM, Friday KJ, Scherlag BJ, et al. Direct endocardial recording from an accessory atrioventricular pathway, localization of the site of block effect of antiarrhythmic drugs and attempt at nonsurgical ablation. Circulation. 1983;68:906-16.
9. Dreifus LS, Nichols H, Morse D, et al. Control of recurrent tachycardia of Wolff-Parkinson-White syndrome by surgical ligature of the A-V bundle. Circulation. 1968;38:1030-6.
10. Edmunds JH, Ellison RG, Crews TL. Surgically induced atrioventricular block as treatment for recurrent atrial tachycardia in Wolff-Parkinson-White syndrome. Circulation. 1969;39:105-11.
11. Durrer D, Roos JP. Epicardial excitation of ventricles in patient with Wolff-Parkinson-White syndrome (type B). Circulation. 1967;35:15-21.
12. Burchell HB, Frye RL, Anderson MW, et al. Atrioventricular and ventriculoatrial excitation in Wolff-Parkinson-White syndrome (type B). Circulation. 1967;36:663-9.
13. Cobb FR, Blumenschein SD, Sealy WC, et al. Successful surgical interruption of the bundle of Kent in a patient with Wolff-Parkinson-White syndrome. Circulation. 1968;38:1018-29.
14. Iwa T, Kazui T, Sugii S, et al. Surgical treatment of Wolff-Parkinson-White syndrom. Kyobu Geka. 1970;23:513-8.
15. Scheinman MM, Morady F, Hess DS, et al. Catheter-induced ablation of the atrioventricular junction to control refractory supraventricular arrhythmias. JAMA. 1982;248:851-5.
16. Morady F, Scheinman MM. Transvenous catheter ablation of a posteroseptal accesory pathway in a patient with the Wolff-Parkinson-White syndrome. N Engl J Med. 1984;310:705-7.
17. Morady F, Scheinman MM, Kou WH, et al. Long-term results of catheter ablation of a posteroseptal accessory atrioventricular connection in 48 patients. Circulation. 1989;79:1160-70.
18. Warin JF, Haissaguerre M, Lemetayer P, et al. Catheter ablation of accessory pathways with a direct approach. Results in 35 patients. Circulation. 1988;78:800-15.

19. Borggrefe M, Budde T, Podczeck A, et al. High frequency alternating current ablation of an accessory pathway in humans. J Am Coll Cardiol. 1987;10:576-82.
20. Jackman WM, Wang XZ, Friday KJ, et al. Catheter ablation of accessory atrioventricular pathways (Wolff-Parkinson-White syndrome) by radiofrequency current. N Engl J Med. 1991;334:1605-11.
21. Kuck KH, Schlüter M, Geiger M, et al. Radiofrequency current catheter ablation of accessory atrioventricular pathways. Lancet. 1991;337:1557-61.
22. Calkins H, Sousa J, El-Atassi R, et al. Diagnosis and cure of the Wolff-Parkinson-White syndrome or paroxysmal supraventricular tachycardias during a single electrophysiologic test. N Engl J Med. 1991;324:1612-8.
23. Scheinman MM. NASPE survey on catheter ablation. Pacing Clin Electrophysiol. 1995;18:1474-8.
24. Hindricks G for the Multicentre European Radiofrequency Survey (MERFS) investigators of the Work Group on Arrhythmias of the European Society of Cardiology. The Multicentre European radiofrequency survey (MERFS): complications of radiofrequency catheter ablation of arrhythmias. Eur Heart J. 1993;14:1644-53.
25. Moe GK, Preston JB, Burlington H. Physiologic evidence for a dual A-V transmission system. Circ Res. 1956;4:357-75.
26. Mendez C, Moe GK. Demonstration of a dual A-V nodal conduction system in the isolated rabbit heart. Circ Res. 1966;19:378-93.
27. Goldreyer BN, Bigger JT. The site of re-entry in paroxysmal supraventricular tachycardia in man. Circulation. 1971;43:15-26.
28. Denes P, Wu D, Dhingra RC, et al. Demonstration of dual A-V nodal pathways in patients with paroxysmal supraventricular tachycardia. Circulation. 1973;48:549-55.
29. Jais P, Haissaguerre M, Shah DC, et al. Successful radiofrequency ablation of a slow atrioventricular nodal pathway on the left posterior atrial septum. Pacing Clin Electrophysiol. 1999;22:525-7.
30. Sousa J, El-Atassi R, Rosenheck S, et al. Radiofrequency catheter ablation of the atrioventricular junction from the left ventricle. Circulation. 1991;84:567-71.
31. Ross D, Johnson D, Denniss A, et al. Curative surgery for atrioventricular junctional ("AV nodal") re-entrant tachycardia. J Am Coll Cardiol. 1985;6:1383-92.
32. Cox J, Holman W, Cain M. Cryosurgical treatment of atrioventricular node re-entrant tachycardia. Circulation. 1987;76:1329-36.
33. Guiraudon GM, Klein GJ, van Hemel N, et al. Anatomically guided surgery to the AV node. AV nodal skeletonization: experience in 26 patients with AV nodal re-entrant tachycardia. Eur J Cardiothorac Surg. 1990;4:464-5.
34. Ruder MA, Mead RH, Smith NA, et al. Comparison of pre- and postoperative conduction patterns in patients surgically cured of atrioventricular node re-entrant tachycardia. J Am Coll Cardiol. 1990;17:397-402.
35. Heidbüchel H, Jackman WM. Characterization of subforms of AV nodal re-entrant tachycardia. Europace. 2004;6:316-29.
36. Haissaguerre M, Warin J, Lemetayer P, et al. Closed-chest ablation of retrograde conduction in patients with atrioventricular nodal re-entrant tachycardia. N Engl J Med. 1989;320:426-33.

37. Epstein LM, Scheinman MM, Langberg JJ, et al. Percutaneous catheter modification of the atrioventricular node: a potential cure for atrioventricular nodal tachycardia. Circulation. 1989;80:757-68.
38. Lee MA, Morady F, Kadish A, et al. Catheter modification of the atrioventricular junction with radiofrequency energy for control of atrioventricular nodal re-entry tachycardia. Circulation. 1991;83:827-35.
39. Jackman WM, Beckman KJ, McClelland JH, et al. Treatment of supraventricular tachycardia due to atrioventricular nodal re-entry by radiofrequency catheter ablation of slow-pathway conduction. N Engl J Med. 1992;327:313-8.
40. Haissaguerre M, Gaita F, Fischer B, et al. Elimination of atrioventricular nodal re-entrant tachycardia using discrete slow potentials to guide application of radiofrequency energy. Circulation. 1992;85:2162-75.
41. Kalbfleisch SJ, Strickberger SA, Williamson B, et al. Randomized comparison of anatomic and electrogram mapping approaches to ablation of the slow pathway of atrioventricular node re-entrant tachycardia. J Am Coll Cardiol. 1994;23:716-23.
42. Morady F. Catheter ablation of supraventricular tachycardia: state of the art. J Cardiovasc Electrophysiol. 2004;15:124-39.
43. Li YG, Gronedfeld G, Bender B, et al. Risk of development of delayed atrioventricular block after slow pathway modification in patients with atrioventricular nodal re-entrant tachycardia and pre-existing prolonged PR interval. Eur Heart J. 2001;22:89-95.
44. Sra JS, Jazayeri MR, Blanck Z, et al. Slow pathway ablation in patients with atrioventricular node re-entrant tachycardia and a prolonged PR interval. J Am Coll Cardiol. 1994;24:1064-8.
45. Lee SH, Chen SA, Tai CT, et al. Atrioventricular node re-entrant tachycardia in patients with a prolonged AH interval during sinus rhythm: clinical features, electrophysiologic characteristics and results of radiofrequency ablation. J Interv Card Electrophysiol. 1997;1:305-10.
46. Skanes AC, Dubuc M, Klein GJ, et al. Cryothermal ablation of the slow pathway for the elimination of atrioventricular nodal re-entrant tachycardia. Circulation. 2000;102:2856-60.
47. Wolff L, Parkinson J, White PD. Bundle-branch block with short P-R interval in healthy young people prone to paroxysmal tachycardia. Am Heart J. 1930;5:685-704.
48. Kent AFS. A conducting path between the right auricle and the external wall of the right ventricle in the heart of the mammal. J Physiol. 1914;48:57.
49. Wood FC, Wolferth CC, Geckeler GD. Histologic demonstration of accessory muscular connections between auricle and ventricle in a case of short P-R interval and prolonged QRS complex. Amer Heart J. 1943;25:454-62.
50. Öhnell RF. Pre-excitation, cardiac abnormality, pathophysiological, pathoanatomical and clinical studies of excitatory spread phenomenon bearing upon the problem of the WPW (Wolff, Parkinson, and White) electrocardiogram and paroxysmal tachycardia. Acta Med Scand. 1944;152:1-167.
51. Timmermans C, Smeets JL, Rodriguez LM, et al. Aborted sudden death in the Wolff-Parkinson-White syndrome. Am J Cardiol. 1995;76:492-4.

52. Munger TM, Parker DL, Hammill SC, et al. A population study of the natural history of Wolff-Parkinson-White syndrome in Olmsted County, Minnesota, 1953-1989. Circulation. 1993;87:866-73.
53. Calkins H, Yong P, Miller JM, et al. Catheter ablation of accessory pathways, atrioventricular nodal re-entrant tachycardia, and the atrioventricular junction: final results of a prospective, multicenter clinical trial. The Atakr Multicenter Investigators Group. Circuilation. 1999;99:262-70.
54. Gaita F, Haissaguerre M, Giustetto C, et al. Safety and efficacy of cryoablation of accessory pathways adjacent to the normal conduction system. J Cardiovasc Electrophysiol. 2003;14:825-9.
55. Calkins H, Langberg J, Sousa J, et al. Radiofrequency catheter ablation of accessory atrioventricular connections in 250 patients. Abbreviated therapeutic approach to Wolff-Parkinson-White syndrome. Circulation. 1992;85:1337-46.
56. Morady F. Catheter ablation of supraventricular arrhythmias: state of the art. Pacing Clin Electrophysiol. 2004;27:125-42.
57. Lesh MD, Van Hare G, Kao AK, et al. Radiofrequency catheter ablation for Wolff-Parkinson-White syndrome associated with a coronary sinus diverticulum. Pacing Clin Electrophysiol. 1991;14:1479-84.
58. Beukema WP, Van Dessel PF, Van Hemel NM, et al. Radiofrequency catheter ablation of accessory pathways associated with a coronary sinus diverticulum. Eur Heart J. 1994;15:1415-8.
59. Langberg JJ, Man KC, Vorperian VR, et al. Recognition and catheter ablation of subepicardial accessory pathways. J Am Coll Cardiol. 1993;22:1100-4.
60. Valderrabano M, Cesario DA, Ji S, et al. Percutaneous epicardial mapping during ablation of difficult accessory pathways as an alternative to cardiac surgery. Heart Rhythm. 2004;1:311-6.
61. Scheinman MM. History of Wolff-Parkinson-White syndrome. Pacing Clin Electrophysiol. 2005;28:152-6.
62. Josephson ME. Clinical cardiac electrophysiology: techniques and interpretations, 3rd edition. Philadelphia, PA: Lippincott Williams and Wilkins; 2002.
63. Garson A Jr, Gillette PC. Electrophysiologic studies of supraventricular tachycardia in children II. Prediction of specific mechanism by noninvasive features. Am Heart J. 1981;102:383-8.
64. Engelstein ED, Lippman N, Stein KM, et al. Mechanism-specific effects of adenosine on atrial tachycardia. Circulation 1994;89:2645-54.
65. Roberts-Thomson KC, Kistler PM, Kalman JM. Atrial tachycardia: mechanisms, diagnosis, and anatomic location. Curr Probl Cardiol. 2005;30:529-73.
66. Kistler PM, Roberts-Thomson KC, Haqqani HM, et al. P-wave morphology in focal atrial tachycardia: development of an algorithm to predict the anatomic site of origin. J Am Coll Cardiol. 2006;48:1010-7.
67. Schmitt C, Zrenner B, Schneider M, et al. Clinical experience with a novel multielectrode basket catheter in right atrial tachycardias. Circulation. 1999;99:2414-22.
68. Natale A, Breeding L, Tomassoni G, et al. Ablation of right and left ectopic atrial tachycardias using a three-dimensional nonfluoroscopic mapping system. Am J Cardiol. 1998;82:989-92.

69. Lesh MD, Kalman JM, Olgin JE. New approaches to treatment of atrial flutter and tachycardia. J Cardiovasc Electrophysiol. 1996;7:368-81.
70. Wong T, Segal OR, Markides V, et al. Cryoablation of focal atrial tachycardia originating close to the atrioventricular node. J Cardiovasc Electrophysiol. 2004;15:838.
71. Tsai CF, Tai CT, Chen SA. Catheter ablation of atrial tachycardia. In: Jalife J, Zipes DP (eds). Cardiac electrophysiology: from cell to bedside, 4th edition. Philadelphia, PA: Saunders; 2004. pp. 1060-8.
72. Natale A, Newby KH, Pisano E, et al. Prospective randomized comparison of antiarrhythmic therapy versus first-line radiofrequency ablation in patients with atrial flutter. J Am Coll Cardiol. 2000;35:1898-904.
73. Waldo AL, MacLean WAH, Karp RB, et al. Entrainment and interruption of atrial flutter with atrial pacing. Studies in man following open heart surgery. Circulation. 1977;56:737-45.
74. Klein GJ, Guiraudon GM, Sharma AD, et al. Demonstration of macrore-entry and feasibility of operative therapy in the common type of atrial flutter. Am J Cardiol. 1986;57:587-91.
75. Saoudi N, Atallah G, Kirkorian G, et al. Catheter ablation of the atrial myocardium in human type I atrial flutter. Circulation. 1990;81:762-71.
76. Chauvin M, Brechenmacher C. A clinical study of the application of endocardial fulguration in the treatment of recurrent atrial flutter. Pacing and Clin Electrophysiol. 1989;12:219-24.
77. Feld GK, Fleck RP, Chen PS, et al. Radiofrequency catheter ablation for the treatment of human type 1 atrial flutter: identification of a critical zone in the re-entrant circuit by endocardial mapping techniques. Circulation. 1992;86:1233-40.
78. Cosio FG, Lopez-Gil M, Goicolea A, et al. Radiofrequency ablation of the inferior vena cava-tricuspid valve isthmus in common atrial flutter. Am J Cardiol. 1993;71:705-9.
79. Olgin JE, Kalman JM, Fitzpatrick AP, et al. Role of right atrial structures as barriers to conduction during human type I atrial flutter. Activation and entrainment mapping guided by intracardiac echocardiography. Circulation. 1995;92:1839-48.
80. Cheng J, Scheinman MM. Acceleration of typical atrial flutter due to double-wave re-entry induced by programmed electrical stimulation. Circulation. 1998;97:1589-96.
81. Cheng J, Cabeen WR, Scheinman MM. Right atrial flutter due to lower loop re-entry; mechanism and anatomic substrates. Circulation. 1999;99:1700-5.
82. Yang Y, Cheng J, Bochoeyer A, et al. Atypical right atrial flutter patterns. Circulation. 2001;103:3092-8.
83. Zhang S, Younis G, Hariharan R, et al. Lower loop re-entry as a mechanism of clockwise right atrial flutter. Circulation. 2004;109:1630-5.
84. Yang Y, Varma N, Keung EC, et al. Re-entry within the cavotricuspid isthmus: an isthmus dependent circuit. PACE. 2005;28:808-18.
85. Yang Y, Varma N, Badhwar N, et al. Prospective Observations in the Clinical and Electrophysiological Characteristics of Intra-Isthmus Re-entry. J Cardiovasc Electrophysiol. 2010;21:1-8.
86. Poty H, Saoudi N, Nair M, et al. Radiofrequency catheter ablation of atrial flutter. Further insights into the various types of isthmus

block: application to ablation during sinus rhythm. Circulation. 1996;94:3204-13.
87. Shah DC, Haissaguerre M, Takahashi A, et al. Differential pacing for distinguishing block from persistent conduction through an ablation line. Circulation. 2000;102:1517-22.
88. Tada H, Oral H, Sticherling C, et al. Double potential along the ablation line as a guide to radiofrequency ablation of typical atrial flutter. J Am Coll Cardiol. 2001;38:750-5.
89. Mangat I, Yang Y, Cheng J, et al. Optimizing the detection of bidirectional block across the flutter isthmus for patients with typical isthmus-dependent atrial flutter. Am J Cardiol. 2003;91:559-64.
90. Kall JG, Rubenstein DS, Kopp DE, et al. Atypical atrial flutter originating in the right atrial free wall. Circulation. 2000;101:270-9.
91. Nakagawa H, Shah N, Matsudaira K, et al. Characterization of re-entrant circuit in macrore-entrant right atrial tachycardia after surgical repair of congenital heart disease: isolated channels between scars allow "focal" ablation. Circulation. 2001;103:699-709.
92. Tai CT, Huang JL, Lin YK, et al. Noncontact three-dimensional mapping and ablation of upper loop re-entry originating in the right atrium. J Am Coll Cardiol. 2002;40:746-53.
93. Jais P, Shah DC, Haissaguerre M, et al. Mapping and ablation of left atrial flutters. Circulation. 2000;101:2928-34.
94. Bochoeyer A, Yang Y, Cheng J, et al. Surface electrocardiographic characteristics of right and left atrial flutter. Circulation. 2003;108: 60-6.
95. Moss AJ, Hall WJ, Cannom DS, et al. Improved survival with an implanted defibrillator in patients with coronary disease at high risk for ventricular arrhythmia. Multicenter Automatic Defibrillator Implantation Trial Investigators. N Engl J Med. 1996;335:1933-40.
96. A comparison of antiarrhythmic-drug therapy with implantable defibrillators in patients resuscitated from near-fatal ventricular arrhythmias. The Antiarrhythmics versus Implantable Defibrillators (AVID) Investigators. N Engl J Med. 1997;337:1576-83.
97. Kuck KH, Cappato R, Siebels J, et al. Randomized comparison of antiarrhythmic drug therapy with implantable defibrillators in patients resuscitated from cardiac arrest: the Cardiac Arrest Study Hamburg (CASH). Circulation. 2000;102:748-54.
98. Connolly SJ, Gent M, Roberts RS, et al. Canadian implantable defibrillator study (CIDS): a randomized trial of the implantable cardioverter defibrillator against amiodarone. Circulation. 2000;101:1297-302.
99. Kamphuis HC, de Leeuw JR, Derksen R, et al. Implantable cardioverter defibrillator recipients: quality of life in recipients with and without ICD shock delivery: a prospective study. Europace. 2003;5:381-9.
100. Irvine J, Dorian P, Baker B, et al. Quality of life in the Canadian Implantable Defibrillator Study (CIDS). Am Heart J. 2002;144: 282-9.
101. Schron EB, Exner DV, Yao Q, et al. Quality of life in the antiarrhythmics versus implantable defibrillators trial: impact of therapy and influence of adverse symptoms and defibrillator shocks. Circulation. 2002;105:589-94.
102. Moss AJ, Greenberg H, Case RB, et al. Long-term clinical course of patients after termination of ventricular tachyarrhythmia by an implanted defibrillator. Circulation. 2004;110:3760-5.

103. Poole JE, Johnson GW, Hellkamp AS, et al. Prognostic importance of defibrillator shocks in patients with heart failure. N Engl J Med. 2008;359:1009-17.
104. Aksoz E, Aksoz T, Bilge SS, et al. Antidepressant-like effects of echo-planar magnetic resonance imaging in mice determined using the forced swimming test. Brain Res. 2008;1236:194-9.
105. Daubert JP, Zareba W, Cannom DS, et al. Inappropriate implantable cardioverter-defibrillator shocks in MADIT II: frequency, mechanisms, predictors, and survival impact. J Am Coll Cardiol. 2008;51:1357-65.
106. Connolly SJ, Dorian P, Roberts RS, et al. Comparison of beta-blockers, amiodarone plus beta-blockers, or sotalol for prevention of shocks from implantable cardioverter defibrillators: the OPTIC study: a randomized trial. JAMA. 2006;295:165-71.
107. Pacifico A, Hohnloser SH, Williams JH, et al. Prevention of implantable-defibrillator shocks by treatment with sotalol. d,l-Sotalol Implantable Cardioverter-Defibrillator Study Group. N Engl J Med. 1999;340:1855-62.
108. Stevenson WG, Wilber DJ, Natale A, et al. Irrigated radiofrequency catheter ablation guided by electroanatomic mapping for recurrent ventricular tachycardia after myocardial infarction: the multicenter thermocool ventricular tachycardia ablation trial. Circulation. 2008;118:2773-82.
109. Kuck K-H, Schaumann A, Eckardt L, et al. Catheter ablation of stable ventricular tachycardia before defibrillator implantation in patients with coronary heart disease (VTACH): a multicentre randomised controlled trial. Lancet. 2010;375:31-40.
110. Reddy VY, Reynolds MR, Neuzil P, et al. Prophylactic catheter ablation for the prevention of defibrillator therapy. N Engl J Med. 2007;357:2657-65.
111. Raymond J-M, Sacher F, Winslow R, et al. Catheter ablation for scar-related ventricular tachycardias. Curr Probl Cardiol. 2009;34:225-70.
112. de Bakker JM, van Capelle FJ, Janse MJ, et al. Re-entry as a cause of ventricular tachycardia in patients with chronic ischemic heart disease: electrophysiologic and anatomic correlation. Circulation. 1988;77:589-606.
113. Haïssaguerre M, Shoda M, Jaïs P, et al. Mapping and ablation of idiopathic ventricular fibrillation. Circulation. 2002;106:962-7.
114. Zipes DP, Camm AJ, Borggrefe M, et al. ACC/AHA/ESC 2006 guidelines for management of patients with ventricular arrhythmias and the prevention of sudden cardiac death: a report of the American College of Cardiology/American Heart Association Task Force and the European Society of Cardiology Committee for Practice Guidelines (Writing Committee to Develop Guidelines for Management of Patients with Ventricular Arrhythmias and the Prevention of Sudden Cardiac Death). J Am Coll Cardiol. 2006;48:e247-e346.
115. Aliot EM, Stevenson WG, Almendral-Garrote JM, et al. EHRA/HRS Expert Consensus on Catheter Ablation of Ventricular Arrhythmias: developed in a partnership with the European Heart Rhythm Association (EHRA), a Registered Branch of the European Society of Cardiology (ESC), and the Heart Rhythm Society (HRS); in collaboration with the American College of Cardiology (ACC) and the American Heart Association (AHA). Europace. 2009;11: 771-817.

116. Carbucicchio C, Della Bella P, Fassini G, et al. Percutaneous cardiopulmonary support for catheter ablation of unstable ventricular arrhythmias in high-risk patients. Herz. 2009;34:545-52.
117. Dandamudi G, Ghumman WS, Das MK, Miller JM. Endocardial catheter ablation of ventricular tachycardia in patients with ventricular assist devices. Heart Rhythm. 2007;4:1165-9.
118. Friedman PA, Munger TM, Torres N, et al. Percutaneous endocardial and epicardial ablation of hypotensive ventricular tachycardia with percutaneous left ventricular assist in the electrophysiology laboratory. Journal of Cardiovascular Electrophysiology. 2007; 8:106-9.
119. Josephson ME, Callans DJ. Using the twelve-lead electrocardiogram to localize the site of origin of ventricular tachycardia. Heart Rhythm. 2005;2:443-6.
120. Berruezo A, Mont L, Nava S, et al. Electrocardiographic recognition of the epicardial origin of ventricular tachycardias. Circulation. 2004;109:1842-7.
121. Stevenson W, Khan H, Sager P, et al. Identification of re-entry circuit sites during catheter mapping and radiofrequency ablation of ventricular tachycardia late after myocardial infarction. Circulation. 1993;88:1647-70.
122. Kocovic DZ, Harada T, Friedman PL, et al. Characteristics of electrograms recorded at re-entry circuit sites and bystanders during ventricular tachycardia after myocardial infarction. J Am Coll Cardiol. 1999;34:381-8.
123. Stevenson WG, Friedman PL, Sager PT, et al. Exploring postinfarction re-entrant ventricular tachycardia with entrainment mapping. J Am Coll Cardiol. 1997;29:1180-9.
124. Stevenson WG, Khan H, Sager P, et al. Identification of re-entry circuit sites during catheter mapping and radiofrequency ablation of ventricular tachycardia late after myocardial infarction. Circulation. 1993;88:1647-70.
125. El-Shalakany A, Hadjis T, Papageorgiou P, et al. Entrainment/mapping criteria for the prediction of termination of ventricular tachycardia by single radiofrequency lesion in patients with coronary artery disease. Circulation. 1999;99:2283-9.
126. Bogun F, Taj M, Ting M, et al. Spatial resolution of pace mapping of idiopathic ventricular tachycardia/ectopy originating in the right ventricular outflow tract. Heart Rhythm. 2008;5:339-44.
127. Brunckhorst CB, Delacretaz E, Soejima K, et al. Identification of the ventricular tachycardia isthmus after infarction by pace mapping. Circulation. 2004;110:652-9.
128. Brunckhorst CB, Stevenson WG, Soejima K, et al. Relationship of slow conduction detected by pace-mapping to ventricular tachycardia re-entry circuit sites after infarction. J Am Coll Cardiol. 2003;41:802-9.
129. Bogun F, Good E, Reich S, et al. Isolated potentials during sinus rhythm and pace-mapping within scars as guides for ablation of post-infarction ventricular tachycardia. J Am Coll Cardiol. 2006;47:2013-9.
130. Soejima K, Stevenson WG, Maisel WH, et al. Electrically unexcitable scar mapping based on pacing threshold for identification of the re-entry circuit isthmus: feasibility for guiding ventricular tachycardia ablation. Circulation. 2002;106:1678-83.
131. Calkins H, Epstein A, Packer D, et al. Catheter ablation of ventricular tachycardia in patients with structural heart disease using cooled

radiofrequency energy: results of a prospective multicenter study. Cooled RF Multi Center Investigators Group. J Am Coll Cardiol. 2000;35:1905-14.

132. Scheinman MM, Huang S. The 1998 NASPE prospective catheter ablation registry. Pacing Clin Electrophysiol. 2000;23:1020-8.

133. Sacher F, Roberts-Thomson K, Maury P, et al. Epicardial ventricular tachycardia ablation a multicenter safety study. J Am Coll Cardiol. 2010;55:2366-72.

134. Sosa E, Scanavacca M, d'Avila A, et al. Nonsurgical transthoracic epicardial catheter ablation to treat recurrent ventricular tachycardia occurring late after myocardial infarction. J Am Coll Cardiol. 2000;35:1442-9.

135. Daniels DV, Lu Y-Y, Morton JB, et al. Idiopathic epicardial left ventricular tachycardia originating remote from the sinus of valsalva: electrophysiological characteristics, catheter ablation, and identification from the 12-lead electrocardiogram. Circulation. 2006;113:1659-66.

136. Vallès E, Bazan V, Marchlinski FE. ECG criteria to identify epicardial ventricular tachycardia in nonischemic cardiomyopathy. Circ Arrhythm Electrophysiol. 2010;3:63-71.

137. Michowitz Y, Mathuria N, Tung R, et al. Hybrid procedures for epicardial catheter ablation of ventricular tachycardia: Value of surgical access. Heart Rhythm. 2010;7:1635-43.

138. Miller JM, Altemose GT, Jayachandran JV. Catheter ablation of ventricular tachycardia in patients with structural heart disease. Cardiol Rev. 2001;9:302-11.

139. Bhakta D, Miller JM. Principles of electroanatomic mapping. Indian Pacing Electrophysiol J. 2008;8:32-50.

140. Brooks R, Burgess JH. Idiopathic ventricular tachycardia. A review. Medicine (Baltimore). 1988;67:271-94.

141. Okumura K, Tsuchiya T. Idiopathic left ventricular tachycardia: clinical features, mechanisms and management. Card Electrophysiol Rev. 2002;6:61-7.

142. Nakagawa M, Takahashi N, Nobe S, et al. Gender differences in various types of idiopathic ventricular tachycardia. J Cardiovasc Electrophysiol. 2002;13:633-8.

143. Altemose GT, Buxton AE. Idiopathic ventricular tachycardia. Annu Rev Med. 1999;50:159-77.

144. Yarlagadda RK, Iwai S, Stein KM, et al. Reversal of cardiomyopathy in patients with repetitive monomorphic ventricular ectopy originating from the right ventricular outflow tract. Circulation. 2005;112:1092-7.

145. Jadonath RL, Schwartzman DS, Preminger MW, et al. Utility of the 12-lead electrocardiogram in localizing the origin of right ventricular outflow tract tachycardia. Am Heart J. 1995;130:1107-13.

146. Coggins DL, Lee RJ, Sweeney J, et al. Radiofrequency catheter ablation as a cure for idiopathic tachycardia of both left and right ventricular origin. J Am Coll Cardiol. 1994;23:1333-41.

147. Dixit S, Gerstenfeld EP, Callans DJ, et al. Electrocardiographic patterns of superior right ventricular outflow tract tachycardias: distinguishing septal and free-wall sites of origin. J Cardiovasc Electrophysiol. 2003;14:1-7.

148. Ito S, Tada H, Naito S, et al. Development and validation of an ECG algorithm for identifying the optimal ablation site for idiopathic ventricular outflow tract tachycardia. J Cardiovasc Electrophysiol. 2003;14:1280-6.

149. Yamauchi Y, Aonuma K, Takahashi A, et al. Electrocardiographic characteristics of repetitive monomorphic right ventricular tachycardia originating near the His-bundle. J Cardiovasc Electrophysiol. 2005;16:1041-8.
150. Timmermans C, Rodriguez LM, Crijns HJ, et al. Idiopathic left bundle-branch block-shaped ventricular tachycardia may originate above the pulmonary valve. Circulation. 2003;108:1960-7.
151. Sekiguchi Y, Aonuma K, Takahashi A, et al. Electrocardiographic and electrophysiologic characteristics of ventricular tachycardia originating within the pulmonary artery. J Am Coll Cardiol. 2005;45:887-95.
152. Moorman AF, Christoffels VM. Cardiac chamber formation: development, genes, and evolution. Physiol Rev. 2003;83:1223-67.
153. Callans DJ, Menz V, Schwartzman D, et al. Repetitive monomorphic tachycardia from the left ventricular outflow tract: electrocardiographic patterns consistent with a left ventricular site of origin. J Am Coll Cardiol. 1997;29:1023-7.
154. Kamakura S, Shimizu W, Matsuo K, et al. Localization of optimal ablation site of idiopathic ventricular tachycardia from right and left ventricular outflow tract by body surface ECG. Circulation. 1998;98:1525-33.
155. Krebs ME, Krause PC, Engelstein ED, et al. Ventricular tachycardias mimicking those arising from the right ventricular outflow tract. J Cardiovasc Electrophysiol. 2000;11:45-51.
156. Lamberti F, Calo L, Pandozi C, et al. Radiofrequency catheter ablation of idiopathic left ventricular outflow tract tachycardia: utility of intracardiac echocardiography. J Cardiovasc Electrophysiol. 2001;12:529-35.
157. Hachiya H, Aonuma K, Yamauchi Y, et al. Electrocardiographic characteristics of left ventricular outflow tract tachycardia. Pacing Clin Electrophysiol. 2000;23:1930-4.
158. Tada H, Nogami A, Naito S, et al. Left ventricular epicardial outflow tract tachycardia: a new distinct subgroup of outflow tract tachycardia. Jpn Circ J. 2001;65:723-30.
159. Dixit S, Gerstenfeld EP, Lin D, et al. Identification of distinct electrocardiographic patterns from the basal left ventricle: distinguishing medial and lateral sites of origin in patients with idiopathic ventricular tachycardia. Heart Rhythm. 2005;2:485-91.
160. Shimoike E, Ohnishi Y, Ueda N, et al. Radiofrequency catheter ablation of left ventricular outflow tract tachycardia from the coronary cusp: a new approach to the tachycardia focus. J Cardiovasc Electrophysiol. 1999;10:1005-9.
161. Sadanaga T, Saeki K, Yoshimoto T, et al. Repetitive monomorphic ventricular tachycardia of left coronary cusp origin. Pacing Clin Electrophysiol. 1999;22:1553-6.
162. Kanagaratnam L, Tomassoni G, Schweikert R, et al. Ventricular tachycardias arising from the aortic sinus of valsalva: an underrecognized variant of left outflow tract ventricular tachycardia. J Am Coll Cardiol. 2001;37:1408-14.
163. Ouyang F, Fotuhi P, Ho SY, et al. Repetitive monomorphic ventricular tachycardia originating from the aortic sinus cusp: electrocardiographic characterization for guiding catheter ablation. J Am Coll Cardiol. 2002;39:500-8.

164. Yang Y, Saenz LC, Varosy PD, et al. Analyses of phase differences from surface electrocardiogram recordings to distinguish the origin of outflow tract tachycardia (abstr). Heart Rhythm. 2005;2:S80.
165. Tanner H, Hindricks G, Schirdewahn P, et al. Outflow tract tachycardia with R/S transition in lead V3: six different anatomic approaches for successful ablation. J Am Coll Cardiol. 2005;45: 418-23.
166. Meininger GR, Berger RD. Idiopathic ventricular tachycardia originating in the great cardiac vein. Heart Rhythm. 2006;3:464-6.
167. Daniels DV, Lu YY, Morton JB, et al. Idiopathic epicardial left ventricular tachycardia originating remote from the sinus of Valsalva: electrophysiological characteristics, catheter ablation, and identification from the 12-lead electrocardiogram. Circulation. 2006;113:1659-66.
168. Doppalapudi H, Yamada T, Ramaswamy K, et al. Idiopathic focal epicardial ventricular tachycardia originating from the crux of the heart. Heart Rhythm. 2009;6:44-50.
169. Buxton AE, Waxman HL, Marchlinski FE, et al. Right ventricular tachycardia: clinical and electrophysiologic characteristics. Circulation. 1983;68:917-27.
170. Lemery R, Brugada P, Bella PD, et al. Nonischemic ventricular tachycardia. Clinical course and long-term follow-up in patients without clinically overt heart disease. Circulation. 1989;79:990-9.
171. Rowland TW, Schweiger MJ. Repetitive paroxysmal ventricular tachycardia and sudden death in a child. Am J Cardiol. 1984;53:1729.
172. Chugh SS, Shen WK, Luria DM, et al. First evidence of premature ventricular complex-induced cardiomyopathy: a potentially reversible cause of heart failure. J Cardiovasc Electrophysiol. 2000;11:328-9.
173. Thiene G, Nava A, Corrado D, et al. Right ventricular cardiomyopathy and sudden death in young people. N Engl J Med. 1988;318:129-33.
174. Marcus FI, Fontaine GH, Guiraudon G, et al. Right ventricular dysplasia: a report of 24 adult cases. Circulation. 1982;65:384-98.
175. Gill JS, Mehta D, Ward DE, et al. Efficacy of flecainide, sotalol, and verapamil in the treatment of right ventricular tachycardia in patients without overt cardiac abnormality. Br Heart J. 1992;68:392-7.
176. Kobayashi Y, Miyata A, Tanno K, et al. Effects of nicorandil, a potassium channel opener, on idiopathic ventricular tachycardia. J Am Coll Cardiol. 1998;32:1377-83.
177. Gepstein L, Hayam G, Ben-Haim SA. A novel method for nonfluoroscopic catheter-based electroanatomical mapping of the heart. In vitro and in vivo accuracy results. Circulation. 1997;95: 1611-22.
178. Fung JW, Chan HC, Chan JY, et al. Ablation of nonsustained or hemodynamically unstable ventricular arrhythmia originating from the right ventricular outflow tract guided by noncontact mapping. Pacing Clin Electrophysiol. 2003;26:1699-705.
179. Joshi S, Wilber DJ. Ablation of idiopathic right ventricular outflow tract tachycardia: current perspectives. J Cardiovasc Electrophysiol. 2005;16:S52-8.
180. Friedman PL, Stevenson WG, Bittl JA, et al. Left main coronary artery occlusion during radiofrequency catheter ablation of idiopathic outflow tract ventricular tachycardia (abstr). Pacing Clin Electrophysiol. 1997;20:1184.

181. Sosa E, Scanavacca M, D'Avila A, et al. Endocardial and epicardial ablation guided by nonsurgical transthoracic epicardial mapping to treat recurrent ventricular tachycardia. J Cardiovasc Electrophysiol. 1998;9:229-39.
182. Cole CR, Marrouche NF, Natale A. Evaluation and management of ventricular outflow tract tachycardias. Card Electrophysiol Rev. 2002;6:442-7.
183. Nogami A. Idiopathic left ventricular tachycardia: assessment and treatment. Card Electrophysiol Rev. 2002;6:448-57.
184. Zipes DP, Foster PR, Troup PJ, et al. Atrial induction of ventricular tachycardia: re-entry versus triggered automaticity. Am J Cardiol. 1979;44:1-8.
185. Belhassen B, Rotmensch HH, Laniado S. Response of recurrent sustained ventricular tachycardia to verapamil. Br Heart J. 1981;46:679-82.
186. Ohe T, Shimomura K, Aihara N, et al. Idiopathic sustained left ventricular tachycardia: clinical and electrophysiologic characteristics. Circulation. 1988;77:560-8.
187. Shimoike E, Ueda N, Maruyama T, et al. Radiofrequency catheter ablation of upper septal idiopathic left ventricular tachycardia exhibiting left bundle branch block morphology. J Cardiovasc Electrophysiol. 2000;11:203-7.
188. German LD, Packer DL, Bardy GH, et al. Ventricular tachycardia induced by atrial stimulation in patients without symptomatic cardiac disease. Am J Cardiol. 1983;52:1202-7.
189. Lin FC, Finley CD, Rahimtoola SH, et al. Idiopathic paroxysmal ventricular tachycardia with a QRS pattern of right bundle branch block and left axis deviation: a unique clinical entity with specific properties. Am J Cardiol. 1983;52:95-100.
190. Klein GJ, Millman PJ, Yee R. Recurrent ventricular tachycardia responsive to verapamil. Pacing Clin Electrophysiol. 1984;7:938-48.
191. Ward DE, Nathan AW, Camm AJ. Fascicular tachycardia sensitive to calcium antagonists. Eur Heart J. 1984;5:896-905.
192. Ohe T, Aihara N, Kamakura S, et al. Long-term outcome of verapamil-sensitive sustained left ventricular tachycardia in patients without structural heart disease. J Am Coll Cardiol. 1995;25:54-8.
193. Kottkamp H, Chen X, Hindricks G, et al. Idiopathic left ventricular tachycardia: new insights into electrophysiological characteristics and radiofrequency catheter ablation. Pacing Clin Electrophysiol. 1995;18:1285-97.
194. Nakagawa H, Beckman KJ, McClelland JH, et al. Radiofrequency catheter ablation of idiopathic left ventricular tachycardia guided by a Purkinje potential. Circulation. 1993;88:2607-17.
195. Nogami A, Naito S, Tada H, et al. Verapamil-sensitive left anterior fascicular ventricular tachycardia: results of radiofrequency ablation in six patients. J Cardiovasc Electrophysiol. 1998;9:1269-78.
196. Akhtar M, Shenasa M, Jazayeri M, et al. Wide QRS complex tachycardia. Reappraisal of a common clinical problem. Ann Intern Med. 1988;109:905-12.
197. Wellens HJ, Bar FW, Lie KI. The value of the electrocardiogram in the differential diagnosis of a tachycardia with a widened QRS complex. Am J Med. 1978;64:27-33.
198. Brugada P, Brugada J, Mont L, et al. A new approach to the differential diagnosis of a regular tachycardia with a wide QRS complex. Circulation. 1991;83:1649-59.

199. Wen MS, Yeh SJ, Wang CC, et al. Successful radiofrequency ablation of idiopathic left ventricular tachycardia at a site away from the tachycardia exit. J Am Coll Cardiol. 1997;30:1024-31.
200. Tsuchiya T, Okumura K, Honda T, et al. Significance of late diastolic potential preceding Purkinje potential in verapamil-sensitive idiopathic left ventricular tachycardia. Circulation. 1999;99:2408-13.
201. Nogami A, Naito S, Tada H, et al. Demonstration of diastolic and presystolic purkinje potentials as critical potentials in a macrore-entry circuit of verapamil-sensitive idiopathic left ventricular tachycardia. J Am Coll Cardiol. 2000;36:811-23.
202. Ouyang F, Cappato R, Ernst S, et al. Electroanatomic substrate of idiopathic left ventricular tachycardia: unidirectional block and macrore-entry within the purkinje network. Circulation. 2002;105:462-9.
203. Chen M, Yang B, Zou J, et al. Non-contact mapping and linear ablation of the left posterior fascicle during sinus rhythm in the treatment of idiopathic left ventricular tachycardia. Europace. 2005;7:138-44.
204. Ma FS, Ma J, Tang K, et al. Left posterior fascicular block: a new endpoint of ablation for verapamil-sensitive idiopathic ventricular tachycardia. Chin Med J (Engl). 2006;119:367-72.
205. Thakur RK, Klein GJ, Sivaram CA, et al. Anatomic substrate for idiopathic left ventricular tachycardia. Circulation. 1996;93:497-501.
206. Lin FC, Wen MS, Wang CC, et al. Left ventricular fibromuscular band is not a specific substrate for idiopathic left ventricular tachycardia. Circulation. 1996;93:525-8.
207. Page RL, Shenasa H, Evans JJ, et al. Radiofrequency catheter ablation of idiopathic recurrent ventricular tachycardia with right bundle branch block, left axis morphology. Pacing Clin Electrophysiol. 1993;16:327-36.
208. Yeh SJ, Wen MS, Wang CC, et al. Adenosine-sensitive ventricular tachycardia from the anterobasal left ventricle. J Am Coll Cardiol. 1997;30:1339-45.
209. Nagasawa H, Fujiki A, Usui M, et al. Successful radiofrequency catheter ablation of incessant ventricular tachycardia with a delta wave-like beginning of the QRS complex. Jpn Heart J. 1999;40: 671-5.
210. Kondo K, Watanabe I, Kojima T, et al. Radiofrequency catheter ablation of ventricular tachycardia from the anterobasal left ventricle. Jpn Heart J. 2000;41:215-25.
211. Tada H, Ito S, Naito S, et al. Idiopathic ventricular arrhythmia arising from the mitral annulus: a distinct subgroup of idiopathic ventricular arrhythmias. J Am Coll Cardiol. 2005;45:877-86.
212. Kumagai K, Yamauchi Y, Takahashi A, et al. Idiopathic left ventricular tachycardia originating from the mitral annulus. J Cardiovasc Electrophysiol. 2005;16:1029-36.
213. Tada H, Tadokoro K, Ito S, et al. Idiopathic ventricular arrhythmias originating from the TA: prevalence, electrocardiographic characteristics, and results of radiofrequency catheter ablation. Heart Rhythm. 2007;4:7-16.

# Cardiac Resynchronization Therapy

CHAPTER 12

*David Singh, Nitish Badhwar*

## Chapter Outline

- CRT: Rationale for Use
- CRT in Practice
  - Miracle Study
  - Companion Study
  - Care-HF
- Summary of CRT Benefit
- Prediction of Response to CRT Therapy
  - Is there Adequate BIV Capture?
  - Optimization of CRT Device
- Role of Dyssynchrony Imaging
  - Septal to Posterior Wall Motion Delay
  - Tissue Doppler Imaging
  - Tissue Synchronization Imaging
  - Strain Rate Imaging
  - Speckled Tracking
  - The Prospect Trial
  - Other Dyssynchrony Imaging Techniques
  - Magnetic Resonance Imaging
  - Nuclear Imaging
  - Real-time Three-dimensional Echocardiography
  - Multidetector Computed Tomography
- Dyssynchrony Summary
- LV Lead Placement
- CRT Complications
  - Phrenic Nerve Simulation
  - Loss of CRT
  - CRT and Ventricular Arrhythmias
- Emerging CRT Indications
  - Narrow QRS
  - Atrial Fibrillation
  - Pacemaker Dependant Patients
  - Minimally Symptomatic Heart Failure
  - CRT for Acute Decompensated Heart Failure
- Guidelines

## INTRODUCTION

Despite major advances in the treatment of systolic heart failure (HF), it continues to enact a large burden on healthcare systems around the world. The estimated direct and indirect cost of HF in the United States alone for 2010 was $39.2 billion. In the United States, 1 out of 5 individuals in the age group of 40 years and above will develop a clinical HF syndrome.[1] Advances in pharmacological therapy, most notably the use of beta-blockers, ace inhibitors, angiotensin receptor blockers and aldosterone antagonists have reduced mortality in this population.[2-7] Despite this, HF patients have a poor prognosis with a 20% annual and nearly 50% five-year mortality rate.[1]

The introduction of device-based therapies, including internal cardiac defibrillators (ICD) and cardiac resynchronization therapy (CRT) also known as biventricular (BIV) pacing have transformed the landscape of HF management. Both of these modalities have been independently shown to improve survival among patients with systolic HF.[8-10] Based on estimates between 2001 and 2005, as many as 500,00 CRT devices have been implanted in the United States.[11]

Despite these advances, the prevalence of HF remains high and is estimated to affect 5.8 million individuals in the United States.[12] Parallelling, this has been a rise in HF related hospitalizations.[12] The complexity of acute HF management has increased considerably over the past decades. In addition to an ever-expanding armamentarium of HF medications, device-based therapies have become more sophisticated with each generation. Thus the need for clinicians who are well versed in all the aspects of HF management has never been greater. Proper management of these patients requires an interdisciplinary approach, including intensivists, cardiologists, HF specialists, nurses and electrophysiologists.

## CRT: RATIONALE FOR USE

The contractile apparatus of the human heart is influenced by a myriad of factors including the highly coordinated electrical activation of the atria and ventricles. Disruption to this activation pattern [for example, in the case of left bundle branch block (LBBB)] can impede ventricular performance. In advanced HF, it is common to see abnormal electrical conduction which promotes asynchronous activation of the ventricles, reduced cardiac output and, in the long-term, adverse ventricular remodeling.[13] The term mechanical dyssynchrony has been used to describe the loss of synchronized contraction both between and within the right and left ventricles. This phenomenon is usually, but not always the result of disorganized electrical activation.

Patients with depressed systolic function are more susceptible to the adverse effects of conduction disturbances and mechanical dyssynchrony. Patients with first-degree heart block have suboptimal contribution of atrial systole, less filling time for the LV, and can have diastolic mitral regurgitation (MR)[14,15] Among HF patients, the most common conduction abnormality is LBBB. In LBBB, the electrical activation of the ventricles occurs first through the right bundle to the RV followed by transseptal conduction that eventually results in activation of the lateral LV myocardium. The delayed contraction of the LV lateral wall occurs during the period of septal relaxation. This results in inefficient LV contraction since the septum and lateral walls fail to move in unison to eject blood. While this may be one of the most common forms of mechanical dyssynchrony, any variation in the timing of regional contraction can impede ventricular performance.

## CRT IN PRACTICE

The CRT typically involves placing pacing leads in the right atrium, right ventricle and a branch of the coronary sinus (CS) **(Figs 1 and 2)**. The CRT implantation is performed using a transvenous approach whereby the CS is cannulated and a pacing lead is advanced into a lateral CS branch. The CS lead

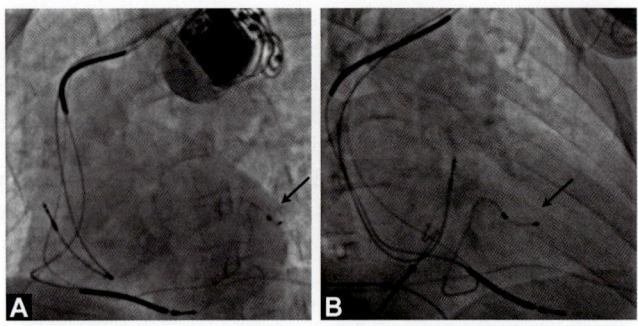

**FIGURES 1A AND B:** (A) Left anterior oblique; (B) Right anterior oblique fluoroscopic views of a biventricular pacemaker-defibrillator with left ventricular lead positioned in a lateral branch of the coronary sinus (arrows)

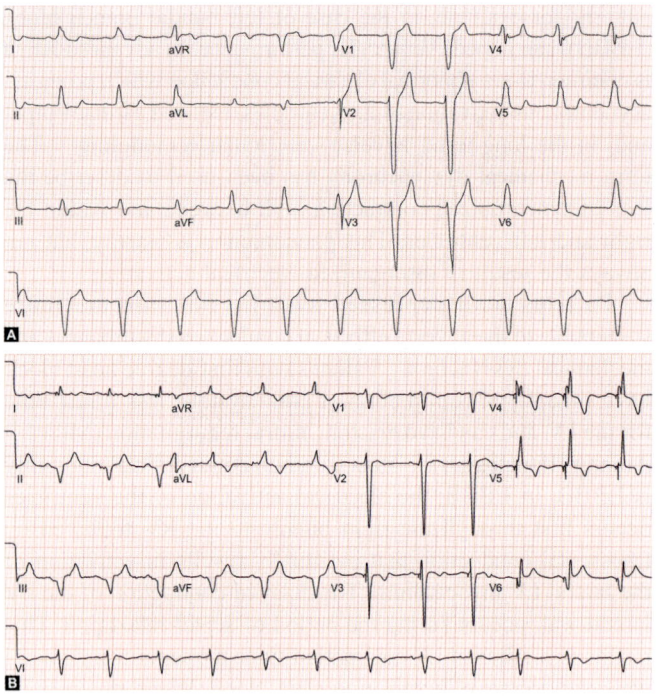

**FIGURES 2A AND B:** (A) ECG before; (B) ECG after implantation of cardiac resynchronization therapy (CRT) device. Note the considerable QRS narrowing and change in QRS morphology with small Q wave in lead I, positive QRS in aVR that is consistent with biventricular pacing

is also known as the LV lead as it activates LV myocardium. Optimal lead placement is the subject of ongoing research and is dependent on many factors including scar location and the regional mechanics of an individual's ventricle. However, it is generally accepted that optimal placement involves maximal separation of the RV and LV leads ideally in a posterolateral branch of the CS.[16]

The CRT can be utilized to influence several key elements of ventricular performance. The AV interval can be adjusted to optimize ventricular preload. The timing of RV and LV pacing can be adjusted to improve interventricular (VV) dyssynchrony. Intraventricular dyssynchrony (LV) can be improved by coordinating the contraction between the LV septum and the lateral wall. Finally, earlier activation of the lateral wall can help to reduce MR, which is likely related to the improved timing of papillary muscle contraction and augmented dP/dt **(Figs 3A and B)**.[17,18] All of these mechanisms contribute to the improvements in myocardial function associated with CRT.

To date, CRT has demonstrated a number of beneficial effects in patients with advanced systolic HF. Several studies have demonstrated its impact on physiologic endpoints such as improved hemodynamics, reduction in MR, increased ejection fraction, increased blood pressure and reverse remodeling.[10,17-23]

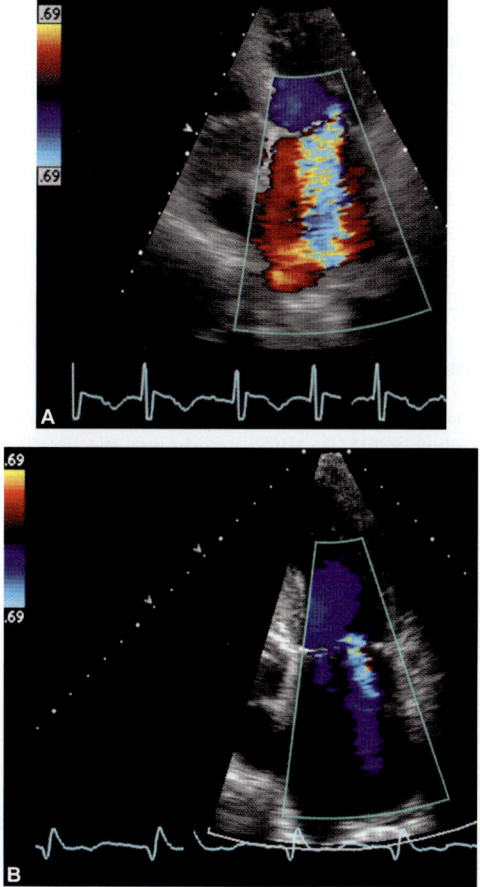

**FIGURES 3A AND B:** Echocardiographic images (A) and (B) showing significant improvement in mitral regurgitation after cardiac resynchronization therapy (CRT)

In addition, randomized and observational studies have shown that CRT favorably impacts clinical endpoints, including, exercise capacity, New York Heart Association (NYHA) functional class, hospitalization rate and quality of life (QOL).[18,21,24-26] More recently, at least one randomized controlled trial (RCT) and a meta-analysis of 14 RCTs have shown that CRT reduces mortality among patients with wide QRS and NYHA class III or IV HF.[26,27] **Table 1** illustrates the results of randomized clinical trials of CRT in patients with advanced HF. The following section will detail the results of three landmark trials involving CRT.

## Miracle Study

The multicenter insync randomized clinical evaluation (MIRACLE) trial was the first large scale, prospective, randomized, double-blind trial of CRT.[28] A total of 453 patients with NYHA class III or IV HF, EF less than or equal to 35%, and QRS duration greater than or equal to 130 millisecond were enrolled. Patients were randomized to have the CRT feature turned on or off. At 6 months, patients randomized to CRT on had significant improvement in QOL, 6-minute walking distance (39 meters vs 10 meters, $p = 0.005$), NYHA functional class ($p < 0.001$) and exercise treadmill time, EF (+ 4.6 vs – 0.2%, $p < 0.001$). Furthermore, patients in the CRT on group had significantly fewer hospitalizations (15 vs 7%, $p = 0.02$) and improved peak oxygen consumption (+ 0.2 vs + 1.1, $p = 0009$).

## Companion Study

The comparison of medical therapy, pacing and defibrillation in heart failure (COMPANION) study was the first large scale, randomized CRT trial to suggest that in addition to symptomatic improvement, CRT may confer mortality benefit.[25] A total of 1,520 patients with NYHA class III and IV HF due to ischemic or nonischemic causes, and QRS duration greater than or equal to 120 milliseconds were randomized to optimal medical therapy, implantation of CRT device or implantation of a CRT device with defibrillator. The mean follow period was 12 months. As with the MIRACLE trial, COMPANION showed that CRT improved HF symptoms based on exercise tolerance testing and QOL surveys. In addition, there was a significant 20% reduction in the primary composite endpoint of all-cause mortality of hospitalization for any cause among those randomized to the CRT arm as compared to medical therapy. Although patients in the CRT-ICD arm experienced a significant reduction in all-cause mortality (HR 0.64, $p = 0.003$), the implantation of CRT

### TABLE 1
Randomized clinical trials of cardiac resynchronization therapy

| Study | Design | No. of patients | Mean follow-up (months) | Results | P value |
|---|---|---|---|---|---|
| MUSTIC (NEJM 2001) | Crossover CRT vs no CRT in patients with CHF NYHA III, EF < 35%, QRS > 150 ms, LVEDD >60 mm, NSR | 58 | 6 | Improved 6 MWT QOL Hospitalization Peak $V_{O_2}$ | < 0.001 < 0.001 < 0.05 < 0.03 |
| MIRACLE (NEJM 2002) | Parallel arms CRT vs no CRT in patients with CHF NYHA III, EF <35%, QRS >130 ms, LVEDD >55 mm, 6 MWT <450 m, NSR | 453 | 6 | Improved 6 MWT NYHA class QOL LVEF Peak $V_{O_2}$ | = 0.005 < 0.001 = 0.001 < 0.001 = 0.009 |
| PATH-CHF (JACC 2002) | Crossover CRT (LV or BiV) vs no CRT in patients with CHF NYHA III-IV, EF <35%, QRS >120 ms, PR >150 ms, NSR | 41 | 12 | Improved 6 MWT Peak $V_{O_2}$ QOL NYHA class LV and BiV had similar improvement | = 0.03 = 0.002 = 0.062 < 0.001 |
| MIRACLE ICD (JAMA 2003) | Parallel arms CRT + ICD vs CRT in patients with CHF NYHA III, EF < 35%, QRS >130 ms, LVEDD >55 mm, | 369 | 6 | Improved NYHA class QOL No change 6 MWT | = 0.007 = 0.02 = 0.36 |

*Contd...*

*Contd...*

| Study | Design | No. of patients | Mean follow-up (months) | Results | P value |
|---|---|---|---|---|---|
| | cardiac arrest due to VT/VF, spontaneous VT or inducible VT/VF, NSR | | | | |
| CONTAK CD (JACC 2003) | Crossover, parallel controlled CRT vs no CRT in patients undergoing ICD implantation with CHF NYHA II-IV, EF <35%, QRS >120 ms, NSR, indications for ICD implantation | 490 | 6 | Improved 6 MWT<br>Peak V$_{O2}$ LVEF<br>LV volumes<br>No significant change<br>NYHA class<br>QOL HF progression | = 0.043<br>= 0.030 < 0.001<br>= 0.02<br>= 0.10<br>= 0.40<br>= 0.35 |
| PATH-CHF II (JACC 2003) | Crossover CRT (LV only) vs no CRT in patients with CHF NYHA II-IV, EF <30%, QRS >120 ms, NSR, Peak V$_{O2}$ <18 ml/min/kg | 86 | 6 | Improved 6 MWT<br>QOL Peak V$_{O2}$<br>No benefit in QRS 120–150 ms | = 0.021<br>= 0.015 < 0.001 |
| COMPANION (NEJM 2004) | Parallel arms<br>Optimal pharmacological therapy (OPT) vs CRT vs CRT + ICD (CRT-D) in patients with CHF NYHA III-IV, EF ≤35%, QRS >120 ms | 1520 | 16 | Death or hospitalization for CHF reduced by 34% in CRT, 40% in CRT-D As compared to OPT | < 0.002<br>< 0.001 |

*Contd...*

*Contd...*

| Study | Design | No. of patients | Mean follow-up (months) | Results | P value |
|---|---|---|---|---|---|
| CARE-HF (NEJM 2005) | Open label, randomized Medical therapy vs Medical therapy + CRT in patients with CHF NYHA III-IV, EF ≤35%, QRS >120 ms with dyssynchrony (aortic pre-ejection >140 ms, interventricular mechanical delay >40 ms, delayed activation of postlateral LV) QRS >150 ms (no dyssynchrony evidence needed) | 814 | Stopped early by DSMB 29.4 | All cause mortality reduced by 36% in CRT-D 24% in CRT All cause mortality/ hospitalization reduction by 37% in CRT All cause mortality reduced by 36% in CRT Improvement in QOL LVEF LVESV NYHA class | = 0.003 = 0.05 < 0.001 < 0.002 < 0.01 |

(*Abbreviations*: 6 MWT: 6-Minute walking test; AF: Atrial fibrillation; CARE-HF: Cardiac resynchronization-heart failure study group; CHF: Congestive heart failure; CONTAK-CD: CONTAK-Cardiac defibrillator; COMPANION: Comparison of medical therapy, resynchronization, and defibrillation therapies in heart failure study group; CRT: Cardiac resynchronization therapy; DSMB: Data safety monitoring board; EF: Ejection fraction; ICD: Implantable cardioverter-defibrillator; JACC: Journal of American College of Cardiology; JAMA: Journal of American Medical Association; LVEDD: LV end diastolic diameter; LVESV: LV end systolic volume; MIRACLE: Multicenter insync randomized clinical evaluation trial; MUSTIC: Multisite stimulation in cardiomyopathies study group; NEJM: New England Journal of Medicine; NSR: Normal sinus rhythm; NYHA: New York Heart Association; QOL: Quality of life; PACE: Pacing and clinical electrophysiology; PATH-CHF: Pacing therapies in heart failure study group; VT: Ventricular tachycardia; VF: Ventricular fibrillation) device alone was associated with a marginally nonsignificant reduction with respect to this endpoint (HR 0.76, p = 0.06).

## Care-HF

The cardiac resynchronization-heart failure (CARE-HF) trial randomly assigned 813 patients with NYHA class III or IV HF (ischemic and nonischemic) with EF less than 35% and QRS prolongation to optimal medical therapy or CRT.[26] Patients with QRS duration of 120–149 milliseconds were required to have echocardiographic evidence of dyssynchrony for enrollment. There was a 37% reduction in the primary endpoint of death from any cause or unplanned hospitalization for a major cardiac event (p < 0.001). The major secondary endpoint in CARE-HF was all-cause mortality. The CRT was associated with a 36% reduction in this endpoint as compared to medical therapy alone (p < 0.002). As per previous studies, CRT was associated with improvements in a number of parameters including ejection fraction, reverse remodeling, systolic blood pressure, MR and QOL.

## SUMMARY OF CRT BENEFIT

These and other trials have provided robust evidence that CRT has a favorable impact on many important physiologic and non-physiologic endpoints in HF. In addition, there is evidence that CRT alone (without back-up defibrillator) reduces mortality. There is some uncertainty about whether CRT coupled to ICD therapy confers additional mortality benefit (in COMPANION, the risk reductions associated with CRT and CRT + ICD were similar). However, due to the wide range of benefits associated with CRT, it is reasonable to combine CRT and ICD therapy in patients who meet criteria for both. In accordance with this, the most recent ACC/AHA/HRS guidelines recommend CRT (with or without ICD) in patients who have left ventricular ejection fraction (LVEF) less than 35%, QRS duration more than 120 millisecond, and NHYA III or IV on optimal medical therapy.[29]

## PREDICTION OF RESPONSE TO CRT THERAPY

One of the great challenges associated with CRT is how to determine which patients are likely to derive the most benefit. Response to CRT is dependent on the endpoint evaluated. When a clinical endpoint, such as NHYA classification, is used to determine response to CRT, there appears to be a consistent 20–30% nonresponder rate. However, when more objective measures, such as echocardiographic parameters, are employed, nonresponder rates may be closer to 40%.[30-32] It remains unknown whether this discrepancy is related to the placebo effect from device implantation or for some other reason.

The reasons for lack of response include suboptimal HF drug therapy, end stage HF, significant MR and other comorbidities such as obesity. Device-related reasons include ineffective biventricular pacing (BiV), suboptimal atrioventricular (AV) and VV timing, suboptimal LV lead position and absence of mechanical dyssynchrony in selected patients. An illustration of a step-by-step approach to CRT nonresponders is given in **Flowchart 1**.

## Is There Adequate BIV Capture?

Optimal delivery of CRT requires continuous ventricular pacing. Although a formal device interrogation may be necessary to assess effective BIV pacing, a great deal can be learned from the surface 12-lead electrocardiogram (ECG). Prior to inspection of the ECG, it is helpful to examine a patient's chest radiograph to determine the position of the RV and LV leads. Several pacing patterns can be observed with: (i) BIV pacing, (ii) complete BIV capture, (iii) isolated RV capture, (iv) isolated LV capture, (v) absence of BIV capture (native QRS), and (vi) BIV capture with fusion.[33]

Traditional RV pacing (with the RV lead in an apical position) activates the myocardium in an inferior-superior and right-left fashion. The surface ECG therefore usually demonstrates a "superior axis" (negative in the inferior leads) and LBBB. Isolated LV pacing can produce a variety of patterns depending on the location of the LV lead. In general, the activation of the ventricle proceeds from left to right producing a "rightward axis" (negative or initial negative QRS complex in leads I and AVL) and a right bundle branch pattern (RBBB).

The pattern of BIV pacing represents the summed vector of RV and LV lead activation. The ECG pattern of BIV capture can vary widely depending on device programming and the placement of the RV and LV leads. In general BIV capture produces a rightward axis (negative or initial negative in leads I, AVL and positive in aVR) and R greater than S in lead $V_1$ **(Figs 3A and B)**. The inferior leads (II, III and aVF) can be positive or negative depending on the location of the LV lead. The absence of this pattern should not however be interpreted to mean the loss of BIV capture. Often there is narrowing of the intrinsic QRS complex with BIV pacing. This has been shown to correlate with clinical benefit.[34,35] However, it has also been shown that wide QRS with BIV pacing also correlates with clinical benefit.[36] Hence, the duration of the BIV paced QRS complex cannot be used to assess presence or absence of BIV

**FLOWCHART 1:** Stepwise algorithm for management of heart failure patients who are nonresponders to CRT

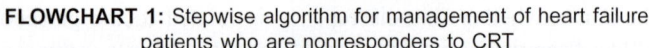

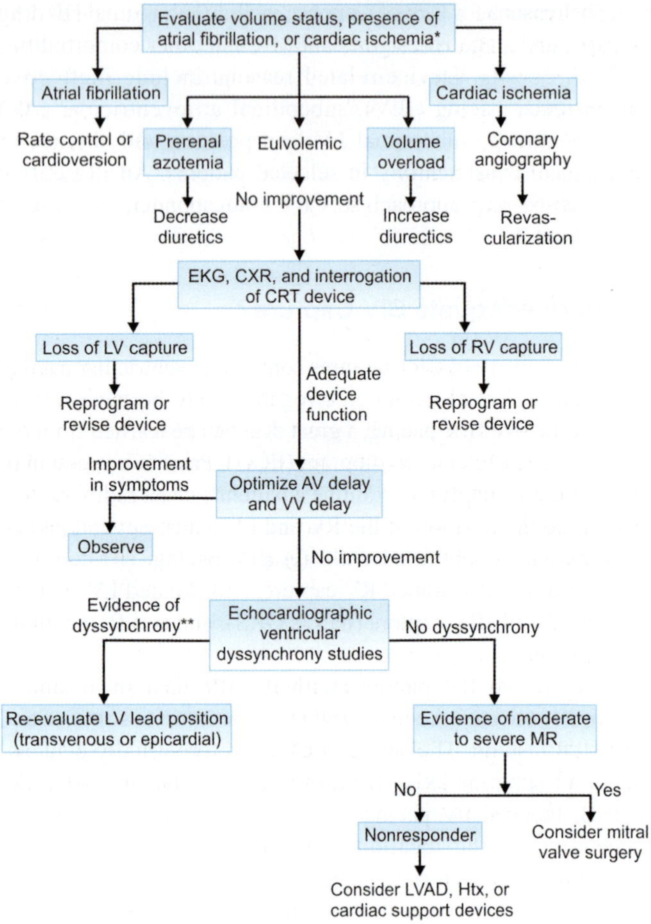

(*Abbreviations:* AV: Atrioventricular; CXR: Chest X-ray; EKG: Electrocardiogram; Htx: Heart transplant; LV: Left ventricular; LVAD: Left ventricular assist device; MR: Mitral regurgitation; RV: Right ventricular; VV: Interventricular).
*Cardiac ischemia is evaluated in patients with ischemic cardiomyopathy.
**Evidence of dyssynchrony includes septal to posterior wall motion delay ≥ 130 ms, intraventricular mechanical delay ≥ 40 ms and tissue Doppler imaging ≥ 65 ms. (*Source:* Modified from Aranda, et al, Management of heart failure after cardiac resynchronization therapy: integrating advanced heart failure treatment with optimal device function. J Am Coll Cardiol. 2005;46:2193-8)

pacing although a narrower BIV paced QRS (when compared to intrinsic QRS) suggests a good prognosis.

Georger and his colleagues analyzed ECG patterns of patients with CRT and observed a Q wave in lead I in 17 out of 18 patients during BIV pacing.[37] A Q wave in lead I with RV pacing alone was found to extremely uncommon. In this series, the absence of a Q wave in lead I was 100% predictive of loss of LV capture. Although this was a small study, the assessment of lead I may be a simple way to assess the presence of LV capture during BIV pacing.

A simple algorithm to assess LV capture among patients with baseline LBBB and RV apical leads was developed by Ammann and his colleagues using leads $V_1$ and I.[38] An R-S ratio greater than or equal to 1 in lead $V_1$ reliably detected left ventricular capture. In the absence of this finding, lead I was analyzed. An R-S ratio of less than or equal to 1 suggested the presence of LV capture. The sensitivity of the algorithm to correctly identify loss of left ventricular capture was 94% (95% CI, 88.2–97.7%), and the specificity was 93% (CI, 86.3–95.8%).

Adequate BIV pacing can only occur if the programmed AV delay is shorter than a patient's native PR interval. When BIV output competes with native AV nodal conduction the result is called fusion. In such cases, both the BIV device and native conduction contribute to the ventricular depolarization. Pseudofusion refers to the phenomena of a pacemaker stimulus that appears to precede ventricular depolarization, but does not contribute to ventricular depolarization. In this case, the QRS complex should be identical to a nonpaced beat. To determine this, it is necessary to compare the ECG in question with a prior ECG that is known to represent nonpaced intrinsic conduction. We recommend performing 12-lead ECG on patients with CRT during their device visit and compare it to previous BIV paced ECG and intrinsic ECG to ensure effective delivery of BIV pacing **(Figs 3A and B)**.

## Optimization of CRT Device

Manipulation of AV and VV delays to achieve an optimal hemodynamic response in patients with CRT devices is known as "optimization". Commonly, noninvasive optimization protocols utilize echocardiography to achieve the desired hemodynamic changes. A number of optimization methods have been developed to establish the optimal AV and VV delays using a variety of different echo parameters. Although alterations in AV delay are more established with respect to their hemodynamic benefits, changes in VV delays are more contentious. At least two randomized trials have failed to show benefit associated with optimization of VV delays.[39,40]

Although echo optimization protocols vary widely, they are usually performed by systematically altering the AV and VV delays to achieve a desired hemodynamic response. The AV and VV delay with the best hemodynamic profile is considered to be "optimal". Some of the hemodynamic parameters used for echo optimization include mitral inflow doppler patterns, time velocity integral of the left ventricular outflow tract (which is proportional to stroke volume), and dP/dt (which can be assessed noninvasively through analysis of continuous wave doppler of an MR jet).[41-43] In addition to echo-guided optimization, a number of other noninvasive optimization techniques have been reported, including impedance cardiography, finger plethysmography and radionuclide ventriculography.[44-46] In addition, some CRT devices possess intracardiac electrogram-based (IEGM) algorithms that can determine optimal AV and VV delays.[47]

Although optimization studies have demonstrated improved ejection fraction, NYHA class, QOL, 6-minute walking distance and cardiac hemodynamics; in general, these studies have been small, nonrandomized, and frequently lack control groups.[43,48-50] There is therefore no consensus regarding the optimal optimization method or universally accepted protocol for patients with CRT devices. Although some practitioners utilize optimization more frequently, in most institutions, only patients who gain suboptimal benefit from CRT undergo optimization as it can be costly, time-consuming and requires specialized skill and expertise.

## ROLE OF DYSSYNCHRONY IMAGING

Cardiac dyssynchrony can occur with respect to atrio-ventricular (A-V) VV delay (RV-LV) or LV. In general, patients with LV dyssynchrony are more likely to response to CRT.[51,52] While QRS duration is a reasonable marker of VV (RV-LV) dyssynchrony; it does not predict LV dyssynchrony (as assessed by echocardiogram) with great accuracy.[53,54]

There have been a number of dyssynchrony criteria that have been shown to predict response to CRT. In general, these trials have been conducted at single centers, with relatively small numbers of patients using a variety of echocardiographic techniques. While a complete review of dyssynchrony parameters is beyond the scope of this paper, a brief review of some of these techniques is provided below.

### Septal to Posterior Wall Motion Delay

The LV dyssynchrony was initially assessed with conventional M-mode echocardiography that measured the delay in contraction between the septum and the posterior wall. This measure is obtained by taking the shortest interval between the maximal posterior displacement of the septum and the maximal displacement of the left posterior wall using an M-mode short-axis view of the left ventricle at the level of the papillary muscle. Several observational studies demonstrated that a SPWMD greater than 130 millisecond predicted a favorable response to CRT.[55-57] However, a recent study showed that this technique was not predictive of response to CRT in a larger study cohort.[58]

## Tissue Doppler Imaging

The tissue doppler imaging (TDI) is an echocardiographic technique that uses ultrasound to image the velocity of cardiac tissue. The TDI is used to assess LV dyssynchrony by comparing the time to peak velocity of various myocardial segments. Measurements are obtained for the time to peak systolic velocity (from the onset of QRS complex) in different segments of the LV and the delay between them is used as a marker of LV dyssynchrony **(Figs 4A and B)**. There have been a number of dyssynchrony indices that have been derived using this technique. Initial studies used a four-segment model (septal, lateral, inferior and anterior) and showed that a delay greater than 65 millisecond predicted response to CRT.[59] Yu et al. used a 12-segment model (6 basal and 6 mid segment) and derived an LV dyssynchrony index from the standard deviation of all 12 intervals.[60,61] An LV dyssynchrony index greater than 31 millisecond yielded a sensitivity and specificity of 96 and 78% to predict LV reverse remodeling.[62]

## Tissue Synchronization Imaging

The tissue synchronization imaging (TSI) builds upon the technique of TDI by transforming the timing of regional peak velocities into color codes **(Figs 5A and B)**. This allows visual

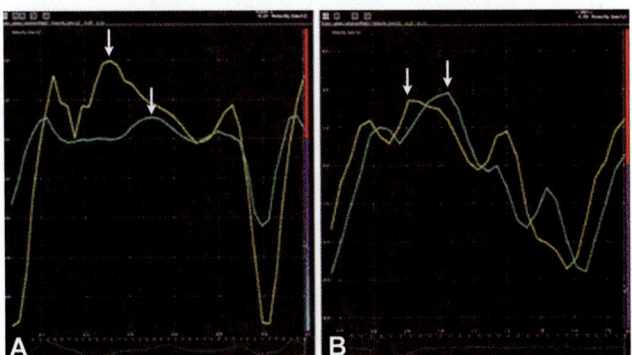

**FIGURES 4A AND B:** Regional myocardial velocity curves obtained by tissue Doppler imaging (TDI) at the basal septal (yellow) and basal lateral (green) segments. (A) In a patient with left bundle branch block with QRS duration of 180 ms, there was delay in peak systolic contraction (arrows) of 95 ms in the lateral wall compared to the septal wall; (B) After biventricular pacing, there was improvement in synchronicity as reflected by the near overlapping of myocardial velocity curves with a difference of only 20 ms. (*Source:* Modified from Yu et al. Comparison of efficacy of reverse remodeling and clinical improvement for relatively narrow and wide QRS complexes after cardiac resynchronization therapy for heart failure. J Cardiovasc Electrophysiol. 2004;15:1058-65)

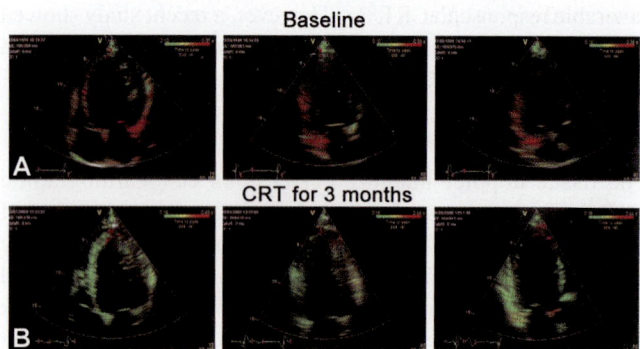

**FIGURES 5A AND B:** Tissue synchronization imaging (TSI) on three apical views showing the presence of extensive regional wall delay in a heart failure patient with prolonged QRS duration. The TSI method was set up to measure the time to peak myocardial systolic velocity (Ts) at ejection phase. The Ts values were then transformed into various color coding depending on the severity of delay, in the sequence of *green, yellow, orange and red*. (A) Before cardiac resynchronization therapy (CRT), this patient had severe delay over the basal to mid-lateral wall and the whole septal wall (*red* color in four-chamber view), severe delay over the whole inferior wall (*red* color in two-chamber view) and moderate to severe delay over the whole posterior wall (*orange* to *red* color in long-axis view); (B) Three months after CRT, corresponding views showed dramatic improvement of these delays, with only mild residual delay over the lateral and inferior wall (*green* to *yellow*). (*Source:* Yu C, et al. A novel tool to assess systolic asynchrony and identify responders of cardiac resynchronization therapy by tissue synchronization imaging. J Am Coll Cardiol. 2005;45:677-84)

identification of regional delay in systole by comparing the color-coding of opposing walls, thus providing rapid identification of dyssynchrony by simple visual evaluation. As with TDI, quantitative assessment of regional delay is possible, and a number of models have been constructed to define dyssynchrony using variable numbers of myocardial segments. Several studies have demonstrated that TSI (including visual and quantitative parameters) is useful in predicting a response to CRT.[62-64]

## Strain Rate Imaging (SRI)

One of the major limitations of TDI is that myocardial velocities may be overestimated or underestimated by translational motion or tethering of the myocardium respectively.[65] Strain imaging overcomes this problem by measuring the actual extent of myocardial deformation in selected regions of the heart. In this manner, it can distinguish between passive motion and active contraction. Myocardial deformation occurs in three dimensions, and strain can therefore be assessed along each axis. Radial strain (RS) represents the myocardial thickening in a short-axis plane; circumferential strain (CS) represents myocardial shortening in a circumferential plane, and longitudinal strain (LS) represents the myocardial shortening in the long-axis plane.[66] Regional differences in strain along any of these axes, can be a marker of LV dyssynchrony. As with other echocardiographic techniques to evaluate dyssynchrony, a number of SRI-derived indices have been shown to predict reverse remodeling and response to CRT.[67-69]

## Speckled Tracking

The speckled tracking (ST) is another echocardiographic technique that takes advantage of acoustic markers produced by reflection, scattering and interference of the echo ultrasound beam to assess regional myocardial motion. Unlike TSI, which relies on tissue Doppler to assess myocardial strain, ST is not limited by angle dependence of the ultrasound transducer. Several studies have validated the used of speckled tracking measured LV dyssynchrony to predict response to CRT.[70-72]

## The Prospect Trial

One of the major limitations of the studies, such as the ones mentioned above, was that they were for the most part, confined to single centers and contained relatively few numbers of patients. The PROSPECT trial was designed to address these limitations. The PROSPECT was a nonrandomized observational study that sought to identify which of the previously published markers of dyssynchrony could predict response to CRT in a multicenter setting in three major regions (United States, Europe, Hong Kong).[32] A total number of 12 echocardiographic markers of dyssynchrony were evaluated in nearly 500 patients with blinded analysis of dyssynchrony in three core laboratories. Dyssynchrony markers were based on both conventional and tissue Doppler based methods, speckled tracking was not evaluated.

The results of PROSPECT raised several concerns with respect to previously published single-center experiences. The feasibility of image acquisition was a major limitation, particularly for TDI measures. Specifically, the percentage of individual parameters deemed interpretable by the core laboratories was between 37% and 82% for TDI-based tests. Intraoperator and interoperator reproducibility was also a major issue ranging from 3.8 to 24.3% and 6.5 to 72.1%, respectively. Most disappointing was that no single parameter appeared to predict response to CRT effectively. Sensitivity for predicting improvement in the clinical composite endpoint ranged from 6 to 74% and specificity ranged from 35 to 92%. For all measured parameters, the area under the ROC curve to predict a positive response was less than or equal to 0.62.

Despite the results of PROSPECT, many investigators and clinicians continue to believe in the utility of dyssynchrony imaging for CRT. In PROSPECT, some investigators have raised concerns about site selection, lack of quality control, lack of adequate training, poor patient selection and poor image acquisition as the factors that might account for the discrepancies between its results and other previously published studies.[73,74] The role of echo-derived dyssynchrony imaging for CRT thus remains uncertain, and future studies will likely be performed to help elucidate this issue.

## Other Dyssynchrony Imaging Techniques

In addition to two-dimensional echocardiography, a myriad of other imaging techniques have been utilized to assess dyssynchrony and/or predict response to CRT.

## Magnetic Resonance Imaging

The magnetic resonance imaging (MRI) is able to offer an integrated assessment of myocardial viability, function, dyssynchrony, anatomy and scar burden making it an attractive modality for patients in whom CRT is being considered. Advantages of CMR include high spatial resolution and reproducibility, accurate assessment of cardiac chamber size and the ability to assess myocardial deformation in three dimensions. However, it is limited by high-cost, long acquisition times and, for the time being, incompatibility with implanted devices.

Three major CMR techniques have been used to assess LV dyssynchrony. The CMR myocardial tagging is similar to speckled tracking analysis whereby a grid is superimposed onto the myocardial image and myocardial strain is assessed via analysis of grid deformation. The CMR phase-contrast tissue velocity mapping (TVM) allows direct myocardial wall motion measurement similar to TDI (i.e. comparing velocity timing obtained in different regions of the myocardium). Unlike TDI, MR TVM is not limited by the acoustical windows of the chest and can acquire three-directional velocity information of the entire myocardium.[75] Displacement-encoded MRI, or DENSE, is a CMR technique that is similar to TVM. However, instead of coding for velocity, DENSE codes for myocardial displacement which are then used to calculate myocardial strain and dyssynchrony. While these MRI techniques are promising, there is currently little data to support their used in predicting response to CRT.

In addition to the quantification of strain, MRI is particularly useful in the assessment of myocardial scar via a technique known as delayed gadolinium enhancement. A high scar burden has been shown to correlate negatively with response to CRT.[76,77] The location of scar is also an important factor in considering CRT. Bleeker and his colleagues demonstrated that patients with ischemic cardiomyopathy, dyssynchrony and posterolateral scar as assessed by MRI do not respond well to CRT.[78]

## Nuclear Imaging

Equilibrium radionuclide angiography (ERNA) and gated SPECT myocardial perfusion imaging (MPI) have also been used to assess VV and LV. The ERNA derived phase image analysis, a functional method based on the first Fourier harmonic fit of the gated blood pool versus radioactivity curve, generates the parameters of amplitude (A) and phase angle (Ø). Amplitude (A) measures the magnitude of regional contraction and phase angle (Ø) represents the timing of regional contraction **(Fig. 6)**. In a healthy heart, all segments of the myocardium should contract during the same phase angle. The mean and standard deviation of LV Ø have been used to characterize LV dyssynchrony[79] and has been used to predict changes in ejection fraction after CRT.[80] More recently, the synchrony (S) [efficiency of contraction in a region of interest (ROI)] and entropy (E) (disorder of contraction in a ROI) parameters have been developed and applied to planar ERNA as a tool for

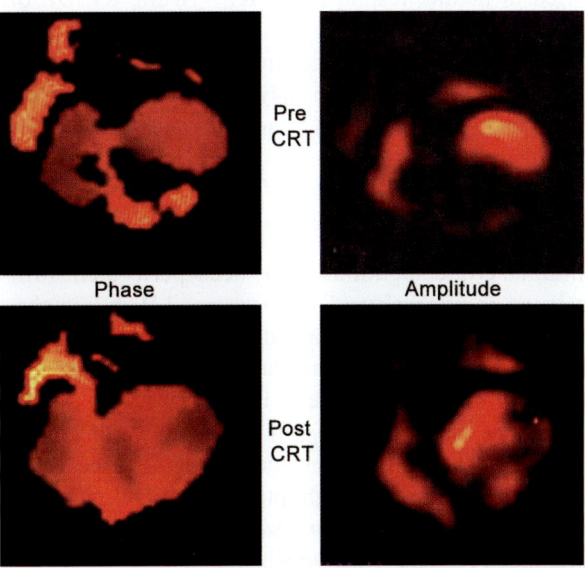

**FIGURE 6:** Equilibrium radionuclide angiogram (ERNA) images showing phase and amplitude analysis that are used to measure left ventricular dyssynchrony before and after cardiac resynchronization therapy (CRT). Phase analysis shows timing of regional contraction that shows apical dyssynchrony that corrects after CRT. Amplitude analysis reflects magnitude of regional contraction that shows improvement after CRT

evaluation of LV dyssynchrony.[81] In one study, these parameters were shown to detect mechanical dyssynchrony with low interobserver and intraobserver variability.[82]

The role of ERNA or MPI dyssynchrony imaging represents a promising advance of a well-established myocardial imaging technology. Whether or not it will be useful in predicting response to CRT will need to be assessed in future studies. The role of nuclear imaging for evaluation of scar burden is somewhat better established. As with CMR, MPI has also been used to assess LV scar burden. Similar to MRI studies, scar burden as assessed by SPECT has been shown to correlate negatively with response to CRT.[83-85]

## Real-time Three-dimensional Echocardiography

Real-time three-dimensional echocardiography (RT3DE) has emerged as a new technique for assessment of LV dyssynchrony based on evaluation of LV regional volumetric changes.[86] This technique is accomplished by dividing the LV into 17 standard subvolumes and assessing the time for each segment to reach the minimum systolic volume (Tmsv). In a normally contracting heart the Tmsv should occur simultaneously for all myocardial segments. The standard deviation of 16 segments is used to create a dyssynchrony index (DI).[87] Preliminary data from a single center suggests that this technique may be used to predict response to CRT.[87] While RT3DE offers several advantages including accurate assessment of chamber size and volume, angle independence and semi-automated measurement, its application may be limited by translational artifacts and suboptimal image quality which may render the data unreadable.[88]

## Multidetector Computed Tomography

Preliminary investigations have been performed utilizing computed tomography (CT) to assess LV dyssynchrony. Truong and his colleagues have derived several dyssynchrony indices with 64 slice CT using changes in wall thickness, wall motion and volume overtime.[89] The global LV dyssynchrony metric using changes in LV wall thickness overtime (average of the SD of 6 segments per slice, using all slices) had the best reproducibility with high interobserver and intraobserver reproducibility. Compared to aged-matched controls, patients with systolic HF and wide QRS had a higher DI. This DI was also moderately well correlated with dyssynchrony as measured by 2D speckled tracking and RT3DE. As with MRI, CT assessment can provide additional information about the heart including its chamber size volumetric analysis, and contractile function. One of the unique features of CT is its ability to visualize CS anatomy, which could help operators to determine optimal lead location prior to implantation.

## DYSSYNCHRONY SUMMARY

Despite extensive research and the multitude of imaging modalities established to assess myocardial dyssynchrony, its role in CRT remains uncertain. No single dyssynchrony parameter has been shown to conclusively predict response to CRT. While many of the above techniques appear promising, conclusive large-scale trials will need to be performed before dyssynchrony assessment can be incorporated into routine clinical practice. Accordingly, neither the ACC/AHA/HRS guidelines for device-based therapy of cardiac rhythm abnormalities nor the ACC/AHA guidelines for the diagnosis and management of HF recommend the use of dyssynchrony imaging to establish candidacy for CRT.[29,90]

## LV LEAD PLACEMENT

The site of placement of the LV lead has also been shown to be an important determinant of the effects of CRT with demonstration of significantly better outcomes with lateral LV pacing as compared to anterior LV pacing.[16] Echocardiogram with TDI has been used to select sites of latest activation in the LV that will be ideal sites for placement of the LV lead.[91] Surgical LV lead placement should be considered when these areas of latest activation do not have a suitable CS branch vein that allows transvenous lead placement.[92-94] We have used multiple gated acquisition scan (MUGA) to identify areas of latest mechanical activation and shown significant improvement in clinical outcomes with imaging guided lead placement as compared to the traditional placement in the lateral LV.[95] Radiographic LV-RV interlead distance has also been shown to predict acute hemodynamic response to CRT as measured by a rise in dP/dt and this can be used to improve the success rate at the time of lead implantation.[96] Placement of the LV lead at areas of LV scarring is unlikely to show response to CRT and can lead to worsening of congestive heart failure (CHF) due to unopposed RV pacing[97] or worsening of ventricular tachycardia.[98] Imaging with PET or contrast enhanced cardiovascular magnetic resonance can identify areas of LV scar preoperatively.

## CRT COMPLICATIONS

As with any implantable device there are a myriad of potential complications associated with CRT. In addition to standard device complications, such as infection and bleeding, there are several complications specific to CRT, including CS dissection and perforation, phrenic nerve stimulation, and LV lead dislodgement. In major trials, LV lead dislodgement occurred in 4–6% of the patients.[10,18] The CS sinus dissection or perforation ranged from 0.3 to 4% and 0.8 to 2%, respectively.[10,18,25] Management for CS dissection or perforation is usually conservative and, in most instances, CS cannulation can safely be performed several weeks later.

## Phrenic Nerve Simulation

The LV lead is frequently positioned in the region of the left phrenic nerve that may in turn lead to diaphragmatic stimulation. Although great care is taken to avoid phrenic nerve stimulation during implantation, subtle changes in lead position as well as postural changes can cause this complication at anytime postimplant. Phrenic nerve stimulation is often easy to recognize by observing the contractions of the abdomen during pacemaker output. Frequently, patients will complain of discomfort that occurs with certain positions or movements. It is sometimes helpful to ask a patient to recreate the setting in which they experience discomfort in order to make the diagnosis. A cardiologist or industry representative should be made immediately aware if the diagnosis of phrenic nerve stimulation is made. Although it is not life threatening, it can be the source of considerable discomfort for a patient. Frequently, a CRT device can be reprogrammed to eliminate phrenic nerve stimulation. At times, however, revision of the LV lead may be necessary.

## Loss of CRT

As discussed above, it is essential to ensure maximum BiV pacing among patients with CRT devices. There are a number of settings in which maximum BiV pacing can be compromised **(Table 2)**. It is important to recognize the loss of BiV capture in order to maximize the benefits of CRT.

### TABLE 2
**Causes of loss of biventricular pacing**

- Atrial undersensing can be caused by a variety of factors including sinus tachycardia with first-degree AV block, atrial fibrillation and lead dislodgement
- Fusion or pseudofusion (discussed above)
- Ventricular oversensing
- Atrial tachyarrhythmias with rapid ventricular conduction (frequently AF)
- Frequent ventricular ectopy
- Loss of LV capture due to LV pacing threshold increase
- Loss of LV capture due to LV lead dislodgement

## CRT and Ventricular Arrhythmias

Several small studies have suggested that CRT may have a role in reducing the incidence of ventricular arrhythmias.[99-101] Although decreasing wall tension and favorable remodeling provide a biological basis for this hypothesis, CRT has not been formally tested in this manner. In many of the large randomized trials of CRT, the impact of CRT on the incidence of VT and VT storm was not systematically studied. However, meta-analysis of five major CRT trials revealed that sudden cardiac death was not significantly reduced by CRT when compared to optimal medical therapy.[102] In all but one of these trials, (CARE-HF) CRT, was not associated with a decreased risk of sudden death.

There is also concern that CRT has been associated with increases in ventricular arrhythmias. There are numerous reported cases of patients developing recurrent ventricular arrhythmias following implantation of a BiV device.[98,103-106] Various mechanisms have been proposed to explain CRT-associated ventricular arrhythmia including LV pacing in close proximity to scar, increases in the transmural dispersion of repolarization (which is proarrhythmic), and alteration of the wavefront of LV activation which is thought to facilitate re-entry.[106,107]

Nayak et al. described a series of 8 out of 191 (4%) patients who developed VT storm (VTS) following BiV implantation.[106] Several observations, such as (i) VTS developed a mean of 16 ± 12.5 days after initiation of BVP, (ii) VTS was refractory to intravenous antiarrhythmic medication and was managed by turning off LV pacing and/or radiofrequency catheter ablation and long-term oral antiarrhythmic therapy, (iii) of the four patients who refused catheter ablation, three had cessation of VTS after turning off the LV lead; the fourth had the LV lead reprogrammed to a lower output resulting in considerable reduction in the burden of VT, and (iv) despite elimination of VT, the presence of VTS carried a poor prognosis in that all eight patients subsequently developed refractory CHF, were made in this single-center case-series.

Although reports, such as these raise the possibility that CRT may facilitate VT in select patients, this concept has not been firmly established. Nonetheless, clinicians should consider this possibility in patients with BIV devices admitted for incessant VT refractory to antiarrhythmic therapy.

## EMERGING CRT INDICATIONS

### Narrow QRS

Although the benefits of CRT have been well established in patients with wide QRS duration, its role in patients with normal

QRS is less clear. A number of small nonrandomized studies have been conducted which suggest that patients with narrow QRS and dyssynchrony benefit from CRT.[108-111] In addition, Jeevanantham and his colleagues performed a meta-analysis of nonrandomized narrow QRS CRT trials. Pooled data from three studies (totaling 98 patients) found that CRT was associated with improvements in NYHA, LVEF and 6-minute walking distance (6MWD).[112]

The role of CRT in narrow QRS patients was evaluated in a prospective randomized fashion in the RethinQ trial.[113] In this study, 172 patients with LVEF less than 35%, NHYA class III HF, QRS interval of less than 130 millisecond and evidence of mechanical dyssynchrony as measured by echocardiography were randomized to CRT + ICD or ICD-alone groups. No significant difference between these groups was found with respect to the primary endpoint of increase in peak oxygen consumption at 6 months. Additionally, no difference with respect of echocardiographic evidence of remodeling was observed between the two groups.

Despite these negative results, many clinicians and investigators continue to believe in the value of CRT in narrow QRS HF populations. Some of the criticisms of RethinQ, include, reliance on nonspecific dyssynchrony criteria (TDI-measured opposing wall delay greater than 65 millisecond, short follow-up time and a primary endpoint that was not studied in the major CRT trials. Further clinical trials are currently underway to determine whether alternative study designs may help to elucidate the discrepancies between RethinQ and previous observational studies among patients with HF and narrow QRS durations.[114]

## Atrial Fibrillation

Multiple observational and at least one randomized trial have demonstrated benefit among patients who meet standard criteria for CRT and coexisting atrial fibrillation (AF).[115-118] In these trials, many of the enrolled patients had well-controlled ventricular rates, or else, had undergone prior AV nodal ablation. This underscores the challenges associated with the optimal delivery of BiV pacing in patients with AF. In AF, high ventricular rates may inhibit consistent BiV pacing. In addition, heart rate irregularity may result in fusion or pseudofusion thereby attenuating or eliminating the effects of BiV capture.

To overcome these challenges AV nodal ablation is increasingly utilized in patients with AF who meet criteria for CRT. Although this procedure renders a patient pacemaker-dependent and eliminates AV synchrony, it serves to regularize ventricular performance, and ensures 100% BiV pacing. The net result may be improvement in patient symptoms and beneficial remodeling in patients with HF. In one study, Gasparini et al. prospectively evaluated 673 patients (162 in AF, 511 in sinus rhythm) with LVEF less than or equal to 35%, QRS greater than or equal to 120 millisecond, and NYHA greater than or equal to II.[119] Patients who were deemed to have inadequate BiV capture (arbitrarily determined to be <85%) underwent AV nodal ablation. Both SR and AF groups showed significant and sustained improvements of all assessed parameters (p < Û0.001 for all parameters). However, within the AF group, only patients who underwent ablation showed a significant increase of ejection fraction (p < 0.001), reverse remodeling effect (p < 0.001) and improved exercise tolerance (p < 0.001); no improvements with respect to these parameters were observed in AF patients who did not undergo ablation. Although this strategy has not been evaluated in a randomized prospective fashion, a clinician should consider AV nodal ablation in AF patients with CRT devices, in whom consistent BIV capture cannot consistently be obtained or reliably assessed.

## Pacemaker-dependent Patients

It is well established that RV pacing is associated with hemodynamic derangement, the promotion of dyssynchrony and worsening of LV function, particularly among patients with decreased ejection fraction.[120-126] Early evidence that RV pacing may impact clinical endpoints came from analysis of the Mode Selection Trial (MOST) a 6-year randomized trial of dual chamber rate adaptive pacemaker (DDDR) versus ventricular rate modulated pacing (VVIR) pacing in patients with sinus node dysfunction. Analysis of the trial suggested that the cumulative percent of ventricular paced beats (Cum%VP) was a strong predictor of HF hospitalization in both DDDR and VVIR modes.[125]

Corroborating evidence came from the dual chamber and VVI implantable defibrillator (DAVID) trial in which patients eligible for ICD were randomized to dual chamber universal, rate responsive (DDD/R) pacing at a lower rate of 70 or VVI, at a lower rate of 40 beats/min. The study was terminated prematurely due to an excess of HF and deaths in the DDD/R arm.[123] Subsequent analysis of the DAVID trial suggested that the lowest risk of HF worsening and death was seen in patients randomized to DDD/R with a low Cum%VP.[127]

The findings from these studies provided a therapeutic rational to investigate whether CRT may attenuate the negative impact of chronic long-term RV apical pacing. The Homburg

Biventricular Pacing Evaluation (HOBIPACE) was a prospective randomized crossover study of patients with LV dysfunction and need for antibradycardiac pacing.[128] When compared with RV pacing, CRT was found to be superior to RV pacing as it induced reverse LV remodeling with significant reductions in LV end-diastolic and end-systolic volumes and an increase in ejection fraction. In addition, BiV pacing was found to impact favorably on the Minnesota living with HF score, NT-proBNP levels and peak oxygen consumption. The HOBIPACE was a small trial (30 patients) but served to reinforce the notion that RV pacing may particularly be detrimental to patients with pre-existing LV dysfunction.

The role of CRT among patients with normal EF and standard indications for pacing was tested in the pacing to avoid cardiac enlargement (PACE) trial.[129] A total of 177 patients with standard pacing indications in whom a BiV pacemaker was implanted were randomized to BiV pacing or RV apical pacing. The primary endpoints were LVEF and ESV at 12 months. In both groups, devices were programmed to ensure maximum pacing. During a follow-up period of 12 months, no effect on the primary endpoints was observed among those randomized to BiV pacing. However, there was a decline of 6.7 percentage points in LVEF and a 25% increase in left ventricular end-systolic volume in patients who were assigned to right ventricular pacing. The LVEF declined to less than 45% in 9% of the patients in the right-ventricular-pacing group.

While thought provoking, the results of PACE by no means suggest that all patients with pacing indications require a CRT device. First, roughly 40% of patients in each group had sinus node dysfunction. Since the study was designed to force RV pacing in the RV apical pacing group, native AV conduction would have been possible with alternative programming. In many patients, particularly those with sinus node dysfunction, RV pacing can be minimized by the use of extended AV delays, hysteresis and device algorithms that promote native AV conduction.

Second, there is a limited body of evidence that selective site right ventricular pacing (for example, from the RVOT) may be less detrimental than RV apical pacing.[130] It is possible that alternative RV pacing sites may sufficient to prevent adverse remodeling in properly selective patients. This could in theory obviate the need for CRT in patients with normal ejection fraction and AV nodal disease. However, larger clinical trials will need to be conducted to establish this as a viable pacing strategy.

Third, the clinical significance of the volumetric changes observed in the RV pacing group remains unknown. Although these changes are intuitively undesirable, they were not associated with concomitant reductions in 6-minute walking test, QOL or hospitalization for HF. It is possible that with longer follow-up duration, these parameters may have also been adversely affected.

Finally, only 9% of patients in the RV apical pacing group experienced significant reductions in LVEF (<45%). This suggests that, at one year, the vast majority of patients with normal EF in whom antibradycardic pacing is indicated will not experience steep declines in their cardiac function with RV apical pacing. Accordingly, clinicians might consider initial implantation of an RV lead in these patients with the addition of an LV lead should a patient's EF decline with serial echocardiographic monitoring.

## Minimally Symptomatic Heart Failure

The beneficial effect of CRT on ventricular remodeling has prompted investigators to evaluate the role of CRT in patients with depressed ejection fraction and minimal symptoms (NYHA I-II). The resynchronization reverse remodeling in systolic left ventricular dysfunction (REVERSE) trial enrolled 610 patients with NYHA I or II HF, QRS greater than or equal to 120 millisecond and LVEF less than or equal to 40%.[131] All enrollees were implanted with CRT devices +/– ICD. Following implantation, patients were randomized in 2:1 fashion to CRT-on or CRT-off. At 12 months, there was no difference between these groups with respect to the trial's primary endpoint, a clinical composite that assessed worsening of HF through a number of measures including all-cause mortality, heart-failure hospitalizations, crossover due to worsening HF, NYHA class and the patient global assessment. The CRT was however shown to improve left ventricular end-systolic volumes (LVESV), left ventricular end-diastolic volumes (LVEDV) and LVEF.

A prospectively planned two-year follow-up of 262 European participants in REVERSE was also conducted. At two years, comparison of the CRT-on versus CRT-off groups demonstrated that a significantly higher percentage of patients in the latter group had a worsening of HF (34 vs 19% p = 0.01). Improvement in volumetric parameters persisted over this time period. In addition, there was a 12% absolute risk reduction in the time to first hospitalization or death, associated with CRT therapy (p = 0.003).

A similar hypothesis was tested in the MADIT-CRT trial.[132] In this study, 1,820 patients with ischemic or nonischemic cardiomyopathy, EF less than or equal to 30%, QRS greater than or equal to 130 millisecond and NYHA class I or II symptoms were randomly assigned to receive CRT + ICD or ICD alone. The primary end point was death from any cause or nonfatal heart-failure event (whichever came first). After a mean follow-up of 2.5 years, CRT therapy was associated with a significant 34% reduction in the risk of death or nonfatal HF (p = 0.001). The difference between CRT and control groups was driven primarily by a reduced incidence of HF events among those randomized to CRT (23 vs 14% p < 0.001). As with many other CRT trials, CRT was associated with significant improvements in LVESV, LVEDV and EF.

The results of REVERSE and MADIT-CRT suggest that CRT may play a role in delaying HF progression in patients with minimal symptoms. Based on MADIT-CRT, FDA has recently been approved CRT therapy in patients with wide QRS, reduced ejection fraction and NYHA class I and II.

## CRT for Acute Decompensated Heart Failure

Patients with systolic HF who are admitted to the ICU may be candidates for CTR. In these instances, patients who meet current guidelines for CRT implantation should be referred for further evaluation. In general, however, it is preferable to wait until a patient is stabilized, before CRT implantation is undertaken. Rarely, is CRT therapy required in an acute setting for patient stabilization. Although there is little evidence to guide the practice, many clinicians advocate acute implantation of CRT device for patients who meet criteria for implantation and cannot be weaned from inotropic therapies or else are responding poorly to aggressive HF management.[133,134] In this setting, there may be a role for acute CRT implantation as carefully selected patients may acutely respond with dramatic improvement in their clinical status. These cases should be evaluated on an individual basis in concert with consulting cardiologists, electrophysiologists or HF specialists.

## SUMMARY

The advent of CRT has been an important development in the management of HF. The results of multiple large-scale clinical trials have consistently demonstrated its favorable impact on symptoms related to HF. In addition, there is mounting evidence that CRT is associated with mortality benefit. Current indications for CRT include patients with wide QRS and ejection fraction less than or equal to 35% with advanced HF despite optimal medical management. Around 30–40% of the patients who are candidates for CRT do not respond. This can be improved by optimizing the device, using imaging to select patients based on dyssynchrony and optimal LV lead placement. In the future, indications for BIV implantation may expand to include select patients with systolic HF and narrow QRS and patients with normal ejection fraction who require chronic RV pacing.

# Modified Summary of Guidelines (ACC/AHA/HRS Guidelines for Device-based Therapy, JACC. 2008;51:e1-62)

*Modified by Kanu Chatterjee*

Class I: Conditions for which there is evidence and/or general agreement that a given procedure/therapy is useful and effective.

Class II: Conditions for which there is conflicting evidence and/or a divergence of opinion about the usefulness/efficacy of performing the procedure/therapy.

Class IIa: Weight of evidence/opinion is in favor of usefulness/efficacy.

Class IIb: Usefulness/efficacy is less well established by evidence/opinion.

Class III: Conditions for which there is evidence and/or general agreement that a procedure/therapy is not useful/effective and in some cases may be harmful.

Level A (highest): Derived from multiple randomized clinical trials.

Level B (intermediate): Data are on the basis of a limited number of randomized trials, nonrandomized studies or observational registries.

Level C (lowest): Primary basis for the recommendation was expert opinion.

## Recommendations for Permanent Pacing in Sinus Node Dysfunction (SND)

Class I:
1. Permanent pacemaker implantation is indicated in symptomatic patients with symptomatic bradycardia including frequent symptomatic sinus pauses (Level of Evidence C).
2. Permanent pacemaker implantation is indicated for symptomatic chronotropic incompetence (Level of Evidence C).

Class IIa:
1. Permanent pacemaker implantation is reasonable for SND with heart rate less than 40/bpm, when a clear association between symptoms consistent with bradycardia and actual presence of bradycardia has not been documented (Level of Evidence C).
2. Permanent pacemaker implantation is reasonable in patients with unexplained syncope when SND is

documented by electrophysiologic studies (Level of Evidence C).

Class IIb:
1. Permanent pacemaker therapy is reasonable in minimally symptomatic patients with chronic awake heart rate of less than 40/bpm (Level of Evidence C).

## Recommendations for Acquired Atrioventricular Block in Adults

Class I:
1. Permanent pacemaker therapy is indicated in patients with third-degree and advanced second-degree AV block even in absence of symptoms (Level of Evidence C).
2. Permanent pacemaker therapy is indicated in patients with third-degree or advanced second-degree AV block in asymptomatic patients with documented period of asystole of 3 seconds or greater (Level of Evidence C).
3. Permanent pacemaker implantation is indicated in patients with third-degree or advanced second-degree AV block developing after AV nodal ablation (Level of Evidence C).
4. Permanent pacemaker therapy is indicated in postcardiac surgery AV block when AV block is unlikely to resolve (Level of Evidence C).
5. Permanent pacemaker treatment is indicated in patients with neuromuscular diseases (e.g. Duchene's and Baker's, limb girdle, peroneal muscular dystrophy) with third-degree or advanced second-degree AV block with or without symptoms (Level of Evidence B).
6. Permanent pacemaker therapy is indicated in symptomatic patients with any type of second degree AV block (Level of Evidence B).
7. Permanent pacemaker therapy is indicated in patients with systolic heart failure with third-degree or infranodal AV block in absence of symptoms related to heart block (Level of Evidence B).
8. Permanent pacemaker implantation is indicated in patients who develop second or third-degree AV block during exercise unrelated to myocardial ischemia (Level of Evidence C).

Class IIa:
1. Permanent pacemaker therapy is reasonable in asymptomatic patients with third-degree, or intra or infra Hisian AV block (Level of Evidence C).
2. Symptom limited exercise test at 3–6 weeks after discharge to assess prognosis, activity prescription or evaluation of medical therapy if early exercise test was submaximal.

## Recommendations for permanent Pacing in Chronic Bifascicular Block

Class I:
1. Permanent pacemaker therapy is indicated in patients with bifascicular block with advanced second-degree, intermittent third-degree or alternating bundle-branch block (Level of Evidence B).

Class IIa:
1. Permanent pacemaker implantation is reasonable in patients with bifascicular block with history of syncope when other causes have been excluded (Level of Evidence B).
2. Permanent pacemaker implantation is reasonable in patients with bifascicular block if HV interval is 100 ms or greater documented during electrophysiologic study (Level of Evidence B).

## Recommendations for Permanent Pacing after the Acute Phase of Myocardial Infarction

Class I:
1. Permanent pacemaker therapy is indicated in post-ST elevation myocardial infarction with intermittent or persistent third-degree, advanced second-degree infranodal AV block or alternating bundle-branch block (Level of Evidence B).
2. Permanent pacemaker therapy is indicated in symptomatic second-degree or third-degree AV block (Level of Evidence C).

## Recommendations for Permanent Pacing in Hypersensitive Carotid Sinus Syndrome and Neurocardiogenic Syncope

Class I:
1. Permanent pacemaker implantation is indicated in patients with recurrent syncope due to hypersensitive carotid sinus syndrome with ventricular asystole of 3 seconds or longer (Level of Evidence C).

Class IIa:
1. Permanent pacing is reasonable in patient with hypersensitive carotid sinus syndrome with cardioinhibitory response of 3 seconds or longer without provocative events (Level of Evidence C).

Class IIb:
1. In patients with neurocardiogenic syncope permanent pacemaker therapy may be considered associated

cardioinhibitory response occurring spontaneously or during tilt-table test (Level of Evidence B).

## Recommendations for Pacing after Cardiac Transplantation

1. Permanent pacemaker implantation is indicated for inappropriate heart rate response and for Class I indications as in non transplant patients (Level of Evidence C).

Class IIb:
1. Permanent pacemaker therapy can be considered in postcardiac transplant patients with recurrent prolonged bradycardia or inappropriate heart rate response that limits rehabilitation (Level of Evidence C).

## Recommendations for Pacing to Prevent Tachycardia

Class I:
1. Permanent pacemaker implantation is indicated for sustained pause-dependent ventricular tachycardia with or without QT prolongation (Level of Evidence C).

Class IIa:
1. Permanent pacemaker implantation is reasonable in high-risk patients with congenital Long QT syndrome (Level of Evidence B).

Class IIb:
1. Permanent pacing may be considered in patients with brady-tachy syndrome with recurrent atrial fibrillation (Level of Evidence B).
2. After discharge for activity counseling and/or exercise training as part of cardiac rehabilitation in patients who have undergone revascularization

Class IIb:
1. In patients with ECG abnormalities of LBBB, pre-excitation syndrome, left ventricular hypertrophy, digoxin therapy, greater than 1 mm resting ST depression. Electronically paced ventricular rhythm.
2. Periodic monitoring in patients who continue to participate in exercise training or cardiac rehabilitation.

Class III:
1. Severe comorbidity likely to limit life expectancy and/or candidacy for revascularization.
2. To evaluate patients with acute myocardial infarction with uncompensated heart failure, cardiac arrythmia or

noncardiac conditions that limit the ability to exercise (Level of Evidence C).
3. Predischarge exercise test in patients who had already cardiac catheterization (Level of Evidence C).

## Asymptomatic Diabetic Patients

Class IIa:
1. Evaluation of asymptomatic patients with diabetes who plan to do vigorous exercise (Level of Evidence C).

Class IIb:
1. Evaluation of patients with multiple risk factors as guide to risk-reduction therapy.
2. Evaluation of asymptomatic men older than 45 years or women older than 55 years who plan to do vigorous exercise or who are involved in an occupation in which exercise impairment may impact public safety or who are at high-risk for CAD.

Class III:
1. Routine screening of asymptomatic men or women.

## Patients with Valvular Heart Disease

Class I:
1. In patients with chronic aortic regurgitation for assessment of symptoms and functional capacity in whom it is difficult assess symptoms.

Class IIa:
1. In patients with chronic aortic regurgitation for evaluation of symptoms and functional capacity before participation in athletic activity.
2. In patients with chronic aortic regurgitation for assessment of prognosis before aortic valve replacement in asymptomatic or minimally symptomatic patients with left ventricular dysfunction.

Class IIb:
1. Evaluation of patients with valvular heart disease (see guide lines in valvular heart disease).

Class III:
1. For diagnosis of CAD in patients with moderate to severe valvular heart disease or with LBBB, electronically paced rhythm, preexcitation syndrome or greater than 1 mm ST depression in the rest ECG.

> **Patients with Rhythm Disorders**
>
> Class I:
> 1. For identification of appropriate settings in patients with rate-adaptive pacemakers.
> 2. For evaluation of congenital complete heart block in patients considering increased physical activity or participation in competitive sports (Level of Evidence C).
>
> Class IIa:
> 1. Evaluation of patients with known or suspected exercise-induced arrhythmias.
> 2. Evaluation medical, surgical or ablation therapy in patients with exercise-induced arrhythmias (including atrial fibrillation).
>
> Class IIb:
> 1. Investigation of isolated ventricular ectopic beats in middle aged patients without other evidence of CAD.
> 2. For investigation of prolonged first-degree atrioventricular block or type I second degree Wenckebach, left bundle-branch block, right bundle-branch block or isolated ectopic beats in young persons considering participation in competitive sports (Level of Evidence C).
>
> Class III:
> 1. Routine investigations of isolated ectopic beats in young patients.

## REFERENCES

1. Lloyd-Jones D, Adams RJ, Brown TM, et al. Heart disease and stroke statistics 2010 update: a report from the American Heart Association. Circulation. 2010;121:e46-215.
2. Rogers WJ, Johnstone DE, Yusuf S, et al. Effect of enalapril on survival in patients with reduced left ventricular ejection fractions and congestive heart failure. N Engl J Med. 1991;325:293-302.
3. CIBIS-II Investigators. The cardiac insufficiency bisoprolol study II: a randomised trial. Lancet. 1999;353:9-13.
4. Packer M, Coats AJ, Fowler MB, et al. Effect of carvedilol on survival in severe chronic heart failure. N Engl J Med. 2001;344:1651-8.
5. Hjalmarson A, Goldstein S, Fagerberg B, et al. Effects of controlled-release metoprolol on total mortality, hospitalizations and well-being in patients with heart failure: the Metoprolol CR/XL Randomized Intervention Trial in congestive heart failure (MERIT-HF). MERIT-HF Study Group. JAMA. 2000;283:1295-302.
6. Pitt B, Remme W, Zannad F, et al. Eplerenone: a selective aldosterone blocker in patients with left ventricular dysfunction after myocardial infarction. N Engl J Med. 2003;348:1309-21.
7. Pitt B, Zannad F, Remme WJ, et al. The effect of spironolactone on morbidity and mortality in patients with severe heart failure. Randomized aldactone evaluation study. N Engl J Med. 1999;341:709-17.

8. Moss AJ, Zareba W, Hall WJ, et al. Prophylactic implantation of a defibrillator in patients with myocardial infarction and reduced ejection fraction. N Engl J Med. 2002;346:877-83.
9. Bardy GH, Lee KL, Mark DB, et al. Amiodarone or an implantable cardioverter-defibrillator for congestive heart failure. N Engl J Med. 2005;352:225-37.
10. Cleland JG, Daubert JC, Erdmann E, et al. The effect of cardiac resynchronization on morbidity and mortality in heart failure. N Engl J Med. 2005;352:1539-49.
11. Reicin G, Miksic M, Yik A, et al. Hospital supplies and medical technology 4Q05. Statistical handbook: growth moderating but outlook remains strong. New York: Morgan Stanley Equity Research; 2005. p. 44.
12. American Heart Association. Heart Disease and Stroke Statistics–2010 Update. Dallas, Texas: American Heart Association; 2010.
13. Auricchio A, Prinzen FW. Update on the pathophysiological basics of cardiac resynchronization therapy. Europace. 2008;10:797-800.
14. Auricchio A, Stellbrink C, Block M, et al. Effect of pacing chamber and atrioventricular delay on acute systolic function of paced patients with congestive heart failure. The pacing therapies for congestive heart failure study group. The guidant congestive heart failure research group. Circulation. 1999;99:2993-3001.
15. Nishimura R, Hayes D, Holmes D, et al. Mechanism of hemodynamic improvement by dual-chamber pacing for severe left ventricular dysfunction: an acute Doppler and catheterization hemodynamic study. J Am Coll Cardiol. 1995;25:281-8.
16. Butter C, Auricchio A, Stellbrink C, et al. Effect of resynchronization therapy stimulation site on the systolic function of heart failure patients. Circulation. 2001;104:3026-9.
17. Breithardt OA, Sinha AM, Schwammenthal E, et al. Acute effects of cardiac resynchronization therapy on functional mitral regurgitation in advanced systolic heart failure. J Am Coll Cardiol. 2003;41:765-70.
18. Abraham WT, Fisher WG, Smith AL, et al. Cardiac resynchronization in chronic heart failure. N Engl J Med. 2002;346:1845-53.
19. Leclercq C, Cazeau S, Le Breton H, et al. Acute hemodynamic effects of biventricular DDD pacing in patients with end-stage heart failure. J Am Coll Cardiol. 1998;32:1825-31.
20. Kass DA, Chen CH, Curry C, et al. Improved left ventricular mechanics from acute VDD pacing in patients with dilated cardiomyopathy and ventricular conduction delay. Circulation. 1999;99:1567-73.
21. St John Sutton MG, Plappert T, Abraham WT, et al. Effect of cardiac resynchronization therapy on left ventricular size and function in chronic heart failure. Circulation. 2003;107:1985-90.
22. Saxon LA, De Marco T, Schafer J, et al. Effects of long-term biventricular stimulation for resynchronization on echocardiographic measures of remodeling. Circulation. 2002;105:1304-10.
23. Yu C, Chau E, Sanderson J, et al. Tissue Doppler echocardiographic evidence of reverse remodeling and improved synchronicity by simultaneously delaying regional contraction after biventricular pacing therapy in heart failure. Circulation. 2002;105:438-45.
24. Cazeau S, Leclercq C, Lavergne T, et al. Effects of multisite biventricular pacing in patients with heart failure and intraventricular conduction delay. N Engl J Med. 2001;344:873-80.

25. Bristow MR, Saxon LA, Boehmer J, et al. Cardiac-resynchronization therapy with or without an implantable defibrillator in advanced chronic heart failure. N Engl J Med. 2004;350:2140-50.
26. Cleland JG, Daubert JC, Erdmann E, et al. The effect of cardiac resynchronization on morbidity and mortality in heart failure. N Engl J Med. 2005;352:1539-49.
27. McAlister FA, Ezekowitz J, Hooton N, et al. Cardiac resynchronization therapy for patients with left ventricular systolic dysfunction: a systematic review. JAMA. 2007;297:2502-14.
28. Abraham W, Fisher W, Smith A, et al. Cardiac resynchronization in chronic heart failure. N Engl J Med. 2002;346:1845-53.
29. Jessup M, Abraham WT, Casey DE, et al. Focused update: ACCF/AHA guidelines for the diagnosis and management of heart failure in adults. J Am Coll Cardiol. 2009;119:1977-2016.
30. Bleeker GB, Bax JJ, Fung JW, et al. Clinical versus echocardiographic parameters to assess response to cardiac resynchronization therapy. Am J Cardiol. 2006;97:260-3.
31. Birnie DH, Tang AS. The problem of non-response to cardiac resynchronization therapy. Curr Opin Cardiol. 2006;21:20-6.
32. Chung ES, Leon AR, Tavazzi L, et al. Results of the Predictors of Response to CRT (PROSPECT) trial. Circulation. 2008;117:2608-16.
33. Sweeney MO. Programming and follow-up of cardiac resynchronization devices. In: Ellenbogen KA (Ed). Clinical Cardiac Pacing, Defibrillation, and Resynchronization Therapy. Philadelphia: Elsevier; 2007. pp. 1087-140.
34. Molhoek SG, VANE L, Bootsma M, et al. QRS duration and shortening to predict clinical response to cardiac resynchronization therapy in patients with end-stage heart failure. Pacing Clin Electrophysiol. 2004;27:308-13.
35. Lecoq G, Leclercq C, Leray E, et al. Clinical and electrocardiographic predictors of a positive response to cardiac resynchronization therapy in advanced heart failure. Eur Heart J. 2005;26:1094-100.
36. Kashani A, Barold SS. Significance of QRS complex duration in patients with heart failure. J Am Coll Cardiol. 2005;46:2183-92.
37. Georger F SC, Collet B. Specific electrocardiographic patterns may assess left ventricular capture during biventricular pacing. Pacing Clin Electrophysiol. 2002;25:37-41.
38. Ammann P, Sticherling C, Kalusche D, et al. An electrocardiogram-based algorithm to detect loss of left ventricular capture during cardiac resynchronization therapy. Ann Intern Med. 2005;142:968-73.
39. Rao RK, Kumar UN, Schafer J, et al. Reduced ventricular volumes and improved systolic function with cardiac resynchronization therapy: a randomized trial comparing simultaneous biventricular pacing, sequential biventricular pacing, and left ventricular pacing. Circulation. 2007;115:2136-44.
40. Boriani G, Müller CP, Seidl KH, et al. Randomized comparison of simultaneous biventricular stimulation versus optimized interventricular delay in cardiac resynchronization therapy. The resynchronization for the hemodynamic treatment for heart failure management II implantable cardioverter defibrillator (RHYTHM II ICD) study. Am Heart J. 2006;151:1050-8.
41. Jansen AH, Bracke FA, van Dantzig JM, et al. Correlation of echo-Doppler optimization of atrioventricular delay in cardiac resynchronization therapy with invasive hemodynamics in patients

with heart failure secondary to ischemic or idiopathic dilated cardiomyopathy. Am J Cardiol. 2006;97:552-7.
42. Barold SS, Ilercil A, Herweg B. Echocardiographic optimization of the atrioventricular and interventricular intervals during cardiac resynchronization. Europace. 2008;10:88-95.
43. Morales MA, Startari U, Panchetti L, et al. Atrioventricular delay optimization by doppler-derived left ventricular dP/dt improves 6-month outcome of resynchronized patients. Pacing Clin Electrophysiol. 2006;29:564-8.
44. Whinnett ZI, Davies JE, Willson K, et al. Haemodynamic effects of changes in atrioventricular and interventricular delay in cardiac resynchronisation therapy show a consistent pattern: analysis of shape, magnitude and relative importance of atrioventricular and interventricular delay. Heart. 2006;92:1628-34.
45. Braun MU, Schnabel A, Rauwolf T, et al. Impedance cardiography as a noninvasive technique for atrioventricular interval optimization in cardiac resynchronization therapy. J Interv Card Electrophysiol. 2005;13:223-9.
46. Burri H, Sunthorn H, Somsen A, et al. Optimizing sequential biventricular pacing using radionuclide ventriculography. Heart Rhythm. 2005;2:960-5.
47. Kamdar R, Frain E, Warburton F, et al. A prospective comparison of echocardiography and device algorithms for atrioventricular and interventricular interval optimization in cardiac resynchronization therapy. Europace. 2010;12:84-91.
48. Hardt SE, Yazdi SH, Bauer A, et al. Immediate and chronic effects of AV-delay optimization in patients with cardiac resynchronization therapy. Int J Cardiol. 2007;115:318-25.
49. Riedlbauchová L, Kautzner J, Frídl P. Influence of different atrioventricular and interventricular delays on cardiac output during cardiac resynchronization therapy. Pacing Clin Electrophysiol. 2005;28:S19-23.
50. Sawhney NS, Waggoner AD, Garhwal S, et al. Randomized prospective trial of atrioventricular delay programming for cardiac resynchronization therapy. Heart Rhythm. 2004;1:562-7.
51. Bax J, Abraham T, Barold S, et al. Cardiac resynchronization therapy: Part 1—issues before device implantation. J Am Coll Cardiol. 2005;46:2153-67.
52. Bax JJ, Bleeker GB, Marwick TH, et al. Left ventricular dyssynchrony predicts response and prognosis after cardiac resynchronization therapy. J Am Coll Cardiol. 2004;44:1834-40.
53. Ghio S, Constantin C, Klersy C, et al. Interventricular and intraventricular dyssynchrony are common in heart failure patients, regardless of QRS duration. Eur Heart J. 2004;25:571-8.
54. Bleeker G, Schalij M, Molhoek S, et al. Relationship between QRS duration and left ventricular dyssynchrony in patients with end-stage heart failure. J Cardiovasc Electrophysiol. 2004;15:544-9.
55. Pitzalis MV, Iacoviello M, Romito R, et al. Ventricular asynchrony predicts a better outcome in patients with chronic heart failure receiving cardiac resynchronization therapy. J Am Coll Cardiol. 2005;45:65-9.
56. Pitzalis MV, Iacoviello M, Romito R, et al. Cardiac resynchronization therapy tailored by echocardiographic evaluation of ventricular asynchrony. J Am Coll Cardiol. 2002;40:1615-22.

57. Sassone B, Capecchi A, Boggian G, et al. Value of baseline left lateral wall postsystolic displacement assessed by M-mode to predict reverse remodeling by cardiac resynchronization therapy. Am J Cardiol. 2007;100:470-5.
58. Marcus GM, Rose E, Viloria EM, et al. Septal to posterior wall motion delay fails to predict reverse remodeling or clinical improvement in patients undergoing cardiac resynchronization therapy. J Am Coll Cardiol. 2005;46:2208-14.
59. Bax J, Bleeker G, Marwick T, et al. Left ventricular dyssynchrony predicts response and prognosis after cardiac resynchronization therapy. J Am Coll Cardiol. 2004;44:1834-40.
60. Yu CM, Chau E, Sanderson JE, et al. Tissue Doppler echocardiographic evidence of reverse remodeling and improved synchronicity by simultaneously delaying regional contraction after biventricular pacing therapy in heart failure. Circulation. 2002;105:438-45.
61. Yu CM, Fung JW, Zhang Q, et al. Tissue Doppler imaging is superior to strain rate imaging and postsystolic shortening on the prediction of reverse remodeling in both ischemic and nonischemic heart failure after cardiac resynchronization therapy. Circulation. 2004;110:66-73.
62. Yu C, Zhang Q, Fung J, et al. A novel tool to assess systolic asynchrony and identify responders of cardiac resynchronization therapy by tissue synchronization imaging. J Am Coll Cardiol. 2005;45:677-84.
63. Gorcsan J, 3rd, Kanzaki H, Bazaz R, et al. Usefulness of echocardiographic tissue synchronization imaging to predict acute response to cardiac resynchronization therapy. Am J Cardiol. 2004;93:1178-81.
64. Van de Veire NR, Bleeker GB, De Sutter J, et al. Tissue synchronisation imaging accurately measures left ventricular dyssynchrony and predicts response to cardiac resynchronisation therapy. Heart. 2007;93:1034-9.
65. Oh JK, Seward JB, Jamil Tajik A. The Echo Manual, 3rd edition. Philadelphia: Lippincott Williams and Wilkins; 2007.
66. Bogaert J, Rademakers FE. Regional nonuniformity of normal adult human left ventricle. Am J Physiol Heart Circ Physiol. 2001;280: H610-20.
67. Porciani MC, Lilli A, Macioce R, et al. Utility of a new left ventricular asynchrony index as a predictor of reverse remodelling after cardiac resynchronization therapy. Eur Heart J. 2006;27:1818-23.
68. Mele D, Pasanisi G, Capasso F, et al. Left intraventricular myocardial deformation dyssynchrony identifies responders to cardiac resynchronization therapy in patients with heart failure. Eur Heart J. 2006;27:1070-8.
69. Dohi K, Suffoletto MS, Schwartzman D, et al. Utility of echocardiographic radial strain imaging to quantify left ventricular dyssynchrony and predict acute response to cardiac resynchronization therapy. Am J Cardiol. 2005;96:112-6.
70. Delgado V, Ypenburg C, van Bommel RJ, et al. Assessment of left ventricular dyssynchrony by speckle tracking strain imaging comparison between longitudinal, circumferential, and radial strain in cardiac resynchronization therapy. J Am Coll Cardiol. 2008;51:1944-52.
71. Gorcsan J 3rd, Tanabe M, Bleeker GB, et al. Combined longitudinal and radial dyssynchrony predicts ventricular response after resynchronization therapy. J Am Coll Cardiol. 2007;50:1476-83.

72. Suffoletto MS, Dohi K, Cannesson M, et al. Novel speckle-tracking radial strain from routine black-and-white echocardiographic images to quantify dyssynchrony and predict response to cardiac resynchronization therapy. Circulation. 2006;113:960-8.
73. Sanderson JE. Echocardiography for cardiac resynchronization therapy selection. J Am Coll Cardiol. 2009;53:1960-4.
74. Bax JJ, Gorcsan J 3rd. Echocardiography and noninvasive imaging in cardiac resynchronization therapy: results of the PROSPECT (Predictors of Response to Cardiac Resynchronization Therapy) study in perspective. J Am Coll Cardiol. 2009;53:1933-43.
75. Delfino JG, Fornwalt BK, Oshinski JN, et al. Role of MRI in patient selection for CRT. Echocardiography. 2008;25:1176-85.
76. Ypenburg C, Roes SD, Bleeker GB, et al. Effect of total scar burden on contrast-enhanced magnetic resonance imaging on response to cardiac resynchronization therapy. Am J Cardiol. 2007;99:657-60.
77. White JA, Yee R, Yuan X, et al. Delayed enhancement magnetic resonance imaging predicts response to cardiac resynchronization therapy in patients with intraventricular dyssynchrony. J Am Coll Cardiol. 2006;48:1953-60.
78. Bleeker GB, Kaandorp TA, Lamb HJ, et al. Effect of posterolateral scar tissue on clinical and echocardiographic improvement after cardiac resynchronization therapy. Circulation. 2006;113:969-76.
79. Botvinick EH, O'Connell JW, Badhwar N. Imaging synchrony. J Nucl Cardiol. 2009;16:846-8.
80. Kerwin WF, Botvinick EH, O'Connell JW, et al. Ventricular contraction abnormalities in dilated cardiomyopathy: effect of biventricular pacing to correct interventricular dyssynchrony. J Am Coll Cardiol. 2000;35:1221-7.
81. O'Connell JW, Schreck C, Moles M, et al. A unique method by which to quantitate synchrony with equilibrium radionuclide angiography. J Nucl Cardiol. 2005;12:441-50.
82. Wassenaar R, O'Connor D, Dej B, et al. Optimization and validation of radionuclide angiography phase analysis parameters for quantification of mechanical dyssynchrony. J Nucl Cardiol. 2009;16:895-903.
83. Ypenburg C, Schalij MJ, Bleeker GB, et al. Impact of viability and scar tissue on response to cardiac resynchronization therapy in ischaemic heart failure patients. Eur Heart J. 2007;28:33-41.
84. Adelstein EC, Saba S. Scar burden by myocardial perfusion imaging predicts echocardiographic response to cardiac resynchronization therapy in ischemic cardiomyopathy. Am Heart J. 2007;153:105-12.
85. Sciagra R, Giaccardi M, Porciani MC, et al. Myocardial perfusion imaging using gated SPECT in heart failure patients undergoing cardiac resynchronization therapy. J Nucl Med. 2004;45:164-8.
86. Marsan NA, Tops LF, Nihoyannopoulos P, et al. Real-time three dimensional echocardiography: current and future clinical applications. Heart. 2009;95:1881-90.
87. Marsan NA, Bleeker GB, Ypenburg C, et al. Real-time three-dimensional echocardiography as a novel approach to assess left ventricular and left atrium reverse remodeling and to predict response to cardiac resynchronization therapy. Heart Rhythm. 2008;5:1257-64.
88. Hawkins NM, Petrie MC, Burgess MI, et al. Selecting patients for cardiac resynchronization therapy. J Am Coll Cardiol. 2009;53:1944-59.

89. Truong QA, Singh JP, Cannon CP, et al. Quantitative analysis of intraventricular dyssynchrony using wall thickness by multidetector computed tomography. JACC Cardiovasc Imaging. 2008;1:772-81.
90. Epstein AE, Dimarco JP, Ellenbogen KA, et al. ACC/AHA/HRS 2008 guidelines for device-based therapy of cardiac rhythm abnormalities. J Am Coll Cardiol. 2008;51:e1-62.
91. Ansalone G, Giannantoni P, Ricci R, et al. Doppler myocardial imaging to evaluate the effectiveness of pacing sites in patients receiving biventricular pacing. J Am Coll Cardiol. 2002;39:489-99.
92. Dekker AL, Phelps B, Dijkman B, et al. Epicardial left ventricular lead placement for cardiac resynchronization therapy: optimal pace site selection with pressure-volume loops. J Thorac Cardiovasc Surg. 2004;127:1641-7.
93. Koos R, Sinha AM, Markus K, et al. Comparison of left ventricular lead placement via the coronary venous approach versus lateral thoracotomy in patients receiving cardiac resynchronization therapy. Am J Cardiol. 2004;94:59-63.
94. Fernandez AL, Garcia-Bengochea JB, Ledo R, et al. Minimally invasive surgical implantation of left ventricular epicardial leads for ventricular resynchronization using video-assisted thoracoscopy. Rev Esp Cardiol. 2004;57:313-9.
95. Badhwar N, Lee BK, Kumar UN, et al. Utility of equilibrium radionuclide angiograms to guide coronary sinus lead placement in heart failure patients requiring resynchronization therapy (abstract). Western Regional Meeting. Carmel, CA; 2006.
96. Heist EK, Fan D, Mela T, et al. Radiographic left ventricular-right ventricular interlead distance predicts the acute hemodynamic response to cardiac resynchronization therapy. Am J Cardiol. 2005;96:685-90.
97. Kanhai SM, Viergever EP, Bax JJ. Cardiogenic shock shortly after initial success of cardiac resynchronization therapy. Eur J Heart Fail. 2004;6:477-81.
98. Guerra JM, Wu J, Miller JM, et al. Increase in ventricular tachycardia frequency after biventricular implantable cardioverter defibrillator upgrade. J Cardiovasc Electrophysiol. 2003;14:1245-7.
99. Nordbeck P, Seidl B, Fey B, et al. Effect of cardiac resynchronization therapy on the incidence of electrical storm. International Journal of Cardiology. 2010;143:330-6 (Epub. 2009).
100. Walker S, Levy TM, Rex S, et al. Usefulness of suppression of ventricular arrhythmia by biventricular pacing in severe congestive cardiac failure. Am J Cardiol. 2000;86:231-3.
101. Zagrodzky JD, Ramaswamy K, Page RL, et al. Biventricular pacing decreases the inducibility of ventricular tachycardia in patients with ischemic cardiomyopathy. Am J Cardiol. 2001;87:1208-10; A7.
102. Rivero-Ayerza M, Theuns DA, Garcia-Garcia HM, et al. Effects of cardiac resynchronization therapy on overall mortality and mode of death: a meta-analysis of randomized controlled trials. European Heart Journal. 2006;27:2682-8.
103. Combes N, Marijon E, Boveda S, et al. Electrical storm after CRT implantation treated by AV delay optimization. Journal of Cardiovascular Electrophysiology. 2010;21:211-3 (Epub. 2009).
104. Kantharia BK, Patel JA, Nagra BS, et al. Electrical storm of monomorphic ventricular tachycardia after a cardiac-resynchronization-therapy-defibrillator upgrade. Europace. 2006;8:625-8.

105. Bortone A, Macia J-C, Leclercq F, et al. Monomorphic ventricular tachycardia induced by cardiac resynchronization therapy in patient with severe nonischemic dilated cardiomyopathy. Pacing Clin Electrophysiol. 2006;29:327-30.
106. Nayak HM, Verdino RJ, Russo AM, et al. Ventricular tachycardia storm after initiation of biventricular pacing: incidence, clinical characteristics, management, and outcome. Journal of Cardiovascular Electrophysiology. 2008;19:708-15.
107. Fish JM, Brugada J, Antzelevitch C. Potential proarrhythmic effects of biventricular pacing. J Am Coll Cardiol. 2005;46:2340-7.
108. Gasparini M, Regoli F, Galimberti P, et al. Three years of cardiac resynchronization therapy: could superior benefits be obtained in patients with heart failure and narrow QRS? Pacing Clin Electrophysiol. 2007;30:S34-9.
109. Yu C-M, Chan Y-S, Zhang Q, et al. Benefits of cardiac resynchronization therapy for heart failure patients with narrow QRS complexes and coexisting systolic asynchrony by echocardiography. J Am Coll Cardiol. 2006;48:2251-7.
110. Achilli A, Sassara M, Ficili S, et al. Long-term effectiveness of cardiac resynchronization therapy in patients with refractory heart failure and "narrow" QRS. J Am Coll Cardiol. 2003;42:2117-24.
111. Bleeker G, Holman E, Steendijk P, et al. Cardiac resynchronization therapy in patients with a narrow QRS complex. J Am Coll Cardiol. 2006;48:2243-50.
112. Jeevanantham V, Zareba W, Navaneethan S, et al. Meta-analysis on effects of cardiac resynchronization therapy in heart failure patients with narrow QRS complex. Cardiology Journal. 2008;15:230-6.
113. Beshai JF, Grimm RA, Nagueh SF, et al. Cardiac-resynchronization therapy in heart failure with narrow QRS complexes. N Engl J Med. 2007;357:2461-71.
114. [cited; Available from: http://clinicaltrials.gov/ct2/show/NCT00683696]
115. Leon AR, Greenberg JM, Kanuru N, et al. Cardiac resynchronization in patients with congestive heart failure and chronic atrial fibrillation: effect of upgrading to biventricular pacing after chronic right ventricular pacing. J Am Coll Cardiol. 2002;39:1258-63.
116. Linde C, Leclercq C, Rex S, et al. Long-term benefits of biventricular pacing in congestive heart failure: results from the MUltisite STimulation in cardiomyopathy (MUSTIC) study. J Am Coll Cardiol. 2002;40:111-8.
117. Molhoek SG, Bax JJ, Bleeker GB, et al. Comparison of response to cardiac resynchronization therapy in patients with sinus rhythm versus chronic atrial fibrillation. Am J Cardiol. 2004;94:1506-9.
118. Dong K, Shen WK, Powell BD, et al. Atrioventricular nodal ablation predicts survival benefit in patients with atrial fibrillation receiving cardiac resynchronization therapy. Heart Rhythm. 2010;7:1240-5.
119. Gasparini M, Auricchio A, Regoli F, et al. Four-year efficacy of cardiac resynchronization therapy on exercise tolerance and disease progression: the importance of performing atrioventricular junction ablation in patients with atrial fibrillation. J Am Coll Cardiol. 2006;48:734-43.
120. Delgado V, Tops LF, Trines SA, et al. Acute effects of right ventricular apical pacing on left ventricular synchrony and mechanics. Circulation: Arrhythm Electrophysiol. 2009;2:135-45.

121. Lieberman R, Padeletti L, Schreuder J, et al. Ventricular pacing lead location alters systemic hemodynamics and left ventricular function in patients with and without reduced ejection fraction. J Am Coll Cardiol. 2006;48:1634-41.
122. O'Keefe JH, Abuissa H, Jones PG, et al. Effect of chronic right ventricular apical pacing on left ventricular function. Am J Cardiol. 2005;95:771-3.
123. Wilkoff BL, Cook JR, Epstein AE, et al. Dual-chamber pacing or ventricular backup pacing in patients with an implantable defibrillator: the Dual Chamber and VVI Implantable Defibrillator (DAVID) trial. JAMA. 2002;288:3115-23.
124. Sweeney MO, Prinzen FW. A new paradigm for physiologic ventricular pacing. J Am Coll Cardiol. 2006;47:282-8.
125. Sweeney MO, Hellkamp AS, Ellenbogen KA, et al. Adverse effect of ventricular pacing on heart failure and atrial fibrillation among patients with normal baseline QRS duration in a clinical trial of pacemaker therapy for sinus node dysfunction. Circulation. 2003;107:2932-7.
126. Thambo J-B, Bordachar P, Garrigue S, et al. Detrimental ventricular remodeling in patients with congenital complete heart block and chronic right ventricular apical pacing. Circulation. 2004;110:3766-72.
127. Sharma AD, Rizo-Patron C, Hallstrom AP, et al. Percent right ventricular pacing predicts outcomes in the DAVID trial. Heart Rhythm. 2005;2:830-4.
128. Kindermann M, Hennen B, Jung J, et al. Biventricular versus conventional right ventricular stimulation for patients with standard pacing indication and left ventricular dysfunction: the Homburg Biventricular Pacing Evaluation (HOBIPACE). J Am Coll Cardiol. 2006;47:1927-37.
129. Yu C-M, Chan JY-S, Zhang Q, et al. Biventricular pacing in patients with bradycardia and normal ejection fraction. N Engl J Med. 2009;361:2123-34.
130. Albouaini K, Alkarmi A, Mudawi T, et al. Selective site right ventricular pacing. Heart. 2009;95:2030-9.
131. Linde C, Abraham WT, Gold MR, et al. Randomized trial of cardiac resynchronization in mildly symptomatic heart failure patients and in asymptomatic patients with left ventricular dysfunction and previous heart failure symptoms. J Am Coll Cardiol. 2008;52:1834-43.
132. Moss AJ, Hall WJ, Cannom DS, et al. Cardiac-resynchronization therapy for the prevention of heart-failure events. N Engl J Med. 2009;361:1329-38.
133. Herweg B, Ilercil A, Cutro R, et al. Cardiac resynchronization therapy in patients with end-stage inotrope-dependent class IV heart failure. Am J Cardiol. 2007;100:90-3.
134. James KB, Militello M, Barbara G, et al. Biventricular pacing for heart failure patients on inotropic support: a review of 38 consecutive cases. Tex Heart Inst J. 2006;33:19-22.

# Ambulatory Electrocardiographic Monitoring

**CHAPTER 13**

Renee M Sullivan, Brian Olshansky, James B Martins, Alexander Mazur

## Chapter Outline

- Holter Monitoring
- Event Recorders
- Mobile Cardiac Outpatient Telemetry
- Implantable Loop Recorders
- Key Considerations in Selecting a Monitoring Modality
- Guidelines

## INTRODUCTION

Ambulatory electrocardiographic (AECG) monitoring, the recording of the electrocardiogram (ECG) over an extended period of time using a portable or implantable recording device, enables the clinician to study dynamic electrocardiographic changes during real-life activities. It is considered to be the cornerstone in the evaluation of patients with suspected cardiac arrhythmias.[2]

The AECG was championed by Norman J Holter in the late 1940s and was introduced into clinical practice in the early 1960s.[1] Early devices consisted of continuous single-lead ECG recordings on a magnetic tape with a storage capacity of only a few hours. The recorded data could be played back at an increased speed for manual review by an operator using a specific analyzer equipped with an oscilloscope. Technological advances in signal recording, processing and transmitting, as well as automatic data analysis, have significantly enhanced the diagnostic capabilities of the ambulatory monitors we use today.

With the advent of digital data acquisition and solid state memory technology, recording devices have been substantially downsized and now capable of recording and storing high fidelity multichannel continuous electrocardiographic data over several days. Modern computer-based analysis systems use advanced diagnostic software algorithms that provide automatic quantification of a large amount of stored ECG data with calculation of multiple electrocardiographic parameters and generation of arrhythmia counters. In addition, areas of interest in the recordings that require operator review are automatically identified.

With the availability of event recorders that store only a few minutes of ECG when activated manually or automatically, based upon programmed parameters, and the capability of transmitting stored information over the phone, the period of monitoring has been extended to weeks and months. The implantable version of event recorders has further expanded the period of monitoring up to three years. More recently, the development of automatic arrhythmia detection algorithms, as well as wireless communication technology, has enabled continuous ambulatory monitoring with real-time data transfer and analysis. Currently available monitoring modalities include: continuous or Holter monitors, external event (postevent and loop) recorders, implantable loop recorders (ILRs), and mobile cardiac outpatient telemetry (MCOT) **(Fig. 1 and Table 1)**.

This chapter reviews the clinical utility and appropriate cost-effective selection of currently available monitoring modalities based on their advantages, limitations and diagnostic yields in specific populations of patients.

## HOLTER MONITORING

Ambulatory Holter monitoring is accomplished with portable battery-operated devices that continuously record multiple electrocardiographic channels, typically over a 24–48 hours period. Some modern devices can store up to two weeks of continuous ECG data. Holter monitors generally record two to three ECG leads. Although devices that allow for the recording of up to 12 ECG leads are currently available, the clinical advantage of multichannel recordings is not well determined. Manually activated events are marked with timestamps which are linked to patient diaries in an attempt to correlate symptoms with an arrhythmia.

Automatic analysis of the full data set is performed using proprietary software, but a manual over read is generally completed to validate accuracy of the automatic arrhythmia diagnosis. The standard Holter monitor analysis summarizes heart rate trends, along with the presence and frequency of tachyarrhythmias and bradyarrhythmias, atrial and ventricular ectopy, as well as asystolic pauses. With the improved quality

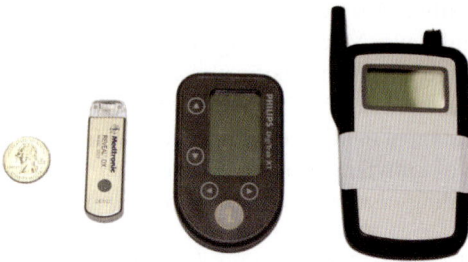

**FIGURE 1:** Examples of current monitoring devices are pictured next to a quarter, shown as a reference for size. From left to right: an implantable loop recorder; a Holter monitor and an event monitor. This model of the event monitor may also be used as mobile cardiac outpatient telemetry when programmed as such

### TABLE 1
**Characteristics of monitoring modalities**

| | Recording type | Monitoring period | Event activation | Transmission | Data analysis |
|---|---|---|---|---|---|
| Holter monitor | Continuous, full disclosure | Typically 24–48 hours | Manual | Typically none | Delayed |
| Loop recorder | Intermittent pre-and post-event | Typically up to 30 days | Manual and automatic | Dial-in trans-telephonic | Delayed |
| Event recorder | Intermittent post-event | | Manual | Dial-in trans-telephonic | Delayed |
| ILR | Intermittent | Up to 3 years | Manual and automatic | Dial-in trans-telephonic or wireless | Delayed |
| MCOT | Continuous, full disclosure | Individualized, up to 30 days | Manual and automatic | Automatic and dial-in wireless | Immediate |

(*Abbreviations*: ILR: Implantable loop recorder; MCOT: Mobile cardiac outpatient telemetry)

of acquired ECG signals in present day recorders, standard ECG measurements including PR and QT intervals, QRS width and variation in ST segments can be assessed accurately. The recorded ECG signals can also be used for more complex analyses including heart rate variability, signal averaged ECG, heart rate turbulence and T wave alternans.

Holter monitoring is generally utilized to detect a cause for symptoms and to diagnose rhythm disturbances that are expected to occur within a 24–48 hour monitoring period **(Fig. 2 and Table 2)**. The major advantage of this modality is the continuous nature of the recording that provides "full disclosure" of the ECG during the monitoring period. This type of information is particularly useful for assessment of ventricular rate response in patients with atrial fibrillation or quantification of arrhythmia burden in patients with frequent ectopy.

Although an effective diagnostic modality in patients with daily symptoms, the diagnostic yield of a Holter monitor is likely to be low, as in the case of infrequent episodes of syncope or palpitations. Depending on patient selection and the length of monitoring, the likelihood of documenting cardiac rhythm during syncope is usually less than 20%; most of the captured events correlate with the absence of significant arrhythmia.[3-5] Patients with palpitations have a diagnosis secured by a Holter monitor more so than patients with syncope but only when the symptoms are frequent and occur during the recording.[6,7] The presence of an asymptomatic arrhythmia in patients with syncope or other symptoms should be interpreted with caution since it may have little clinical meaning. Rarely, an asymptomatic arrhythmia such as Mobitz II or complete atrial-ventricular block, prolonged sinus pauses, or significant QT interval prolongation, among others, may provide a diagnostic clue as to the cause of symptoms **(Fig. 3)**. However, it is always critical to know whether an arrhythmia is temporally linked to a symptom or unnecessary therapy that may fail to provide benefit and cause harm. For instance, although sinus bradycardia may be documented during sleep, sinus bradycardia

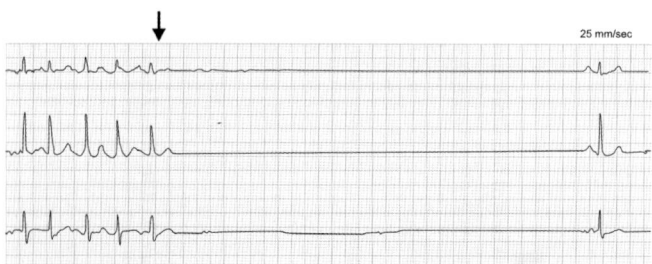

**FIGURE 2:** A 3-ECG channel Holter monitor recording obtained in a patient with recurrent palpitations and syncope. Note a 6-second sinus pause following termination of atrial fibrillation (arrow)

## TABLE 2
### Selection of monitoring modalities for common clinical indications

| | Holter monitor | Event monitor Looping | Event monitor Non-looping | Implantable loop recorder | Mobile cardiac outpatient telemetry |
|---|---|---|---|---|---|
| **Syncope** | | | | | |
| ≥1 episode/week | + | | | | |
| ≥1 episode/month | | +* | | | + |
| <1 episode/month | | | | + | + |
| **Palpitations** | | | | | |
| ≥1 episode/week | + | | | | |
| ≥1 episode/month | | + | + | | + |
| <1 episode/month | | | + | + | + |
| **Risk assessment (HCM, CAD, LQTS)** | + | +* | + | | |
| **Atrial fibrillation** | | | | | |
| Burden | + | | | | + |
| Rate control | + | +*# | | | + |
| **AAD monitoring** | | +* | | | + |

* With auto-trigger capability
# With an automatic atrial fibrillation algorithm
*Abbreviations:* AAD: Antiarrhythmic drug; MCOT: Mobile cardiac outpatient telemetry; HCM: Hypertrophic cardiomyopathy; CAD: Coronary artery disease; LQTS: Long QT syndrome

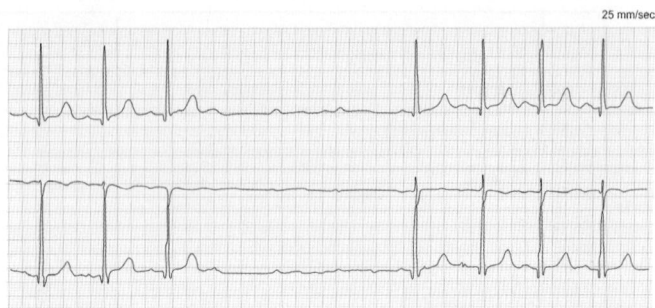

**FIGURE 3:** Paroxysmal AV block recorded during an episode of dizziness in a young patient with recurrent exertional syncope. The patient has became asymptomatic following placement of a permanent pacemaker

may not necessarily correlate with symptoms while the patient is awake.

In addition to its diagnostic indications, Holter monitoring may be useful as a screening tool in identifying patients at increased risk for sudden death. The presence of asymptomatic nonsustained ventricular tachycardia may aid in risk stratification in patients with hypertrophic cardiomyopathy or ischemic heart disease and impaired left systolic ventricular function.[8,9] Transient QT interval prolongation and macroscopic T wave alternans are recognized markers of risk for life-threatening ventricular arrhythmias in patients with long QT syndrome.[10] While possible to ascertain from high quality Holter monitor recordings, the clinical utility of heart rate variability, SAECG, heart rate turbulence, microscopic T wave alternans and other complex methods of analyzing continuous ECG recordings remains controversial.

Holter monitoring has a limited role in assessing the efficacy and proarrhythmic response to antiarrhythmic medications due to the relatively short period of recording. Furthermore, conventional Holter recordings require off-line stored ECG processing and, therefore, do not provide immediate notification to the prescribing physician and patient about serious arrhythmic events. In this regard, the MCOT has recently emerged as a promising modality that permits extended periods of monitoring with automatic wireless transmission of arrhythmia events and real-time ECG analysis.[11] Short-term recording periods covered by conventional 24–48 hours Holter monitors are usually insufficient to quantify atrial fibrillation burden following initiation of antiarrhythmic therapy or a catheter ablation procedure.[12]

Detection of myocardial ischemia based upon variations in ST segments is another potential application of Holter monitoring. The ST segment changes on Holter recordings have been associated with adverse outcomes in patients with

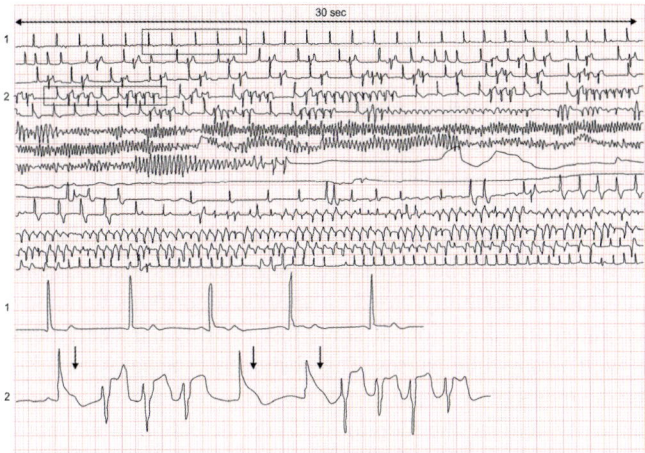

**FIGURE 4:** Continuous single channel Holter monitor ECG recorded during a prolonged syncopal spell in a 60-year-old patient with recurrent episodes of chest pressure and syncope. The bottom tracings (1 and 2) show selected portions of the ECG (boxes 1 and 2) in an expanded scale. Note transient ST-segment elevation (arrows, bottom tracing 2) followed by a 90-second episode of ventricular fibrillation and a 50-second asystolic pause. Coronary angiography showed mild (40%) narrowing of the right coronary artery

coronary artery disease in some series[13,14] but not in others.[15] Despite ongoing clinical interest in this area, the role of Holter monitoring in managing patients with coronary artery disease remains ill defined. This is at least in part due to the low specificity of ST segment changes on Holter recordings to predict ischemia as a number of technical and physiological factors may limit interpretation of the ST segment including body position, nonstandard lead position, medications and changes in autonomic tone. Rarely, Holter monitoring may be useful in the evaluation of patients with suspected variant angina **(Fig. 4)**.

## EVENT RECORDERS

Similar to Holter monitors, event recorders are used to correlate symptoms to arrhythmias but over longer periods of time, usually up to one month. Unlike continuous monitors, these devices have limited memory capacity and are capable of storing only short intervals of ECG recordings related to manually or automatically activated events and, therefore, do not provide "full disclosure" data. Data from event recorders are usually transmitted from a patient transtelephonically to a central location for interpretation and dispersal, by either internet transmission or facsimile, to the prescribing physician.

Event recorders are differentiated mainly by the presence or absence of the memory loop recording capability which allows for the storage of ECG recording immediately preceding a triggered event. A loop recorder continuously stores in internal loop memory several minutes of the most recent ECG by overwriting earlier data. Similarly to a Holter monitor, it is connected to the chest with leads and adhesive electrodes. When the patient develops symptoms and activates the device, pre- and postactivation ECG data are stored for several seconds to minutes depending on specific programmable recording time intervals. Some advanced loop monitors also have an auto-trigger capability based on specific algorithms that allow for automatic detection of slow, fast or irregular heart rates, as well as asystolic pauses. This feature is especially useful in the diagnosis of syncopal events when the patient is not able to manually activate the monitor or in the detection of asymptomatic arrhythmia **(Fig. 5)**. It has been shown that the automatic arrhythmia detection capability improves the diagnostic yield of monitoring devices primarily by detecting asymptomatic arrhythmias.[16] A retrospective analysis of a large AECG database showed that clinically significant arrhythmias were detected in 36% of patients using auto-triggered loop recorders as compared to 17% and 6% using manually activated loop recorders and Holter monitors, respectively.[16] The relatively high diagnostic yield of auto-triggered recorders in this study was due to better documentation of asymptomatic atrial fibrillation or transient bradycardia.

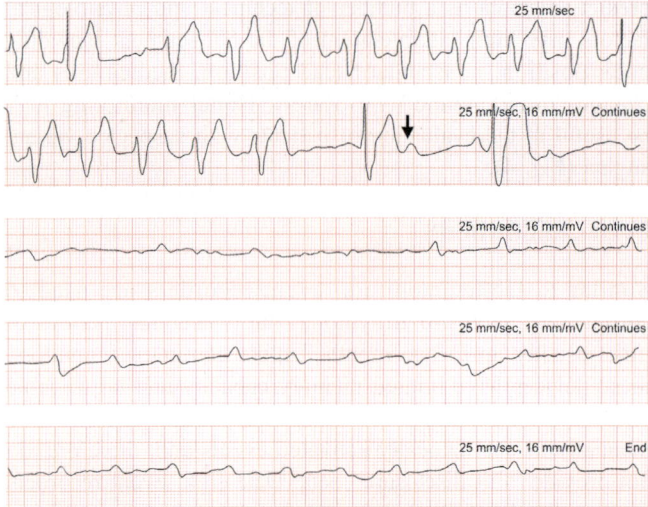

**FIGURE 5:** A continuous single channel recording automatically captured during an episode of syncope in a patient with recurrent palpitations and syncope. In this case, the patient was not able to activate the loop monitor manually. Note paroxysmal complete AV block with prolonged asystolic pause following termination of atrial fibrillation (arrow)

genic cardiac conditions who are at high risk for sudden death (hypertrophic and right ventricular cardiomyopathies, myotonic dystrophy, long and short QT syndromes, Brugada syndrome, etc.) as well as long-term management of atrial fibrillation.[17] Automatic algorithms for the diagnosis of atrial fibrillation have been recently introduced, although accuracy of the derived information requires further validation.

With future advent of multiple physiological sensors (such as blood pressure, oxygen saturation, drug concentrations, etc.) implantable monitoring technology may become an invaluable tool in the management of a variety of cardiac and non-cardiac conditions.

Disadvantages of ILRs exist. The major issue remains inappropriate detection of arrhythmia episodes secondary to either undersensing, most commonly due to loss of electrode contact within the device pocket or oversensing due to "noise" **(Fig. 7)**. This may compromise automatic arrhythmia detection either by undersensing of tachyarrhythmia episodes or by saturation of the device memory with inappropriately sensed ECG recordings and thereby precluding storage of true arrhythmia episodes. In a recent study that analyzed a large database of automatically stored ECGs by an ILR, inappropriately detected events were found in 71.9% of all recordings from 88.6% of patients.[33] An ILR is the most expensive of all available monitoring modalities, although some data suggest that it may be more cost effective compared to conventional diagnostic approaches in selected patients with rare, unexplained syncope.[34] The device requires surgical implantation with inherent risk of pocket complications.

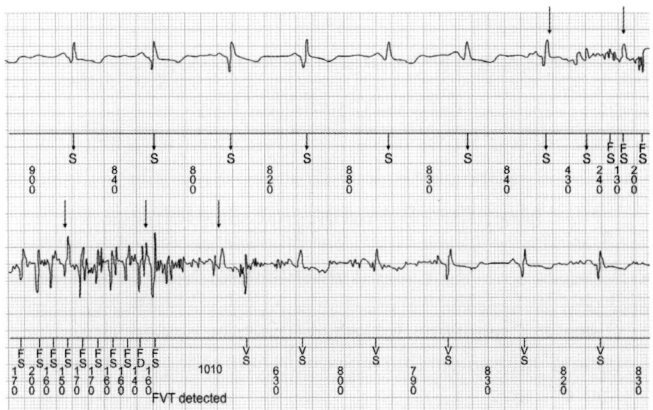

**FIGURE 7:** An example of inappropriate automatic detection by an implantable loop recorder due to "noise" oversensing. Note normal QRS complexes (arrows) "marching through" nonphysiologic high frequency signals

## KEY CONSIDERATIONS IN SELECTING A MONITORING MODALITY

- Major considerations in the selection of a monitoring modality are type and frequency of symptoms **(Table 2)**. In patients with daily symptoms, a Holter monitor remains a preferred device. External event recorders are useful in patients with less frequent (but at least monthly) symptoms. An ILR is generally reserved for evaluation of rare symptoms, usually unexplained syncope, suggestive of an arrhythmic cause.
- In the case of syncope or brief palpitations, loop memory monitors that allow for capturing of ECG data preceding the activation are required. Recorders with auto-triggered capability or MCOT that do not rely solely on the patient's ability to activate the device during symptoms may help to better secure the diagnosis.
- For detection of asymptomatic arrhythmia recurrences or quantification of arrhythmia burden, as well as in patients who are not able to properly activate the device or transmit recorded ECG data, devices providing continuous, "full disclosure", type of information (Holter monitor or MCOT) or event recorders with auto-triggered capability and special arrhythmia algorithms are the most appropriate choice.
- The MCOT offers continuous wireless live ECG monitoring with automatic arrhythmia recognition and immediate notification of the prescribing physician and patient regarding significant arrhythmia events.

---

**(AAA/AHA Guidelines for Ambulatory Electrocardiography:
Executive Summary And Recommendations,
Circulation. 1999;100:886-93)
Modified Summary of Guidelines**

*Modified by Kanu Chatterjee*

Class I: Conditions for which there is evidence and/or general agreement that a given procedure/therapy is useful and effective

Class II: Conditions for which there is conflicting evidence and/or a divergence of opinion about the usefulness/efficacy of performing the procedure/therapy

"Non-looping" or postevent monitors do not have internal loop memory and record ECG only prospectively following manual activation by the patient. These small handheld or wrist worn monitors that have "built in" electrodes are applied directly to the skin for recording. Since no continuous application of adhesive electrodes is required, long-term compliance is usually better compared to loop recorders. Also, these monitors are ideal for use in patients with sensitivity or allergy to adhesives. However, given the technical aspects of the device, its clinical application is limited to situations when prompt activation of the device during symptoms is feasible. It is, therefore, not practical for patients with brief symptoms or syncope. Furthermore, lack of internal memory in these devices does not allow for the capturing of the ECG during the onset of arrhythmia, this information may be helpful in understanding the mechanism of some tachyarrhythmias. Finally, "non-looping" monitors do not provide information about asymptomatic arrhythmias. As mentioned earlier, in rare circumstances, detection of asymptomatic arrhythmias may help in guiding appropriate therapy in symptomatic patients. In addition, detection of asymptomatic recurrences plays an important role in assessing the efficacy of medical or interventional rhythm control strategies for atrial fibrillation.

Event recorders are useful in the diagnosis of symptoms (such as palpitations, syncope or pre-syncope) suspected to be directly caused by an arrhythmia that occur at least monthly **(Table 2)**. Extended surveillance with event recorders provides higher diagnostic yield than conventional short-term Holter monitoring and allows symptom-ECG correlation in up to two-thirds of patients with frequent palpitations.[6,7] Similarly, loop recorders have been shown to be superior to Holter monitors in the evaluation of patients with frequent syncope **(Fig. 5)**.[4] However, the relative diagnostic utility of these devices for evaluation of syncope as compared to palpitations is lower because of the unpredictable course of syncopal events.[17] The optimal duration of monitoring remains unclear. Although some studies suggest that 70–90% of arrhythmias are usually diagnosed within the first two weeks of surveillance,[7,18] the extent of the required monitoring period should be individualized depending on the type and frequency of symptoms.

Loop recorders with auto-trigger capability may be useful in the surveillance of patients undergoing medical therapy for arrhythmias or following catheter ablation procedures. In this regard, management of atrial fibrillation is one of the major applications for these devices, since they allow for the detection of asymptomatic recurrences. It is well recognized that atrial fibrillation recurrences are commonly asymptomatic, particularly following catheter ablation procedures.[19,20] Newer devices can

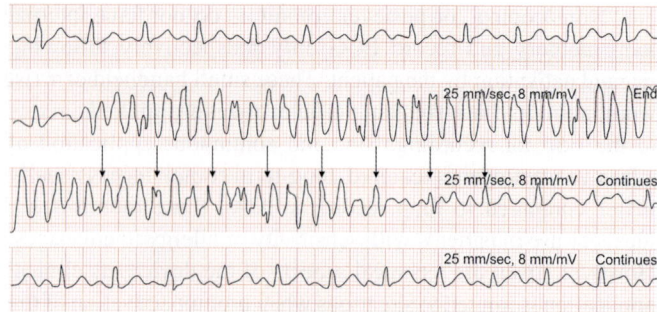

**FIGURE 6:** An automatically logged asymptomatic event consistent with recording artifact mimicking ventriculartachycardia. Note normal QRS complexes (arrows) "marching through" the artifacts

detect atrial fibrillation automatically and display daily atrial fibrillation burden, although accuracy of this information needs further validation. However, this modality does not offer live monitoring with real-time data analysis and, therefore, may not be optimal for patients who are at high risk for serious proarrhythmia.

While event recorders have inherent advantages, they have limitations. Recorders without auto-trigger capability rely upon the patient to activate the device at the time of, or directly after, an episode. In one study, using patient activated loop recorders, one quarter of patients were unable to activate the device properly despite previous education and test transmissions.[4] Long-term compliance in wearing loop recorders is usually limited due to the need for application of adhesive electrodes.[5] As noted previously, patients with extremely rare symptoms may not be good candidates for external event recording. Correct interpretation of recording artifacts may be challenging when only a single lead ECG recording is available **(Fig. 6)**.

## MOBILE CARDIAC OUTPATIENT TELEMETRY

More recently, MCOT has been introduced into clinical practice. The devices used for MCOT are similar in size to conventional loop recorders and are capable of transmitting ECG data wirelessly, either directly or via a portable data manager (an external cellular telephone-sized device). Some providers use the same recorder which can be programmed either as a loop or MCOT monitor **(Fig. 1)**. Potential advantages of MCOT over other modalities include continuous live ECG monitoring with automatic arrhythmia recognition and real-time ECG transmission to a central location that operates 24 hours a day and provides immediate notification of the ordering physician and patient about significant events based on prespecified notification criteria, in addition to daily summary reports. Since the ECG is transmitted continuously to a receiver system, there are practically no memory constraints and "full disclosure" data are available for analysis, including quantification of heart rate and arrhythmia burden.

Initial observational experience suggests that MCOT may offer an improved diagnostic yield in patients with symptoms concerning for arrhythmia and may also potentially be useful for outpatient initiation of antiarrhythmic medications, thereby obviating the need for hospitalization.[7,11] A randomized prospective evaluation of patients with symptoms suggestive of arrhythmia showed that symptom-ECG correlation could be obtained in 88% of patients randomized to MCOT, compared to 75% randomized to a patient activated loop recorder.[18] Higher diagnostic yield of MCOT in this study was due to detection of asymptomatic arrhythmia deemed to be clinically significant with atrial fibrillation and nonsustained ventricular tachycardia accounting for the majority of automatically captured events. However, outcome data show that asymptomatic arrhythmia could be used as a surrogate to guide appropriate therapy for symptomatic events are limited.[21]

Although there are potential advantages to this real-time approach to monitoring, more data are needed to define its role in the management of arrhythmia patients. The MCOT is a technologically and operationally demanding modality and, therefore, is substantially more expensive than conventional event recorders. The system requires not only well trained technical personnel but also a physician available 24 hours a day to manage large amounts of ECG data. It remains unclear whether or not this method can provide cost-effective benefit over older technologies, particularly auto-triggered loop recording.

## IMPLANTABLE LOOP RECORDERS

Implantable loop recorders (ILRs) have extended diagnostic capabilities not afforded by external loop recorders and are generally indicated in patients with infrequent symptoms suspicious for cardiac arrhythmia. These small leadless devices are implanted subcutaneously, usually in the left pectoral area, and provide up to three years of continuous monitoring. They record a single bipolar ECG lead from a pair of electrodes embedded into the shell of the device. Despite relatively closely spaced electrodes, P waves and QRS complexes are generally visible. A subcutaneous wire antenna utilized in some new devices may improve quality of recording by providing more flexible electrode configurations and a larger inter-electrode distance.[22] Similar to external loop recorders, ILRs have limited memory capacity (42–48 minutes of compressed ECG signals) and store only short intervals (seconds to minutes) of both patient and automatically activated ECG recordings based on prespecified parameters. Stored data can be retrieved either manually during interrogation with a standard pacemaker programmer or remotely over the phone. Devices with wireless transmission capability (similar to MCOT) have recently become available,[22] although the clinical utility of live monitoring using ILRs has yet to be confirmed because of inherent problems with inappropriate sensing in these devices.

The longer periods of monitoring afforded by ILRs compared to external loop recorders allow for better correlation of events with arrhythmias in patients with rare but serious symptoms. The devices are usually well tolerated and there are no long-term compliance issues. Most data support their use in patients with recurrent unexplained syncope in whom a conventional invasive and noninvasive evaluation has been unrevealing. The reported diagnostic yield in symptom-ECG correlation is 30–88% depending on studied population of patients.[23-26] The diagnostic yield is directly proportional to the frequency of syncope while likelihood of the diagnosis of significant arrhythmia underlying syncopal events is higher in patients with structural heart disease and/or conduction abnormalities.[17] Some data suggest that selected patients with unexplained syncopal events clinically suspicious for arrhythmia may benefit from relatively early utilization of these devices before embarking on the conventional diagnostic, particularly invasive, techniques provided that cardiac conditions associated with high risk of life-threatening arrhythmia are carefully excluded.[27-29]

The ILRs play a relatively limited role in the evaluation of patients with palpitations as compared to syncope. Palpitations usually are a less severe symptom and are more likely to be diagnosed with external recorders or electrophysiologic testing.[17] In one study which randomized 50 patients with infrequent and sustained palpitations to either ILR or a conventional diagnostic approach including external monitoring and electrophysiology testing, the diagnostic yield of ILR was 73% compared to only 21% using the conventional strategy.[30] The ILR guided therapy yielded symptomatic benefit in the majority of patients.

Some observational data suggest that ILRs may be useful in guiding pacemaker therapy in patients with severe and frequent episodes of neurocardiogenic syncope caused by significant bradycardia.[29] However, prospective randomized studies are warranted before this approach can be adopted in routine clinical practice. The ILRs may potentially be helpful in establishing arrhythmic cause of recurrent nonaccidental falls,[31] as well as unexplained episodes of loss of consciousness.[32]

Other emerging areas of application for ILRs are risk stratification of patients with structural or primary arrhythmo-

Class IIa: Weight of evidence/opinion is in favor of usefulness/efficacy

Class IIb: Usefulness/efficacy is less well established by evidence/opinion

Class III: Conditions for which there is evidence and/ or general agreement that a procedure/therapy is not useful/ effective and in some cases may be harmful

Level A (highest): Derived from multiple randomized clinical trials

Level B (intermediate): Data are on the basis of a limited number of randomized trials, nonrandomized studies or observational registries

Level C (lowest): Primary basis for the recommendation was expert opinion

## Indications for Ambulatory Electrocardiography (AECG) to Assess Symptoms Possibly Related to Rhythmic Disturbances

Class I:
1. Patients with unexplained syncope, presyncope or episodic dizziness
2. Patients with unexplained recurrent palpitation

Class IIb:
1. Patients with unexplained episodic shortness of breath, chest pain, or fatigue
2. Patients with neurological events when transient atrial fibrillation or flutter is suspected
3. Patients with persistent symptoms after non-arrhythmogenic cause of syncope, presyncope dizziness or palpitation have been detected and treated

Class III:
1. Patients with symptoms of syncope, presyncope episodic dizziness or palpitation in whom other causes have been established
2. Patients with cerebrovascular accidents without other evidence of arrhythmia

## Indications for AECG Arrhythmia Detection to Assess Risk for Future Cardiac Events in Patients without Symptoms from Arrhythmia

Class I: None

Class IIb:
1. Post-MI patients with LV systolic dysfunction (ejection fraction of 40% or less)

2. Patients with congestive heart failure
3. Patients with idiopathic hypertrophic cardiomyopathy

Class III:
1. Patients who have sustained myocardial contusion
2. Systemic hypertensive patients with LV hypertrophy
3. Post-MI patients with normal LV function
4. Preoperative arrhythmia evaluation of patients for noncardiac surgery
5. Patients with sleep apnea
6. Patients with valvular heart disease

## Indications for Measurement of Heart Rate Variability (HRV) to Assess Risk for Future Cardiac Events in Patients without Symptoms from Arrhythmia

Class I: None

Class IIb:
1. Post-MI patients with LV dysfunction
2. Patients with congestive heart failure
3. Patients with idiopathic hypertrophic cardiomyopathy

Class III:
1. Post-MI patients with normal LV function
2. Diabetic subjects to evaluate for diabetic neuropathy
3. Patients with rhythmic disturbances that preclude HRV analysis (i.e. atrial fibrillation)

## Indications for AECG to Assess Antiarrhythmic Therapy

Class I: To assess antiarrhythmic drug response if required

Class IIa: To detect proarrhythmic response in patients at high risk

Class IIb:
1. To assess rate control during atrial fibrillation
2. To document recurrent or asymptomatic nonsustained arrhythmias in outpatients

## Indications for AECG to Assess Pacemaker and ICD Function

Class 1:
1. In patients with frequent palpitation, syncope, or presyncope to assess device malfunction
2. To assess the response to adjunctive pharmacotherapy

Class IIb:
1. Evaluation of device function immediately after implantation
2. Evaluation of the rate of supraventricular arrhythmias

Class III:
1. Assessment of device malfunction when it's diagnosis has been already established
2. For routine follow-up in asymptomatic patients

## Indications for AECG for Ischemia Monitoring

Class IIa: Patients with suspected variant angina

Class IIb:
1. Evaluation of patients with chest pain who cannot exercise
2. Preoperative evaluation for vascular surgery who cannot exercise
3. Patients with known CAD and atypical chest pain syndrome

Class III:
1. Initial evaluation of patients with chest pain who are able to exercise
2. Routine screening of asymptomatic subjects

## REFERENCES

1. Holter NJ. New method for heart studies. Science. 1961;134:1214-20.
2. Crawford MH, Bernstein SJ, Deedwania PC, et al. ACC/AHA Guidelines for Ambulatory Electrocardiography. A report of the American College of Cardiology/American Heart Association Task Force on Practice Guidelines (Committee to Revise the Guidelines for Ambulatory Electrocardiography). Developed in collaboration with the North American Society for Pacing and Electrophysiology. J Am Coll Cardiol. 1999;34:912-48.
3. Gibson TC, Heitzman MR. Diagnostic efficacy of 24-hour electrocardiographic monitoring for syncope. Am J Cardiol. 1984;53:1013-7.
4. Sivakumaran S, Krahn AD, Klein GJ, et al. A prospective randomized comparison of loop recorders versus Holter monitors in patients with syncope or presyncope. Am J Med. 2003;115:1-5.
5. Linzer M, Yang EH, Estes NA, et al. Diagnosing syncope. Part 2: Unexplained syncope. Clinical Efficacy Assessment Project of the American College of Physicians. Ann Intern Med. 1997;127:76-86.
6. Kinlay S, Leitch JW, Neil A, et al. Cardiac event recorders yield more diagnoses and are more cost-effective than 48-hour Holter monitoring in patients with palpitations. A controlled clinical trial. Ann Intern Med. 1996;124:16-20.

7. Zimetbaum PJ, Kim KY, Josephson ME, et al. Diagnostic yield and optimal duration of continuous-loop event monitoring for the diagnosis of palpitations. A cost-effectiveness analysis. Ann Intern Med. 1998;128:890-5.
8. Maron BJ, Spirito P. Implantable defibrillators and prevention of sudden death in hypertrophic cardiomyopathy. J Cardiovasc Electrophysiol. 2008;19:1118-26.
9. Moss AJ, Hall WJ, Cannom DS, et al. Improved survival with an implanted defibrillator in patients with coronary disease at high risk for ventricular arrhythmia. Multicenter Automatic Defibrillator Implantation Trial Investigators. N Engl J Med. 1996;335:1933-40.
10. Zareba W, Moss AJ, le Cessie S, et al. T wave alternans in idiopathic long QT syndrome. J Am Coll Cardiol. 1994;23:1541-6.
11. Olson JA, Fouts AM, Padanilam BJ, et al. Utility of mobile cardiac outpatient telemetry for the diagnosis of palpitations, presyncope, syncope and the assessment of therapy efficacy. J Cardiovasc Electrophysiol. 2007;18:473-7.
12. Kottkamp H, Tanner H, Kobza R, et al. Time courses and quantitative analysis of atrial fibrillation episode number and duration after circular plus linear left atrial lesions: trigger elimination or substrate modification: early or delayed cure? J Am Coll Cardiol. 2004;44: 869-77.
13. Rocco MB, Nabel EG, Campbell S, et al. Prognostic importance of myocardial ischemia detected by ambulatory monitoring in patients with stable coronary artery disease. Circulation. 1988;78:877-84.
14. Scirica BM, Morrow DA, Budaj A, et al. Ischemia detected on continuous electrocardiography after acute coronary syndrome: observations from the MERLIN-TIMI 36 (Metabolic Efficiency with Ranolazine for Less Ischemia in Non-ST-Elevation Acute Coronary Syndrome-Thrombolysis in Myocardial Infarction 36) trial. J Am Coll Cardiol. 2009;53:1411-21.
15. Nair CK, Khan IA, Esterbrooks DJ, et al. Diagnostic and prognostic value of Holter-detected ST-segment deviation in unselected patients with chest pain referred for coronary angiography: a long-term follow-up analysis. Chest. 2001;120:834-9.
16. Reiffel JA, Schwarzberg R, Murry M. Comparison of auto-triggered memory loop recorders versus standard loop recorders versus 24-hour Holter monitors for arrhythmia detection. Am J Cardiol. 2005;95:1055-9.
17. Brignole M, Vardas P, Hoffman E, et al. Indications for the use of diagnostic implantable and external ECG loop recorders. Europace. 2009;11:671-87.
18. Rothman SA, Laughlin JC, Seltzer J, et al. The diagnosis of cardiac arrhythmias: a prospective multi-center randomized study comparing mobile cardiac outpatient telemetry versus standard loop event monitoring. J Cardiovasc Electrophysiol. 2007;18:241-7.
19. Joshi S, Choi AD, Kamath GS, et al. Prevalence, predictors, and prognosis of atrial fibrillation early after pulmonary vein isolation: findings from 3 months of continuous automatic ECG loop recordings. J Cardiovasc Electrophysiol. 2009;20:1089-94.
20. Pontoppidan J, Nielsen JC, Poulsen SH, et al. Symptomatic and asymptomatic atrial fibrillation after pulmonary vein ablation and the impact on quality of life. Pacing Clin Electrophysiol. 2009;32: 717-26.

21. Krahn AD, Klein GJ, Yee R, et al. Detection of asymptomatic arrhythmias in unexplained syncope. Am Heart J. 2004;148:326-32.
22. Jacob S, Kommuri NV, Zalawadiya SK, et al. Sensing performance of a new wireless implantable loop recorder: a 12-month follow up study. Pacing Clin Electrophysiol. 2010;33:834-40.
23. Krahn AD, Klein GJ, Yee R, et al. Use of an extended monitoring strategy in patients with problematic syncope. Reveal investigators. Circulation. 1999;99:406-10.
24. Brignole M, Menozzi C, Moya A, et al. Mechanism of syncope in patients with bundle branch block and negative electrophysiological test. Circulation. 2001;104:2045-50.
25. Menozzi C, Brignole M, Garcia-Civera R, et al. Mechanism of syncope in patients with heart disease and negative electrophysiologic test. Circulation. 2002;105:2741-5.
26. Solano A, Menozzi C, Maggi R, et al. Incidence, diagnostic yield and safety of the implantable loop-recorder to detect the mechanism of syncope in patients with and without structural heart disease. Eur Heart J. 2004;25:1116-9.
27. Krahn AD, Klein GJ, Yee R, et al. Randomized assessment of syncope trial: conventional diagnostic testing versus a prolonged monitoring strategy. Circulation. 2001;104:46-51.
28. Farwell DJ, Freemantle N, Sulke N. The clinical impact of implantable loop recorders in patients with syncope. Eur Heart J. 2006;27:351-6.
29. Brignole M, Sutton R, Menozzi C, et al. Early application of an implantable loop recorder allows effective specific therapy in patients with recurrent suspected neurally mediated syncope. Eur Heart J. 2006;27:1085-92.
30. Giada F, Gulizia M, Francese M, et al. Recurrent unexplained palpitations (RUP) study comparison of implantable loop recorder versus conventional diagnostic strategy. J Am Coll Cardiol. 2007;49:1951-6.
31. Armstrong VL, Lawson J, Kamper AM, et al. The use of an implantable loop recorder in the investigation of unexplained syncope in older people. Age Ageing. 2003;32:185-8.
32. Pezawas T, Stix G, Kastner J, et al. Implantable loop recorder in unexplained syncope: classification, mechanism, transient loss of consciousness and role of major depressive disorder in patients with and without structural heart disease. Heart. 2008;94:e17.
33. Brignole M, Bellardine Black CL, Thomsen PE, et al. Improved arrhythmia detection in implantable loop recorders. J Cardiovasc Electrophysiol. 2008;19:928-34.
34. Krahn AD, Klein GJ, Yee R, et al. Cost implications of testing strategy in patients with syncope: randomized assessment of syncope trial. J Am Coll Cardiol. 2003;42:495-501.

# CHAPTER 14

# Cardiac Arrest and Resuscitation

Christine Miyake, Richard E Kerber

## Chapter Outline

- Overview or Background
  - Evolution of Cardiac Resuscitation
  - Cardiopulmonary Arrest
  - Emergency Medical Services
- Basic Life Support
  - Role of Bystanders
  - Emergency Medical Services Activation
  - Dispatcher Assisted Cardiopulmonary Resuscitation
  - Compression only Cardiopulmonary Resuscitation
  - Chest Compressions or Airway Management
  - Mechanical Devices for Cardiopulmonary Resuscitation
  - Use of Automatic External Defibrillators
  - Pacemaker or Automatic Implantable Cardioverter Defibrillator Patient in Cardiac Arrest
  - Complications of Cardiopulmonary Resuscitation
- Advanced Cardiac Life Support
  - Overview-Statistics of Success
  - Advanced Airway Management
  - Pharmaceutical Interventions
  - Defibrillation or Cardioversion
- Cessation of Resuscitation
- Post-resuscitation Care
  - Cardiopulmonary Support
  - Cardiac Interventions
  - Therapeutic Hypothermia

## OVERVIEW OR BACKGROUND

Cardiopulmonary resuscitation (CPR) guidelines are continuously changing as new evidence and techniques are developed and researched. Despite these advances overall survival from sudden cardiac arrest remains low.[1] The return of spontaneous circulation (ROSC) is directly related to adequate coronary perfusion, while good clinical outcomes are more closely related to adequate vital organ perfusion during and immediately after resuscitation. This chapter covers the most recent understanding of blood flow, current emergency medical services (EMS) out of hospital resuscitation efforts, standard techniques of CPR, basic life support (BLS) and advanced life support (ALS) as well as post-resuscitative care. This chapter does not cover the details of pediatric resuscitation.

### Evolution of Cardiac Resuscitation

Cardiopulmonary resuscitation is a relatively new concept. The idea of artificial blood flow with artificial respirations as a means to restore life, after what appears to be death, was not a concept that came easily. It took many years of trial and error and research to develop the idea that blood can flow without opening the chest and directly massaging the heart. In the 1700s mouth-to-mouth resuscitation was recommended for drowning victims; however, there was no formal training for physicians or any type of EMS.[1] The Society for the Recovery of Drowned Persons became the first organization to deal with sudden and unexpected death; however, it was not until the 1950s that James Elam proved expired air was sufficient to maintain adequate oxygenation until blood flow was restored or other mechanical means of ventilation could be established.[1] In the late 1800s to early 1900s, external chest compressions were being used sporadically in humans with little scientific research or impact on survival rates. The first documented successful use of external chest compressions was reported by Dr George Crile in 1903.[1] After the landmark paper by Kouwenhoven and Jude in the 1960s, who showed adequate circulation could be achieved with closed chest cardiac massage, CPR guidelines were developed and the American Heart Association (AHA) started a program to train physicians and the general public in the techniques of closed-chest cardiac massage.[2] Prior to this landmark paper, it was believed that the only way to artificially circulate blood was to open the chest and perform direct cardiac massage. Since then the AHA has established the standards of care for CPR. The AHA re-evaluates its recommendations and updates the guidelines as new information and research become available.

The mechanism of forward blood flow during CPR has been the subject of much debate and research. There have been two proposed mechanisms most widely recognized: (1) the "thoracic pump model" and (2) the "cardiac pump model". The thoracic pump model postulates that blood flows during closed chest cardiac massage, or CPR, due to an increase in intrathoracic vascular pressure that exceeds extrathoracic vascular pressures. This theory was postulated by Weale and Rothwell-Jackson in 1962. They showed almost equivalent increases in arterial and venous pressures in animals during closed chest CPR.[3] Blood flow is in the proper direction due to venous valves that prevent retrograde flow. The heart is essentially passive with the valves remaining open due to equal pressure from all sides during compression **(Fig. 1)**. The cardiac pump model, proposed by Kouwenhoven et al., states that flow occurs due to compression of the heart between the sternum and the spine.[2] Flow is maintained in the proper direction due to the mitral valve staying closed during systole or the compression phase. With release of the chest compression the heart expands, the mitral and tricuspid valves open and the heart fills with blood **(Fig. 2)**. In 1993, Redberg et al. performed transesophageal echocardiography on 20 patients during resuscitation in an attempt to determine if cardiac size changed and if the mitral valve opened during diasystole or the release phase of CPR. They found a reduction of ventricular cavity size with compression and mitral valve opening during cardiac release, supporting the cardiac pump theory.[4] Multiple other

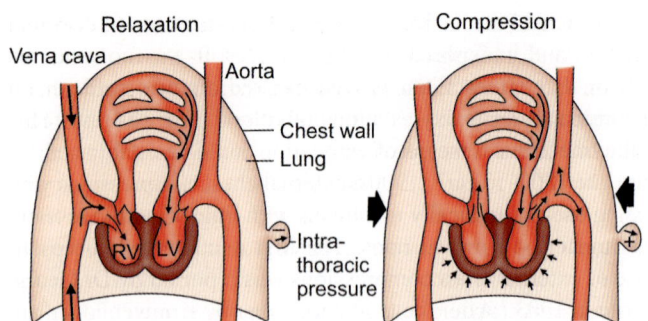

**FIGURE 1:** Cardiac pump model of cardiopulmonary resuscitaton. (*Source:* Modified from Luce JM, Cary JM, Ross BK, et al. New developments in cardiopulmonary resuscitation. JAMA. 1980;244: 1366-70)

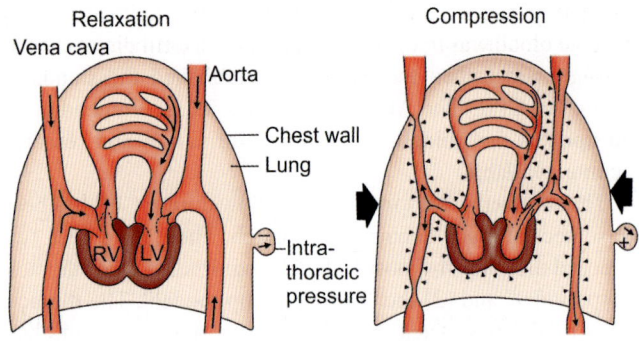

**FIGURE 2:** Thoracic pump model of cardiopulmonary resuscitation. (*Source:* Modified from Luce JM, Cary JM, Ross BK, et al. New developments in cardiopulmonary resuscitation. JAMA. 1980;244: 1366-70)

studies have used ultrasound in an attempt to determine the mechanism of blood flow with mixed results. Studies done on animal models are difficult to extrapolate to humans and many of the human studies show conflicting results without much correlation with survival rates. Both mechanisms may occur, especially during prolonged resuscitation, when the myocardium becomes edematous and stiffer, which might make the heart less compressible favoring the thoracic pump mechanism. While it is still uncertain exactly which mechanism is correct or if it is a combination of both, it is known that the quality and rate of chest compression is extremely important for increasing rates of survival from sudden cardiac arrest.

Using electricity for cardioversion and defibrillation in order to terminate dysrhythmias has been used worldwide for many years. Its origins began in 1775 when Abildgaard demonstrated that stunned chickens can be revived by electrical shocks to the head and heart.[5] In 1899, Prevost and Batelli showed that dogs in ventricular fibrillation (VF) could be restored to a normal rhythm by electric shocks.[6] Then in the 1900s the Consolidated Edison Company of New York began funding research on the mechanisms and treatments of electrical accidents.[6] This allowed extensive research into the electrical mechanisms of heart function and, as a consequence, ways in which electricity can be used to restore proper function in the setting of electrical dysfunction. Human defibrillation began in an operating room in 1947 when Beck administered electrical shocks to the exposed heart (open chest defibrillation) to terminate VF.[7] The first closed chest defibrillation was not achieved until 1965 by Zoll et al.; prior to this a few successful attempts with direct cardiac defibrillation were reported.[8] After Kouwenhoven et al. developed closed-chest cardiac massage, the AHA started a program to train physicians with the techniques of advanced CPR that included CPR as well as electrical defibrillation.[2,6] Types of defibrillators and mechanism of defibrillation are discussed later in this chapter.

## Cardiopulmonary Arrest

Sudden cardiac arrest is still a major public health problem and a leading cause of death in the United States. It is important when discussing resuscitation strategies from sudden cardiac arrest to define what is meant by "sudden", "cardiac arrest" and "death". The epidemiologic definition of "sudden" is usually defined by less than 1 hour from onset of symptoms to terminal clinical event which could include: death, loss of detectable pulse or cessation of breathing. This definition does not take into account unwitnessed events. The World Health organization has included in its definition of "sudden" unwitnessed deaths that occur less than 24 hours prior to discovery of the victim. "Death" is defined as an absolute irreversible event; this is a biologic, legal and literal definition. "Cardiac arrest" can be reversible and is defined as the cessation of pump function. If the patient is unable to be resuscitated or resuscitation is not performed then the event becomes irreversible and is considered sudden cardiac death. Cardiovascular collapse is defined as loss of effective blood flow either due to cardiac dysfunction or loss of vascular function.

Approximately 10% of all emergency department visits are cardiac related.[6] Cardiac causes are by far the most common cause of sudden cardiac arrest. Ventricular tachycardia (VT) and VF account for the vast majority of sudden death cases. Other causes of sudden death include intracranial hemorrhage, pulmonary embolism, drug overdose, lung disease, aortic dissection or rupture, trauma and drowning. Atherosclerotic disease is the leading cause of death in the United States and in addition it carries significant morbidity, disability and loss of productivity. Atherosclerotic disease is present to some extent in all adult patients, but genetic background and lifestyle risk factors, such as cigarette smoking, sedentary lifestyles and high fat diets, put patients at higher risk for developing significant atherosclerotic disease complications or death. The initiating event is usually injury to the vascular endothelium leading to accumulation of macrophages and lipids at the site of injury. Plaque formation then occurs which can subsequently rupture leading to clot formation compromising blood flow through the arterial lumen. Decreased blood flow or complete occlusion leads to myocardial ischemia, hypoxia, acidosis and infarction. The consequences of arterial occlusion depend upon the availability of collateral blood flow and the size of the area of myocardium supplied by the occluded vessel. Risk factors for increasing plaque formation include: genetic predisposition, hypercholesterolemia, diabetes, hypertension, smoking, male gender and postmenopausal status in women.

The AHA estimates that there are 300,000 out of hospital cardiac arrests each year.[1] Unfortunately, survival rates are extremely low. Of those that do survive, more than half have poor neurologic outcomes.[9] According to the AHA, an average of 31.4% of out of hospital cardiac arrest patients receive bystander CPR, of which 60% are treated by EMS.[1] Immediate CPR can substantially improve a victim's chance of survival. Increasing survival rates involves many factors including measuring outcomes and improving "weak links" in the chain of survival. The AHA uses four "links" in the "chain of survival" for victims of sudden cardiac arrest: (1) early recognition of the emergency and activation of emergency medical services; (2) early bystander CPR; (3) early defibrillation if indicated and (4) early ALS and post-resuscitative care. Quality rescuer education and frequency of retraining are also critical factors.

The cardiopulmonary resuscitation is highly accessible and requires little medical training and no equipment but out of hospital cardiac arrest survival rates remain low. Many studies have documented that only 15–30% of victims receive bystander CPR. The EMS often takes 6–7 minutes or longer to arrive at the scene; each minute that passes without blood flow to vital organs decreases the victims' chance of survival. Many theories have been proposed to explain the hesitation to perform CPR by bystanders even when they are trained. One explanation is the concern bystanders have performing mouth-to-mouth resuscitation. The complexity of the guidelines and instructional materials may prevent bystanders from performing CPR. The fear of poor performance or failure may also prevent many from even attempting. Legal liability is also a concern of many individuals because they may not be aware of "good Samaritan" laws that provide some protection for rescuers.

Every 5 years the AHA revises its guidelines for resuscitation care. The 2005 revision placed greater emphasis on compressions and de-emphasized ventilations. This change was studied after its initiation and found an increase in survival rates suggesting less interruption of chest compressions improves outcomes.[10,11] The bigger problem and concern was the need to increase bystander performed CPR. Given the concerns bystanders have of performing mouth-to-mouth resuscitation several studies were conducted to evaluate the outcome differences between compression only CPR and standard CPR with ventilations. Svensson et al. conducted a randomized prospective study comparing the two groups.[12] They found no significant difference in survival rates of compression only CPR and standard CPR for witnessed out of hospital cardiac arrest victims. This along with other studies prompted the AHA to change its recommendations for the 2010 guidelines to compression only CPR for witnessed out of hospital cardiac arrest victims. Two main conclusions can be currently drawn from these studies: (1) first and foremost CPR needs to be performed as quickly as possible and (2) quality compressions with minimal interruption need to occur. The 2010 AHA guidelines for pulseless arrest are discussed later in this chapter.

Implementation of new guidelines does improve outcomes; however, expediting guideline implementation is challenging. It can take several years for new guidelines to be implemented; barriers to implementation include: delays in instruction, technology upgrades, and difficulties with coordination of medical direction, government agencies and participation in research.

## Emergency Medical Services

The credit for developing the first EMS system seems to go to Napoleon's Surgeon-in-Chief, Barron Jean Larrey. He noted that wounded soldiers were left unattended until the fighting ceased, after which rescue teams would enter the battlefield and care for the wounded. He was convinced that if the wounded were attended to sooner, mortality rates would improve. He positioned medical transport teams closer to the battlefield to remove the injured soldiers. During the American Civil War the medical director of the Army of the Potomac, Jonathan Letterman, organized horse-drawn trains to rapidly remove wounded soldiers from the battlefield to field hospitals set-up nearby.[13]

During the first half of the 20th century most civilian transports were performed by morticians. There were no government regulations or financial support. The poor state of emergency medical response and treatment along with recommendations for improvement were first published by the National Safety Council's Traffic Conference after surveying several cities across the United States on how injuries from traffic accidents were handled.[14] In 1966, a second national survey by the National Academy of Sciences-National Research Council was used to complete the "White Paper" entitled Accidental Death and Disability: the Neglected Disease of Modern Society.[15] The issues brought forth in this paper along with public concern pushed congress to draft legislation that enabled the US Department of Transportation and the National Highway Traffic Safety Administration (NHTSA) to develop a national program to improve emergency medical care.[16] In 1973, the US Senate passed the EMS Systems Act which gave federal funding to improve regional EMS systems.[17] Currently, all states have an administrative department that governs EMS activities and assists in the planning, licensing and development of standards of practice.

Each state is responsible to ensure that their citizens receive prompt emergency medical care. The National Highway Safety Administration has made recommendations and guidelines for the implementation of an EMS system and course curriculum training for care providers. At a minimum EMS, programs should have these ten components: (1) Regulation and Policy—this should include comprehensive legislation, regulations and operational policies and procedures to provide emergency medical and trauma care services; (2) Resource Management—the state should establish a central lead agency that identifies, categorizes and coordinates resources necessary for the overall system implementation and operation; (3) Human Resources and Training—the EMS system must have trained persons to perform the required tasks, including first responders, emergency medical technicians (EMTs), communications, physicians, nurses, hospital administrators and planners; (4) Transportation—reliable and safe ambulance transportation is critical for an effective EMS system; (5) Facilities—proper facilities that are accessible are required to ensure high-quality care and these must be available in a timely manner. Hospital resources and capabilities need to be designated and known in advance and agreements need to be established between facilities to ensure patients receive treatment at the closest, most appropriate facility; (6) Communications—an effective communication system is needed to provide a means for persons to access resources, mobilize units, manage and coordinate those resources to provide care in a timely fashion; (7) Trauma Systems—a designated trauma system needs to be established to provide high quality, effective patient care; (8) Public Information and

Education—public awareness and education of the EMS system is essential for proper access to occur; (9) Medical Direction—physician involvement is critical for a system to provide high quality and proper care. Physicians must delegate responsibilities to non-physician providers; protocols need to be developed, implemented and have continuous oversight with audits and evaluations. Immediate medical direction by a physician should be available at all times to ensure quality patient care and (10) Evaluation—evaluation is required to provide improvement in the system as new medical knowledge is obtained.[18] **Table 1** summarizes all these ten components.

Not all EMS systems operate in similar fashion. System designs need to accommodate the needs of the local community or jurisdiction. The system may be served by a private or public agency, it may provide BLS services only or both BLS and ALS. Responses to 911 calls may be in the form of single-tiered, multi-tiered or first responder only. Currently not all systems incorporate record keeping, data collection or auditing programs. This puts these systems at a disadvantage due to their inability to improve patient care through changes in protocol, tracking those changes to ensure proper implementation and demonstrating a true impact on patient care.

Public agencies are the responsibility of local governments; many of these systems use fire departments to provide EMS services. Providers are then cross trained as firefighter or paramedics. Some public EMS systems are separate entities

### TABLE 1
**Components needed for EMS system operation**

| | |
|---|---|
| Regulation and policy | Legislation, regulations and operational policies and procedures |
| Resource management | Lead agency identifies categorizes and coordinates resources |
| Human resources and training | Trained persons to perform required tasks |
| Transportation | Reliable and safe ambulances |
| Facilities | Proper and accessible with known hospital capabilities |
| Communication | Communication system for resource allocation |
| Trauma system | Predetermined for timely access |
| Public information and education | Public awareness/education for proper utilization of resources |
| Medical direction | Physician directed protocols and oversight |
| Evaluation | Provide improvement and implementation of new medical knowledge |

referred to as municipal third-service systems. Some communities may combine public agencies such as EMS, fire and police services with one director or administrator. Funding public EMS systems may be tax based or a combination of use fees plus government funding. Medical oversight is usually provided by an appointed physician or medical control board.

Private agencies that provide EMS services may be locally owed, hospital based or operated by large corporations. Most private agencies are funded by user fees; some government subsidies may be provided; however, depending on the local needs and percent of uninsured population. Some systems have multiple agencies, public or private, providing services to the same area. These systems may have varying ways of allocating calls, such as rotational coverage or zone coverage.

Single-tier systems provide the same level of personnel and equipment irrelevant of call types, for example, all BLS or all ALS. A multi-tiered system dispatches different levels, BLS or ALS depending on the nature of the call. Cost differences are debated and may depend on the community being served. It may prove to be cost effective to have a single ALS system providing consistent advanced care avoiding the potential to under-triage a call and send a lower BLS unit when an ALS unit is actually required. This can be difficult to fund as paramedics are more costly to staff than EMTs. First responders are usually part of any system and consist of police or firefighter who may arrive at the scene prior to the ambulance.

Individual states are responsible for drafting recognized provider levels, testing and recertification requirements. The NHTSA provides recommendations for a national standard curriculum. Basic provider levels include: First responder, emergency medical technician-basic (EMT-B), emergency medical technician-intermediate (EMT-I) and paramedic. The BLS and ALS refer to type of emergency care provided. The BLS or BLS services involve life-saving skills such as bag-valve-mask ventilation, oral and nasal airway use, CPR training, bleeding control techniques, basic fracture care or splinting and childbirth assistance. The use of an automated external defibrillator (AED) is also often included in BLS training. The ALS includes BLS training but also incorporates more advanced airways such as intubation, laryngeal mask airway (LMA) use and the use of rapid sequence intubation (RSI) medications. They also include cardiac medications for the resuscitation of cardiac arrest victims. Details about BLS and ALS are to be discussed later in this chapter.

First responder training is typically done for all personnel, such as firefighters and police officers, who might be the first to respond to an emergency. Many bystanders may also be first responder trained. First responders can provide limited life-saving procedures such as CPR, Heimlich maneuver, spinal immobilization and basic bleeding control measures. Most first responders receive 40 hours of didactic instruction and 16–36 hours required for refresher training as recommended by the NHTSA.[19]

The EMT-B is the minimal level of training required to staff an ambulance. It incorporates the skill of a first responder as well as some training in patient assessment, immobilization procedures and the use of oxygen administration. Depending on the state, some incorporate the use of life saving cardiac medications such as epinephrine given subcutaneously with auto injection, albuterol via nebulizer and intravenous (IV) fluids. The NHTSA recommends 100–125 hours of training that includes laboratory training.[20] The EMT-B usually also includes training in the use of AED.

Emergency Medical Technician-Intermediate (EMT-I) is trained in the skills of first responder and EMT-B plus more advanced techniques of care such as laryngeal mask airway, endotracheal intubation, IV line with fluid resuscitation and are trained in the use of a defibrillator. The EMT-I approaches the level of training of a paramedic but with less education and in most cases less cost. The NHTSA recommends 300–400 hours of initial education that combines classroom education as well as clinical training through hospital and field rotations.[21] Some states are removing the EMT-I level of training to simplify the training process.

Paramedics have the highest level of prehospital training. Their training includes all BLS training plus advanced cardiac life support (ACLS) training that includes the use of RSI medications, use of 12 lead ECG and its basic interpretation, defibrillator use, many medications including cardiac medications, pain medications and antiseizure medications. Paramedics are also trained in the use of alternative access lines such as the use of intraosseous access. Training also includes more advanced emergent delivery techniques such as neonatal resuscitation for emergency deliveries. The NHTSA recommends 1100–1200 hours of training that include 500–600 hours of classroom and laboratory time, 250–300 clinical hours in the hospital and 250–300 hours of field training.[22] All levels of training require some level of continuing medical education and refresher training. **Table 2** summarizes prehospital personnel training levels.

## BASIC LIFE SUPPORT

Basic life support is the first step and the foundation for saving lives from sudden cardiac arrest. It involves immediate

**TABLE 2**
Summary of EMS provider levels

| Training level | Personnel | Skills | Education |
|---|---|---|---|
| First responder | Personnel first on scene: Police/Firefighters/Bystanders | Limited life-saving procedures: CPR, Heimlich | 40 hours didactic training |
| EMT-B | Minimum level of training to staff ambulance | First responder + immobilization techniques and use of oxygen, auto injection medications, albuterol | 100–125 hours training including laboratory |
| EMT-I | EMT-B plus some advanced resuscitation techniques | EMT-B + LMA, IV line with fluid resuscitation, defibrillator use | 300–400 hours training with clinical/field rotations |
| Paramedic | Most advanced prehospital training | EMT-I + ACLS, RSI, ECG interpretation, limited medication administration, IO | 1100–1200 hours training: didactic + laboratory<br>250–300 hours clinical<br>250–300 hours field |

recognition, activation of emergency response system, performing high-quality CPR and rapid defibrillation when appropriate.[23] The BLS also involves basic trauma and other medical techniques that have not been covered in this chapter. The more people who are trained in BLS the better survival rates can be since witnessed arrest victims would receive CPR sooner. It is not the intention of this chapter to provide complete instruction on the performance of CPR; interested individuals should seek a certified AHA BLS or ACLS classes for complete training and instruction.

## Role of Bystanders

The role of bystanders in sudden cardiac arrest is extremely important in the "chain of survival". Bystanders can perform 3 of the 4 links in the chain of survival and greatly impact a victim's chance of survival with good neurologic outcome. Bystanders are important for recognition and EMS activation, perform immediate CPR and apply and use an AED for defibrillation. Kitamura et al.[24] conducted a prospective, observational study in Japan which evaluated the effects of nationwide dissemination of public-access AEDs on the rate of survival of out of hospital cardiac arrest victims. Nationwide access to AEDs resulted in earlier administration of shocks by laypersons and an increase in survival at one month with minimal neurologic impairment. The new 2010 AHA guidelines for BLS emphasize immediate chest compressions without delay for rescue breathing and application of an AED as soon as it is available.

## Emergency Medical Services Activation

Immediate activation of EMS is extremely important for the survival of sudden cardiac arrest victims. In many communities, the time interval for EMS arrival is 7–8 minutes. This means that for the first several minutes the chances of survival for the victim is in the hands of bystanders. The sooner EMS can arrive at the scene the sooner victims can receive ACLS and post-resuscitative care. The BLS algorithm has been simplified: immediate activation of the emergency response system and initiate chest compressions for any unresponsive adult victim who is not breathing normally.[23] The previous recommendation of "Look, Listen and Feel" step was too time consuming and inconsistent between rescuers. Lay rescuers should not attempt to check for a pulse as even trained healthcare providers often incorrectly assess the presence or absence of a pulse especially if blood pressure is extremely low.[23] Chest compression performed on a patient with a heart beat is rarely associated with significant injury.[25] Many arrest victims will have gasping respirations or appear to be having a seizure. Lay rescuers should be instructed to start CPR immediately on any unresponsive victim who appears to be struggling to breathe given the unusual presentations of sudden cardiac arrest.

## Dispatcher Assisted Cardiopulmonary Resuscitation

When dispatchers receive a 911 call of a witnessed sudden cardiac arrest they are encouraged to instruct the lay rescuer to perform "hands-only" (compression only) CPR. Compression only CPR is much easier to perform and prevents delays in providing chest compressions. Positioning the head, attaining a seal for mouth-to-mouth or assembling a bag mask can take time and as such this delay has been found to decrease overall survival rates. In 2005, the AHA published new guidelines for cardiac arrest victims in which chest compressions were emphasized before first defibrillation. Placing electrode pads and analyzing rhythm all take time during which the victim is not receiving CPR. It was believed that the lack of immediate chest compressions and the delay in CPR during the "3 stacked shocks" that were previously advised for cardiac arrest due to VF or unstable or pulseless VT was decreasing vital organ blood flow and contributing to poor outcomes. After implementation of the 2005 AHA guidelines multiple studies found significant improvement in survival rates.[10,26,27]

## Compression Only Cardiopulmonary Resuscitation

Changing chest compression ratios, emphasizing more compressions and less ventilation was a controversial topic at the 2005 International Consensus Conference on Resuscitation and a major change to the AHA 2005 guidelines for CPR. Recent studies have demonstrated an increase in survival from out of hospital cardiac arrest resulting from improved quality of CPR, with adequate rate and depth and minimizing interruptions by avoiding excessive ventilations and "stacking shocks".[26-30] During the time between 2005 and 2010 the AHA has been studying ways to simplify CPR and increase its use by laypersons due to the fact that survival rates of out of hospital cardiac arrest remain low.[1] Compression only CPR for most adults for out of hospital cardiac arrest has been shown to achieve similar outcomes to those who receive standard CPR with rescue breathing.[12,31-34] Thus, bystanders are encouraged and directed by dispatch to perform compression only CPR until the arrival of EMS. Starting the procedure with compressions only eliminates the step that most laypersons have difficulty with: opening the airway and giving rescue breaths. However, children rarely arrest from a primary cardiac cause and thus for the pediatric cardiac arrest victim, rescue breathing may be much more important. Pediatric cardiac resuscitation has not been covered in this chapter.

## Chest Compressions or Airway Management

The newest change to the AHA guidelines for CPR will be changing from "A-B-C" (airway, breathing, circulation) to "C-A-B" (chest compressions, airway, breathing).[23] The change will not be an easy one to make. Everyone who has ever learned CPR will have to be re-educated. The vast majority of sudden cardiac arrest patients arrest from VF or pulseless VT, and these patients have improved survival from immediate high quality CPR with early defibrillation.[23] When a collapse is witnessed by a lone rescuer the AHA now advises to confirm unresponsiveness, activate the emergency response system and then begin chest compressions with a rate of 100 per minute for adults with a depth of at least 2 inches, allowing complete recoil of the chest after each compression. Proper hand position is two fingerbreadths above the xiphoid-sternal notch. For the lay person, it is easier to understand "center of the chest between the nipples". The first compression cycle should be 30 compressions in length. Early application of an AED or defibrillator should be done as soon as another rescuer is available. Ventilations should be given with 2 breaths after 30 compressions with minimal interruption of compressions. Bystander or dispatcher assisted CPR should be performed with compressions only if a barrier device is not available and there is concern about exposure. Once a "shockable" rhythm is identified and a defibrillator or AED has been applied a shock should be delivered. After the shock is delivered CPR should be started immediately without checking for a pulse or rhythm. After 2 minutes of CPR, there should be a pause for a rhythm and pulse check. Healthcare providers should tailor the sequence of actions to the most likely cause of the arrest. In a known drowning, for example, conventional CPR with rescue breathing would be more important. Children and newborn infants are more likely to suffer a respiratory cause of arrest and would also benefit from the conventional A-B-C sequence of resuscitation. Children and infant CPR is performed slightly differently and is not covered in this chapter.

Rescue breathing should be performed with a head tilt chin lift maneuver or jaw thrust if trauma is suspected. A bag valve mask or other barrier should be used if available. More advanced airway techniques have been discussed below.

Rescue breaths should be given quickly, minimizing interruption of chest compressions.

## Mechanical Devices for Cardiopulmonary Resuscitation

Mechanical devices for adult CPR have been developed for several reasons. It is possible that mechanical devices may perform CPR better than standard CPR. Mechanical devices are also useful for long transports to prevent rescuer fatigue. The longer a rescuer performs CPR, the more the quality decreases as the rescuer becomes tired. Several studies have shown improved coronary perfusion pressures with mechanical devices. However no studies have shown improved survival with any mechanical device compared to standard CPR.[35-41] Decreased quality of chest compressions with time is however well recognized.[42] It is recommended that rescuers performing CPR change often during the resuscitation and if a long transport is to occur a mechanical device should be considered to be used to prevent rescuer fatigue as well as free up the rescuer to perform other duties. More study is needed to effectively evaluate the use of mechanical CPR devices.

## Use of Automatic External Defibrillators

Automatic external defibrillators are small, portable, battery operated devices that allow providers to defibrillate cardiac arrest victims without interpretation of an electrocardiographic waveform. Laypersons can use the device with minimal or no training, as audible and visual prompts are incorporated into the machine. It is well known that the earlier a victim is defibrillated the better chance of survival. Studies of placement of the devices where it was easily accessible have shown improvement in the survival rate of out of hospital cardiac arrest victims.[24] Most in-hospital defibrillators are now equipped with AED technology to help improve the time to first shock by allowing those untrained or uncertain to use the device.

When a victim is recognized and the emergency medical system is activated CPR should be started. Once the AED is available, the device should be placed by the patient's head for easy access and operation. The AED's power should be turned on which initiates a self check by the machine. The machine then instructs the user in its use. It will begin by advising to attach the electrode pads to the patient. There are pictures on the electrode pads for proper placement on the chest wall. The electrode pads should be placed on the right upper chest just below the clavicle and the left lower lateral chest below the nipple. The chest wall should be dry when the electrode pads are applied. The AED will then analyze the rhythm; some devices require a button to be pushed to analyze the rhythm. The CPR should continue uninterrupted as much as possible during set-up and application of the electrode pads. During analysis of the rhythm the patient should not be touched. The AED will then advise if a shock is indicated or if CPR should be continued. If shock is advised then the machine will indicate to "charge" then will wait for the user to push the "shock" button. Before the shock is initiated everyone must be clear of the patient. After the shock is initiated CPR should be started immediately and not be delayed to analyze the rhythm and check for a pulse. This is a deviation from past AHA guidelines and if the AED is old it may not be programmed appropriately, but CPR should be initiated after the shock in any case. If return of circulation occurs then airway or breathing assistance should be maintained until more definitive care arrives. The cycle of shock, 2 minutes of CPR then rhythm or pulse check should continue until advanced interventions are available.

## Pacemaker or Automatic Implantable Cardioverter Defibrillator Patient in Cardiac Arrest

If a patient has an implantable device such as a defibrillator or pacemaker, care should be taken to avoid placing electrode paddles or pads on the device; placement should be at least 1 inch away to avoid any potential artifact interference during rhythm analysis and potential damage to the device from defibrillation. There is little danger to a rescuer performing CPR from an implantable automatic internal defibrillator. If the defibrillator fires during CPR the rescuer may feel a slight electrical shock; however, it is not harmful. Pacemaker problems can occur from defibrillation or cardioverson, although this is rare, including damage to the circuitry resulting in complete dysfunction or inappropriate pacing or defibrillation. It is not necessary to turn off the devices during CPR; however, if it is indicated or to alleviate fears of the resuscitation team a circular magnet is needed. For an automatic implantable cardioverter defibrillator (AICD), the magnet is placed over the upper right-hand corner, if left in place for 30 seconds the AICD is turned off. The magnet is then removed. Placing the magnet back for 30 seconds will turn the AICD back on. If a magnet is placed on a pacemaker, it changes the pacemaker to a set predetermined rate in an asynchronous mode. It will pace at the predetermined rate without trying to sense an intrinsic rhythm or coordinate its pacing. All patients who present with a cardiac dysrhythmia and have a pacemaker or AICD in place should receive an interrogation of the device by a cardiologist, trained nurse or technician following the resuscitation. This can be helpful in determining the cause of the arrest as well as ensuring proper pacemaker or AICD function and reprogramming.

## Complications of Cardiopulmonary Resuscitation

Cardiopulmonary resuscitation is performed to save the life of the victim; however, it is not without complications. Resuscitation teams need to be aware of the potential complications in order to provide better care to a patient whose resuscitation is not going well or for those who deteriorate post-resuscitation. Many studies have been conducted in an attempt to determine the rate of complication from CPR. The most common complication found was sternal and rib fractures with a rate of 25–30%. Other complications include: anterior mediastinal hemorrhage, upper airway complications, abdominal organ injuries and lung injuries.[43-48]

Rib and sternal fractures are the most common complications of even well performed CPR.[45] The sternum may become separated from the ribs during the first several compression cycles and the anterior ribs during this time can be fractured. Most commonly this does not cause permanent problems for the patient, but will be painful during recovery. Sometimes, the fractured ribs can puncture or injure the lung leading to a tension pneumothorax. If a patient during resuscitation becomes difficult to bag it is important to consider pneumothorax as the cause. If a pneumothorax occurs this can quickly lead to tension for patients who are receiving positive pressure ventilation. Tension pneumothorax interferes with cardiac filling during diastole or the release phase of CPR and can lead to pulseless electrical activity (PEA). Providers need to decompress the chest quickly. To decompress the chest, a chest tube needs to be placed, but this can take time during which the patient may continue to deteriorate. While obtaining the supplies and placing a chest tube a needle thoracotomy should be performed. This is performed by placing a large gauge (14-G is preferred) angiocatheter in the second intercostal space on the anterior chest in the midclavicular line. A rush of air should be heard releasing the pressure on the heart allowing for proper filling. The needle can then be removed leaving the plastic catheter in place until a chest tube is in place. If a patient is very large with a thick chest wall a spinal needle may be used instead. While sternal and rib fractures are common complications of CPR they rarely cause any problems for the resuscitation. However, better CPR technique leads to a lower rate of complication.

During a resuscitation attempt many patients will develop gastric distention from either mouth-to-mouth or other rescue breathing prior to intubation. Patients may vomit from this gastric distention or from the resuscitation itself, which can lead to aspiration, especially before a definitive airway can be achieved. An endotracheal tube does not completely occlude the airway and fluid may still pass the endotracheal tube balloon; however, intubation does prevent large amounts of fluid and food particles from being aspirated. If aspiration occurs the patient is at risk for pneumonitis from the acidic contents.

Abdominal organ injury from CPR is not a common complication, but can occur. The most common organ injured is the liver; liver lacerations can occur due to fractured ribs and from CPR performed with the hands too low on the anterior chest. Splenic lacerations have also been known to occur. If a post-resuscitation patient becomes acutely hypotensive, one must consider liver or splenic lacerations as a potential source of hemorrhage. Bowel injury or perforation can also occur.

Fatal bleeding following CPR and the initiation of thrombolytic therapies is not a common complication. Even despite multiple rib fractures most patients do not suffer fatal hemorrhage.[49] Concern over the possible bleeding risks of thrombolytic agents should not preclude providers from thrombolysis following post-cardiac arrest if it is medically indicated for the treatment of acute coronary syndrome (ACS).

## ADVANCED CARDIAC LIFE SUPPORT

### Overview-Statistics of Success

Advanced cardiac life support includes high quality BLS and interventions that can prevent cardiac arrest in the setting of ACSs, treat cardiac arrest and improve outcomes after the cardiac arrest patient is resuscitated. During any resuscitation the healthcare provider must recognize and treat reversible causes of cardiac arrest. The "4 Hs and 4 Ts" are known causes and possible complications of cardiac arrest. These are listed in **Table 3**. The new AHA 2010 guidelines for the "Chain of Survival" includes "Part 9: Post-Cardiac Arrest Care" emphasizing comprehensive multidisciplinary care beginning with the recognition of cardiac arrest and concludes with hospital discharge, but may carry beyond to prevent future cardiac complications.[23]

### Advanced Airway Management

The new 2010 AHA guidelines for ACLS have new recommendations for airway management that include: the use of quantitative waveform capnography for confirmation and continuous monitoring of endotracheal tube placement for adults, the use of supraglottic advanced airways as alternative

### TABLE 3
**Reversible causes of cardiac arrest**

| Five Hs | Five Ts |
|---|---|
| Hypoxia | Toxins |
| Hypovolemia | Tamponade |
| Hydrogen ion (acidosis) | Tension pneumothorax |
| Hypokalemia/hyperkalemia | Thrombosis, pulmonary |
| Hypothermia | Thrombosis, coronary |

(*Source:* Neumar RE, Ottto CS, Link MS, et al. Guidelines for cardiopulmonary resuscitation and emergency cardiovascular care. Part 8: Adult advanced cardiovascular life support. Circulation. 2010;122(Suppl. 3):S729-67)

to endotracheal intubation and they no longer recommend the routine use of cricoid pressure during airway management.[23]

Endotracheal intubation is the recommended airway of choice for all patients needing invasive ventilation management or airway protection due to alteration in mentation, airway swelling or any other injuring that may compromise the upper airways. The possibility of spinal injury is a relative contraindication of direct laryngoscopy orotracheal intubation; however, if a patient requires endotracheal intubation for life-saving reasons then one must perform the procedure with as much spinal immobilization as possible without interfering with the intubation procedure. Interested individuals should seek a certified ACLS course for instruction on endotracheal intubation.

The difficult airway where intubation fails can be due to prominent upper incisors, inability to extend the neck, extremely large tongue, swelling, blood or secretions in the airway, small lower jaw, inability to completely open the mouth, tumors or any other unusual anatomy. Some patients despite normal-appearing anatomy without complicated history may pose an unexpected challenge to intubate. For those experienced and skilled in the practice of endotracheal intubation, it is still the best airway; however, intubation is also a motor skill that requires practice and for those providers in the EMS system who may not be very experienced airway adjuncts are extremely important. Good bag-valve-mask (BVM) ventilation is the first technique and probably the most important to know. BVM, however, can cause gastric distention and does not protect the airway from aspiration. The most common adjunct supraglottic airways include the LMA, Combitube and King LT. An ideal airway should be rapidly and reliably inserted with minimal training, control ventilation, protect against aspiration and be able to be inserted with ongoing chest compressions. No one device provides all of these things but the LMA, Combitube and King

LT are easy to use and found to be useful as adjunct airways for the inexperienced.[50] Multiple mannequin studies have found that most out of hospital providers prefer the King LT as easier and faster to insert.[51-54]

## Pharmaceutical Interventions

Pharmaceutical interventions for cardiac arrest victims are a controversial subject. There is no evidence that any medications given during cardiac arrest that have lead to any improvement in survival to hospital discharge.[55,56] The most important factors in survival are high quality CPR and early defibrillation. When pharmaceutical therapies are to be used it is important to continue high quality CPR with minimal interruptions in compressions for line placement and intubation. The medications used during CPR should assist in potentiating the return of circulation, enhance cardiac function, support blood pressure and shunt blood toward vital organs.[57] Drug therapy regimens should ultimately increase survival to discharge and not just increase initial resuscitation rates which may only result in unsalvageable patients with transient cardiac activity.[55,56] The AHA includes several medications in their Advanced Cardiovascular Life Support Pulseless Algorithm.

### Epinephrine

Epinephrine is a mixed alpha and beta adrenergic receptor agonist. Alpha agonists are potent vasopressors and increase systemic vascular resistance which results in elevated aortic diastolic pressure which then increases coronary and carotid blood flow. However, epinephrine may increase myocardial workload and decrease endocardial perfusion, compromising cardiac tissue. The recommendation of the AHA is 1 mg of epinephrine on every 3–5 minutes for pulseless VT or VF and asystole. High-dose epinephrine does not enhance long-term outcome, can be detrimental and should not be used.[58]

### Vasopressin

Vasopressin is a naturally occurring polypeptide produced from cells within the hypothalamus. When administered in pharmacologic dosages it acts as a peripheral vasoconstrictor. Vasopressin has also a longer half-life than epinephrine, about 10–20 minutes. Studies on the use of vasopressin in sudden cardiac arrest have not shown any greater benefit than epinephrine on long-term survival in sudden cardiac arrest.[59-63] If vasopressin is used, the AHA recommendation is 40 units IV in lieu of epinephrine for the first or second dose.

## Lidocaine

Lidocaine increases uniformity of the action-potential duration and refractory period and can terminate reentrant rhythms. Lidocaine is given as a bolus of 1 mg/kg. The AHA recommends its administration after epinephrine, vasopressin and amiodarone have been tried. A second loading dose of 1 mg/kg can be given 10–15 minutes after the first one.

## Amiodarone

Amiodarone is an antiarrhythmic agent that lengthens the cardiac action potential, prolongates refractoriness of the myocytes and decreases cardiac oxygen consumption. Amiodarone also improves cardiac pump performance, dilates coronary arteries, causes peripheral arterial vasodilatation and increases coronary blood supply. It does have the side effect of decreasing systemic vascular resistance and causing hypotension. The ALIVE trial showed significant improvement in terminating VF and increasing survival to hospital admission in the prehospital setting over the use of lidocaine; however, it is unclear if there is any long-term benefit from its use.[64,65] The AHA does have amiodarone in its guidelines for the use of refractory VF or VT in their pulseless arrest algorithm. The recommended dosing is 300 mg IV bolus with a second dose of 150 mg IV if needed.

## Procainamide

Procainamide is a sodium channel-blocking antiarrhythmic medication that prolongs the refractory period and slows conduction through the myocardial conduction system. There are very few studies addressing the use of procainamide during pulseless cardiac arrest. Procainamide must be infused slowly; therefore, its practical use during cardiac arrest is limited. Procainamide must be avoided in patients with torsades de pointes. More study is needed to determine if procainamide has any long-term benefits for pulseless cardiac arrest. The AHA recommends procainamide for stable wide-QRS tachycardia but not for pulseless cardiac arrest.

## Atropine

Atropine is a competitive antagonist of acetylcholine at muscarinic receptors. For cardiac arrest patients or those with symptomatic bradycardia, parasympathetic tone is increased due to vagal stimulation. Atropine blocks the depressant effect of the vagus nerve at the sinus and AV nodes. The AHA, however, has removed atropine from its 2010 guidelines for PEA or asystole ACLS Cardiac Arrest Algorithm due to a lack of any evidence showing therapeutic benefit.[66,67]

## Magnesium Sulfate

If reversible causes of cardiac arrest are identified and treated patient outcomes improve. Magnesium deficiency can precipitate refractory VF. No prospective clinical trials have been published that show any change in long-term outcome from the routine use of magnesium in cardiac arrest patients except those with hypomagnesemia and patients in trosades de pointes. The recommended dose in cardiac arrest is 1–2 g diluted in 10 mL D5W given IV.

## Calcium Chloride

Calcium chloride is not recommended for routine use in pulseless cardiac arrest unless there is an identifiable reason such as known hypocalcemia, hyperkalemia or calcium channel blocker toxicity. Increases in intracellular calcium potentiate myocardial ischemic injury. Canine myocardial cells were investigated and found to have increased uptake of calcium after myocardial ischemia followed by reperfusion, the mechanism of this uptake has not been established but could be a concern for ischemic cellular injury.[68] For hyperkalemia and calcium channel blocker overdose, the recommended dosing is 500–1,000 mg IV.

## Morphine and Oxygen

The new 2010 ACLS guidelines from the AHA do not recommend the routine administration of oxygen for all ACS patients. The recommendation is to use oxygen to keep saturations greater than 94%. For cardiac arrest patients, 100% oxygen should be used during resuscitation but weaned down in the post-resuscitation period. Morphine is indicated for ST elevation myocardial infarction and when a patient's chest pain is unresponsive to nitrates. Caution should be used when administering to unstable angina or non-ST elevation myocardial infarction patients.

## Defibrillation or Cardioversion

Defibrillation is a procedure where controlled electrical energy is applied to the myocardium either through the chest wall or through directly on the heart and is designed to terminate an unstable or pulseless rhythm. Defibrillators are divided into two main types: (1) Manual and (2) Automatic. The AEDs were discussed earlier. Manual defibrillators require the provider to obtain and interpret an electrocardiogram and determine if: (1) defibrillation is necessary; (2) select an energy level and (3) decide if synchronization should be used. The goal of defibrillation is to uniformly depolarize a majority of the myocardium and terminate the abnormal dysrhythmia. Once electrical activity is reset and the myocardium regains its excitability, the SA node will presumptively reinitiate normal pacing and the myocardium can begin coordinated rhythmic contractions.

There are currently three basic theories of defibrillation: (1) The "critical mass" theory hypothesizes that the electrical current depolarizes a "critical mass" of myocardium and that the remaining myocardium that is not depolarized is inadequate to sustain the dysrhythmia. Zipes et al.[69] studied VF in dogs and found that successful defibrillation occurred most often when the electrodes were placed in the right ventricular apex and the posterior base of the left ventricle and least often when delivered between the two right ventricular electrodes;[69] (2) The theory of "upper limit of vulnerability" hypothesizes that the most important factor in defibrillation is achieving a critical current density throughout the ventricular myocardium that not only stops the fibrillation fronts but also does not reinitiate fibrillation by the same mechanism by which a shock during the vulnerable period of sinus or paced rhythm initiates fibrillation;[70] (3) It was found that the strength of shocks that stopped inducing VF were similar to the strength of shocks at the defibrillation threshold.[70] The theory of "extension of refractoriness" hypothesizes the shock prolongs the refractoriness in most of the myocardium so the fibrillation wavefronts cannot propagate.[71] Post-shock response durations, from shock to repolarization, were significantly longer in successful defibrillations than for unsuccessful defibrillation.[71] The "critical mass" theory may be important in both the "extension of refractoriness" and "upper limit of vulnerability" hypothesizes.

Defibrillators are classified by the type of waveforms they produce. Monophasic defibrillators send an electrical wave from one electrode to the other in only one direction. With biphasic defibrillators the electrode potential is reversed in midshock so the current reverses direction **(Fig. 3)**. The waveform can also be classified by the way the current flows and terminates during the "discharge" period with respect to time on an x-y Cartesian plot—rapid or gradual at onset and termination. If the wave has a rapid rise to a peak and then a gradual decline to baseline it is considered a damped sinusoidal waveform. If the waveform has a rapid rise and then a rapid descent, it is considered a truncated waveform **(Fig. 3)**. In 1962, Lown introduced the monophasic damped sinusoidal waveform for defibrillation and it remained the standard waveform for defibrillation for 30 years.[72] More recently studies have shown improved rates of terminating VF using biphasic truncated exponential (BTE) waveforms

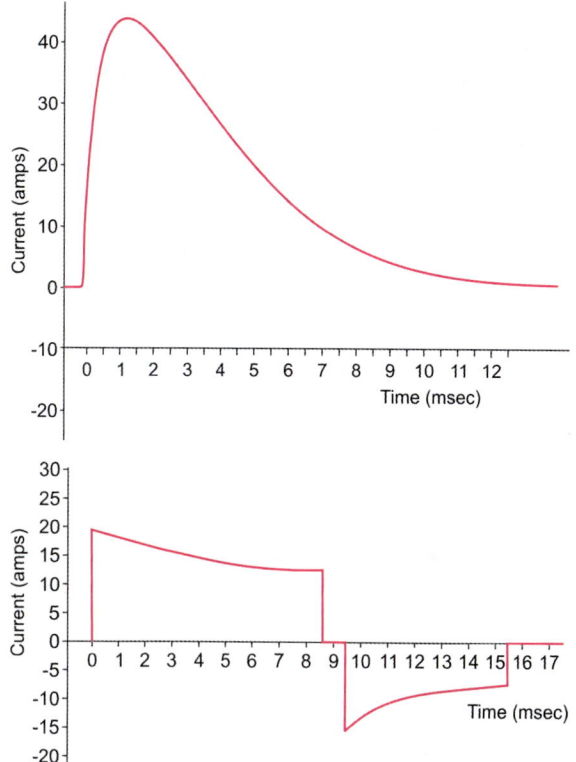

**FIGURE 3:** Monophasic vs Biphasic waveforms. The monophasic waveform is damped and the biphasic waveform is truncated. (*Source:* Modified from Deakin CD, Nolan JP, Sunde KJ, et al. European resuscitation council guidelines for resuscitation 2010 Section 3. Electrical therapies: automated external defibrillation, cardioversion and pacing. Resuscitation. 2010;81:1293-304)

compared to monophasic damped sinusoidal waveforms. The BTE waveform also causes fewer post-shock arrhythmias with less myocardial damage.[73-78] The second phase of the biphasic shock also removes excess charge left on the myocardial cells after the first shock, a process called "charge burping".[79]

Biphasic defibrillators require lower energy levels to terminate VF or VT; the lower energy levels result in a decreased chance of myocardial damage and also allow smaller machines to be built. This technology has allowed portable battery operated AEDs to be developed and placed in public areas for use by lay rescuers. Public access to AEDs has increased survival rates for out of hospital cardiac arrest by decreasing time to first shock as discussed earlier in the chapter.[24,80]

Minimizing transthoracic impedance increases success rates of defibrillation. Paddle force, paddle orientation, a couplant, such as conductive gel use, shaving the chest and lung volumes, can all affect the transthoracic impedance. It is important that gel is applied when using hand-held paddles; without conductive gel the transthoracic impedance is extremely high resulting in poor current flow and decreased defibrillation success. Care must be taken not to smear the gel across the chest between the paddles as this could cause the current to flow through the low impedance path of the gel away from the heart.[81] Breast tissue can also increase impedance; therefore, the paddles or pads should be placed adjacent to or under the breast. Commercially available biphasic defibrillators are able to automatically adjust voltages and pulse duration to compensate for high or low transthoracic impedance. Monophasic defibrillators do not have this capability and result in considerable variability in delivered waveform depending on transthoracic impedance. "Smart" biphasic defibrillators can alter the waveform duration and/or voltage of the two pulses individually and instantaneously to optimize performance.

While defibrillation is a life-saving procedure, it is not without risk. There are three basic risks: (1) risk to the patient; (2) risk to the user and (3) risk to equipment or environment. Care must be taken to minimize these risks.

## Risk to the Patient

Defibrillators are equipped with "synchronization" mode which allows the user to avoid unintended delivery of a shock to the "T" wave of the ECG [which can induce VF, **(Fig. 4)**]. The user must be sure, when using a manual defibrillator, to appropriately use the synchronization for cardioversion of stable dysrhythmias but must avoid its use in unstable or pulseless rhythms, such as VF or polymorphic VT, where there will not be a discrete "R" wave to synchronize on and the device will not fire, thereby causing a delay until the synchronizer is disabled.

The energy level applied to the myocardium from defibrillation can cause damage manifest as myocardial necrosis and functional damage evident as atrioventricular conduction disturbances.[82] Chest wall impedance also posses challenges to delivery of safe defibrillation; high impedance results in a wide waveform with lower current, low impedance results in a narrow waveform with high current. Excessive current runs the risk of myocardial damage and low current may be inadequate to achieve defibrillation. As discussed earlier electrode orientation, electrode placement, chest hair and lung volumes can all affect impedance.

Incorrectly displayed asystole can occur when paddles or gel pads are used to display the ECG tracing due to electrical voltage "offset". This "false" asystole display can last long enough to mislead rescuers. If asystole is displayed it must be confirmed immediately by attaching the standard ECG

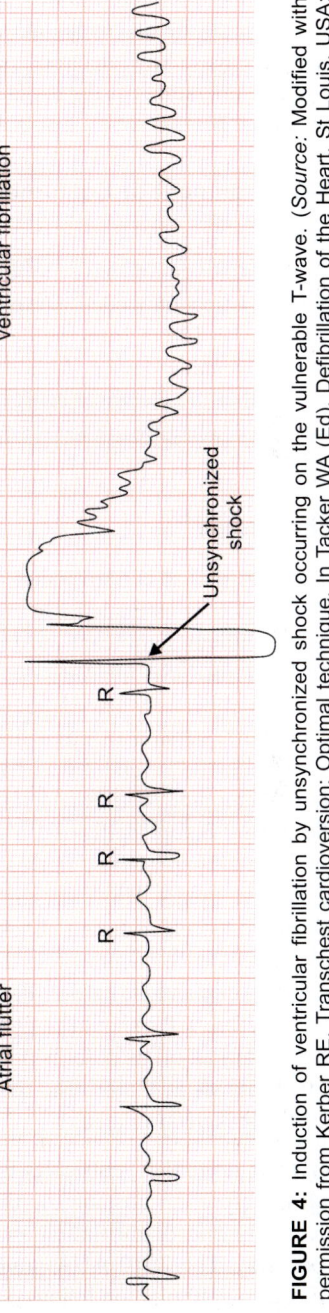

**FIGURE 4:** Induction of ventricular fibrillation by unsynchronized shock occurring on the vulnerable T-wave. (*Source*: Modified with permission from Kerber RE. Transchest cardioversion: Optimal technique. In Tacker WA (Ed). Defibrillation of the Heart. St Louis, USA: Mosby-Year Book; 1994. pp. 46-81)

electrodes; therefore, whenever possible gel pads or paddles should not be used for ECG display or monitoring.[83]

After transthoracic shocks first-degree skin burns are common.[84] Current flows preferentially to the edges of the electrodes increasing the thermal temperature causing skin burns mostly around the edges. New electrode pads are designed to decrease thermal temperatures at the contact site and therefore decrease skin burns.[85]

## Risk to the Environment or Equipment

An increasing number of patients requiring CPR and defibrillation or cardioversion have implantable pacemakers or defibrillators. It is recommended that placement of paddles or pads are at least 12–15 cm from the device to help prevent damage to the implantable device. Although this risk is very low, it is possible that the electrical energy can be transferred down the lead wires damaging the wires and preventing proper defibrillation.[86-88]

## Risks to the Rescuer

Risks to the rescuer during cardioversion or defibrillation are difficult to quantify. It is estimated that injury to paramedics was 1 per 1,700 without significant morbidity or mortality.[89] Recent clinical studies have measured current flow through rescuers who have deliberately placed themselves in the current pathway during cardioversion of atrial fibrillation; measured current flow through the rescuers was trivial and no injury occurred.[90] It has been suggested that the traditional admonition to "clear" the patient before delivering a defibrillating shock is therefore unnecessary, providing gloves are worn and self-adhesive paddles are used.[91] However, the AHA recommends that all personnel stand clear during shock delivery. There is also no significant evidence documenting that performing CPR on a patient with an implantable cardiac defibrillator is harmful to the rescuer; therefore, CPR should be performed as usual but the device may be turned off as described earlier in the chapter when a magnet becomes available. The most common rescuer injuries are musculoskeletal related.

*VF or VT-stable vs unstable*: Ventricular fibrillation is always pulseless and, therefore, an unstable rhythm; CPR should be started immediately and defibrillation should be performed as soon as possible. The 2010 AHA guidelines recommend immediate CPR then defibrillation followed by more immediate CPR, minimizing pauses and "hands off" time. The monophasic energy recommendation is 360 J. Biphasic waveform defibrillator configurations differ among manufacturers and none have been directly compared to humans, therefore, the AHA recommends using the manufacturer's recommendations for energy dose on biphasic machines. Since no optimal biphasic energy level has been determined the AHA does not make any definitive recommendation for the selected energy for subsequent biphasic defibrillation attempts. If subsequent shocks are required then the energy levels should be equivalent or higher than the first shock.

Ventricular tachycardia can be either unstable or stable. If the rhythm does not produce a blood pressure sufficient to maintain mentation then CPR should be started with defibrillation as soon as possible. If the patient with VT has a pulse with tolerable blood pressures then elective cardioversion is recommended, as soon as possible to prevent deterioration to a pulseless rhythm. Cardioversion is always performed using synchronized shocks, which avoids shock delivery during the relative refractory period of the cardiac cycle when the shock is most likely to induce VF. Monomorphic VT has been found to convert with lower energy and current during cardioversion than polymorphic VT which behaves more like VF.[92] Amiodarone administration for persistent VF or VT has been found to be relatively safe but ineffective for the acute termination of sustained VT.[93] Amiodarone is administered at a dose of 300 mg IV bolus.

*Atrial fibrillation or supraventricular tachycardia*: Atrial dysrhythmias and other tachycardias may require urgent cardioversion due to hypotension or pulmonary edema. Treatment of the cause of the arrhythmia may restore sinus rhythm or prevent recurrence. Causes include hyperthyroidism, pulmonary embolism, congestive heart failure and valve disorders. **Table 4** lists some of the causes of wide and narrow tachycardia. Factors that may influence immediate and long-term success of cardioversion include the type of tachycardia, underlying cause of the dysrhythmia, and duration of the dysrhythmia. The AHA recommends initial biphasic energy doses of 120–200 J for atrial fibrillation; the monophasic energy dose should be 200 J. Atrial flutter and other supraventricular rhythms require less energy and a starting dose of 50–100 J of monophasic or biphasic with increasing step wise dosing is often sufficient. Synchronization is required for cardioversion. **Table 4** summarizes classification of tachycardia.

*Defibrillation or cardioversion in pacemaker or AICD patients*: As described earlier in the chapter, defibrillation with pacemakers and AICDs in place pose little risk to the patient and the rescuer; however, some precautions are indicated. Placement of the pads or paddles should be in an anterior-posterior or anterior-lateral locations avoiding overlap with

> **TABLE 4**
> **Classification of tachycardias**
>
> - Narrow–QRS-complex (SVT) tachycardias (QRS <0.12 second), listed in order of frequency
>   — Sinus tachycardia
>   — Atrial fibrillation
>   — Atrial flutter
>   — AV nodal re-entry
>   — Accessory pathway–mediated tachycardia
>   — Atrial tachycardia
>   — Multifocal atrial tachycardia (MAT)
>   — Junctional tachycardia
>
> - Wide–QRS-complex tachycardias (QRS $\geq$0.12 second)
>   — Ventricular tachycardia (VT) and ventricular fibrillation (VF)
>   — SVT with aberrant conduction
>   — Pre-excitation tachycardias [Wolff-Parkinson-White (WPW) syndrome]
>   — Ventricular paced rhythms
>
> (*Source:* Neumar RW, Otto CS, Link MS, et al. American Heart Association guidelines for cardiopulmonary resuscitation and emergency cardiovascular care. Part 8: Adult advanced cardiovascular life support. Circulation. 2010;122(Suppl. 3):S729-67)

the device if at all possible. Delay in defibrillation should not occur due to an implantable device. Pacemaker spikes with unipolar pacing may confuse AED software and prevent VF from being recognized.[67] As soon as more advanced equipment and personnel are available the rhythm should be analyzed.

## CESSATION OF RESUSCITATION

Termination of resuscitation is difficult for all providers of cardiac arrest patients but it can become especially difficult for emergency medical personnel in the prehospital setting. There are ethical, legal and cultural factors that need to be taken into consideration when deciding the need for termination of resuscitation. Initiation of resuscitation may conflict with a patient's desires or may not be in the best interest of the patient and in some instances resuscitation may not be the best use of limited resources. The public in general overestimates the probability of survival from cardiac arrest and even most physicians cannot accurately predict mortality rates of sudden cardiac arrest.

The 2010 AHA guidelines give some guidance for termination of resuscitation efforts in adults without of hospital cardiac arrest. The AHA has developed the "BLS termination of resuscitation rule"; if all the following criteria are met then there is no indication for ambulance transport: (1) arrest was not witnessed by and EMS provider or first responder; (2) no ROSC after three complete rounds of CPR and AED analysis and (3) no AED shocks delivered. The "ALS termination of resuscitation rule" states if all the following criteria are met then termination of resuscitation before transport is indicated: (1) arrest not witnessed by anyone; (2) no bystander CPR provided; (3) no ROSC after complete ALS care in the field and (4) no shocks delivered. Implementation of these rules usually includes contacting the EMS medical control. EMS providers should be trained in sensitive communication with family members about outcomes. Collaboration with hospital EDs, medical coroner's office, online medical directors and the police are necessary. These rules have been validated for adult out of hospital cardiac arrest victims in multiple EMS settings across the United States, Canada and Europe.[67,94-96] Implementation of these rules reduces the rate of unnecessary hospital transports which can place providers and the public at risk from road traffic hazards, decrease unnecessary exposures from potential biohazards and decrease costs.

The decision to terminate resuscitative efforts is never an easy one. While guidelines are available, all providers including first responders, EMS personnel and physicians need to take into consideration many factors when deciding to terminate efforts. It is clear that resuscitation should not be started for victims who have a valid do not resuscitate (DNR), newborn premature infants less than 23 weeks gestation and victims who present with obvious signs of death such as: rigor mortis, decapitation or dependent lavidity.

## POST-RESUSCITATION CARE

Post-resuscitative care is a new section in the 2010 AHA guidelines for CPR. The goal is to emphasize an organized multidisciplinary program that focuses on optimizing neurologic, hemodynamic and metabolic function that may provide an increase in survival to hospital discharge.[67] Patients should be cared for in a multidisciplinary environment with angiography and interventional capabilities and a team capable of caring for patients with multiorgan dysfunction.

### Cardiopulmonary Support

Once circulation is restored oxygen should be weaned down to lowest required to maintain oxygen greater than 94%, to avoid hyperoxia. The 2010 international consensus on CPR and ECC science with treatment recommendations found harmful effects of hyperoxia after ROSC.[67,97] An oxyhemoglobin of 100% can correspond to a $PaO_2$ anywhere between 80 and 500. With return of circulation a "post-arrest syndrome" often presents that requires proper inotropic support and monitoring. Standard vasopressor treatment is indicated to improve patient hemodynamics. Many patients will develop multisystem organ dysfunction and this must be anticipated and treated in a timely fashion.

## Cardiac Interventions

The goal of interventions is to prevent further myocardial necrosis and left ventricular dysfunction leading to heart failure. Percutaneous coronary intervention (PCI) will provide patients with ST elevation myocardial infarction the most favorable outcomes, even in out of hospital cardiac arrest where overall survival remains low.[98-101] For those patients without obvious ST elevations an ischemic cardiac etiology may still be a reasonable assumption given the insensitivity or possible misleading electrocardiogram following cardiac arrest; more study is needed to determine if this subgroup of patients would benefit from intervention.[100] Patients should be transferred or taken from the scene to a facility that is capable of providing a comprehensive post-cardiac arrest treatment system of care that includes advanced cardiac interventions even if they have received thrombolytic therapy at a less capable institution. Early PCI following thrombolysis is associated with reduced recurrence of ischemia and reinfarction without increased risk of major hemorrhage.[102,103]

## Therapeutic Hypothermia

Predicting neurologic outcomes post-cardiac arrest is challenging. Therapeutically induced mild hypothermia after successful initial resuscitation has been studied and found to improve neurologic outcome in comatose cardiac arrest survivors and to decrease overall mortality.[104-107] Methods of inducing hypothermia to 33°C for 24 hours include: external ice, blankets through which cold water continuously circulates, cold IV saline and endovascular coils.[108,109] Interventions to rapidly reduce body temperature during CPR have been shown to improve defibrillation success and ROSC in animal models; however, limited clinical experience has been favorable.[110-113] Improving patients' final functional outcome also involves early recognition and treatment of treatable neurologic disorders such as seizures. Seizures may be difficult to diagnose in the hypothermia induced patient due to neuromuscular blockade use that is designed to prevent shivering. It is therefore important to have electroencephalographic monitoring during this coma induced state. To date there have not been any diagnostic evaluations or tools found that consistently predict neurologic outcome in post-cardiac arrest patients. There is limited evidence to guide clinical decisions and, therefore, best clinical judgment with family discussion should be used to make decisions regarding withdraw of life support.

## SUMMARY

Cardiopulmonary resuscitation from sudden cardiac arrest is challenging. Since 1960, our impact on patient survival from out of hospital cardiac arrest has changed very little. What we do know is that high quality, minimally interrupted CPR that is started immediately upon recognition is extremely important for improving patient survival. Rate and depth of compressions is the key to high quality CPR. Providers who do not perform CPR often should have frequent retraining. The AHA guidelines for 2010 have been simplified to help improve compliance and to emphasize compressions over other interventions in the field to help improve the number of patients who receives bystander CPR; this includes compression only CPR for those who may be unwilling or unable to perform conventional CPR. Many fears that potential rescuers have may be alleviated with education and debriefing which may improve first responder or bystander initiation of resuscitation. Continuous evaluations and re-education should be implemented in all levels from the prehospital setting to hospital resuscitation teams, to improve resuscitation performance with an emphasis on CPR quality and team work. Continued research is required in all areas of resuscitation from prehospital to hospital discharge to improve survival and neurologic outcomes.

## REFERENCES

1. CPR and Sudden Cardiac Arrest Fact Sheet. American Heart Association. Available from www.heart.org [Accessed October, 2010].
2. Kouwenhoven WB, Jude JR, Knickerbocker CG. Closed chest cardiac massage. JAMA. 1960;173:1064-7.
3. Weale FE, Rothwell-Jackson RL. Cardiac massage. Lancet. 1962; 7237:990-2.
4. Redberg RF, Tucker KJ, Cohen TJ, et al. Physiology of blood flow during cardiopulmonary resuscitation. A transesophageal echocardiographic study. Circulation. 1993;88:534-42.
5. Driscoll TE, Ratnoff OD, Nygard OF. The remarkable Dr. Abildgaard and countershock. Ann Internal Med. 1975;83:878-82.
6. American Heart Association. History of CPR. Available from www.heart.org [Accessed October, 2010].
7. Beck CS, Pritchard WH, Feil HS. Ventricular fibrillation of long duration abolished by electric shock. J Am Med Assoc. 1947;135:985-6.
8. Zoll P, Linenthal A, Gibson W, et al. Termination of ventricular fibrillation in man by externally applied electrical countershock. NEJM. 1956;254:727-32.
9. Young BG. Neurologic prognosis after cardiac arrest. NEJM. 2009;361:605-11.
10. Sayre MR, Cantrell SA, White LJ, et al. Impact of the 2005 American Heart Association cardiopulmonary resuscitation and emergency cardiovascular care guidelines on out-of-hospital cardiac arrest survival. Prehosp Emerg Care. 2009;13:469-77.

11. Hinchey PR, Myers JB, Lewis R, et al. Improved out-of-hospital cardiac arrest survival after the sequential implementation of 2005 AHA guidelines for compressions, ventilations and induced hypothermia: the Wake County experience. Ann Emerg Med. 2010;56:358-61.
12. Svensson L, Bohm K, Castren M, et al. Compression-only CPR or standard CPR in out of hospital cardiac arrest. NEJM. 2010;363: 434-42.
13. Boyd DR. The conceptual development of EMS systems in the United States. Emerg Med Serv. 1982;11:19-23.
14. Hampton OP. Transportation of the injured: a report. Bull Am Coll Surg. 1960;45:55-9.
15. Division of Medical Sciences, National Academy of Sciences-National Research Council: accidental death and disability: the neglected disease of modern society. Washington DC: US Government Printing Office; 1966.
16. National Highway Safety Act of 1966, Public Law 89-564, Washington DC: 89th US Congress; 1966.
17. Emergency Medical Services Systems Act of 1973, Public law 93-154, Washington DC: 93rd US Congress; 1973.
18. US Department of Transportation-National Highway Traffic Safety Administration: Highway Safety Program Guideline No 11 Emergency Medical Services, 2009 [www.nhtsa.gov].
19. US Department of Transportation-National Highway Traffic Safety Administration: First responder: National standard curriculum, 1997 [www.nhtsa.gov].
20. US Department of Transportation-National Highway Traffic Safety Administration: Emergency Medical Technician-Basic: National standard curriculum, 1997 [www.nhtsa.gov].
21. US Department of Transportation-National Highway Traffic Safety administration: Emergency Medical Technician-Intermediate: National Standard Curriculum, 1999 [www.nhtsa.gov].
22. US Department of Transportation-National Highway Traffic Safety Administration: Emergency Medical Technician-Paramedic: National Standard Curriculum, 2004 [www.nhtsa.gov].
23. Executive Summary: 2010 American Heart Association Guidelines for Cardiopulmonary resuscitation. Circulation. 2010. Available from www.circ.ahajournals.org [Accessed October, 2010]
24. Kitamura T, Iwami T, Kawamura T, et al. Nationwide public-access defibrillation in Japan. NEJM. 2010;362:994-1004.
25. White L, Roger J, Bloomingdale M, et al. Dispatcher-assisted cardiopulmonary resuscitation: risks for the patients not in cardiac arrest. Circulation. 2010;121:91-7.
26. Iwami T, Nichol G, Hiraide A, et al. Continuous improvements in "chain of survival" increased survival after out-of-hospital cardiac arrests: a large-scale population-based study. Circulation. 2009;119:728-34.
27. Rea TD, Helbock M, Perry S, et al. Increasing use of cardiopulmonary resuscitation during out-of-hospital ventricular fibrillation arrest: survival implications of guideline changes. Circulation. 2006;114:2760-5.
28. Hollenberg J, Herlitz J, Lindqvist J, et al. Improved survival after out-of-hospital cardiac arrest is associated with an increase in proportion of emergency crew-witnessed cases and bystander cardiopulmonary resuscitation. Circulation. 2008;118:389-96.

29. Lund-Kordahl I, Olasveengen TM, Lorem T, et al. Improving outcome after out-of-hospital cardiac arrest by strengthening weak links of the Chain of Survival: quality of advanced life support and post-resuscitation care. Resuscitation. 2010;81:422-6.
30. Bobrow BJ, Clark LL, Ewy GA, et al. Minimally interrupted cardiac resuscitation by emergency medical services for out-of-hospital cardiac arrest. JAMA. 2008;299:1158-65.
31. Iwami T, Kawamura T, Hiraide A, et al. Effectiveness of bystander-initiated cardiac-only resuscitation for patients with out-of-hospital cardiac arrest. Circulation. 2007;116:2900-7.
32. Ong ME, Ng FS, Anushia P, et al. Comparison of chest compression only and standard cardiopulmonary resuscitation for out-of-hospital cardiac arrest in Singapore. Resuscitation. 2008;78:119-26.
33. Bohm K, Rosenqvist M, Herlitz J, et al. Survival is similar after standard treatment and chest compression only in out-of-hospital bystander cardiopulmonary resuscitation. Circulation. 2007;116:2908-12.
34. Rea TD, Fahrenbruch C, Culley L, et al. CPR with chest compression alone or with rescue breathing. NEJM. 2010;363:423-33.
35. Dickinson ET, Verdile VP, Schneider RM, et al. Effectiveness of mechanical versus manual chest compressions in out-of-hospital cardiac arrest resuscitation: a pilot study. Am J Emerg Med. 1998;16:289-92.
36. Ward KR, Menegazzi JJ, Zelenak RR, et al. A comparison of chest compressions between mechanical and manual CPR by monitoring end-tidal $PCO_2$ during human cardiac arrest. Ann Emerge Med. 1993;22:669-74.
37. Larsen AI, Hjornevik AS, Ellingsen CL, et al. Cardiac arrest with continuous mechanical chest compression during percutaneous coronary intervention. A report on the use of the LUCAS device. Resuscitation. 2007;75:454-9.
38. Wik L, Bircher NG, Safar P. A comparison of prolonged manual and mechanical external chest compression after cardiac arrest in dogs. Resuscitation. 1996;32:241-50.
39. Niemann JT, Rosborough JP, Kassabian L, et al. A new device producing manual sternal compression with thoracic constraint for cardiopulmonary resuscitation. Resuscitation. 2006;69:295-301.
40. Timerman S. Cardoso LF, Ramires JA, et al. Improved hemodynamic performance with a novel chest compression device during treatment of in-hospital cardiac arrest. Resuscitation. 2004;61:273-80.
41. Axelsson C, Nestin J, Svensson L, et al. Clinical consequences of the introduction of mechanical chest compression in the EMS system for the treatment of out-of-hospital cardiac arrest—a pilot study. Resuscitation. 2006;71:47-55.
42. Hightower D, Thomas SH, Stone CK, et al. Decay in quality of closed-chest compressions over time. Ann Emerg Med. 1995;26:300-3.
43. Fitchet A, Neal R, Bannister P. Splenic trauma complicating cardiopulmonary resuscitation. BMJ. 2001;322:480-1.
44. Bedell SE, Fulton EJ. Unexpected findings and complications at autopsy after cardiopulmonary resuscitation (CPR). Arch Intern Med. 1986;146:1725-8.
45. Lederer W, Mair D, Rabi W, et al. Frequency of rib and sternum fractures associated with out-of-hospital cardiopulmonary resuscitation is underestimated by conventional chest X-ray. Resuscitation. 2004;60:157-62.

46. Hoke RS, Chamberlain D. Skeletal chest injuries secondary to cardiopulmonary resuscitation. Resuscitation. 2004;63:327-38.
47. Boz B, Erdur B, Acar K, et al. Frequency of skeletal chest injuries associated with cardiopulmonary resuscitation: forensic autopsy. Ulus Trauma Acil Cerrahi Derg. 2008;14:216-20.
48. Krisher JP, Fine EG, Davis JH, et al. Complications of cardiac resuscitation. Chest. 1987;92:287-91.
49. Scholz KH, Tebbe U, Hermann C, et al. Frequency of complications of cardiopulmonary resuscitation after thrombolysis during acute myocardial infarction. Am J Cardiol. 1992;69:724-8.
50. Cook TM, Hommers C. New airways for resuscitation? Resuscitation. 2006;69:371-87.
51. Russi CS, Hartley MJ, Buresh CT. A pilot study of the King LT supralaryngeal airway use in a rural Iowa EMS system. Int J Emerg Med. 2008;1:135-8.
52. Burns JB, Branson R, Barnes SL, et al. Emergency airway placement by EMS providers: comparison between the King LT supralaryngeal airway and endotracheal intubation. Prehosp Disaster Med. 2010;25:92-5.
53. Tumpach EA, Lutes M, Ford D, et al. The King LT versus the combitube: flight crew performance and preference. Prehosp Emerg Care. 2009;13:324-8.
54. Murray MJ, Vermeulen MJ, Morrison LJ, et al. Evaluation of prehospital insertion of the laryngeal mask airway by primary care paramedics with only classroom mannequin training. CJEM. 2002;4:338-43.
55. Herlitz J, Ekstrom L, Wennerblom B, et al. Adrenaline in out-of-hospital ventricular fibrillation: does it make any difference? Resuscitation. 1995;29:195-201.
56. Olasveengen TM, Sunde K, Brunbarg C, et al. Intravenous drug administration during out-of-hospital cardiac arrest: a randomized trial. JAMA. 2009;302:2222-9.
57. White SJ, Himes D, Rouhani M, et al. Selected controversies in cardiopulmonary resuscitation. Semin Respir Crit Care Med. 2001;22:35-50.
58. Brown CG, Martin DR, Pepe PE, et al. A comparison of standard-dose and high-dose epinephrine in cardiac arrest outside the hospital the multicenter high-dose epinephrine study group. NEJM. 1992;327:1051-5.
59. Guevniaud PY, David JS, Chamzy E, et al. Vasopressin and epinephrine vs epinephrine alone in cardiopulmonary resuscitation. NEJM. 2008;359:21-30.
60. Mentzelopoulos SD, Zakynthinos SG, Tzoufi M, et al. Vasopressin, epinephrine and corticosteroids for in-hospital cardiac arrest. Arch Intern Med. 2009;169:15-24.
61. Morris DC, Dereczyk BE, Grzybowski M, et al. Vasopressin can increase coronary perfusion pressure during human cardiopulmonary resuscitation. Acad Emerge Med. 1997;4:878-83.
62. Lindner KH, Prengel AW, Brinkmann A, et al. Vasopressin administration in refractory cardiac arrest. Ann Intern Med. 1996;124:1061-4.
63. Stiell IG, Hebert PC, Wells GA, et al. Vasopressin versus epinephrine for in-hospital cardiac arrest: a randomized controlled trial. Lancet. 2001;358:105-9.

64. Gonzalez ER, Kannewurf BS, Ornato JP. Intravenous amiodarone for ventricular arrhythmias: overview and clinical use. Resuscitation. 1998;39:33-42.
65. Kudenchuk PJ, Cobb LA, Copass MK, et al. Amiodarone for resuscitation after out-of-hospital cardiac arrest due to ventricular fibrillation. NEJM. 1999;341:871-8.
66. Coon GA, Clinton JE, Ruiz E. Use of atropine for brady-asystolic prehospital cardiac arrest. Ann Emerg Med. 1981;10:462-7.
67. Highlights of the 2010 American Heart Association Guidelines for CPR and ECC. Available from www.static.org/eccguidelines [Accessed December 2010].
68. Shen AC, Jennings RB. Kinetics of calcium accumulation in acute myocardial ischemic injury. Am J Pathol. 1972;67:441-52.
69. Zipes DP, Fischer J, King RM, et al. Termination of ventricular fibrillation in dogs by depolarizing a critical amount of myocardium. Am J Cardiol. 1975;36:37-44.
70. Malkin RA, Souza JJ, Ideker RE. The ventricular defibrillation and upper limit of vulnerability dose-response curves. J Cardiovasc Electrophysiol. 1997;8:895-903.
71. Tovar OH, Jones JL. Relationship between "extension of refractoriness" and probability of successful defibrillation. Am J Physiol. 1997;272:H1011-9.
72. Lown B, Neuman J, Amarasinghem R. Comparison of alternating current with direct current electroshock across closed chest. Am J Cardiol. 1962;10:223-7.
73. Behr JC, Hartley LL, York DK, et al. Truncated exponential versus damped sinusoidal waveform shocks for transthoracic defibrillation. Am J Cardiology. 1996;78:1242-5.
74. Bardy GH, Marchlinski F, Sharma A, et al. Multicenter comparison of truncated biphasic shocks and standard damped sine wave monophasic shocks for transthoracic ventricular fibrillation. Circulation. 1996;94:2507-14.
75. Schneider JT, Martens PR, Paschen H, et al. Multicenter, randomized controlled trial of 150-J biphasic shocks compared with 200 to 360-J monophasic shocks in the resuscitation of out-of-hospital cardiac arrest victims. Circulation. 2000;102:1780-7.
76. Van Alem AP, Chapman FW, Lank P, et al. A prospective, randomized and blinded comparison of first shock success of monophasic and biphasic waveforms in out-of-hospital cardiac arrest. Resuscitation. 2003;58:17-24.
77. Tang W, Weil MH, Sun S, et al. The effects of biphasic and conventional monophasic defibrillation on postresuscitation myocardial function. J Am Coll Cardiol. 1999;34:815-22.
78. Martens PR, Russel JK, Wolcke B, et al. Optimal response to cardiac arrest study: defibrillation waveform effects. Resuscitation. 2001;49:233-43.
79. White RD, Kerber RE. Ventricular fibrillation and defibrillation: experimental and clinical experience with waveforms and energy. Textbook of Emergency Cardiovascular Care CPR. Philadelphia Lippincott, Williams and Wilkins; 2009. pp. 222-31.
80. The Public Access Defibrillation Trial Investigators. Public access defibrillation and survival after out-of-hospital cardiac arrest. NEJM. 2004;351:637-46.
81. Caterine MR, Yoerger DM, Spencer KT, et al. Effect of electrode position and gel-application technique on predicted transcardiac

current during transthoracic defibrillation. Ann Emerg Med. 1997;29:588-95.
82. Kerber RE. Transthoracic cardioversion and defibrillation. Cardiac Electrophysiology Zipes and Jalife, 4th edition. Philadelphia: Saunders; 2004. pp. 966-9.
83. Chamberlain D. Gel pads should not be used for monitoring ECG after defibrillation. Resuscitation. 2000;43:159-60.
84. Pagan-Carlo LA, Stone MS, Kerber RE. Nature and determinants of skin burns after transthoracic cardioversion. AM J Cardiol. 1997;79:689-91.
85. Meyer PF, Gadsby PD, Van Sickle D, et al. Impedance-gradient electrode reduces skin irritation induced by transthoracic defibrillation. Med Biol Eng Comput. 2005;43:225-9.
86. Waller C, Callies F, Langenfeld H. Adverse effects of direct current cardioversion on cardiac pacemakers and electrodes. Is external cardioversion contraindicated in patients with permanent pacing systems? Europace. 2004;6:165-8.
87. Aylward P, Blood R, Tonkin A. Complications of defibrillation with permanent pacemaker *in situ*. Pacing Clin Elctrophysiol. 1979;2:462-4.
88. Manegold JC, Israel CW, Ehrlich JR. et al. External cardioversion of atrial fibrillation in patients with implanted pacemaker or cardioverter-defibrillator systems: a randomized comparison of monophasic and biphasic shock energy application. Eur Heart J. 2007;28:1668-9.
89. Gibbs W, Eisenberg M, Damon SK. Dangers of defibrillation: injuries to emergency personnel during patient resuscitation. Am J Emerg Med. 1990;8:101-4.
90. Lloyd MS, Heeke BS, Walter PF, et al. Hands-on defibrillation. An analysis of electrical current flow through rescuers in direct contact with patients during biphasic external defibrillation. Circulation. 2008;117:2510-4.
91. Kerber RE. "I'm clear, you're clear, everybody's clear" A tradition no longer necessary for defibrillation? Circulation. 2008;117:2435-6.
92. Kerber RE, Olshansky B, Waldo AL, et al. Ventricular tachycardia rate and morphology determine energy and current requirements for transthoracic cardioversion. Circulation. 1992;85:158-63.
93. Marill KA, DeSouza IS, Nishijima DK, et al. Amiodarone is poorly effective for the acute termination of ventricular tachycardia. Ann Emerg Med. 2006;47:217-24.
94. Ong ME, Jaffey J, Stiell I, et al. Comparison of termination-of-resuscitation guidelines for basic life support: defibrillator providers in out-of-hospital cardiac arrest. Ann Emerg Med. 2006;47:337-43.
95. Morrison LJ, Verbeek PR, Vermeulen MJ, et al. Derivation and evaluation of a termination of resuscitation clinical prediction rule for advanced life support providers. Resuscitation. 2007;74:266-75.
96. Sherbino J, Keim SM, Davis DP. Clinical decision rules for termination of resuscitation in out-of-hospital cardiac arrest. J Emerg Med. 2010;38:80-6.
97. Kilgannon JH, Jones AE, Shapiro NI, et al. Association between arterial hyperoxia following resuscitation from cardiac arrest and in-hospital mortality. JAMA. 2010;303:2165-71.
98. Reynolds JC, Callaway CW, Khoudary SR, et al. Coronary angiography predicts improved outcome following cardiac arrest: propensity-adjusted analysis. J Intensive Care Med. 2009;24:179-86.

99. Lettieri C, Savonitto S, De Servi S, et al. Emergency percutaneous coronary intervention in patients with ST-elevation myocardial infarction complicated by out-of-hospital cardiac arrest: early and medium-term outcome. Am Heart J. 2009;157:569-75.
100. Garot P, Lefevre T, Eltchaninoff H, et al. Six-month outcome of emergency percutaneous coronary intervention in resuscitated patients after cardiac arrest complicating ST-elevation myocardial infarction. Circulation. 2007;115:1354-62.
101. Kern KB, Rahman O. Emergent percutaneous coronary intervention for resuscitated victims of out-of-hospital cardiac arrest. Catheter Cardiovasc Interv. 2010;75:616-24.
102. D'Souza SP, Marnas MA, Fraser DG, et al. Routine early coronary angioplasty versus ischemia-guided angioplasty after thrombolysis in acute ST-elevation myocardial infarction: a meta-analysis. Eur Heart J. 2010; online publication October 28, 2010.
103. Sanchez P, Fernandez-Aviles F. Routine early coronary angioplasty after thrombolysis in acute ST-elevation myocardial infarction: lysis is not the final step. Eur Heart J. 2010; online publication December 22, 2010.
104. The Hypothermia Cardiac Arrest Study Group. Mild therapeutic hypothermia to improve the neurologic outcome after cardiac arrest. NEJM. 2002;346:549-56.
105. Bernard SA, Gray TW, Buist MD, et al. Treatment of comatose survivors of out of hospital cardiac arrest with induced hypothermia. NEJM. 2002;346:557-63.
106. Arrich J, Holzer M, Herkner H, et al. Hypothermia for neuroprotection in adults after cardiopulmonary resuscitation. Conchrane Database Syst Rev. 2009;4:CD004128.
107. Lee R, Asare K. Therapeutic hypothermia for out-of-hospital cardiac arrest. Am J Health Syst Pharm. 2010;67:1229-37.
108. Cheung KW, Green RS, Magee KD. Systematic review of randomized controlled trials of therapeutic hypothermia as a neuroprotectant in post cardiac arrest patients. CJEM. 2006;8:329-37.
109. Boddicker KA, Zhang Y, Zimmerman B, et al. Hypothermia improves defibrillation success and resuscitation outcomes from ventricular fibrillation. Circulation. 2005;111:3195-201.
110. Staffey KS, Dendi R, Kerber RE, et al. Liquid ventilation with perfluorocarbons facilitates resumption of spontaneous circulation in a swine cardiac arrest model. Resuscitation. 2008;78:77-84.
111. Riter HG, Brooks LA, Kerber RE, et al. Intra-arrest hypothermia: Both cold liquid ventilation with perfluorocarbons and cold intravenous saline rapidly achieve hypothermia, but only cold liquid ventilation improves resumption of spontaneous circulation. Resuscitation. 2009;80:561-6.
112. Boller M, Lampe JW, Katz JM, et al. Feasibility of intra-arrest hypothermia induction: a novel nasopharyngeal approach achieves preferential brain cooling. Resuscitation. 2010;81:1025-30.
113. Busch HJ, Eichwede F, Fodisch M, et al. Safety and feasibility of nasopharyngeal evaporative cooling in the emergency department setting in survivors of cardiac arrest. Resuscitation. 2010;81:943-9.

# Risk Stratification for Sudden Cardiac Death

**CHAPTER 15**

*Dwayne N Campbell, James B Martins*

## Chapter Outline

- Healthy Athletes
- Brugada Syndrome
- Long QT Interval Syndrome
- Early Repolarization
- Short QT Syndrome
- Catecholamine Polymorphic Ventricular Tachycardia
- Wolff-Parkinson-White Syndrome
- Arrhythmogenic Right Ventricular Cardiomyopathy
- Hypertrophic Cardiomyopathy
- Marfan Syndrome
- Noncompaction
- Congenital Heart Disease
- Nonischemic Cardiomyopathy
- Coronary Artery Disease

## INTRODUCTION

Sudden death is overwhelmingly a cardiac etiology [sudden cardiac death (SCD)] defined as unexpected death occurring within one hour of symptoms.[1] The heart rhythm causing SCD is most frequently ventricular tachycardia (VT) or ventricular fibrillation (VF). Assessing risk is not trying to predict the future, but to plan for the possibility of this disaster with cost effective strategies; the most comprehensive discussion of individual assessment was published almost a decade ago and yet is still largely valid.[1] As Myerberg and Castellanos[2] have so aptly depicted this problem, the groups of patients where we think we know what to do to prevent SCD account for a small fraction of the total number; hence our need to continue looking to improve risk assessment. Our approach is to aid the clinician with the most recent scientifically based risk assessment to guide clinical judgment. The most frustrating aspects of this effort are that the first symptom of many entities may be SCD and that after a decade of reports little better risk assessment is available.

Many noninvasive as well as invasive procedures are available to help the clinician to evaluate the risk of SCD. A partial list includes signal averaged electrocardiogram (ECG), heart rate variability and turbulence, and T-wave alternans.[3] Unfortunately while these assessments have excellent basic background in theory, they do not pan out in studies of even the most common ischemic heart disease with congestive heart failure. However, we find studies revealing important simple clinical ways to help patients in specific categories. In general most SCDs involve patients with previous cardiac arrest or syncope or a family history of SCD.

## HEALTHY ATHLETES

We start with those most healthy athletes which would apriori be the least likely to succumb to SCD. Surprisingly there was a 2.5-fold increase in SCD in an Italian athletic population, compared to non-athletes.[4] Some high profile deaths in professional sports have lead international groups to make recommendations which disagree.[5] The Bethesda conference recommends careful physical and history to include family occurrence of syncope and SCD. Symptoms of palpitations, syncope or seizure, especially during exercise, require further work up to identify the cause. Parenthetically, simple orthostatic dizziness due to vasodilatation is very common in elite athletes and needs to be separated from other symptoms. However, the European Society of Cardiology Consensus recommends routine ECGs, based on a reduction in SCD in athletes in Italy, but which is now reduced to the frequency observed in most other countries including the US.[6,7] However, this reduction was an uncontrolled observation.[8] Clearly standard evaluations based on good scientific criteria must be developed to protect athletes, without being overly invasive. Here we hope to give evidence based data that can inform all parties in this endeavor including the patient (athlete) and family.

Athletes may by virtue of their training actually develop a unique set of cardiac findings, different from untrained persons of the same age, which need to be appreciated as normal for sport. These include asymptomatic slower heart rates and second degree AV block at rest, but normal rate and rhythm with exercise.[9] Although we are only beginning to understand that ECG and echocardiographic (ECHO) criteria may have to be altered in athletes such criteria change may depend on race.[10] A recent controlled study involving ECG screening in 510 athletes using ECHO as the standard increased specificity of screening from 5 to 10 of 11 documented ECHO abnormalities, but at the expense of increasing false positives from 5.5% to 16.9%. This study used standard ECG criteria published by European society of cardiology.[11] As expected such screening will significantly add to the cost of case finding.[12] One of six athletes would be expected to have an abnormal screening ECG, but only about 1% of all would have a cardiac abnormality capable of causing SCD. Annual cost estimates for ECG screening and follow-up exceeded $126 million. False-positive ECGs accounted for 98.8% of follow-up costs. Similar evaluation schemes have been used for soccer with similar outcomes.[13] While 4.8% soccer players had potentially abnormal ECGs, only 1% had clearly abnormal evaluations preventing participation in the 2006 world cup.

Recently an international group recommended an entity specific ECG screening approach that may deal with the risk-cost ratio.[14] They made recommended limits on the ECG changes attributable to training in athletes, including incomplete right bundle branch block (RBBB), differentiating it from Brugada and early repolarization (pre-SCD findings, see below), eliminating incomplete RBBB as a risk factor. On the contrary ST segment depression, right atrial disease, ventricular hypertrophy, bundle branch block, QRS greater than 110 millisecond, QTc greater than 500 and lesser than 380 millisecond should prompt a further work up since these findings are not physiological in athletes. Thus the ECG will identify inherited electrical as well as structural heart diseases, which will apply to screening anyone with a family history of SCD including athletes.

## BRUGADA SYNDROME

Electrocardiogram (ECG) screening has the potential for identification of electrical disorders which could lead to SCD, although there is not much evidence that athletes would be affected by these entities. Brugada syndrome is one such disorder;[15] it is an autosomal dominantly inherited disease producing a high-risk-associated ECG with a coved appearance of ST segment elevation (2 mm) in the right precordial leads; this pattern has (4.7-fold increase risk) predictive accuracy for SCD in meta analysis of 947 patients (Veltmann 09). However, in a patient with this ECG finding, syncope may predict SCD with a hazard ratio (HR) = 2.5–6.4 for SCD.[16] Males are 3.5 times at more risk than females and there is clustering of patients with the syndrome from South-east Asia. No other factors are reproducibly predictive including positive family history or abnormal EPS.[17]

## LONG QT INTERVAL SYNDROME

Long QT interval syndrome has been well reviewed in general population[18] as well as in athletes. New data collectively suggest that the magnitude of the QT prolongation is highly predictive.[16] The younger age, the shorter the QT of risk, so in the first decade risk occurs at QTc >500: HR = 2.12, the second decade QTc >530: HR = 2.3. In adults the diagnosis is made with a QTc >480, but the risk in adults is incrementally greater the larger the QTc: QTc 500–549: HR = 3.3 while QTc >550: HR = 6.4. In children males are at higher risk, while after puberty females are at higher risk.[16] Genotyping, to confirm ECG types, may be abnormal in 70% of cases; some subtypes of LQT1, LQT2 and LQT3 may have more risk than others.[16] LQT1 should not compete particularly in swimming. Similar to Brugada, the addition of syncope to the QT increases risk of SCD to 2.7–18 times those without syncope. A recent report suggests that a simple standing QT measurement will confirm a diagnosis of long QT especially in LQT2 patients.[19] Therapy with ICD is indicated if symptoms cannot be controlled with beta blocker therapy and or cervicothoracic sympathectomy.

ICD may be appropriate if there is a strong family history of SCD or inability to take beta blockers.[20]

## EARLY REPOLARIZATION

Early repolarization on the ECG has been implicated in the etiology of SCD although the mechanism is unclear.[21] However, the occurrence of this finding in normal athletes with an apparently good prognosis make it a rather difficult risk assessment tool except perhaps when it is localized in the inferior leads in middle aged subjects.[22] Recently it is suggested that early repolarization localized to the lateral ECG leads is benign associated with athletic training, while localization to inferior leads is intermediate and localization to inferior, lateral and right precordial ECG is more high-risk for SCD.[23]

## SHORT QT SYNDROME

Short QT syndrome[16] known only since 2000 has little patient data from which to guide assessment. Certainly a patient with QTc less than 350 and resuscitated SCD should have a defibrillator, but even syncope is not predictive of SCD unless in the presence of a markedly positive family history.

## CATECHOLAMINE POLYMORPHIC VENTRICULAR TACHYCARDIA

Catecholamine polymorphic VT[16] is a well studied and strikingly reproducible syndrome of exercise facilitated VT with multiple morphologies associated with mutations in ryanodine (autosomal dominant) or calsequestrin (autosomal recessive). Exercise induced bidirectional ventricular ectopy and bidirectional VT makes the diagnosis and therapy with beta adrenergic blockers is indicated with ICD implantation if VT cannot be suppressed. Left upper thoracic chain sympathectomy may prevent ICD shocks. Genetic testing showing mutations in a family member are at risk of syncope and SCD, which may be prevented by beta blockers, has been shown to significantly reduce them. Flecainide may also be protective of VT. Family history of the syndrome does not predict SCD.

## WOLFF-PARKINSON-WHITE SYNDROME

Pre-excitation of the ventricle by non-atrioventricular nodal structures when producing symptoms is called Wolff-Parkinson-White (WPW) syndrome. WPW is a well known entity found in 1% of individuals recently reviewed. It is clear that patients with rapid tachycardia symptoms need therapeutic ablation. However, the recommendation for ablation in order for asymptomatic persons with pre-excitation alone to participate in competitive athletics seems inappropriately invasive when the risk of pre-excitation alone is unclear. Prospective studies have not identified the presence of asymptomatic pre-excitation as a risk[24,25], although in children multiple pathways and rapidly conducting pathways (causing deterioration of supraventricular tachycardia to VT and VF) may suggest danger on follow-up.[26] Even electrophysiological study (EPS) identifying the functional characteristics of rapidly conducting pathways do not predict the occurrence of SCD due to rarity of SCD (0.1%) in asymptomatic pre-excitation. Therefore, asymptomatic pre-excitation does not warrant EPS and ablation. Expert opinions on certain sports, such as sky-diving or occupations such as airline piloting suggest EPS and ablation for rapidly conducting pathways.

Electrical syndromes, however, do not account for most of sudden deaths in young people. In Italy, the top three causes of SCD in athletes were arrhythmogenic right ventricular cardiomyopathy (ARVC) (22%), coronary atherosclerosis (18%) and anomalous origin of a coronary artery (12%). In the US predominant structural cardiac abnormalities identified in the military population were coronary artery abnormalities (61%), myocarditis (20%) and hypertrophic cardiomyopathy (HCM) (13%). An anomalous coronary artery accounted for one-third (21 of 64 recruits) of the cases in this cohort, and, in each, the left coronary artery arose from the right sinus of Valsalva, coursing between the pulmonary artery and the aorta. So the populations are different. Recommendations include attention to young persons who complain of angina.

## ARRHYTHMOGENIC RIGHT VENTRICULAR CARDIOMYOPATHY

Arrhythmogenic right ventricular cardiomyopathy (ARVC) is an autosomal dominantly inherited disease involving molecular regulation of basic cell to cell adhesion, and production of fat, fibrous tissue and apoptosis, which results in VT and VF.[27] A specific list including major and minor criteria has been published including fibrofatty replacement of the RV, ECG depolarization (epsilon wave)/repolarization changes (with T-wave inversions in the right precordium), VT with a LBBBM and atrial fibrillation, and family history.[28,29] Of interest is the fact that the disease can be quiescent for years in youth but progresses to symptomatic disease with time. The highest risk occurs in patients having been resuscitated from SCD, with syncope at a young age or extreme involvement of the RV and LV. The primary prevention management of patients and their families is complex because of variable penetrance.[30] It was hoped that genetic testing would be an effective way to stratify risk.[31] Although eight causative genes have been identified, up to 50% of cases do not have genetic markers. Thus good clinical judgment is necessary since risk factors may be hidden, but progressive.[32,33] Since competitive sports may provoke VT/VF, exercise must be curtailed and beta blockers and sotalol may effectively prevent VT.

## HYPERTROPHIC CARDIOMYOPATHY

Hypertrophic cardiomyopathy (HCM) is the most common cause of SCD in people in the US below the age of 25 years, particularly in athletes, with incidence of about 1%.[34] Clearly interdiction of competitive sports for such a patient would make the best sense. However, even though the arrhythmogenic substrate of myocardial disarray is clear, our ability to predict SCD is flawed due to apparent dormancy despite the presence of substrate. After years of attempts to prevent SCD with drugs although improving symptoms, it is concluded that this approach has never proved effective. Clearly a survivor of SCD can be treated with an ICD, but indications for primary prevention are not clear, particularly since 60% of ICD treated patient do not have VT/VF in 5 years after resuscitation from SCD.[29,35] Risk factors commonly cited have not been well proven until recently, particularly because the evidence was taken from "appropriate" ICD shocks, which are not the same as SCD (a VT which is shocked is only seconds long and may (had no ICD intervention taken place) have only been non-sustained and not cause syncope or SCD.[36] Syncope has clearly been shown on follow-up of untreated patients to predict SCD.[37] Combinations of studies taken together have recently suggested that the aforementioned risk stratification data may be true[38] including VT nonsustained: HR = 2.2–3.6 (95% confidence); syncope: HR = 0.97–4.4; extreme LVH (wall thickness >3 cm): HR = 1.8–4.4; hypotension on exercise HR = 0.6–2.0 and family history of SCD = 1.2–1.4. Also combinations of risk factors produced a higher HR than individual ones. Aggravating factors including atrial fibrillation, ischemia, genotype and exercise do not show risk, but LVOT gradient may increase risk.

## MARFAN SYNDROME

Marfan syndrome is a connective tissue disorder caused by mutations in genes encoding supporting scaffold for elastin. Its prevalence is between 1 in 5,000 or 10,000. It causes SCD, produced by aortic dissection, rupture and pericardial tamponade.[39] Risk stratification for this outcome is based on ECHO measurement of aortic root greater than 50–55 mm.[40] SCD without dissection are reported and ventricular arrhythmias are thought to be the cause;[41] mitral valve prolapse is a common component of Marfan syndrome but LV dilatation is the associated finding suggesting risk in Marfan patients with ventricular ectopy and VT. Unfortunately LV dilatation is not commonly found in sporadic cases of SCD with only MVP studied at autopsy.[42] However, SD can occur without cardiac cause due to the elongated odontoid causing pressure on the cerebellum and medulla owing to atlantoaxial hypermobility.[43]

## NONCOMPACTION

Noncompaction of the left ventricle is a potentially arrhythmogenic condition diagnosable with ECHO which is presently not well understood. It is a cause of SCD which may be appropriately treated with ICD but follow-up shocks occur as frequently in primary as secondary prevention cases,[44] EPS does not predict ICD therapy for VT/VF. It is not clear which patients in absence of resuscitated VT/VF would benefit from ICD implantation.[20] There is simply no information to risk stratify this group at this time, since many patients with the disorder have a benign course.[45]

## CONGENITAL HEART DISEASE

Congenital heart disease (CHD) afflicts approximately 75 of 1,000 live births. Significant advances in the treatment of CHD over past 50 years have allowed the majority of afflicted children to reach adulthood. The number of adults with CHD now exceeds that of children and is expected to increase with further advances.[46] SCD is the most common cause of death in these patients, occurring usually by the third or forth decade of life. The incidence, estimated at 0.09% per year, represents up to a 100-fold increased risk compared to age matched controls.[47] SCD is especially likely in patients with repaired cyanotic and left heart obstructive lesions. In a recent report of over 8,000 patients with CHD, the majority of SCD occurred out of hospital (62%), at rest (only 7% SCD occurred during exercise), and demonstrated seasonal variation with the nadir occurring in summer (22%) and peak in the fall (33%).[48] Predictors of mortality in patients with CHD include New York Heart Association (NYHA) functional class greater than one, cyanosis, age (postsurgical repair) and complexity of malformations.[49,50] There are virtually no prospective or randomized clinical trials of risk stratification. Patients with CHD presenting with cardiac arrest, sustained symptomatic VT or syncope and have significant systemic ventricular dysfunction are stratified as high-risk and ICD therapy is generally recommended.[20] Three special groups are associated with the highest risk of SCD including patients treated with Mustard/Senning procedures, Fontan procedures and repaired tetralogy of Fallot (TOF).[51]

In the latter incidence of SCD is approximately 0.15% per year, risk factors include QRS greater than 180 millisecond, older age at repair, transannular right ventricular outflow tract patch, left ventricular dysfunction, frequent ectopy and inducible sustained VT, with poor specificity. Although reduced left ventricular (EF <35%) function is the strongest risk factor for SCD in ischemic heart disease (see below), there is debate as to whether such findings should be extended to patients with CHD.[52,53] Systemic ventricular dysfunction (such as in corrected transposition) has been demonstrated in numerous observational studies and registries to identify CHD patients at risk for SCD. Studies evaluating the efficacy of ICD therapy in CHD patients have usually been observational and retrospective.[54,55] Beyond secondary prevention scenarios risk stratification in the CHD population must be done on an individual basis combining available diagnostic data with sound clinical judgment and weighing the risk and benefits of a particular intervention.[56]

## NONISCHEMIC CARDIOMYOPATHY

Nonischemic cardiomyopathy (NICM) is the primary etiology in 10–15% of SCDs and accounts for the second largest number of SCDs from cardiac causes behind coronary artery disease (CAD).[57] NICM is characterized by biventricular dilatation and impaired ventricular contractility without CAD. Mortality rates in this patient population range from 12–13% at three years to 20% at five years.[29,58,59] Like ischemic heart disease, numerous diagnostic techniques exist to risk stratify without careful data to identify high-risk patients would benefit from existing interventions (ICDs); in fact the majority of the major primary prevention trials enrolling patients with NICM failed to demonstrate definitive benefit to ICD therapy.[59] The primary risk stratifying approach utilized was a combination of quantitative left ventricular function assessment and functional status based on NYHA functional class. Cardiomyopathy Trial (CAT), Amiodarone versus Implantable Cardioverter-defibrillator Study (AMIOVIRT), Defibrillators in Non-Ischemic Cardiomyopathy Treatment Evaluation (DEFINITE) Trial and SCD in Heart Failure Trials (SCD-HEFT) failed to demonstrate significant mortality benefit of ICDs in patients with NICM.[59] The Comparison of Medical, Therapy, Pacing and Defibrillation in Heart Failure (COMPANION) did demonstrate benefit (significantly lower risk of death from any cause compared to medical therapy HR = 0.5 P = 0.015 but no significant benefit in the primary end point of the study which was a composite of death from any cause or hospitalization for any cause) of ICDs in patients with NICM treated with biventricular pacing.[60] However, the Cardiac resynchronization-Heart Failure (Care-HF) trial which enrolled 813 patients, a majority with NICM,

EF <35%, NYHA class III to IV, QRS >149 or QRS 120–149 with evidence of dyssynchrony did demonstrate a significant benefit of CRT alone in decreasing mortality (36% p <0.003) to the same extent as in the CRT-D arm of the COMPANION trial (36% p <0.003).[61] Interestingly in Care-HF the presence of NICM predicted a better outcome. However, 36% of the deaths in the pacing only arm of the COMPANION trial were attributed to SCD similar to the 35% in the CRT arm CARE-HF trial; these deaths might have been prevented with an ICD.[62] A meta-analysis of the pooled data from the ICD trials (1854 patients) demonstrated 31% reduction in all cause mortality with ICD therapy compared to medical therapy.[63] Thus current guidelines recommend placement of ICDs in patients with NICM, EF less than 35% and NYHA Class II–III symptoms.[20] A number of noninvasive diagnostic methods for risk stratifying patients with NICM have recently been reviewed[64,65] with the exception of syncope, EF and NYHA functional class, no significant data exists for other methods of identifying patients that would benefit from available therapies.[65,66]

## CORONARY ARTERY DISEASE

Coronary artery disease (CAD) accounts for (or is the underlying condition in) 65–80% of patients presenting with SCD. Depending on age group 13–50% of CAD deaths are SCD with coronary occlusion the most common.[57] In general the incidence of SCD in a population depends on the incidence of CAD.[57] In recent years, the steady decrease in mortality due to CAD has correlated with the decrease in SCD, although the prevalence of CAD has increased.[67] Medical therapy directed at treating CAD, in particular beta blockers and renin angiotensin system modifiers (ACE inhibitors ARBs, Aldosterone antagonist) have been demonstrated to decrease the incidence of SCD.[68] Traditional risk factors of CAD (HTN, DM, smoking, hypercholesterolemia) identify patients at risk for ischemic heart disease and hence SCD (with obesity, DM and smoking showing an increased proportion of deaths that are sudden).[29] However, these factors do not discriminate among CAD patients, those at high-risk.[69]

Risk stratification in patients immediately postmyocardial infarction is particularly difficult, especially in patients with impaired ventricular function.[29] These patients may have a particularly high-risk of SCD despite optimal medical therapy. To date no risk stratification strategy exists that utilizes invasive or noninvasive diagnostic testing that can identify patients that would benefit from advanced therapies (ICDs) within the first 40 days of infarction.[70] Both the Defibrillator in Acute Myocardial Infarction Trial (DINAMIT) and the Immediate

Risk-Stratification Improves Survival (IRIS) Trial failed to demonstrated any overall mortality benefit with ICD use.[70] Similarly in a substudy of the MADIT 2 Trial, no survival benefit was found in this population if the time interval from index infarct was within 18 months.[71] In the home use of external automated defibrillators for SCD trial (HAT) no mortality benefit was noted over conventional resuscitation methods in high-risk post-MI patients.[72] Current ICD guidelines reflect these results.[20]

The previously mentioned factors stratifying patients at high risk with NICM also identify CAD patients. So, NYHA functional class higher than 2, EF less than 35 and syncope identify patients who benefit from ICDs.[20] Unlike NICM, there may be utility of invasive electrophysiologic testing in patients with CAD. In the MUSTT and MADIT clinical trials sustained VT/VF during EPS, in addition to LV dysfunction (EF $\leq 40\%$) and presence of NSVT on ambulatory testing, identified patients that benefited from prophylactic ICD therapy.[20] In addition, invasive EP testing is recommended in patients with remote history of myocardial infarction and symptoms suggestive of VT, including palpitations syncope and presyncope.[29] If sustained VT/VF is induced then ICD therapy is usually recommended.[20] Although a positive EPS identifies patients with CAD at high-risk of SCD who benefit from advanced therapies a negative study in patients with severe LV dysfunction less than 30% does not necessarily indicate a good prognosis.[29,73]

Although LV function can identify patients at high-risk of sudden death it does not discriminate well between those at high-risk and those at low-risk of sudden death.[64] The risk of sudden death or cardiac arrest increased by 21% for every 5% decrease in left ventricular function. However, this is a U shaped relationship, in that as EF decreases deaths due to pump failure increase as arrhythmic deaths decrease.[64,74] Recently risk stratification test studied in the Cardiac Arrhythmias and Risk Stratification after Acute Myocardial Infarction (CARISMA) study and the Alternans Before Cardioverter Defibrillator (ABCD) trial were compared to a "coin toss" high-risk (heads) and low-risk (tails).[75] Compared to LVEF (NPV 94% PPV 9% HR 1.3) in CARISMA study population the coin toss performed only mildly worst (NPV 92% PPV 8%).[75,76] It was noted that the coin toss (HR 1 in both populations) has no role in risk stratification as net correct reclassification would always be zero. This comparison demonstrates the limitation of individual risk stratification test for SCD to adequately stratify patients. An alternative approach is a multi-tiered risk stratification strategy similar to that utilized by the SHAPE task force for atherosclerosis.[77,78] To date there have been no prospective studies utilizing such an approach.

## SUMMARY

We have reviewed the available data on risk stratification in major entities encountered by clinicians. Generally good clinical judgment can be enhanced by the published knowledge; in addition to a history of SCD, symptoms of syncope or arrhythmogenic dizziness predict likely risk of SCD in most disorders. A family history of SCD may also inform such an evaluation. An ECG abnormality coupled with the above may focus additional evaluation such as exercise testing in appropriate patients. Documentation of structural heart disease by ECHO confirms risk in various population groups. Normal ECG and ECHO may exclude risk in many groups. Further research is needed to clarify the specific risk in more rare diseases. Unfortunately based on our experience with CAD, it is not likely we will find easy risk stratifiers in many disease states.

## REFERENCES

1. Priori SG, Aliot E, Blomstrom-Lundqvist C, et al. Task force on sudden cardiac death of the european society of cardiology. Eur Heart J. 2001;22:1374-450.
2. Myerberg RJ, Castellanos A. Sudden cardiac death In: Zipes DP, Jalife J (Eds). Cardiac Electrophysiology from Cell to Bedside, 5th edition. Philadelphia: Saunders, Elsevier; 2009.
3. Zipes DP, Jalife J (Editors). Cardiac Electrophysiology from Cell to Bbedside, 5th edition. Philadelphia: Saunders, Elsevier; 2009.
4. Corrado D, Basso C, Rizzoli G, et al. Does sports activity enhance the risk of sudden death in adolescents and young adults? J Am Coll Cardiol. 2003;42:1959-63.
5. Pelliccia A, Zipes DP, Maron BJ. Bethesda conference #36 and the European Society of Cardiology consensus recommendations revisited a comparison of US and European criteria for eligibility and disqualification of competitive athletes with cardiovascular abnormalities. J Am Coll Cardiol. 2008;52:1990-6.
6. Maron BJ, Doerer JJ, Haas TS, et al. Sudden deaths in young competitive athletes: analysis of 1866 deaths in the United States, 1980-2006. Circulation. 2009;119:1085-92.
7. Perez M, Fonda H, Le VV, et al. Adding an electrocardiogram to the pre-participation examination in competitive athletes: a systematic review. Curr Probl Cardiol. 2009;34:586-662.
8. Maron BJ, Haas TS, Doerer JJ, et al. Comparison of US and Italian experiences with sudden cardiac deaths in young competitive athletes and implications for preparticipation screening strategies. Am J Cardiol. 2009;104:276-80.
9. Link MS, Mark Estes NA. Athletes and arrhythmias. J Cardiovasc Electrophysiol; 2010.
10. Rawlins J, Carre F, Kervio G, et al. Ethnic differences in physiological cardiac adaptation to intense physical exercise in highly trained female athletes. Circulation. 2010;121:1078-85.
11. Baggish AL, Hutter AM Jr, Wang F, et al. Cardiovascular screening in college athletes with and without electrocardiography: a cross-sectional study. Ann Intern Med. 2010;152:269-75.

12. O'Connor DP, Knoblauch MA. Electrocardiogram testing during athletic preparticipation physical examinations. J Athl Train. 2010;45:265-72.
13. Thunenkotter T, Schmied C, Dvorak J, et al. Benefits and limitations of cardiovascular pre-competition screening in international football. Clin Res Cardiol. 2010;99:29-35.
14. Corrado D, Pelliccia A, Heidbuchel H, et al. Section of Sports Cardiology, European Association of Cardiovascular Prevention and Rehabilitation, Working Group of Myocardial and Pericardial Disease, European Society of Cardiology. Recommendations for interpretation of 12-lead electrocardiogram in the athlete. Eur Heart J. 2010;31:243-59.
15. Brussey T, Brugada R, Brugada J, et al. The brugada syndrome. In: DP Zipes, J Jalife (Eds). Cardiac Electrophysiology from Cell to Bedside, 5th edition. Philadelphia: Saunders; 2009.
16. Veltmann C, Schimpf R, Borggrefe M, et al. Risk stratification in electrical cardiomyopathies. Herz. 2009;34:518-27.
17. Probst V, Veltmann C, Eckardt L, et al. Long-term prognosis of patients diagnosed with brugada syndrome: results from the FINGER Brugada syndrome registry. Circulation. 2010;121:635-43.
18. Schwartz P, Crotti L. Long QT and short QT syndromes. In: DP Zipes, J Jalife (Eds). Cardiac Electrophysiology from Cell to Bedside, 5th edition. Philadelphia: Saunders, Elsevier; 2009.
19. Viskin S, Postema PG, Bhuiyan ZA, et al. The response of the QT interval to the brief tachycardia provoked by standing: a bedside test for diagnosing long QT syndrome. J Am Coll Cardiol. 2010;55:1955-61.
20. Epstein AE, DiMarco JP, Ellenbogen KA, et al. ACC/AHA/HRS 2008 guidelines for device-based therapy of cardiac rhythm abnormalities: a report of the American College of Cardiology/American Heart Association Task Force on Practice Guidelines (writing committee to revise the ACC/AHA/NASPE 2002 guideline update for implantation of cardiac pacemakers and antiarrhythmia devices) developed in collaboration with the american association for thoracic surgery and society of thoracic surgeons. J Am Coll Cardiol. 2008;51:e1-e62.
21. Haissaguerre M, Derval N, Sacher F, et al. Sudden cardiac arrest associated with early repolarization. N Engl J Med. 2008;358:2016-23.
22. Tikkanen JT, Anttonen O, Junttila MJ, et al. Long-term outcome associated with early repolarization on electrocardiography. N Engl J Med. 2009;361:2529-37.
23. Antzelevitch C, Yan GX. J wave syndromes. Heart Rhythm. 2010;7:549-58.
24. Tischenko A, Fox DJ, Yee R, et al. When should we recommend catheter ablation for patients with the Wolff-Parkinson-White syndrome? Curr Opin Cardiol. 2008;23:32-7.
25. Santinelli V, Radinovic A, Manguso F, et al. Asymptomatic ventricular preexcitation: a long-term prospective follow-up study of 293 adult patients. Circ Arrhythm Electrophysiol. 2009;2:102-7.
26. Santinelli V, Radinovic A, Manguso F, et al. The natural history of asymptomatic ventricular pre-excitation a long-term prospective follow-up study of 184 asymptomatic children. J Am Coll Cardiol. 2009;53:275-80.
27. Fontaine G, Charron P. Arrhythmogenic right ventricular cardiomyopathies. In: DP Zipes, J Jalife (Eds). Cardiac Electrophysiology from Cell to Bedside, 5th edition. Philadelphia: Saunders, Elsevier; 2009.

28. Muthappan P, Calkins H. Arrhythmogenic right ventricular dysplasia. Prog Cardiovasc Dis. 2008;51:31-43.
29. European Heart Rhythm Association, Heart Rhythm Society, Zipes DP, et al. ACC/AHA/ESC 2006 guidelines for management of patients with ventricular arrhythmias and the prevention of sudden cardiac death: a report of the American College of Cardiology/ American Heart Association Task Force and the European Society of Cardiology Committee for Practice Guidelines (Writing Committee to Develop Guidelines for Management of Patients with Ventricular Arrhythmias and the Prevention of Sudden Cardiac Death). J Am Coll Cardiol. 2006;48:e247-e346.
30. Sen-Chowdhry S, Morgan RD, Chambers JC, et al. Arrhythmogenic cardiomyopathy: etiology, diagnosis, and treatment. Annu Rev Med. 2010;61:233-53.
31. Hershberger RE, Cowan J, Morales A, et al. Progress with genetic cardiomyopathies: screening, counseling, and testing in dilated, hypertrophic, and arrhythmogenic right ventricular dysplasia/ cardiomyopathy. Circ Heart Fail. 2009;2:253-61.
32. Boldt LH, Haverkamp W. Arrhythmogenic right ventricular cardiomyopathy: diagnosis and risk stratification. Herz. 2009;34:290-7.
33. Basso C, Corrado D, Marcus FI, et al. Arrhythmogenic right ventricular cardiomyopathy. Lancet. 2009;373:1289-300.
34. Maron BJ. Contemporary insights and strategies for risk stratification and prevention of sudden death in hypertrophic cardiomyopathy. Circulation. 2010;121:445-56.
35. Maron BJ, Spirito P, Shen WK, et al. Implantable cardioverter-defibrillators and prevention of sudden cardiac death in hypertrophic cardiomyopathy. JAMA. 2007;298:405-12.
36. Ellenbogen KA, Levine JH, Berger RD, et al. Are implantable cardioverter defibrillator shocks a surrogate for sudden cardiac death in patients with nonischemic cardiomyopathy? Circulation. 2006;113:776-82.
37. Spirito P, Autore C, Rapezzi C, et al. Syncope and risk of sudden death in hypertrophic cardiomyopathy. Circulation. 2009;119: 1703-10.
38. Christiaans I, van Engelen K, van Langen IM, et al. Risk stratification for sudden cardiac death in hypertrophic cardiomyopathy: systematic review of clinical risk markers. Europace. 2010;12:313-21.
39. Pearson GD, Devereux R, Loeys B, et al. Report of the National Heart, Lung, and Blood Institute and National Marfan Foundation Working Group on research in Marfan syndrome and related disorders. Circulation. 2008;118:785-91.
40. Stout M. The Marfan syndrome: implications for athletes and their echocardiographic assessment. Echocardiography. 2009;26:1075-81.
41. Yetman AT, Bornemeier RA, McCrindle BW. Long-term outcome in patients with marfan syndrome: is aortic dissection the only cause of sudden death? J Am Coll Cardiol. 2003;41:329-32.
42. Anders S, Said S, Schulz F, et al. Mitral valve prolapse syndrome as cause of sudden death in young adults. Forensic Sci Int. 2007;171:127-30.
43. MacKenzie JM, Rankin R. Sudden death due to atlantoaxial subluxation in Marfan syndrome. Am J Forensic Med Pathol. 2003;24:369-70.
44. Kobza R, Steffel J, Erne P, et al. Implantable cardioverter-defibrillator and cardiac resynchronization therapy in patients with left ventricular noncompaction. Heart Rhythm. 2010.

45. Lofiego C, Biagini E, Pasquale F, et al. Wide spectrum of presentation and variable outcomes of isolated left ventricular non-compaction. Heart. 2007;93:65-71.
46. Khairy P. EP challenges in adult congenital heart disease. Heart Rhythm. 2008;5:1464-72.
47. Yap SC, Harris L. Sudden cardiac death in adults with congenital heart disease. Expert Rev Cardiovasc Ther. 2009;7:1605-20.
48. Zomer AC, Uiterwaal CSPM, Velde ETvd, et al. Circumstances of death in adult congenital heart disease. J Am Coll Cardiol. 2010;55:A41.E393.
49. Trojnarska O, Grajek S, Katarzynski S, et al. Predictors of mortality in adult patients with congenital heart disease. Cardiol J. 2009; 16:341-7.
50. Oechslin EN, Harrison DA, Connelly MS, et al. Mode of death in adults with congenital heart disease. Am J Cardiol. 2000;86:1111-6.
51. Triedman JK. Arrhythmias in adults with congenital heart disease. Heart. 2002;87:383-9.
52. Silka MJ, Bar-Cohen Y. Should patients with congenital heart disease and a systemic ventricular ejection fraction less than 30% undergo prophylactic implantation of an ICD? Patients with congenital heart disease and a systemic ventricular ejection fraction less than 30% should undergo prophylactic implantation of an implantable cardioverter defibrillator. Circ Arrhythm Electrophysiol. 2008;1:298-306.
53. Triedman JK. Should patients with congenital heart disease and a systemic ventricular ejection fraction less than 30% undergo prophylactic implantation of an ICD? Implantable cardioverter defibrillator implantation guidelines based solely on left ventricular ejection fraction do not apply to adults with congenital heart disease. Circ Arrhythm Electrophysiol. 2008;1:307-16; discussion 316.
54. Khairy P, Harris L, Landzberg MJ, et al. Implantable cardioverter-defibrillators in tetralogy of fallot. Circulation. 2008;117:363-70.
55. Khairy P, Harris L, Landzberg MJ, et al. Sudden death and defibrillators in transposition of the great arteries with intra-atrial baffles: a multicenter study. Circ Arrhythm Electrophysiol. 2008;1:250-7.
56. Walsh EP. Practical aspects of implantable defibrillator therapy in patients with congenital heart disease. Pacing Clin Electrophysiol. 2008;31:S38-40.
57. Lee KK, Al-Ahmad A, Wang PJ, et al. Epidemiology and etiologies of sudden cardiac death. In: PJ Wang, A Al-Ahmad, HH Hsia, PC Zei (Eds). Ventricular arrhythmias and sudden cardiac death. Oxford, UK: Blackwell Futura; 2009.
58. Dec GW, Fuster V. Idiopathic dilated cardiomyopathy. N Engl J Med. 1994;331:1564-75.
59. Cevik C, Nugent K, Perez-Verdia A, et al. Prophylactic implantation of cardioverter defibrillators in idiopathic nonischemic cardiomyopathy for the primary prevention of death: a narrative review. Clin Cardiol. 2010;33:254-60.
60. Bristow MR, Saxon LA, Boehmer J, et al. Cardiac-resynchronization therapy with or without an implantable defibrillator in advanced chronic heart failure. N Engl J Med. 2004;350:2140-50.
61. Cleland JG, Daubert JC, Erdmann E, et al. The effect of cardiac resynchronization on morbidity and mortality in heart failure. N Engl J Med. 2005;352:1539-49.
62. Ellenbogen KA, Wood MA, Klein HU. Why should we care about CARE-HF? J Am Coll Cardiol. 2005;46:2199-203.

63. Desai AS, Fang JC, Maisel WH, et al. Implantable defibrillators for the prevention of mortality in patients with nonischemic cardiomyopathy: a meta-analysis of randomized controlled trials. JAMA. 2004;292:2874-9.
64. Goldberger JJ, Cain ME, Hohnloser SH, et al. American Heart Association/American College of Cardiology Foundation/Heart Rhythm Society scientific statement on noninvasive risk stratification techniques for identifying patients at risk for sudden cardiac death: a scientific statement from the American Heart Association Council on Clinical Cardiology Committee on Electrocardiography and Arrhythmias and Council on Epidemiology and Prevention. Circulation. 2008;118:1497-518.
65. Okutucu S, Oto A. Risk stratification in nonischemic dilated cardiomyopathy: current perspectives. Cardiol J. 2010;17:219-29.
66. Grimm W, Christ M, Sharkova J, et al. Arrhythmia risk prediction in idiopathic dilated cardiomyopathy based on heart rate variability and baroreflex sensitivity. Pacing Clin Electrophysiol. 2005;28:S202-6.
67. Chugh SS, Reinier K, Teodorescu C, et al. Epidemiology of sudden cardiac death: clinical and research implications. Prog Cardiovasc Dis. 2008;51:213-28.
68. Das MK, Zipes DP. Antiarrhythmic and nonantiarrhythmic drugs for sudden cardiac death prevention. J Cardiovasc Pharmacol. 2010;55:438-49.
69. El-Sherif N, Khan A, Savarese J, et al. Pathophysiology, risk stratification, and management of sudden cardiac death in coronary artery disease. Cardiol J. 2010;17:4-10.
70. Estes NA, 3rd. The challenge of predicting and preventing sudden cardiac death immediately after myocardial infarction. Circulation. 2009;120:185-7.
71. Wilber DJ, Zareba W, Hall WJ, et al. Time dependence of mortality risk and defibrillator benefit after myocardial infarction. Circulation. 2004;109:1082-4.
72. Bardy GH, Lee KL, Mark DB, et al. Home use of automated external defibrillators for sudden cardiac arrest. N Engl J Med. 2008;358: 1793-804.
73. Lopera G, Curtis AB. Risk stratification for sudden cardiac death: current approaches and predictive value. Curr Cardiol Rev. 2009;5: 56-64.
74. Solomon SD, Zelenkofske S, McMurray JJ, et al. Sudden death in patients with myocardial infarction and left ventricular dysfunction, heart failure, or both. N Engl J Med. 2005;352:2581-8.
75. Goldberger JJ. The coin toss: implications for risk stratification for sudden cardiac death. Am Heart J. 2010;160:3-7.
76. Huikuri HV, Raatikainen MJ, Moerch-Joergensen R, et al. Prediction of fatal or near-fatal cardiac arrhythmia events in patients with depressed left ventricular function after an acute myocardial infarction. Eur Heart J. 2009;30:689-98.
77. Naghavi M, Falk E, Hecht HS, et al. From vulnerable plaque to vulnerable patient—part III: executive summary of the screening for heart attack prevention and education (SHAPE) task force report. Am J Cardiol. 2006;98:2H-15H.
78. Bailey JJ, Berson AS, Handelsman H, et al. Utility of current risk stratification tests for predicting major arrhythmic events after myocardial infarction. J Am Coll Cardiol. 2001;38:1902-11.

# Cardiocerebral Resuscitation for Primary Cardiac Arrest

**CHAPTER 16**

*Jooby John, Gordon A Ewy*

## Chapter Outline

- Etiology and Pathophysiology of Cardiac Arrest
  - Primary Cardiac Arrest in Children and Adolescents
  - Pathophysiology of Primary Cardiac Arrest
  - Coronary Perfusion Pressures during Resuscitation Efforts
  - Assisted Ventilation in Primary Cardiac Arrest
  - Not Following Guidelines for Primary Cardiac Arrest
  - The Public Has Made Up Its Mind
  - Increasing the Prevalence of Bystander Resuscitation Efforts
  - Increasing the Ability to Promptly Identify Primary Cardiac Arrest
  - The Three Phases of Ventricular Fibrillation (VF)
  - Cardiocerebral Resuscitation: Prehospital Component
- Drug Therapy in Cardiac Resuscitation
- Cardiac Resuscitation Centers
  - Therapeutic Mild Hypothermia
  - Myocardial Ischemia Causing Cardiac Arrest
- Ending Resuscitative Efforts

## INTRODUCTION

Out-of hospital cardiac arrest (OHCA) claims hundreds of thousands of lives each year.[1,2] Despite this enormous public health problem, and the promulgation of standards and guidelines, with also numerous updates of guidelines **(Fig. 1)**, the aggregate survival rate of OHCA of 7.6% has not significantly changed in almost three decades.[3] Likewise there is little data to indicate that guidelines have improved in-hospital survival. A large medicare database of over 400,000 elderly patients revealed that even in-hospital survival after cardiopulmonary resuscitation (CPR) has been unchanged for the period from 1992 to 2005.[4]

**FIGURE 1:** Reported survival rates of out-of-hospital cardiac arrests 7.6% unchanged over the past 30 years

The major limitation of the chain of survival **(Fig. 2)** was the necessity for "early" initiation of each link.[5] However, in retrospect its failure to recognize that gasping was common during the first few minutes of primary cardiac arrest often prevented "early" recognition, its insistence on mouth-to-mouth ventilation as the initial intervention precluded most bystanders from providing "early" CPR, and many factors, including traffic congestion in larger cities, precluded "early" defibrillation. In addition, the prescription of endotracheal intubation as the initial step in advanced cardiac lift support (ACLS), and the use of automated external defibrillators for the proscribed "stacked" shocks further delayed or interrupted essential chest compressions; all contributed to poor survival rates.[6]

Survival was especially poor in large metropolitan cities. A 2005 report revealed a rate of neurologically intact survival from OHCA of 1.4% in Los Angeles, similar to survival rates previously reported from Chicago, New York City and Detroit.[7,8] Survival is better in areas where the incidence of bystander CPR is high and the emergency medical system (EMS) response times are short.[9] Unfortunately these "links" are rare, so new approaches were needed.

In this chapter we have discussed recent insights into the physiology of cardiac arrest and resuscitation and present a new approach to the management of cardiac arrest developed by our University of Arizona Sarver Heart Center Resuscitation Research Group, called Cardiocerebral Resuscitation (CCR) **(Fig. 3)**. The CCR has been shown to markedly improve survival of patients with OHCA with survival rates of 38% or better in the subset of patients who have greatest chance of

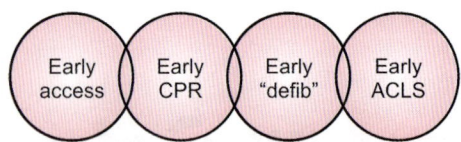

**FIGURE 2:** Cardiopulmonary resuscitation "chain of survival" (*Source:* Modified from Cummings, et al. Circulation. 1991;83(5):1832)

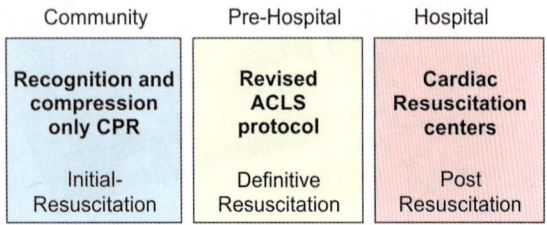

**FIGURE 3:** Cardiocerebral resuscitation "The New CPR" for primary cardiac arrest

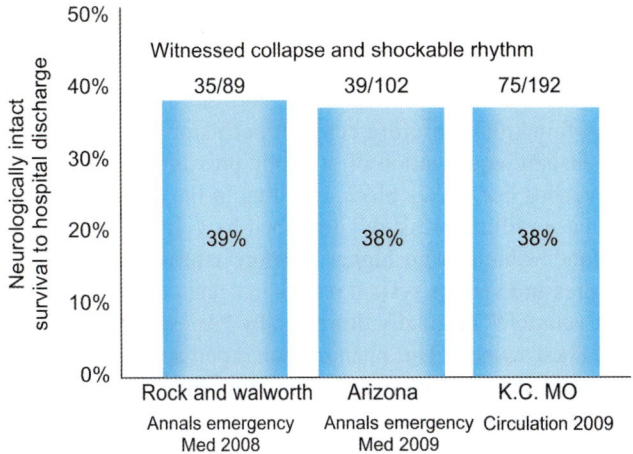

**FIGURE 4:** Cardiocerebral resuscitation for OHCA. Witnessed collapse and shockable rhythm

survival; those with witnessed arrest and a shockable rhythm **(Fig. 4)**.[10-12]

## ETIOLOGY AND PATHOPHYSIOLOGY OF CARDIAC ARREST

Cardiac arrest is either primary or secondary. The emphasis of this chapter is on primary cardiac arrest. The CCR is not recommended for arrests secondary to hypoxia from drowning or respiratory failure. However, it must be emphasized that not all cardiac arrest in individuals under the age of 18 years are respiratory.

### Primary Cardiac Arrest in Children and Adolescents

Respiratory arrests are the reason for the majority of cardiac arrests in children, and should be treated according to guidelines.[13] However, primary cardiac arrest also occurs in children and adolescents in the presence of diverse pathologies such as the prolonged QT syndrome, the cardiomyopathy of arrhythmogenic right ventricular dysplasia (ARVD), Brugada syndrome, anomalous coronary arteries or commotio cordis [where a blow to the precordium at the peak of the electrocardiographic T wave can result in ventricular fibrillation (VF)]. Accordingly, chest compression only (CCO) CPR is the best approach to bystander CPR in children and adolescence when the cardiac arrest is primary.

Primary cardiac arrest is recognized by an unexpected, witnessed (seen or heard) collapse in an individual who is not responsive. As emphasized in this chapter, many individuals with primary cardiac arrest continue to have spontaneous ventilation, and checking for the presence or absence of an arterial pulse by bystanders is no longer recommended.[13]

## Pathophysiology of Primary Cardiac Arrest

One minute into persistent VF or pulseless VT, coronary blood flow comes to a standstill, and minutes later carotid blood flow (and therefore cerebral perfusion) becomes nil.[14] The ensuing equalization of systemic pressures in the arterial and venous beds takes place, resulting in the so-called "mean circulatory filling pressure" described by Guyton.[15] This shift in blood volume from higher pressure arterial system to the low pressure venous system results in "acute distension of the right ventricle" originally described by Professor Stig Steen in open chest swine.[16] This phenomenon is perhaps best illustrated by closed chest imaging techniques **(Fig. 5)**.[17,18] The resultant pericardial restraint produces a constrictive pericardial condition, and even with defibrillation poor contractility occurs due to the lack of stretch of the myocardial fibers.[19] This sequence helps to explain why chest compressions (to decompress the heart to relieve pericardial restraint) are often necessary for successful generations of an arterial pressure following defibrillation.

Untreated VF results in a progressive decrease in left ventricular volumes until a state of extreme myocardial contraction develops (referred to as stone heart). This sequence has also been demonstrated by magnetic resonance imaging (MRI) in closed chest animal models of VF **(Fig. 5)**.[17]

## Coronary Perfusion Pressures during Resuscitation Efforts

Several observations lead the University of Arizona Sarver Heart Center Resuscitation Research Group almost two decades ago to advocate of CCO CPR. One of the most important was the

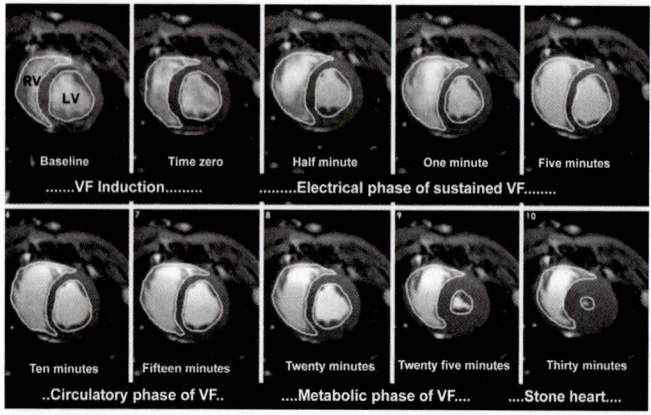

**FIGURE 5:** Development of the "Stone Heart" after prolonged VF (*Source:* Sorrell VL, Altbach MI, Kern KB, et al. Images in cardiovascular medicine. Continuous cardiac magnetic resonance imaging during untreated ventricular fibrillation. Circulation. 2005;111(19):e294)

observation during our early (1980s) experimental studies that survival from prolonged VF arrest was related to the coronary perfusion pressure (CPP) generated by chest compressions **(Fig. 6)**.[20] The CPP is defined as the difference between the aortic and the right atrial pressures during the release phase "diastole" of closed chest cardiac compression **(Fig. 7)**.[21] The CPP is the pressure gradient that is responsible for antegrade coronary flow. Similar to normal sinus rhythm where most blood flow through the coronary arteries occurs in diastole, during chest compressions for cardiac arrest, most coronary blood flow occurs during "compression diastole" or the release phase of chest compressions. During life the pressures in the myocardium, the ventricle and the aorta are similar during ventricular systole and therefore there is very little coronary blood flow. But during diastole, the aortic valve closes, the pressure is higher in the aorta than in the myocardium, and thus most of the coronary

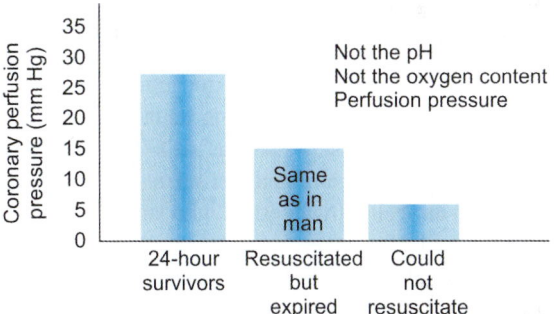

**FIGURE 6:** Survival from prolonged VF arrest in experimental studies was related to pressure generated by chest compressions. (*Source:* Kern, Ewy, Voorhees, Babbs, Tacker. Resuscitation. 1988;16:241-50; Paradis, et al. JAMA .1990;263:1106–13)

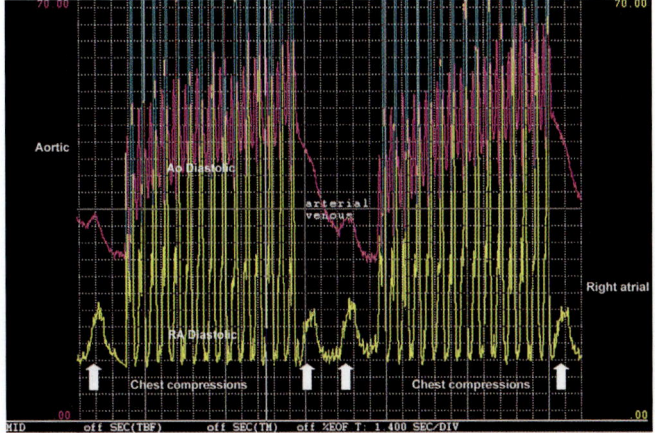

**FIGURE 7:** The difference between the aortic and the right atrial pressures during the released phase "diastole" of closed chest cardiac compression is termed as coronary perfusion pressure

blood flow to the myocardium occurs during diastole. During chest compression for cardiac arrest, the pressure in the heart and aorta are similar. But during the release phase the aortic valve closes[22] and the pressure in the aorta is higher than the right atrium, so antegrade coronary blood flow occurs during the release phase of chest compressions. During cardiac arrest and resuscitation efforts, the amount of coronary flow is predominantly related to the amount of arterial pressure generated by chest compressions. We found that one had to generate a minimal CPP of 15 mm Hg for return of spontaneous circulation (ROSC) in our experimental studies.[20] Of interest is the fact that this was the same value found by Norman Paradis in his measurements during resuscitation efforts in man **(Fig. 6)**.[21] Of note in their report there were no survivors![21]

The CPP is built up slowly with the initiation of chest compressions during resuscitation efforts for cardiac arrest, such that the first few compressions often do not generate a significant CPP **(Fig. 8)**. The CPP is a surrogate for myocardial cellular perfusion and has been shown to be a determinant of survival in prolonged VF.[20,23-25] Obviously to generate an adequate coronary perfusion pressure, one is usually generating an adequate cerebral perfusion pressure as the CPP produced during resuscitation efforts for cardiac arrest relates to neurologically intact survival as well. However, all who survive for 24 hours are not neurologically intact.[26]

When chest compressions are interrupted for even 4 seconds, the CPP decays and with it myocardial perfusion. The coronary perfusion gradient has to be re-established once again when

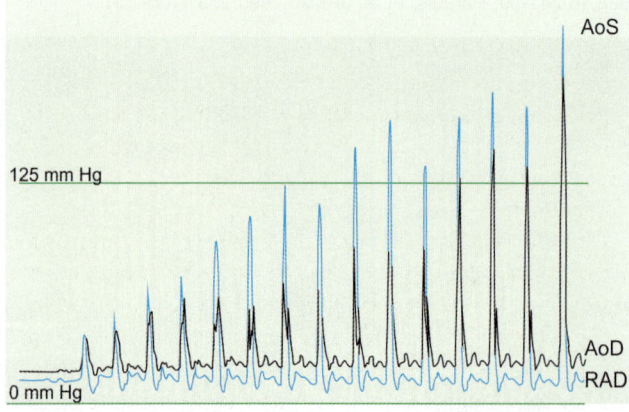

**FIGURE 8:** Simultaneous recording of aortic and right atrial pressures during first 15 external chest compressions in swine in cardiac arrest due to ventricular fibrillation. Note how initial compressions do not generate much of a pressure (*Source:* Modified from Ewy GA. Circulation. 2005;111:2134-42)

compressions are restarted **(Fig. 7)**. This is a key element in understanding why continuous chest compression (CCC) or CCO CPR results better survival. The CPP gradient, once established, is not interrupted and therefore cellular perfusion is maintained.[23,24,27]

## Assisted Ventilation in Primary Cardiac Arrest

The technique of closed chest "cardiac massage" for cardiac arrest was first published in 1960 and, since the survival rate in this initial report was 70% (14/20), it quickly became the preferred technique for both in-hospital and prehospital treatment of cardiac arrest.[28] In their early teachings, the authors, Kouwenhoven, Knickerbocker and Jude said that, "assisted ventilation was not necessary as the victim gasped" during closed chest compression.[29] Seven of their initial 20 reported patients received chest-compression only without assisted ventilation.[28]

Nevertheless, influential individuals advocated bystander mouth-to-mouth assisted ventilations for cardiac as well as for respiratory arrests.[30,31] The possible justifications for this view were: (1) the emphasis of therapy for out-of-hospital arrests was historically based on the resuscitation of drowning victims; (2) it was thought that lay individuals could not reliably differentiate between respiratory and cardiac arrest; (3) gasping was not appreciated as a common event in primary cardiac arrest and (4) studies by Safer and his associates in volunteers given drugs to produce temporary paralysis showed that without assisted ventilation, their blood gases rapidly deteriorated.[31] Unfortunately, it was not recognized that the explanation was that these subjects were not in cardiac arrest, and had normal cardiac outputs; therefore, their arterial blood quickly became unsaturated without assisted ventilation. In contrast, with sudden onset VF in primary cardiac arrest, the arterial blood is fully oxygenated and remains so for several minutes, because it is not circulating. Nevertheless closed chest cardiac massage was merged with mouth-to-mouth to form what became known as cardiopulmonary resuscitation or CPR.[32]

By 1966, standardized methods of training and performance criteria for the administration of CPR had been advocated and published. A few years later the American Heart Association (AHA) adopted CPR as one of its main focus areas, developed standards for CPR and emergency cardiac care (ECC) and spearheaded a campaign to disseminate the techniques of CPR to the public and both CPR and ECC to professionals. Until the 2008 scientific advisory, all previous guidelines recommended mouth-to-mouth or so-called "rescue breathing" as the initial step for bystander initiated CPR.[33]

Years of defibrillation and resuscitation research led the senior author and his associates to advocate CCO CPR decades ago.[27,34-37] In our experimental laboratory, survival was better with CCO CPR than not doing anything until the simulated arrival of the paramedics and definite treatment. As our studies progressed, we were somewhat surprised to find that survival with CCO CPR was equivalent to "ideal" guidelines advocated CPR where each series of chest compressions were interrupted by 4 seconds to deliver the "two quick breaths" of "rescue breathing" **(Fig. 9)**.[27,36]

In 2000, it was documented that recently certified lay individuals interrupted chest compressions for 16 seconds to deliver the two recommended "rescue breaths".[38] Subsequent experimental studies from our laboratories showed that survival was better with CCO CPR than with guidelines recommend CPR when each set of 15 compressions were interrupted a realistic 16 seconds to simulate the interruptions necessary for mouth-to-mouth ventilations.[24] Subsequently the AHA and International Liaison Committee on Resuscitation (ILCOR) changed the recommended compression to ventilation ratio to 30:2.[13] This recommendation was based on "consensus" as there was no data to support this recommendation. Accordingly, in our experimental laboratory, we then compared survival from simulated OHCA with CCO CPR and guidelines (30:2) CPR, and found that survival to be better with continuous chest compression CPR **(Figs 10 to 12)**.[39] If lay individuals interrupted chest compressions an average of 16 seconds to provide the two recommended rescue breaths, could medical students who were younger provide these breaths quicker? How about the professionals, the paramedics? As shown in **(Fig. 13)** no one can provide mouth-to-mouth rescue breaths rapidly.[40,41]

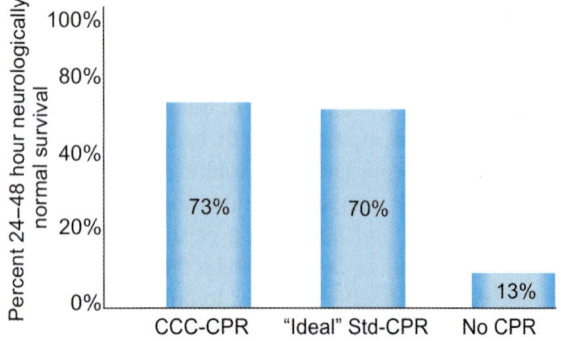

**FIGURE 9:** "CPR" for VF arrest, 6 different publications, in 169 non-paralyzed swine, between 1993 and 2002 (*Source:* University of Arizona Sarver Heart Center Resuscitation Research Group)

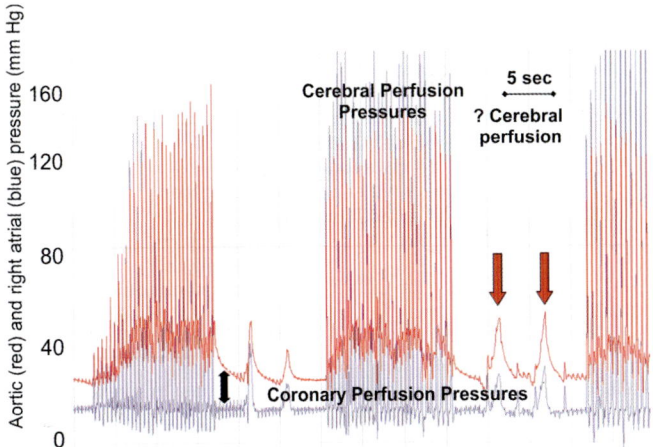

**FIGURE 10:** Hemodynamics of simulated single rescuer performing 30:2 compression: ventilations in experimental animal with realistic 16 sec. interruption of chest compressions for mouth-to-mouth ventilations

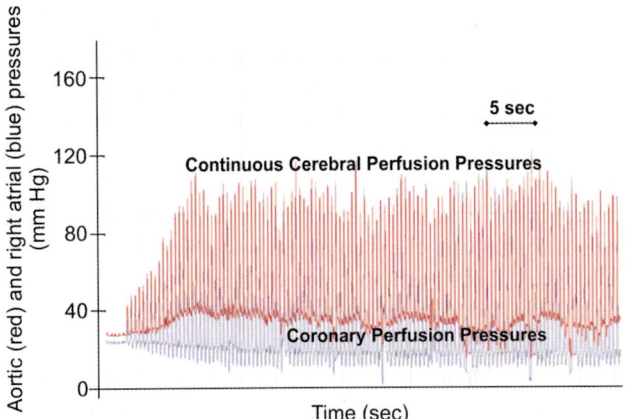

**FIGURE 11:** Hemodynamics of simulated single rescuer performing continuous chest compressions in experimental animal

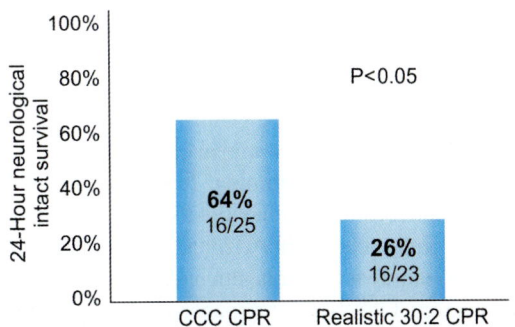

**FIGURE 12:** Survival following simulated single lay rescuer scenario of primary cardiac arrest (4–6 minutes untreated VF followed by bystander CPR; at 12 min, all received ACLS) (*Source:* Ewy GA, Zuercher M, Hilwig RW, et al. Circulation. 2007;116:2525–30)

> **Lay public:**
> **16 ± 1 seconds**
> Assar, et al. Resuscitation
> 2000;45:7-15

> **Medical students:**
> **14 ± 1 seconds**
> Heidenreich, et al. Resuscitation
> 2004;62:283-9

> **Paramedics:**
> **10 ± 1 seconds**
> Higdon et al. Resuscitation
> 2006;71:34-9

**FIGURE 13:** Interruptions of chest compressions by single rescuer CPR for guidelines recommended 2 quick mouth-to-mouth ventilations

## Not Following Guidelines for Primary Cardiac Arrest

Years of defibrillation and resuscitation research and continuing analysis of the resuscitation literature and our failure to be able to influence guidelines led the senior author to conclude in 2003 that we could no longer in good conscience follow the AHA and ILCOR guidelines of 2000.[36,37] We announced our intensions[37] and explained our approach.[36,42] Our new approach was called CCR, due to its focus on the maintaining blood flow to the heart (cardio) and the brain (cerebral) during primary cardiac arrest by near continuous chest compressions prior to defibrillation, a necessary components of neurologically intact survival.[36] The CCR de-emphasizes the early ventilation or "pulmonary" component of traditional CPR and attempts to minimize other possible detrimental aspects of positive pressure ventilation.[36]

The community **(Fig. 3)** component of CCR included CCO CPR. In 2003, CCO CPR was advocated in Tucson, AZ, with free training, local radio spots and newspaper interviews, and inserts into utility bills **(Fig. 14)**. In 2004, with Dr Benjamin Bobrow and his statewide SHARE program a statewide effort was initiated in Arizona to encourage CCO CPR.[43] This was a multiple facet approach to training and information was dissemination in multiple venues that included websites (www.azshare.gov) and (www.heart.arizona.edu), celebrity endorsement, online video training, free in-person training in many setting and locations throughout the state, training kits sent to all 6–12th grade schools in Arizona, inserts mailed in utility bills **(Fig. 15)** tables set up at health and safety fairs by various departments (e.g. Fire, etc.), newspaper articles, editorials, local radio spots and interviews. To determine the effect of this effort, a statewide reporting system was developed.[44] In October 2008, the AHA published a science advisory, recommending CCO (Hands-Only CPR) for lay individuals untrained in CPR.[33]

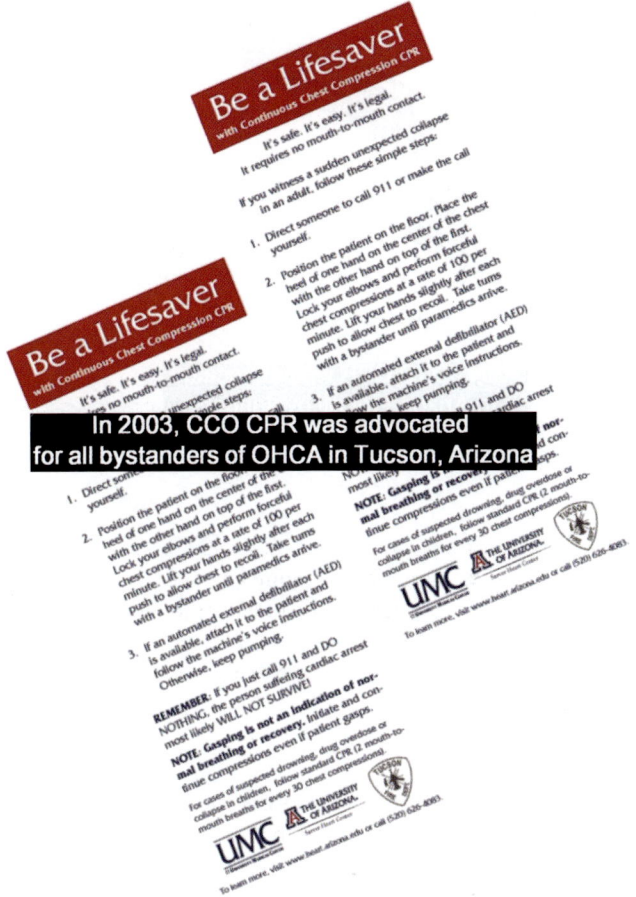

**FIGURE 14:** In 2003, CCO CPR was advocated for all bystanders of OHCA in Tucson, Arizona

Between 1 January 2005 and 31 December 2009 we analyzed the results of this five years effort, published as "chest compression—Only CPR by Lay Rescuers and Survival from Out-of-hospital cardiac arrest" in the Journal of the American Medical Association in 2010.[45] After excluding bystander CPR provided by health care professionals or arrests that occurred in a medical facility, 4,415 patients with OHCA were analyzed.[45] Survival to hospital discharge was 5.2% for no bystander CPR group, 7.8% for conventional CPR and 13.3% for CCO CPR **(Fig. 16)**.[45] Survival to hospital discharge of those most likely to survive, those with witnessed arrest and a shockable rhythm on arrival of the emergency medical services (EMS) personnel, was 17.6% in the no CPR group, 17.7% for conventional CPR and 33.7% for CCO CPR **(Fig. 17)**.[45] There was no adverse effect of CCO CPR in the subgroup less likely to survive, witnessed arrest but without a shockable rhythm.[45]

**FIGURE 15:** 2004, CCO CPR was advocated in Arizona for all bystanders

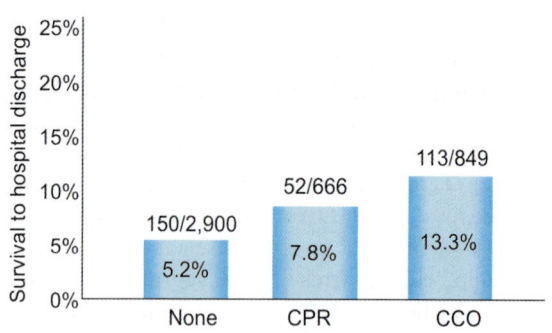

**FIGURE 16:** Effect of bystander CPR for OHCA on survival in Arizona (2005–2010) (*Source:* Modified from Bobrow, et al. JAMA. 2010)

This is the first prospective observational study to show that CCO CPR resulted in improved survival of patients with OHCA.[45] However, the first observational study to find that survival was better with lay individuals performing CCO CPR was the SOS-KANTO study which found that the survival of those individuals most likely to survive, witnessed arrest and a shockable rhythm was 11% for those receiving chest

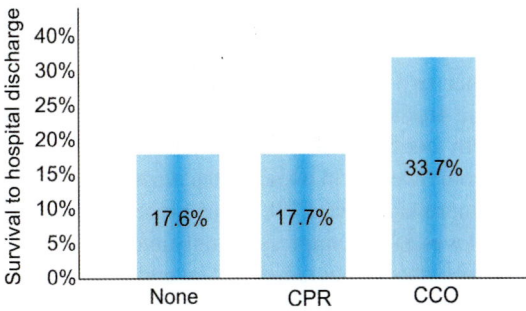

**FIGURE 17:** Bystander CPR for OHCA in Arizona (2005–2010) Witnessed/Shockable OHCA (*Source:* Modified from Bobrow, et al. JAMA. 2010)

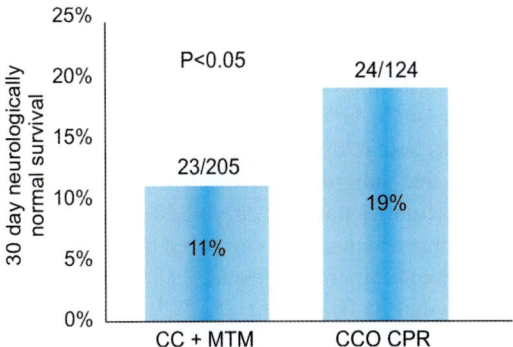

**FIGURE 18:** SOS-KANTO: Subset of patients with witnessed arrest and shockable rhythm (*Source:* Modified from Nagao et al. for the SOS-KANTO. The Lancet. 2007:369;920)

compressions plus mouth-to-mouth ventilations versus 19% for those receiving CCO **(Fig. 18)**.[46] This study is of interest, for several reasons, including the fact that a bystander technique that had not been advocated nor taught was more effective than one that has been guidelines advocated for decades, and in which untold thousands of man-hours had been spent teaching and untold thousands of dollars spent advocating over the past several decades.

To date (2010), there are three reported randomized trials of dispatcher instructed CPR where one group was instructed in guidelines CPR of chest compressions plus mouth-to-mouth ventilations, and the other in CCO CPR. The first was by Hallstrom and his associates who reported in 2000 that the survival was 14.6% in those receiving chest compressions only CPR and 10.4% in those receiving instructions in chest compression and ventilations.[47] Since there were only 520 patients in the study, the difference were not significant and therefore the guidelines were not changed. However, this finding, published in 2000, encouraged the University of Arizona Sarver Heart Center Resuscitation Research Group in our decision in 2003 not to follow the AHA and ILCOR 2000 guidelines.

The other two recently reported studies of patients with OHCA, one from Sweden and the other from Seattle and London, both found no statistically difference in survival.[48,49] In the study from Sweden, in which 620 patients received dispatch directed compression-only CPR and 656 received standard CPR, the survival was 8.7% compression-only and 7.0% for standard CPR. These authors concluded that overall "this study lends further support to the hypothesis that compression-only CPR, which is easier to learn and to perform, should be considered the preferred method for CPR for patients with cardiac arrest."[48]

Rea and his associates from Seattle also reported on dispatch directed chest compression alone (981 patients) versus chest compression plus rescue breathing (960 patients) and found that survival to hospital discharge with favorable neurological outcome was 14.4% with chest compression alone and 11.5% in the CC plus rescue breathing group ($p = 0.13$).[49] However, their prespecified subgroup analysis showed a trend toward a higher proportion of patients survived to hospital discharge with chest compressions alone compared to chest compressions plus rescue breathing (15.5% vs 12.3%) and for those with a shockable rhythm 31.0% versus 25.7%.[49]

## The Public Has Made Up Its Mind

In our statewide Arizona study, it was of interest to find that from 2005 to 2010, lay rescuer CPR only increased from 28.2% to 39.9% **(Fig.19)**.[45] This was rather disappointing. However, of those performing bystander CPR, the proportion of CPR that was CCO CPR increased from 19.6% to 75.9% **(Fig. 20)**.[45] This indicates to us that the public has made up its mind. When encouraged they are much more likely to perform CCO CPR.

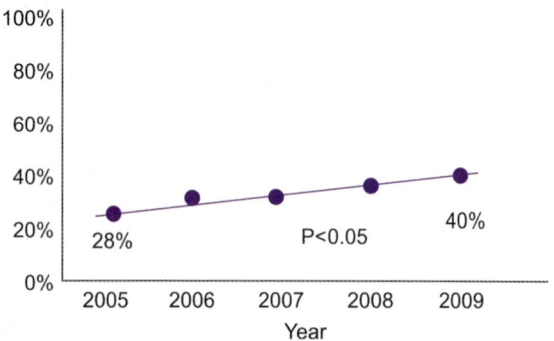

**FIGURE 19:** Incidence of bystander CPR for OHCA in Arizona (2005–2010) (*Source:* Modified from Bobrow, et al. JAMA. 2010)

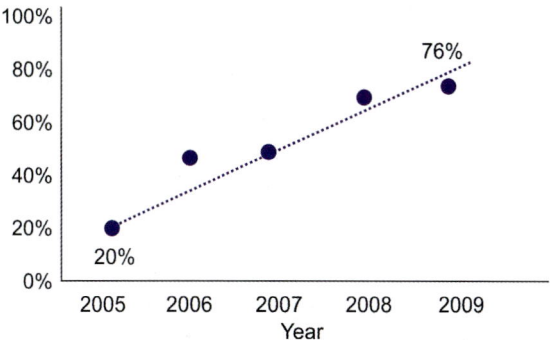

**FIGURE 20:** Bystander CPR for OHCA in Arizona (2005–2010): Percent of lay CPR providers who performed CCO-CPR (*Source:* Modified from Bobrow, et al. JAMA. 2010)

- Fear/concern: mouth-to-mouth contact   ~20%
- Fear/concern: harming person   ~20%
- Fear/concern: legal consequences   ~20%
- Fear/concern: not performing properly   ~20%
- Physically unable to perform CPR   ~20%

**FIGURE 21:** Is fear or concern about MTM contact the only deterrent to bystander CPR? (*Source:* Coons SJ, Guy MC. Resuscitation. 2009;80: 334-40) This study was designed and funded by the Sarver Heart Center University of Arizona College of Medicine and SHARE

## Increasing the Prevalence of Bystander Resuscitation Efforts

However, the fear or concern about mouth-to-mouth breathing is not the only reason a bystander may not initiate resuscitation efforts. We must address all of their concerns **(Fig. 21)** if we are to increase the prevalence of bystander resuscitation.[50]

## Increasing the Ability to Promptly Identify Primary Cardiac Arrest

The prompt recognition of primary cardiac arrest is essential to any program to improve survival of patients with out-of-hospital cardiac arrest. The recommendation used by cardiocerebral resuscitation is the unexpected, witnessed (seen or heard) collapse in an individual who is not responsive. Note that this recommendation does not say anything about spontaneous ventilations. The reason for this is that one of the major impediments to the prompt recognition of primary cardiac arrest is the fact that subjects with cardiac arrest, rats, swine and humans have a high frequency of gasping after cardiac arrest **(Fig. 22)**. Failure to recognize this fact delays the recognition of primary cardiac arrest **(Fig. 23)** lists the

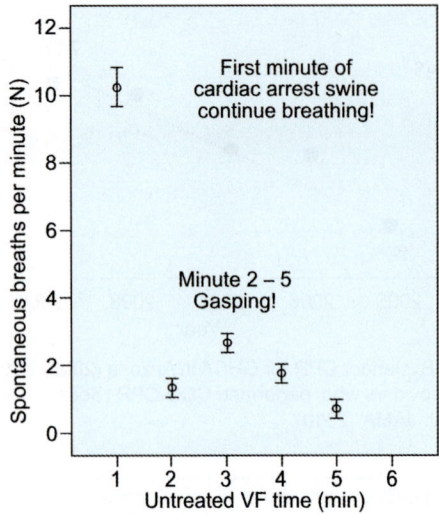

**FIGURE 22:** Recent unexpected finding (*Source:* Modified from Zuercher Ewy, Hillwig, et al. BioMedCentral Cardiovascular Disorders. 2010)

| | |
|---|---|
| • Gasping | • Agonal breathing |
| • Snoring | • Barely breathing |
| • Gurgling | • Labored breathing |
| • Moaning | • Noisy breathing |
| • "Normal" breathing | • Heavy breathing |
| | • Abnormal breathing |

**FIGURE 23:** Spontaneous ventilatory activity in patients with primary cardiac arrest

description of breathing abnormalities that have been described by witnesses of patients with cardiac arrest. Continued normal breathing for the first minute following cardiac arrest **(Fig. 22)** has only recently been observed in swine by the University of Arizona Sarver Heart Center Resuscitation Research Group.[51] However, gasping is well know and has been described in animals to begin during the second minute of VF arrest, and has a classical crescendo-decrescendo frequency pattern, of about 1,3,2,1 gasps per minutes **(Fig. 22)** and is no longer present six or more minutes after the onset of VF arrest, unless chest compressions provides enough blood flow to the brainstem to continue gasping.[52] Physician scientists refer to this phenomenon as gasping or agonal breathing, but lay people may refer to this phenomenon with other terms **(Fig. 23)**.[53]

One of the reasons that assisted ventilation is often not necessary even during prolonged CCO CPR is that effective chest compression provides enough flood flow to the brainstem to maintain this primitive respiratory response.

## Three Phases of Ventricular Fibrillation (VF)

To better understand some of the rationale for the second or pre-hospital phase of cardiocerebral resuscitation, one needs to appreciate not only the pathophysiology of cardiac arrest outlined above but also the electrophysiology of VF.

It has been known for decades that survival rates decrease by about 7–10% for every minute that a patient spends in untreated VF.[6] In the absence of chest compressions, after roughly 12 minutes, defibrillation for VF is rarely effective.[54] Our understanding of the therapy for VF was helped by the three-phase time sensitive concept of untreated VF, articulated by Weisfeldt and Becker in 2002 **(Fig. 24)**.[54] This concept divides VF into an electrical phase (around 0–4 minutes), circulatory phase (roughly 4–10 minutes) and a metabolic phase (> 10 minutes). This model helps us to understand why specific therapies need to be tailored to timelines.

### Electrical Phase (0–4 minutes)

In this initial phase of VF, there is enough myocardial adenosine triphosphate (ATP) and other energy stores that defibrillation alone is adequate to restore a perfusing rhythm.[55] Patients defibrillated within seconds by an ICD or minutes by an AED often return to a perfusing stable rhythm since they are in the electrical phase of VF arrest. Chest compressions can prolong this so-called electrical phase of VF. As a prototype, the city of Seattle, WA, USA, has an average EMS response time of about 5 minutes, with a bystander CPR rate of over 60%. This translates into patients being defibrillated in the electrical phase and consequently having some of the best survival rates from OHCA anywhere in the United States.[9]

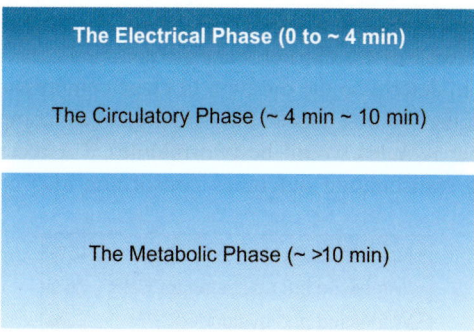

**FIGURE 24:** 3-Phase time-sensitive model of cardiac arrest due to ventricular fibrillation (*Source:* Modified from Weisfeldt ML, Becker LB. JAMA. 2002;288:3035-8)

## Circulatory Phase (4–10 minutes)

Conversely, once the markedly underperfused fibrillating ventricle uses up a significant portion of its energy stores, defibrillation, even if successful results in pulseless electrical activity (PEA) or asystole.[56] Prolonged untreated VF is manifested electrocardiographically by decreasing amplitude of the VF wave form with a transition to "fine" fibrillation waves on the electrocardiogram (ECG). Defibrillation in the absence of chest compressions is rarely successful in this phase of VF. However, if the CPP is re-established by chest compressions, the resultant perfusion of the myocardium allows the formation of new myocardial energy which makes the myocardium more responsive to defibrillation **(Fig. 24)** and less likely to deteriorate to PEA or asystole **(Fig. 25)**. Thus, chest compressions can prolong the electrical phase of VF.

Most OHCA patients are found in the circulatory phase of VF upon arrival of EMS personnel. For instance, the average time to response in the city of Tucson, Arizona, was 6 minutes 34 seconds,[42] placing the patient precisely in the circulatory phase. A defibrillation first strategy in these patients is likely to result in a nonperfusing rhythm like PEA or asystole.[56] In this situation, preshock chest compressions have been demonstrated to increase the likelihood of successful defibrillation in a swine model.[57] A study from Seattle in humans showed increased survival when chest compressions were uniformly performed for 90 seconds prior to defibrillation.[58] In a similar study, Wik and his associates found survival to be improved with 3 minutes of chest compressions prior to defibrillation.[59] Based on these studies, for individuals encountered in the circulatory phase of untreated VF arrest, we prescribed, as part of cardiocerebral resuscitation, a period of 2 minutes of chest compressions prior to the first defibrillation attempt.[36] This was done first due to a compromise between the duration of chest compressions studied by Cobb et al. and by Wik et al. and because the senior author as well as we did not want the paramedics to use their watches to determine the duration of chest compressions prior to delivering the first shock. Two hundred chest compressions at 100 per minutes would be two minutes.[36] A recent report from the resuscitation outcomes consortium also confirmed

| | |
|---|---|
| • Not witnessed | 16% |
| • Witnessed but no bystander CPR | 36% |
| • Witnessed and bystander CPR | 52% |

**FIGURE 25:** Prevalence of VF on arrival of EMS in out-of-hospital cardiac arrest in Arizona (*Source:* Data from 1,296 cardiac arrest in Arizona Voluntary reporting SHARE Program: Data collected October 2004 to April 2006 Bobrow, Clark, Ewy, Kern, Sanders)

improved survival in humans with preshock chest compressions, and interestingly they found that survival was greatest in the subgroup who received 2 minutes of chest compressions prior to the first attempted defibrillation.[60]

## Metabolic Phase (>10 minutes from Onset of Untreated Cardiac Arrest)

This third and terminal phase of untreated VF is universally associated with diminishing odds of successful defibrillation and neurologically intact survival. End organ damage has already set in with irreversible cellular impairment. Ischemic and reperfusion injuries are believed to predominate at this stage. Strategies that may delay the onset of irremediable damage during this phase of untreated VF include hypothermia.

## Cardiocerebral Resuscitation; PreHospital Component

The initial approach to the prehospital component of CCR was implemented in Tucson, AZ, in late 2003 **(Figs 26 to 28)**.[36,37,42] We announced our intensions and gave our rational for no longer following the National and International CPR and

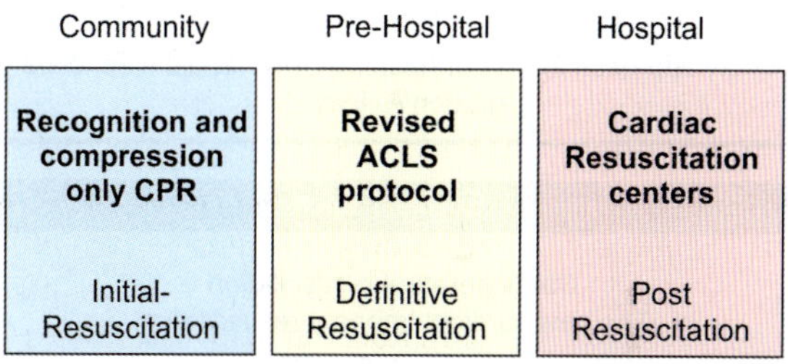

| Community | Pre-Hospital | Hospital |
|---|---|---|
| Recognition and compression only CPR | Revised ACLS protocol | Cardiac Resuscitation centers |
| Initial-Resuscitation | Definitive Resuscitation | Post Resuscitation |

**FIGURE 26:** Cardiocerebral resuscitation "The New CPR" for primary cardiac arrest

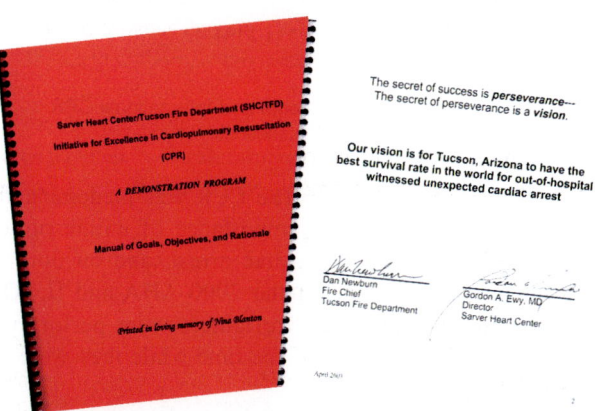

**FIGURE 27:** Manual of Goals, Objective, and Rationale

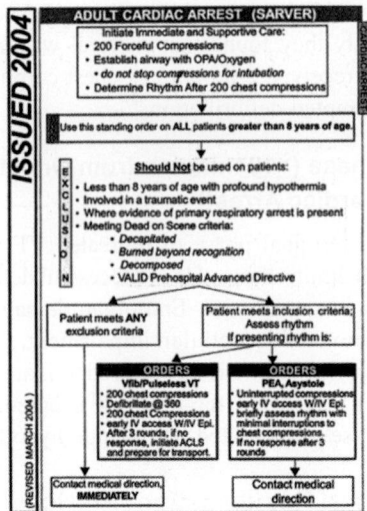

**FIGURE 28:** Tucson Fire Department

 Resuscitation. 2003;58:271

A new approach for out-of-hospital CPR: a bold step forward
gordon A. Ewy [1*]

Special Report

Cardiocerebral resuscitation
The new cardiopulmonary resuscitation

Gordon A. Ewy, MD

"Why is that every time I press n his chest he opens his eyes, every time I stop to breathe for him he goes back to sleep?"

Circulation. 2005;111:2134-42

**FIGURE 29:** Sources of National and International CPR and EMS guidelines

EMS guidelines **(Fig. 29)**.[36,37] The CCR was predominantly based on our findings of the importance of uninterrupted chest compressions during cardiac arrest, and on ours or others findings that following the 2000 AHA and ILCOR guidelines, EMS paramedics/firefighters were performing chest compressions only half of the time while they were on the scene.[61,62]

We initially did not allow endotracheal intubation based our clinical observations that chest compressions were often

- Interruptions in CPR for paramedic endotracheal intubations in 100 OHCA
- Median duration of 1st endotracheal Intubation was 47 seconds
- Almost one-third exceeded 1 minute
- One-fourth exceeded 3 minutes

**FIGURE 30:** Interruptions of chest compressions for endotracheal intubation (*Source:* Modified from Wang, et al. Annals of Emergency Medicine. 2009;54:645)

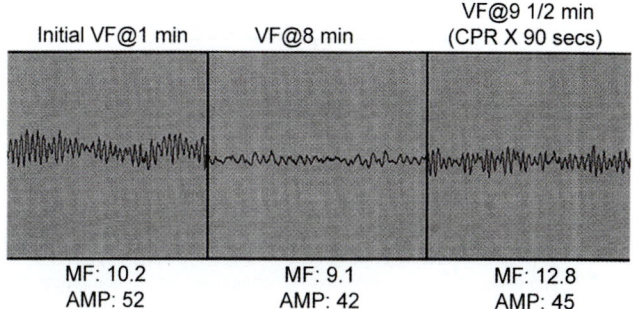

**FIGURE 31:** Untreated, the VF waveform decreased in amplitude with time. Following 90 seconds of chest compressions, the frequency and amplitude of the VF waveform increases

interrupted for prolonged periods of time, even by well trained individuals attempting endotracheal intubation. This assumption proved to be correct as documented in a recent study by Wang and his associates **(Fig. 30)**.[63]

We advocated chest compressions prior to defibrillation based on our experimental finding **(Fig. 31)**,[64] and on the observations by Cobb et al. and Wik et al. that in patients with prolonged VF arrest, chest compressions prior to defibrillation improved survival.[58,59,65]

A single defibrillator shock was recommended based on the long interruptions of chest compressions for "stacked shocks". This approach also proved to be correct. Rae and his associates found increased survival in humans with single rather than stacked shocks, and subsequently this recommendation was made in the guidelines.[66]

In our experimental laboratory, we found that chest compressions immediately after defibrillation shocks improved survival, and this approach was also subsequently advocated in 2005 guidelines.[13]

In 2004, the adverse affects of hyperventilation were reported by Aufderheide and his colleagues **(Fig. 32)**. We had previously reported that during in-hospital cardiac arrests, physicians, in their excitement, were ventilating at an average

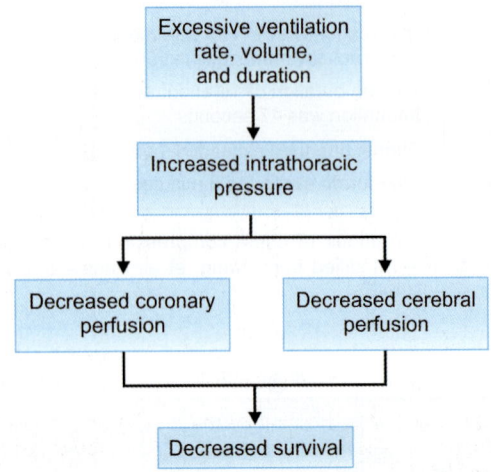

**FIGURE 32:** Consequence of excessive ventilation

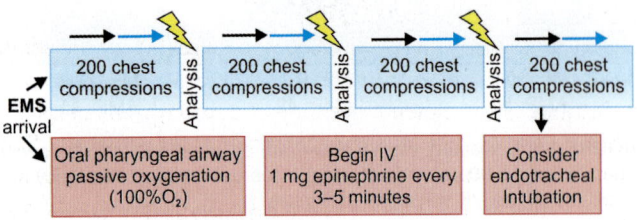

**FIGURE 33:** Cardiocerebral resuscitation. Goal: minimally interrupted chest compressions, avoiding hyperventilation, and early administration of epinephrine

rate of 37 per minute.[67] Aufderheide and his associates documented that paramedics were ventilating at this same rapid rate.[68] Accordingly we eliminated bag-mouth-ventilation and substituted "passive ventilation" as part of cardiocerebral resuscitation.[69,70] In 2004, after visiting us, Dr Mike Kellum and his associates instituted CCR **(Fig. 33)** in Rock and Walworth Counties of Wisconsin.[10,36,71,72] The CCR **(Figs 26 and 33)**, although advocating CCO CPR for bystanders, in the reports from Kellum et al. the intervention was essentially the "prehospital **(Fig. 26)** portion". Its initial component consisted of two hundred chest compressions. While this was being preformed, endotracheal intubation was not allowed, but rather an oral-pharyngeal airway, a non-rebreather mask, and high flow oxygen was administered.[72] A single shock was followed immediately by another 200 continuous chest compressions. After the single shock the EMS personnel were instructed not to feel for a pulse, nor evaluate the rhythm by looking at an ECG. These analyses were allowed, only after the first 400 chest compressions (4 minutes into ACLS). Equally important was the fact that only a single defibrillator shock, at maximal defibrillator

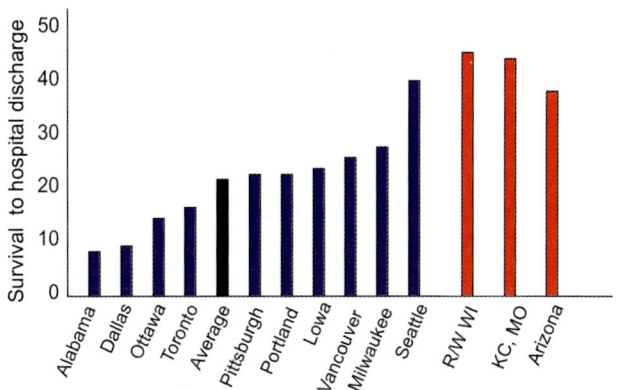

**FIGURE 34:** Survival to hospital discharge (%) of patients with VF arrest ROC 2005 Guidelines (all VF) vs cardiocerebral resuscitation (Witnessed VF)

output, was allowed.[10,36,72] Intravenous (IV) epinephrine was administered as soon as possible.[36,72]

In 2005, the rational for CCR was presented by the senior author to EMS medical directors in Arizona, and some chose to institute CCR in their cities. In part due to the enthusiasm of EMS personnel and in part in response to the statewide Save Hearts in Arizona Research and Education Program (SHARE) spearheaded by Dr Benjamin Bobrow, Director of the Arizona Department of Health Emergency Care and Trauma Service, CCR was expanded to a statewide effort.[11] In 2006, the senior author was invited to Kansas city, MO, to advocate cardiocerebral resuscitation.[12,73]

As shown in Figure 4, survival of the subset of patients more likely to survive, those with witnessed arrest and a shockable rhythm on arrival of EMS personnel, averaged 38% with CCR in all three of these areas. Then compared to arguably some of the better EMS systems in the United States and Canada, the resuscitation outcomes consortiums (ROC), survival with CCR was better than that advocated by the 2005 national and international guidelines for CPR and EMS care **(Fig. 34)**.[1] There is a caveat with this comparison in that the results of CCR are for witnessed arrest and a shockable rhythm, whereas the results reported by ROC was for VF arrest.[1]

## DRUG THERAPY IN CARDIAC RESUSCITATION

Epinephrine is a first line agent for a cardiac arrest and is used in all forms of cardiac arrest. Epinephrine causes immediate peripheral vasoconstriction by its α-adrenergic effect, and thereby during chest compressions for cardiac arrest, increases coronary and cerebral perfusion.[74-76] An IV epinephrine dose of 1 mg is administered ever 3–5 minutes till ROSC has been recommended. Since the time from onset of cardiac arrest to IV administration of epinephrine in the field is prolonged, there is a trend to the increasing use of intraosseous (IO) injections.

Vasopressin, which causes peripheral vasoconstriction through V1a receptors on vascular smooth muscle, is a more controversial drug. A large randomized trial of epinephrine or vasopressin showed no difference in survival between the two groups.[77] A 2005 meta analysis also failed to show any benefit of vasopressin over epinephrine.[78] Since the half-life of vasopressin is 10–20 minutes, it is administered as on one time 40U IV dose. Some have recommended epinephrine as the first vasopressor, followed by vasopressin as the second vasopressors in an effort to decrease the number of epinephrine doses.[10] Since it is the alpha adrenergic effect of epinephrine that is beneficial during resuscitation efforts, this approach could decrease the theoretical adverse effects of excessive beta adrenergic effects of frequent epinephrine doses. Others have recommended epinephrine alone, in efforts to simplify the regimen of EMS personnel. Experimental studies have suggested that vasopressin contributes to postresuscitation myocardial dysfunction, but not survival.[79]

Amiodarone and to a lesser extent lidocaine are anti-arrhythmic drugs of choice for pulseless VT/VF, especially of presumed ischemic etiology. Amiodarone was superior to lidocaine in the ALIVE trial, and had been the recommended first line antiarrhythmic agent for VF/VT arrest.[80,81] Amiodarone is administered as a single 300 mg IV push, followed by, if necessary, another 150 mg IV push.

The largest controlled trial (TROICA) of thrombolytics in cardiac arrest, involving 1,050 patients, was prematurely terminated due to a lack of benefit.[82]

For postresuscitation hypotension, dopamine should be considered. Intra-aortic balloon pumping was not found to be helpful in our experimental laboratory.[83]

## CARDIAC RESUSCITATION CENTERS

A more recently advocated third component of CCR is cardiac resuscitation centers **(Fig. 35)**. Cardiac resuscitations centers are proposed to improve therapy of resuscitated but comatose patients following cardiac arrest. Although neurologic function after prolonged cardiac arrest is of major concern, there is almost always other organ dysfunction as well. In its 2008 consensus statement, ILCOR termed this as the "post cardiac arrest syndrome".[84]

The clinical manifestation of the post cardiac arrest syndrome can include hypoxic encephalopathy, myocardial dysfunction, aspiration pneumonia, ischemic gut injury,

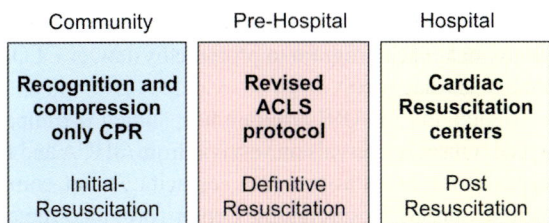

**FIGURE 35:** Cardiocerebral resuscitation "The New CPR" for primary cardiac arrest

ischemic hepatopathy, renal dysfunction as well as peripheral limb ischemia. Generalized activation of immunological and coagulation cascades occur. Relative adrenal insufficiency,[85] glycemia,[86] and ventilator associated pneumonia are common. Seizures/myoclonus, consequent to hypoxic brain injury, is seen in up to 40% of resuscitated patient and should be treated promptly with anticonvulsants. Coronary flow reserve remains below normal for at least 4 hours in animal models **(Fig. 36)**.

In Arizona, hospitals were encouraged to become designated as Cardiac Resuscitation Centers.[87] To be designated, a hospital has 24/7 capability for therapeutic hypothermia, early cardiac catheterization and indicated percutaneous intervention, glucose management protocols, provide hemodynamic optimization, prophylaxis therapy for stress ulcers, infection and venous thrombosis prophylaxis, and assessment for relative adrenal insufficiency, and be willing to submit their outcomes (which are subsequently not identified in statewide outcome reporting). In an effort to further improve survival from cardiac arrest, requirements for evidence based termination of resuscitation including at least a 72 hour moratorium for termination of care following therapeutic hypothermia, a protocol to address

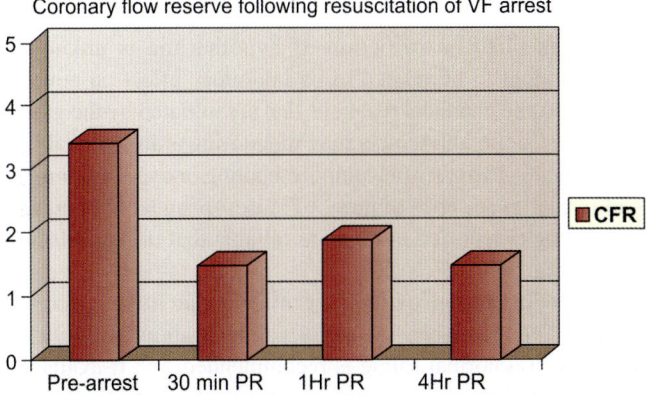

**FIGURE 36:** Coronary flow reserve (CFR) remained significantly below normal (ratio of 2:4) throughout the 4-hr postresuscitation period (*Source:* Modified from Kem KB, et al. Univ. AZ College Medicine)

organ donation, and these hospitals are encouraged to have a community out-reach program to promote bystander CCO CPR for primary cardiac arrest.

In Arizona, the SHARE program has studied the impact of prehospital transport intervals on survival from OHCA and found that bypassing a hospital to bring resuscitated but comatose patient to a Cardiac Resuscitation Center (provided the added transport time is <15 minutes) is justifiable.[88]

Neurologically intact survival is the ultimate goal of resuscitation treatment strategies and is reflected in the inclusion of "cerebral" in "cardiocerebral resuscitation". Resuscitated patients require close monitoring in an intensive care unit setting as they are prone to repeat hemodynamic instability as well as recurrent cardiac arrhythmias.

## Therapeutic Mild Hypothermia

One of the most encouraging approaches to resuscitated cardiac arrest patients with coma is therapeutic controlled mild hypothermia [89.6–93.2°F (32–34°C)]. Two decisive randomized trials have established hypothermia as an integral part of postresuscitation care.[89,90] The hypothermia after cardiac arrest (HACA) study group performed the largest randomized clinical trial of hypothermia to date. Barnard et al. also studied adult patients with out-of-hospital cardiac arrest from VF. Hypothermia has been shown to be safe in patient with cardiogenic shock,[91] where the benefit appears to be even more robust. In 2007, a Norwegian report emphasized aggressive early hypothermia and early coronary angiography in patients who were comatose after OHCA, and demonstrated an increase in survival increased from 26% to 56% with the implementation of this protocol.[92] Based on randomized trial data, approximately 6 patients need to be treated with therapeutic hypothermia to gain one neurologically intact survivor.

The mechanism of action of hypothermia is unknown, but is thought to be related to its inhibitory effect on adverse enzymatic and chemical reactions that are initiated by the global ischemia. Continuous temperature monitoring is an essential part of therapeutic hypothermia as undershooting or overshooting can lead to malignant arrhythmias.[93] Transesophageal temperature monitoring is reported to be more reliable that urinary bladder monitoring.

Early institution of hypothermia in the field is recommended.[94] If available, the rapid administration of 2,000 ml of cold (4°C) normal saline is recommended.[95,96] If available, especially in hot environments, the application of ice packs to the groin, axilla and neck are considered.

The ILCOR, taking into account the increasing evidence, issued an advisory statement in 2003 recommending that unconscious adult patients with spontaneous circulation after out-of-hospital cardiac arrest should be cooled to 89.6–93.2°F (32–34°C) for 12–24 hours when the initial rhythm was VF. Similar recommendations were echoed in more recent guidelines.[13]

Electrolyte abnormalities, coagulation disturbances and alteration of drug metabolism have all been described as complications of therapeutic hypothermia. There is no data supporting one method of cooling over the other.

There are, at present, no reliable predictive tools that can be used in comatose patients to distinguish who will or will not wake up. The best prognostic sign postresuscitation recovery is the return of consciousness. Recent unpublished observation suggests that one should wait at least 72 hours after therapeutic hypothermia before making the decision to discontinue therapy.

## Myocardial Ischemia Causing Cardiac Arrest

The recommendation for early catheterization and possible early percutaneous coronary intervention is based on the fact that about 50% of adult patients with VF arrest may have an acute myocardial infarction as the underlying etiology.[97] Unfortunately, in the postresuscitation state, neither clinical nor electrocardiographic findings are predictors of an acute coronary occlusion. In one study, 48% of patients who had no obvious noncardiac cause and had undergone coronary angiography after resuscitation from out-of-hospital cardiac arrest were found to have had an acute coronary occlusion.[98]

This has led to the concept of bundled postresuscitation care, with standardized protocol for patients with OHCA, including hypothermia and emergent coronary angiography.[87]

## ENDING RESUSCITATIVE EFFORTS

For services delivering advanced cardiac life support in England, the Recognition of Life Extinct (ROLE) guidelines state that "resuscitation attempts should be terminated when the patient remains in asystole despite full advanced life support procedures for more than 20 minutes". The AHA guidelines state that "resuscitation efforts should be continued" until "reliable criteria indicating irreversible death are present".[99] Morrison et al. found that only 0.5% of arrest victims survived if: (1) there was no ROSC; (2) no shocks were administered; (3) the arrest was not witnessed by EMS personnel; (4) when response time greater than 8 minutes was retrospectively added to the prediction rule, the survival rate was 0.3% and (5) when not bystander witnessed no one survived.[100] However, practice patterns vary widely and no single consensus has been established as the gold standard for ending resuscitation efforts.

## SUMMARY

The classic "chain of survival" identifies five fundamental links in resuscitation: early warning, early cardiopulmonary resuscitation by witnesses, early defibrillation, early advanced life support and care of the post-arrest patient. Despite all the advances, until recently there has only been a weak trend toward improved survival to hospital discharge.[3] The CCR is a new approach that has been shown to improve survival. It consists of a community approach, which emphasizes early recognition, including the frequency of gasping post witnessed cardiac arrest and early CCO CPR.[36] A prehospital approach for primary cardiac arrest prohibits early endotracheal intubation, requires prompt initiation of minimally interrupted chest compressions before rhythm analysis or after an indicated single shock and the prompt administration of epinephrine.

The CCR now has human survival outcome data and will likely succeed traditional CPR as the preferred management of primary cardiac arrest.[10-12]

## REFERENCES

1. Nichol G, Thomas E, Callaway CW, et al. Regional variation in out-of-hospital cardiac arrest incidence and outcome. JAMA. 2008; 300:1423-31.
2. Atwood C, Eisenberg MS, Herlitz J, et al. Incidence of EMS-treated out-of-hospital cardiac arrest in Europe. Resuscitation. 2005;67:75-80.
3. Sasson C, Rogers MA, Dahl J, et al. Predictors of survival from out-of-hospital cardiac arrest: a systematic review and meta-analysis. Circ Cardiovasc Qual Outcomes. 2010;3:63-81.
4. Ehlenbach WJ, Barnato AE, Curtis JR, et al. Epidemiologic study of in-hospital cardiopulmonary resuscitation in the elderly. N Engl J Med. 2009;361:22-31.
5. Cummins RO, Ornato JP, Thies WH, et al. Improving survival from sudden cardiac arrest: the "chain of survival" concept. A statement for health professionals from the Advanced Cardiac Life Support Subcommittee and the Emergency Cardiac Care Committee, American Heart Association. Circulation. 1991;83:1832-47.
6. American Heart Association Guidelines 2000 for cardiopulmonary resuscitation and emergency cardiovascular care: international consensus on science. Circulation. 2000;102:I1-I348.
7. Eckstein M, Stratton SJ, Chan LS. Cardiac arrest resuscitation evaluation in Los Angeles: CARE-LA. Ann Emerg Med. 2005;45: 504-9.
8. Dunne RB, Compton S, Zalenski RJ, et al. Outcomes from out-of-hospital cardiac arrest in Detroit. Resuscitation. 2007;72:59-65.

9. Becker L, Gold LS, Eisenberg M, et al. Ventricular fibrillation in King County, Washington: A 30-year perspective. Resuscitation. 2008;79:22-7.
10. Kellum MJ, Kennedy KW, Barney R, et al. Cardiocerebral resuscitation improves neurologically intact survival of patients with out-of-hospital cardiac arrest. Ann Emerg Med. 2008;52:244-52.
11. Bobrow BJ, Ewy GA, Clark L, et al. Passive oxygen insufflation is superior to bag-valve-mask ventilation for witnessed ventricular fibrillation out-of-hospital cardiac arrest. Ann Emerg Med. 2009;54:656-62.
12. Garza AG, Gratton MC, Salomone JA, et al. Improved patient survival using a modified resuscitation protocol for out-of-hospital cardiac arrest. Circulation. 2009;119:2597-605.
13. International consensus on cardiopulmonary resuscitation and emergency cardiovascular care science with treatment recommendations. Resuscitation. 2005;67:181-341.
14. Andreka P, Frenneaux MP. Hemodynamics of cardiac arrest and resuscitation. Curr Opin Crit Care. 2006;12:198-203.
15. Guyton AC, Polizio D, Armstrong GG. Mean circulatory filling pressure measured immediately after cessation of heart pumping. Am J Physiol. 1954;179:261-7.
16. Steen S, Liao Q, Pierre L, et al. The critical importance of minimal delay between chest compressions and subsequent defibrillation: a hemodynamic explanation. Resuscitation. 2003;58:249-58.
17. Sorrell VL, Altbach MI, Kern KB, et al. Images in cardiovascular medicine. Continuous cardiac magnetic resonance imaging during untreated ventricular fibrillation. Circulation. 2005;111(19):e294.
18. Sorrell VL, Bhatt RD, Berg RA, et al. Cardiac magnetic resonance imaging investigation of sustained ventricular fibrillation in a swine model with a focus on the electrical phase. Resuscitation. 2007;73:279-86.
19. Frenneaux M. Cardiopulmonary resuscitation-some physiological considerations. Resuscitation. 2003;58:259-65.
20. Kern KB, Ewy GA, Voorhees WD, et al. Myocardial perfusion pressure: a predictor of 24-hour survival during prolonged cardiac arrest in dogs. Resuscitation. 1988;16:241-50.
21. Paradis NA, Martin GB, Rivers EP, et al. Coronary perfusion pressure and the return of spontaneous circulation in human cardiopulmonary resuscitation. JAMA. 1990;263:1106-13.
22. Higano ST, Oh JK, Ewy GA, et al. The mechanism of blood flow during closed chest cardiac massage in humans: transesophageal echocardiographic observations. Mayo Clin Proc. 1990;65:1432-40.
23. Kern KB, Hilwig RW, Berg RA, et al. Efficacy of chest compression-only BLS CPR in the presence of an occluded airway. Resuscitation. 1998;39:179-88.
24. Kern KB, Hilwig RW, Berg RA, et al. Importance of continuous chest compressions during cardiopulmonary resuscitation: improved outcome during a simulated single lay-rescuer scenario. Circulation. 2002;105:645-9.
25. Berg RA, Sanders AB, Kern KB, et al. Adverse hemodynamic effects of interrupting chest compressions for rescue breathing during cardiopulmonary resuscitation for ventricular fibrillation cardiac arrest. Circulation. 2001;104:2465-70.
26. Kern KB, Ewy GA, Sanders AB, et al. Neurologic outcome following successful cardiopulmonary resuscitation in dogs. Resuscitation. 1986;14:149-55.

27. Berg RA, Kern KB, Hilwig RW, et al. Assisted ventilation does not improve outcome in a porcine model of single-rescuer bystander cardiopulmonary resuscitation. Circulation. 1997;95:1635-41.
28. Kouwenhoven WB, Jude JR, Knickerbocker GG. Closed-chest cardiac massage. JAMA. 1960;173:1064-7.
29. Kouwenhoven WB, Jude JR, Knickerbocker GB. Demonstration of the technique of CPR for New York Society of Anesthesiologist 1960s (Copy of demonstration provided on CD by JR Jude).
30. Safar P. Ventilatory efficacy of mouth-to-mouth artificial respiration; airway obstruction during manual and mouth-to-mouth artificial respiration. J Am Med Assoc. 1958;167:335-41.
31. Safar P, Brown TC, Holtey WJ, et al. Ventilation and circulation with closed-chest cardiac massage in man. JAMA. 1961;176:574-6.
32. Standards for cardiopulmonary resuscitation (CPR) and emergency cardiac care (ECC). II: Basic life support. JAMA. 1974;227:833-68.
33. Sayre MR, Berg RA, Cave DM, et al. Hands-only (compression-only) cardiopulmonary resuscitation: a call to action for bystander response to adults who experience out-of-hospital sudden cardiac arrest: a science advisory for the public from the American Heart Association Emergency Cardiovascular Care Committee. Circulation. 2008;117:2162-7.
34. Ewy GA. Cardiopulmonary resuscitation-strengthening the links in the chain of survival. N Engl J Med. 2000;342:1599-601.
35. Kern K, Hilwig R, Berg R, et al. Assisted ventilation during "bystander" CPR in a swine acute myocardial infarction model does not improve outcome. Circulation. 1997;96:4364-71.
36. Ewy GA. Cardiocerebral resuscitation: the new cardiopulmonary resuscitation. Circulation. 2005;111:2134-42.
37. Ewy GA. A new approach for out-of-hospital CPR: a bold step forward. Resuscitation. 2003;58:271-2.
38. Assar D, Chamberlain D, Colquhoun M, et al. Randomized controlled trials of staged teaching for basic life support. 1. Skill acquisition at bronze stage. Resuscitation. 2000;45:7-15.
39. Ewy GA, Zuercher M, Hilwig RW, et al. Improved neurological outcome with continuous chest compressions compared with 30:2 compressions-to-ventilations cardiopulmonary resuscitation in a realistic swine model of out-of-hospital cardiac arrest. Circulation. 2007;116:2525-30.
40. Heidenreich JW, Higdon TA, Kern KB, et al. Single-rescuer cardiopulmonary resuscitation: 'two quick breaths'—an oxymoron. Resuscitation. 2004;62:283-9.
41. Higdon TA, Heidenreich JW, Kern KB, et al. Single rescuer cardiopulmonary resuscitation: Can anyone perform to the guidelines 2000 recommendations? Resuscitation. 2006;71:34-9.
42. Kern KB, Valenzuela TD, Clark LL, et al. An alternative approach to advancing resuscitation science. Resuscitation. 2005;64:261-8.
43. Bobrow BJ, Spaite DW, Mullins T, et al. The impact of state and national efforts to improve bystander CPR rates in Arizona. Circulation. 2009;120(18):S1443.
44. Bobrow BJ, Vadeboncoeur TF, Clark L, et al. Establishing Arizona's statewide cardiac arrest reporting and educational network. Prehosp Emerg Care. 2008;12:381-7.
45. Bobrow B, Spaite D, Berg R, et al. Chest compression-only CPR by lay rescuers and survival from out-of-hospital cardiac arrest. JAMA. 2010 (In press).

46. SOS-KANTO. Cardiopulmonary resuscitation by bystanders with chest compression only (SOS-KANTO): an observational study. The Lancet. 2007;369:920-6.
47. Hallstrom A, Cobb L, Johnson E, et al. Cardiopulmonary resuscitation by chest compression alone or with mouth-to-mouth ventilation. N Engl J Med. 2000;342:1546-53.
48. Svensson L, Bohm K, Castren M, et al. Compression-only CPR or standard CPR in out-of-hospital cardiac arrest. N Engl J Med. 2010;363:434-42.
49. Rea TD, Fahrenbruch C, Culley L, et al. CPR with chest compression alone or with rescue breathing. N Engl J Med. 2010;363:423-33.
50. Coons SJ, Guy MC. Performing bystander CPR for sudden cardiac arrest: behavioral intentions among the general adult population in Arizona. Resuscitation. 2009;80:334-40.
51. Zuercher M, Ewy GA, Hilwig RW, et al. Continued breathing followed by gasping or apnea in a swine model of ventricular fibrillation cardiac arrest. BMC Cardiovasc Disord. 2010;10:36.
52. Zuercher M, Ewy GA. Gasping during cardiac arrest. Curr Opin Crit Care. 2009;15:185-8.
53. Bobrow BJ, Zuercher M, Ewy GA, et al. Gasping during cardiac arrest in humans is frequent and associated with improved survival. Circulation. 2008;118:2550-4.
54. Weisfeldt ML, Becker LB. Resuscitation after cardiac arrest: a 3-phase time-sensitive model. JAMA. 2002;288:3035-8.
55. Kern KB, Garewal HS, Sanders AB, et al. Depletion of myocardial adenosine triphosphate during prolonged untreated ventricular fibrillation: effect on defibrillation success. Resuscitation. 1990;20:221-9.
56. Ewy GA. Defining electromechanical dissociation. Ann Emerg Med. 1984;13:830-2.
57. Berg RA, Hilwig RW, Ewy GA, et al. Precountershock cardiopulmonary resuscitation improves initial response to defibrillation from prolonged ventricular fibrillation: a randomized, controlled swine study. Crit Care Med. 2004;32:1352-7.
58. Cobb LA, Fahrenbruch CE, Walsh TR, et al. Influence of cardiopulmonary resuscitation prior to defibrillation in patients with out-of-hospital ventricular fibrillation. JAMA. 1999;281:1182-8.
59. Wik L, Hansen TB, Fylling F, et al. Delaying defibrillation to give basic cardiopulmonary resuscitation to patients with out-of-hospital ventricular fibrillation: a randomized trial. JAMA. 2003;289:1389-95.
60. Bradley SM, Gabriel EE, Aufderheide TP, et al. Survival increases with CPR by emergency medical services before defibrillation of out-of-hospital ventricular fibrillation or ventricular tachycardia: observations from the Resuscitation Outcomes Consortium. Resuscitation. 2010;81:155-62.
61. Valenzuela TD, Kern KB, Clark LL, et al. Interruptions of chest compressions during emergency medical systems resuscitation. Circulation. 2005;112:1259-65.
62. Wik L, Kramer-Johansen J, Myklebust H, et al. Quality of cardiopulmonary resuscitation during out-of-hospital cardiac arrest. JAMA. 2005;293:299-304.
63. Wang HE, Simeone SJ, Weaver MD, et al. Interruptions in cardiopulmonary resuscitation from paramedic endotracheal intubation. Ann Emerg Med. 2009;54:645-52.

64. Berg RA, Hilwig RW, Kern KB, et al. Precountershock cardiopulmonary resuscitation improves ventricular fibrillation median frequency and myocardial readiness for successful defibrillation from prolonged ventricular fibrillation: a randomized, controlled swine study. Ann Emerg Med. 2002;40:563-70.
65. Valenzuela TD. Priming the pump—can delaying defibrillation improve survival after sudden cardiac death? JAMA. 2003;289:1434-6.
66. Rea TD, Helbock M, Perry S, et al. Increasing use of cardiopulmonary resuscitation during out-of-hospital ventricular fibrillation arrest: survival implications of guideline changes. Circulation. 2006;114:2760-5.
67. Milander MM, Hiscok PS, Sanders AB, et al. Chest compression and ventilation rates during cardiopulmonary resuscitation: the effects of audible tone guidance. Acad Emerg Med. 1995;2:708-13.
68. Aufderheide TP, Lurie KG. Death by hyperventilation: a common and life-threatening problem during cardiopulmonary resuscitation. Crit Care Med. 2004;32:S345-51.
69. Hayes MM, Ewy GA, Anavy ND, et al. Continuous passive oxygen insufflation results in a similar outcome to positive pressure ventilation in a swine model of out-of-hospital ventricular fibrillation. Resuscitation. 2007;74:357-65.
70. Steen S, Liao Q, Pierre L, et al. Continuous intratracheal insufflation of oxygen improves the efficacy of mechanical chest compression-active decompression CPR. Resuscitation. 2004;62:219-27.
71. Ewy GA, Kern KB, Sanders AB, et al. Cardiocerebral resuscitation for cardiac arrest. Am J Med. 2006;119:6-9.
72. Kellum MJ, Kennedy KW, Ewy GA. Cardiocerebral resuscitation improves survival of patients with out-of-hospital cardiac arrest. Am J Med. 2006;119:335-40.
73. Ewy GA. Do modifications of the American Heart Association guidelines improve survival of patients with out-of-hospital cardiac arrest? Circulation. 2009;119:2542-4.
74. Redding JS, Pearson JW. Evaluation of drugs for cardiac resuscitation. Anesthesiology. 1963;24:203-7.
75. Otto CW, Yakaitis RW, Ewy GA. Effect of epinephrine on defibrillation in ischemic ventricular fibrillation. Am J Emerg Med. 1985;3:285-91.
76. Attaran RR, Ewy GA. Epinephrine in resuscitation: curse or cure? Future Cardiology. 2010;6:473-82.
77. Wenzel V, Krismer A, Arntz H, et al. A comparison of vasopressin and epinephrine for out-of-hospital cardiopulmonary resuscitation. N Engl J Med. 2004;350:105-13.
78. Aung K, Htay T. Vasopressin for cardiac arrest: a systematic review and meta-analysis. Arch Intern Med. 2005;165:17-24.
79. Kern KB, Heidenreich JH, Higdon TA, et al. Effect of vasopressin on postresuscitation ventricular function: unknown consequences of the recent guidelines 2000 for cardiopulmonary resuscitation and emergency cardiovascular care. Crit Care Med. 2004;32:S393-7.
80. Dorian P, Cass D, Schwartz B, et al. Amiodarone as compared with lidocaine for shock-resistant ventricular fibrillation. N Engl J Med. 2002;346:884-90.
81. Kudenchuk PJ, Cobb LA, Copass MK, et al. Amiodarone for resuscitation after out-of-hospital cardiac arrest due to ventricular fibrillation. N Engl J Med. 1999;341:871-8.

82. Bottiger BW, Arntz HR, Chamberlain DA, et al. Thrombolysis during resuscitation for out-of-hospital cardiac arrest. N Engl J Med. 2008;359:2651-62.
83. Kern KB. Postresuscitation myocardial dysfunction. Cardiol Clin. 2002;20:89-101.
84. Neumar RW, Nolan JP, Adrie C, et al. Post-cardiac arrest syndrome: epidemiology, pathophysiology, treatment, and prognostication. A consensus statement from the International Liaison Committee on Resuscitation (American Heart Association, Australian and New Zealand Council on Resuscitation, European Resuscitation Council, Heart and Stroke Foundation of Canada, InterAmerican Heart Foundation, Resuscitation Council of Asia, and the Resuscitation Council of Southern Africa); the American Heart Association Emergency Cardiovascular Care Committee; the Council on Cardiovascular Surgery and Anesthesia; the Council on Cardiopulmonary, Perioperative, and Critical Care; the Council on Clinical Cardiology; and the Stroke Council. Circulation. 2008;118:2452-83.
85. Hekimian G, Baugnon T, Thuong M, et al. Cortisol levels and adrenal reserve after successful cardiac arrest resuscitation. Shock. 2004;22:116-9.
86. Calle PA, Buylaert WA, Vanhaute OA. Glycemia in the post-resuscitation period. The Cerebral Resuscitation Study Group. Resuscitation.1989;17:S181-8; discussion S199-206.
87. Bobrow BJ, Kern KB. Regionalization of postcardiac arrest care. Curr Opin Crit Care. 2009;15:221-7.
88. Spaite DW, Bobrow BJ, Vadeboncoeur TF, et al. The impact of prehospital transport interval on survival in out-of-hospital cardiac arrest: implications for regionalization of post-resuscitation care. Resuscitation. 2008;79:61-6.
89. Bernard SA, Gray TW, Buist MD, et al. Treatment of comatose survivors of out-of-hospital cardiac arrest with induced hypothermia. N Engl J Med. 2002;346:557-63.
90. HACA Study Group. Mild hypothermia to improve the neurologic outcome after cardiac arrest. N Engl J Med. 2002;346:549-56.
91. Skulec R, Kovarnik T, Dostalova G, et al. Induction of mild hypothermia in cardiac arrest survivors presenting with cardiogenic shock syndrome. Acta Anaesthesiol Scand. 2008;52:188-94.
92. Sunde K, Pytte M, Jacobsen D, et al. Implementation of a standardised treatment protocol for post resuscitation care after out-of-hospital cardiac arrest. Resuscitation. 2007;73:29-39.
93. Merchant RM, Abella BS, Peberdy MA, et al. Therapeutic hypothermia after cardiac arrest: unintentional overcooling is common using ice packs and conventional cooling blankets. Crit Care Med. 2006;34:S490-4.
94. Bernard S, Buist M, Monteiro O, et al. Induced hypothermia using large volume, ice-cold intravenous fluid in comatose survivors of out-of-hospital cardiac arrest: a preliminary report. Resuscitation. 2003;56:9-13.
95. Kim F, Olsufka M, Carlbom D, et al. Pilot study of rapid infusion of 2 L of 4°C normal saline for induction of mild hypothermia in hospitalized, comatose survivors of out-of-hospital cardiac arrest. Circulation. 2005;112:715-9.

96. Kim F, Olsufka M, Longstreth WT, et al. Pilot randomized clinical trial of prehospital induction of mild hypothermia in out-of-hospital cardiac arrest patients with a rapid infusion of 4 degrees C normal saline. Circulation. 2007;115:3064-70.
97. Engdahl J, Abrahamsson P, Bang A, et al. Is hospital care of major importance for outcome after out-of-hospital cardiac arrest? Experience acquired from patients with out-of-hospital cardiac arrest resuscitated by the same emergency medical service and admitted to one of two hospitals over a 16-year period in the municipality of Goteborg. Resuscitation. 2000;43:201-11.
98. Spaulding SM, Joly L-M, Rosenberg A, et al. Immediate coronary angiography in survivors of out-of-hospital cardiac arrest. N Engl J Med. 1997;336:1629-33.
99. ROLE. http://www.asancep.org.uk/JRCALC/publications/doc/ROLE_Most_Final_March2003.pdf
100. Morrison LJ, Visentin LM, Kiss A, et al. Validation of a rule for termination of resuscitation in out-of-hospital cardiac arrest. N Engl J Med. 2006;355:478-87.

# Index

Page numbers followed by *f* refer to figure and *t* refer to table

## A

Airway management  463, 467
Aldosterone antagonist  496
Ambulatory electrocardiography  444, 445
Amiodarone  53, 228, 470, 495
Angiotensin receptor blockers, use of  187
Anomalous coronary arteries  505
Antiarrhythmic drug  28, 29, 34, 35*t*, 39, 43, 46, 48, 66, 184-186
   device interactions  74
   emerging  70
   in pregnancy and lactation  72, 72*t*
   selection  71
   selection in atrial fibrillation  70
   therapy  30, 71
Antiarrhythmic trial with dronedarone  63
Aortic and right atrial pressures  507*f*
Arrhythmia  297
   hereditary  322
   mechanisms  1, 29
   triggering oscillations in cell membrane potential  15*f*
Arrhythmic right ventricular dysplasia (ARVD)  251*f*
Arrhythmogenic cardiomyopathy, left dominant  286
Arrhythmogenic right ventricular cardiomyopathy  281, 370, 492
   dysplasia  281, 505
Atherosclerotic disease  453
Atrial fibrillation  30, 93, 61*t*, 169, 397, 412, 477
   ablation  188*f*
   classification of  169
   etiology of  172
   factors predisposing to  173*t*
   genetic causes of  178*t*
   incidence of  170
   investigators  191
   maintenance of  181
   pathogenesis of  172
   prevalence of  170
Atrial flutter  207, 209, 342
   nomenclature of  347
Atrial inflammation, causes of  189
Atrial tachyarrhythmias  174
Atrial tachycardia  175, 211, 212
   focal  211, 339
   monomorphic  207
Atrioventricular block  30, 418
Atrioventricular conduction disease, evaluation of  113
Atrioventricular nodal dependent SVT  214
Atrioventricular nodal independent SVT  230
Atrioventricular nodal re-entry tachycardia  207, 214, 215, 333
Atrioventricular node disease  272
Atropine  470
Automated external defibrillator (AED)  458
   use of  464
Automatic implantable cardioverter defibrillator  465
Autosomal dominant disease  286
AV block
   first-degree  272
   second-degree  273
   third-degree  274
AVNRT
   electrophysiology of  333
   surgical ablation of  334
Azimilide  65
   post-infarct survival evaluation  65

## B

Beta-adrenoceptor blockers  46
Bidirectional tachycardia  254

Biventricular pacing 399
   causes of loss of 410*t*
Blood tests 138
Bradycardia and heart block 266
   causes of 268*t*
   treatment of 277
Bradycardia
   syndromes/diseases 267
   asymptomatic 269
Brain natriuretic peptide levels 138
Brugada syndrome 257, 258, 310, 322, 323, 443, 490, 505
   signs of 324
Bundle branch block 276
Bundle branch re-entry 120, 248
   ablation of 251*f*
Bystander
   CPR for OHCA in arizona (2005–2010) 515*f*, 517*f*
   role of 461

## C

Calcium channel antagonists 66
Cardiac access and catheterization 91
Cardiac arrest 517
   and resuscitation 450
   etiology of 505
   guidelines for primary 512
   pathophysiology of 505
   reversible causes of 468*t*
Cardiac arrhythmia suppression trial 44
Cardiac catheterization 149
Cardiac conduction system, development of 268
Cardiac defibrillator 148
   implantable 151
   internal 390
Cardiac electrophysiology for evaluation of drug therapy 123
Cardiac electrophysiology study 85
   fundamentals of 93
Cardiac interventions 480
Cardiac life support, advanced 467, 504
Cardiac nervous system dysfunction 175
Cardiac pump model of cardiopulmonary resuscitation 452*f*
Cardiac refractory, determination of 104*f*
Cardiac resuscitation
   drug therapy in 525
   evolution of 450
Cardiac resynchronization
   heart failure (CARE-HF) trial 398
   therapy 390, 395*t*, 404*f*, 407*f*
Cardiac ryanodine receptor 288
Cardiac sympathetic denervation, left 319
Cardiac transplantation 269
Cardiac-surgical ablation 332
Cardiac-surgical contribution 336
Cardioactive agents 121
Cardiocerebral resuscitation 504, 504*f*, 521, 521*f*, 524*f*, 527*f*
   for OHCA 505*f*
   for primary cardiac arrest 503
Cardiomyocytes 284
Cardiomyopathy, dilated 290
Cardiopulmonary arrest 453
Cardiopulmonary resuscitation 450, 464
   chain of survival 504*f*
   complications of 466
   compression only 462
   dispatcher assisted 462
   thoracic pump model of 452*f*
Cardiopulmonary support 479
Cardioversion 471
Cardioverter defibrillator
   automatic implantable 465
   implantable 120
   therapy, implantable 318
   trial, alternans before 497
Carotid sinus massage (CSM) 137
Carvajal syndrome 299
Catecholamine
   polymorphic ventricular tachycardia 491
   stimulation 220
Catecholaminergic PVT 259, 286
Catheter ablation 188, 233, 332, 370, 373, 375
   complications of 339

development of 336
efficacy of 341
for SVT, complications of 235*t*
of AVNRT 334
techniques of 340
Cavotricuspid isthmus (CTI) 347
Cerebral perfusion 506
Chest compressions 463
Commotio cordis 505
COMPANION study 394
Conduction system disease 175
Congenital heart disease 155, 252, 494
Congenital long QT interval syndrome 254
Congestive heart failure 54, 63, 397
Consciousness and syncope, classification of 131*fc*
CONTAK-cardiac defibrillator (CONTAK-CD) 397
Continuous pacing 101
Coronary artery disease (CAD) 245, 352, 495, 496
Coronary perfusion pressure (CPP) 506, 507
Coronary sinus 91, 175
branch of 391
ostium of 335
CRT
and ventricular arrhythmias 411
benefit 398
complications 409
device, optimization of 401
for acute decompensated heart failure 416
in practice 391
indications, emerging 411
CTI dependent AFLs, ablation of 343
Cusp VT 367

## D

Defibrillation 471
Defibrillator in
acute myocardial infarction trial (DINAMIT) 496
nonischemic cardiomyopathy treatment evaluation (DEFINITE) trial 154

Desmoplakin gene (DSP) 282
Desmosomal dysfunction 284
Desmosome function 283
Desmosome structure 283
Digoxin 228
Diltiazem 227
Disease severity, different phases of 290*t*
Disopyramide 40
Dofetilide 50
renal dosing algorithm 51*t*
Dronedarone 57
Drug therapy, electrophysiology study 90*t*
Dual atrioventricular node 112*f*
Dysplasia 281
triangle of 291
Dyssynchrony imaging, role of 402
Dyssynchrony index 408
summary 409
techniques 406

## E

Ectopy, atrial 30
Electroanatomic three-dimensional mapping 357
Electrocardiogram 84
Electrocardiographic monitoring 49
Electrophysiological abnormalities 175
Emergency medical services 455
activation 461
Emergency medical technician-basic (EMT-B) 458
Emergency medical technician-intermediate (EMT-I) 458
EMS provider levels 460*t*
Endomyocardial biopsy 295
Epicardial VT 362, 368
Epinephrine 469
Equilibrium radionuclide angiogram (ERNA) 407, 407*f*
Esmolol 228
Excessive ventilation, consequence of 524*f*

## F

Fallot, tetralogy of 494
Fascicular VT 371
Flecainide 43

## G

Genetic arrhythmia syndromes 19*t*
Genotype-specific therapy 319

## H

Heart failure 61*t*, 174
　candesartan in 172
Heart transplant 400
Hematocrit 138
Hemiblock 274, 275
Hemoglobin 138
His bundle electrogram 335
　in atrioventricular block 115*f*
His-Purkinje disease 113*f*
His-Purkinje system 99, 107*f*, 266, 275
Holter monitoring 432
Hypersensitive carotid sinus syndrome 419
Hypertrophic cardiomyopathy 135, 140*f*, 153, 174, 492, 493
Hypotension, symptomatic 190
Hypothermia
　after cardiac arrest (HACA) 528
　therapeutic 480, 528
Hypoxia 130

## I

Iatrogenic VT 254
ICD shock and defibrillation 247*f*
Idiopathic ventricular tachycardia 253, 253*f*, 364, 364*t*
Implantable cardioverter-defibrillator study 495
Intra-atrial re-entrant tachycardia 212
Intracardiac channels 104*f*
Intraventricular conduction defect 248
Intraventricular dyssynchrony 393

Ion channel and cellular properties, consequence of 10
Ion channel proteins, mutation in 19*t*
Ischemia, active 258
Ischemic heart disease 174, 495
Ischemic-related PVT-VF 258

## J

Junctional ectopic tachycardia 207, 220
J-wave syndromes 260

## K

Kaplan-Meier curves 172*f*

## L

Lange-Nielsen syndrome 310
Laryngeal mask airway 458
Lethal arrhythmia syndrome 326
Lidocaine 41, 470
Long QT syndrome 256, 310
Lower pulmonary vein, left 348
Lung disease 453
LV systolic dysfunction 173

## M

Mahaim fiber 219
Marfan syndrome 493
Metoprolol 227
Mexiletine 42, 319
Minimally symptomatic heart failure 415
Miracle study 394
Mitral annular VT 374
Mitral cusp VT 254
Mitral regurgitation 400
Mitral valve prolapse 493
Mobile cardiac outpatient telemetry 432, 440
Molecular genetic 311, 320
　analysis 301
Multielectrode catheters, diagnostic 92*f*
Multifocal atrial tachycardia 207, 212
Multiple gated acquisition scan 409

Myocardial adenosine triphosphate 519
Myocardial cells 7f
Myocardial hypertrophy 259
Myocardial infarction 89, 211, 419
Myocardial ischemia 529
  detection of 436
Myocardial tissue 12f

## N

Naxos disease 299
Neurocardiogenic syncope 419
  asystole in 119f
New task force criteria 292t
New-onset atrial fibrillation 181
Noncardiac causes 176, 267
Nonischemic cardiomyopathy 154, 495
Normal cardiac electrophysiology 95
North American Azimilide Cardioversion Maintenance Trial 66

## O

Obstructive sleep apnea 176
Outflow tract-ventricular tachycardia 365

## P

Pace mapping 358
Paroxysmal AV block 274
Paroxysmal SVT 332
  cause of 333
Permanent junctional reciprocating tachycardia 219, 221f
Phrenic nerve simulation 410
Pill-in-the-pocket approach 185
Polymorphic ventricular tachycardia 254, 259f
Posterior fascicular block, left 374
Post-resuscitation care 479
Postural orthostatic tachycardia syndrome 145f, 209
Pre-excitation syndromes 217, 219
Primary cardiac arrest 505
  assisted ventilation in 509
  pathophysiology of 506
Proarrhythmia 30
Procainamide 40, 470
Propafenone 45
Psychogenic reaction, response of 146
Pulmonary artery 366
Pulmonary disease or hypertension 169
Pulmonary veins 211
Pulmonary venous activity 175

## Q

QRS complex tachycardias, electrophysiology study 87t
QRS tachycardia, differential diagnosis of wide 224t
QT intervals, electrophysiology study 88t
QT syndrome 436, 491
Quinidine 39

## R

RAAS system, modulators of 187
Ranolazine 70
Real-time three-dimensional echocardiography 408
Re-entrant tachycardia 354
Renin-angiotensin-aldosterone system 176
Repolarization, early 491
Rescue breathing 509
Resuscitation, cessation of 478
Rhythm disorders 422
Right atrial disease 490
Right bundle branch
  block 248, 276
  pattern 399
Right lower pulmonary vein 348
Right ventricular outflow tract 252f, 365

## S

Saw tooth 210f
Selective serotonin receptor inhibitors 152
Shockable rhythm 513
Sicilian gambit for classifying antiarrhythmic drugs 33f
Sick sinus syndrome 271
Sinoatrial conduction time 110

Sinoatrial node 3, 14
Sinoatrial re-entry tachycardia 212
Sinus bradycardia 30
Sinus node disease 271
Sinus node function 86$t$
　evaluation of 109
Sinus node recovery time 109
　abnormal of 147$f$
　measurement of 109
Sinus rhythm 360$f$
　maintenance of 60, 184, 187
　restoration of 183
Sinus tachycardia 208
Situational syncope 134
Sodium calcium exchanger 311
Sodium channel blockers 34
Sodium channel gene 319
Sotalol renal dosing algorithm 48$t$
Spontaneous circulation, return of 450
Spontaneous ventricular tachycardia 246$f$
SQT syndrome 320
Strain rate imaging 404
Structural heart disease 173, 498
Sudden cardiac arrest, survivors of 119
Sudden cardiac death 488
　heart failure trial 155
Sudden unexplained death in sleep 322
Sudden unexplained nocturnal deaths 322
Superior vena cava 175
Supraventricular tachycardia 105, 175, 206, 208, 226, 227$t$, 235, 331
　diagnosis 222
　of classification 207, 207$t$
　of treatments 226
　induction of 107$f$
Syncope 130, 157
　causes of 120$f$, 133
　classification of 133
　diagnostic evaluation of 150$f$c
　economic burden of 132
　evaluation of 135, 149
　incidence of 131
　prevalence of 131
　trial, prevention of 152
　unexplained 89$t$
Systolic blood pressure 145$f$
Systolic heart failure, treatment of 390

# T

Tachycardia
　antidromic 217
　classification of 478$t$
　focal 353
　supraventricular 477
　wide complex 104$f$
Three-dimensional mapping systems, role of 123
Thromboembolism, prevention of 191
Tissue synchronization imaging 403, 404$f$
Trauma systems 456
Tricuspid annular VT 375

# U

Upper pulmonary vein, left 348

# V

Vagal tone 268
Valsalva, aortic sinus of 254
Valvular heart disease 174, 421
Vasopressin 469
Vasovagal reactions 269
Vasovagal syncope 133, 134, 149, 151
　international study 153
Vaughan-Williams classification 31
　of antiarrhythmic drugs 32$t$
Ventricular assist device 400
Ventricular dysfunction 172
Ventricular ejection fractions 248
Ventricular end-diastolic volumes 415
Ventricular end-systolic volumes 415
Ventricular fibrillation 364, 505
　induction of 475$f$
Ventricular outflow tract 365
Ventricular pre-excitation and AVNRT, evolution of 336

Ventricular tachycardia 93, 104*f*, 364, 453, 488
    clinical spectrum of 243, 244*t*
    monomorphic 245
    with structural cardiac disease, ablation of 348
Voltage map of
    left ventricle 359*f*
    posterior aspect of left ventricle 362*f*
Voltage-gated calcium channel 314

## W

Wide QRS tachycardia 223
Wolff-Parkinson-White (WPW) syndrome 88*t*, 107*f*, 139*f*, 179, 336, 337, 491